Mediated Geographies and Geographies of Media

Susan P. Mains • Julie Cupples
Chris Lukinbeal
Editors

Mediated Geographies and Geographies of Media

Springer

Editors
Susan P. Mains
Geography
University of Dundee School
of the Environment
Dundee, UK

Julie Cupples
Institute of Geography,
School of Geosciences
University of Edinburgh
Edinburgh, UK

Chris Lukinbeal
School of Geography and Development
The University of Arizona
Tucson, AZ, USA

ISBN 978-94-017-7928-9 ISBN 978-94-017-9969-0 (eBook)
DOI 10.1007/978-94-017-9969-0

Springer Dordrecht Heidelberg New York London

Softcover re-print of the Hardcover 1st edition 2015

Printed on acid-free paper

Springer Science+Business Media B.V. Dordrecht is part of Springer Science+Business Media (www.springer.com)

Dedicated to Fredy Humberto Patiño Durango

Contents

Section I Introducing Mediated Geographies and Geographies of Media

1 **Introducing Mediated Geographies and Geographies of Media** 3
Julie Cupples, Chris Lukinbeal, and Susan P. Mains

2 **Film Geography: A Review and Prospectus** 21
Laura Sharp and Chris Lukinbeal

3 **TV and the Spaces of Everyday Life** 37
Lynn Spigel

4 **Cultural Industries and the (Geographical) Political Economy of the Media** 65
Brett Christophers

5 **A Brief History of Mediated, Sensational and Virtual Geographies** 81
Stuart C. Aitken and James W. Craine

Section II Media Production and Place

6 **Chinese Cinema Cities: From the Margins to the Middle Kingdom** 95
Michael Curtin

7 **US Television Travels Abroad: Global TV and the Formatting Trend** 111
Ann Fletchall

8 **Geographies of the News** 127
Mike Gasher

9 **Remember the Alamo: A Place of Cinematic Experience** 141
Leo Zonn

Section III Transforming Geospatial Technologies and Media Cartographies

10 Vertical Mediation: Geospatial Imagery and the US Wars in Afghanistan and Iraq 159
Lisa Parks

11 The Digital Globe and New Mediations of the Environment 177
Leon Gurevitch

12 GIS as Media Through the Geoweb 191
Daniel Sui and Bo Zhao

13 GIS as Media? 209
Matthew W. Wilson and Monica Stephens

14 Crowd-Sourced Augmented Realities: Social Media and the Power of Digital Representation 223
Matthew Zook, Mark Graham, and Andrew Boulton

Section IV Placing New Media Platforms

15 Science Blogging Below-the-Line: A Progressive Sense of Place? 243
Jonathan Mendel and Hauke Riesch

16 From Webcams to Facebook: Gay/Queer Men and the Performance of Situatedness-in-Displacement 261
Ken Hillis and Michael Petit

17 Placing Satellite Radio in the Contemporary Radio Landscape 273
Sara Beth Keough

Section V Image/Imagining Media Events and Spaces

18 Visibility, Media Events and Convergence Culture: Struggles for the Meaning of 9/11 293
Kevin Glynn

19 The Case of the Missing Laureate: The Communication Geography of the 2010 Nobel Peace Prize 313
Paul C. Adams

20 From Bolt to Brand: Olympic Celebrations, Tourist Destinations and Media Landscapes 329
Susan P. Mains

Section VI Transnational Media Production and Consumption

21 **Development Communication, Popular Pleasure and Media Convergence** 351
Julie Cupples

22 **Indigenous Media: Linking the Local, Translocal, Global and Virtual** 367
Pamela Wilson

23 **Salsation: Affective Encounters and the Right to the City** 385
Tamara M. Johnson

Section VII Media Practices and Pedagogies

24 **Mediated Geographies Across Arizona: Learning Literacy Skills Through Filmmaking** 401
Chris Lukinbeal, John C. Finn, John Paul Jones III, Christina Beal Kennedy, and Keith Woodward

25 **Indigenous Media and Postcolonial Pedagogy** 417
Laurel C. Smith

26 **Owning and Sharing Experiences of Adventure: Tourism, Video and Editing Practices** 433
Eric Laurier

About the Authors 447

Index 457

Section I

Introducing Mediated Geographies and Geographies of Media

Introducing Mediated Geographies and Geographies of Media

1

Julie Cupples, Chris Lukinbeal, and Susan P. Mains

While "media geography" has coalesced in recent years as an identifiable subdiscipline of human geography, media geography did not emerge from a linear history, nor does it have a clearly defined or singular focus. Compiling this edition, participating in media geography networks at conferences and elsewhere, and teaching media at our respective institutions have all abundantly revealed that media geography is a subdiscipline with many different routes and trajectories. People come to identify as media geographers as a result of an interest in a particular medium such as film, television or radio, through the literature on the Internet and geographies of cyberspace, through critical and popular geopolitics, through questions of development and the digital divide, through media and cultural studies, through communication studies, through scholarship on the city and urban studies, and through GIS, the geoweb and geospatial technologies. Media geography intersects with social and cultural geography, development geography, political geography, feminist geography, economic geography and GIS. One of the major contributions of media productions, spaces and analyses are the opportunities they offer for providing an entryway into understanding places and communities that we may otherwise rarely, if ever encounter—but this can be problematic when the identity and places that are being marked as yours, no longer appear recognisable,

J. Cupples
Institute of Geography, School of Geosciences, University of Edinburgh, Drummond Street, Edinburgh EH8 9XP, UK
e-mail: julie.cupples@ed.ac.uk

C. Lukinbeal
School of Geography and Development, The University of Arizona, Harvill Building 437B, 85721 Tucson, AZ, USA
e-mail: chris.lukinbeal@arizona.edu

S.P. Mains (✉)
Geography, School of the Environment, University of Dundee, Dundee, Scotland, DD1 4HN, UK
e-mail: s.p.mains@dundee.ac.uk

S.P. Mains et al. (eds.), *Mediated Geographies and Geographies of Media*,
DOI 10.1007/978-94-017-9969-0_1

representative and/or desirable. The contributions in this collection pay close attention to such opportunities and challenges posed by a range of media formats, contexts and methods. These diverse entry points make for a rich emerging field in which a number of voices and perspectives are present. The field is further complicated and enriched by scholars in media studies who have turned to human geography and human geographic concepts, in order to take space, place and scale seriously in their analyses of media texts, industries and audiences. Given the diversity of the field, we thought it valuable as editors to write three position pieces that situate our work, and us personally, within the broader project represented by the scholars in this volume.

Susan P. Mains, University of Dundee

Although media permeates our lives, the ways in which mediated spaces become embedded in our personal narratives are not always apparent until much later or when a specific place, conversation or television programme acts as a prompt, reminding us of something that reappears and coalesces with other times and places. When I was a child, one of my favourite activities involved spending an evening watching a collection of homemade Super 8 mm cine films—an emotive occasion that brought our immediate family together and various friends who popped by. These film shows transported us to summer holidays in blustery wind-breaker-dotted North Yorkshire beaches, silly walks on our driveway (including those performed by younger versions of my parents when they had first moved into their house—and before the distraction of me and my sister!), and our annual living room birthday party balloon fights. Prior to each "screening," living room chairs would be rearranged, then snacks and drinks dispensed. We all waited with anticipation as my Dad set up the projector on the dining table and hoped that the bulb wouldn't blow or that the film didn't jam and turn into a molten monster projected onto the wall in front of us. There was usually a quick call of "OK, here goes . . . !" and one of us would run to turn off the lights and jostle back into a prime viewing location. The films that unfurled made us giggle: speeding up our movements ever so slightly, while also mercifully editing out the never-ending car journey and Christmas dinner meltdowns, and highlighting instead a series of funny faces, parents in deckchairs pretending to relax while their children fought over buckets, and slightly blurry close-ups of hands waving maniacally in front of the camera.

Our childhood birthday parties provide the most vivid memories of watching these home movies. Screenings became an annual ritual when my sister and I, along with our school friends, would all find a spot on the living room carpet and my parents would show the film from the previous year's party. We couldn't believe how silly our younger selves were—a whole year ago!—and after watching the film through once, we would then watch it in reverse to even greater hilarity (which was also a great opportunity to enjoy gravity defying balloons, backwards dance moves and undone food spillages). As we sat shouting out comments, we would point out each other's outfits, spot people trying to hide in the corner while smuggling extra sausages (or cocktail sticks for puncturing balloons), and make every effort to plan something even more dramatic for the current year's film. My Dad would then be required to show the film several more times, until, in the face of much boisterous resistance, he would have to turn off the projector before it really did combust.

Although it seems obvious now, it was only years later that I realised these humorous gatherings acted as a key catalyst for my curiosity with the creation, viewing and retelling of media geographies (also demonstrated recently through a highly popular screening series of Scottish-based cine film and edited archival footage (Aitken 2014)). These moving image stories produced socio-spatial cartographies that are interwoven through memories and material landscapes, and have become part of a collective experience of wonder, a sense of occasion, and, at times, even frustration. While the films themselves were quirky and fleetingly short—and probably not very exciting unless you were actually in them—it was the shared and heightened sense of expectation, fun and camaraderie of a public viewing that propelled them into the stuff of legend. I relished my friends' enjoyment and the chance to relive past festivities, and was secretly proud of my parents for being savvy enough to capture a nugget of our smaller selves and magically bring it back to life before our very eyes.

As I grew older, I paid more attention to how the places and people with whom I felt most familiar were depicted through a range of media formats and became increasingly aware of media representations with which I felt a connection. I also found myself pondering the many media portrayals that overgeneralised or undermined popular stereotypes (for example, in the Scottish context: the "idyllic" Highland retreat, the "violent" and reactionary Glasgwegian, the "chatty" female tenement resident who lived in a poor, but warm community), and became more interested in how different media forms used their own kind of place shorthand (for example, through dialogue, scenery, clothing, framing, colour, etc). This interest became more formalised through further exploration: as an undergraduate student at the University of Glasgow, I studied Geography and Theatre Studies, and although not initially realising the connections between the two, their affinities soon became apparent, as well as the opportunities they offered for examining identity and place.

As a first year student I had the unusual opportunity to work as a research assistant in the Glasgow University Media Unit (GUMU (now also known as Glasgow Media Group)), headed by Greg Philo in Sociology. This innovative research centre had been investigating a range of controversial issues: mainstream television news coverage of the 1984–1985 miners' strike, the Falklands/Malvinas conflict, and more widespread bias in television and print media. Although only working for a few hours each month, as a new student I had a fascinating introduction to the possibilities, questions and (pain-staking) methods that could be utilised to interrogate our assumptions about the reliability and neutrality of news media in particular. The GUMU's work also illustrated the political context in which academic research functions and its subversive possibilities: for example, ongoing discussions between the unit's researchers and national media organisations in relation to state intervention and control of mainstream media were highlighted through the BBC's censorship of a broadcast addressing the GUMU's publication, *War and Peace News* (1985) investigating the Falklands conflict (Quinn 2014).

While studying at Glasgow, the work of urban and feminist geographers inspired an interest in the ways in which power, representation, gender and space are interrelated (Women and Geography Study Group 1984). The city was going through a process of "re-imagineering:" hosting the 1988 Glasgow Garden Festival and being promoted as the 1990 European City of Culture. Our discussions (including within the student run Geography journal, *Drumlin* (Philo 1998)), interrogated what such civic boosterism meant in the face of high unemployment, displaced tenement residents, and the highly contentious Poll Tax being introduced by Margaret Thatcher's Conservative government (Paddison 1989, 1993). How did advertisements promoting a beautified urban space intersect with concepts of regeneration, inequality and problematic social relations? These questions also informed my Theatre Studies classes where I was investigating the role of community theatre groups as a form of grassroots agency at a city level, as well as the role of drama in challenging exclusionary concepts of identity and nationhood, for example, through the work of the 7:84 theatre group in Scotland, Ntozake Shange's poetic monologues, playwriting and site-specific performances (see, for example, McGrath 1974 (in Davidson 2014), Shange 1980; Bryden and Bailey 1990). My geographic horizons were simultaneously brought into focus and broadened by the poetic writings of playwrights from the US and South Africa, combined with geographic studies challenging neo-colonial development strategies in Latin America and Sub-Saharan Africa. This led to my dissertation examining the work of South African writer, Athol Fugard, whose play, *Sizwe Bansi is Dead* (Fugard et al. 1976)—written and produced in collaboration with John Kani and Winston Ntshona—strikingly illustrates the urgency to understand the ways in which identities and spaces are policed and resisted within racist systems of governance. The central characters in *Sizwe* negotiate the nefarious pitfalls of the pass laws circumscribing the mobility of residents, enacted through an amalgam of insidious racial criteria. This depiction of mapping and placing as being bound up in stories of terror, violence and courage, and the broader implications of international political activities, including those of the UK government and NGO protest movements, led me to an ongoing concern that continues to run through my research: borders, and how they are lived, enforced, mobilised and contested.

While my academic interests emerged from a range of theatre, cultural studies and feminist literature, it was during my postgraduate studies that I attempted to make more explicit connections between these bodies of work with that of geographers. Geography had encouraged me to think critically about urban spaces, economies and political participation (Rogerson et al. 1989). Theatre and cultural studies had provided me with an insight into theoretical concepts about language, movement and identity, specifically the ways in which deixis, semiotics and the implied relationships and hierarchies with which they are associated order, reflect and frame spatial and temporal relationships (Elam 1980). It was while studying for a master's degree in Geography at San Diego State University and then a PhD at the University of Kentucky that a growing discussion about media geographies peaked my desire to probe the consanguinity of power, media and narratives of nationality. Although focusing on urban design, gay and lesbian identities and gentrification—the role of media representations was apparent during my Masters thesis research, and combined with the work of emerging critical cultural geographies that Chris discusses in greater detail below. It was during my studies at Kentucky, however, that I engaged more directly with how media representations could be reproduced, challenged and deployed in relation to specific social groups and places.

The Geography Department and Social Theory programme at the University of Kentucky provided an exciting and engaging environment: staff such as John Paul Jones, III and Wolfgang Natter were exploring the connections between identity, space and representation and taught seminars on these topics (Natter and Jones 1993a, b), Sue Roberts guided us on the exploration of development/anti-development discourses and introduced me to the influential writings of critical theorists, including Escobar (1995); while Karl Raitz and Rich Schein re-visited traditional cultural landscape studies with a keen attention to culture and race, respectively (Peake and Schein 2000 (see also, Cresswell 1996; Sibley 1997)). Geography was actively involved in the interdisciplinary Social Theory graduate certificate and the Women's Studies program, which meant an ongoing encouragement to engage in interdisciplinary research and was also reflected in collaborative faculty projects. This dynamic and diverse scholarly landscape embraced interdisciplinary thinking and built on the work of the Frankfurt School, French poststructuralist thinkers, feminist and postcolonial theories, as we interrogated the ways in which power, inequality and representation were interwoven through space.

A significant component of this research directly involved analysing the processes through which media geographies are produced, particularly through film and television. Returning to the theme of borders mentioned above, I initiated research into ongoing debates about immigration policy in mainstream media, particularly the controversial California vote on Proposition 187, which sought to monitor undocumented migrants through a network of public agencies, Border Patrol policing and strategic use of popular media (Mains 1999). Building on Foucauldian analyses of discipline, power and representation, in conjunction with the work of Gloria Anzaldúa, bell hooks and Stuart Hall, I sought to more closely integrate my understandings of social and material practices and landscapes (focusing on the US-Mexico border). This work was partly borne out of a concern about grounding media analyses in specific experiences of place, discrimination and protest and also an effort to highlight the ways in which border identities, media representations and places were bound up in problematic concepts of race, gender and nation (Mains 2000, 2002, 2004a). (A parallel activist aspect of this was reflected in a lively postgraduate student group, Geographers for Justice (of which many of us in the department were a part), which was challenging and utilising local and national media contacts and images in conjunction with specific campus and city sites to illustrate inequalities around health care accessibility and minimum wage industries.)

Following on from my PhD, a research position at the British Film Institute (BFI) in London, provided the opportunity to explore the role, practices and resources of national media archives. It was through working at the BFI—a key charitable organisation for preservation, production and education related to film in the UK—that I became more aware of the complexities involved in making media collections publicly available. Media archives are part of ongoing conversations about representation and memory, and demonstrate the pedagogical importance of moving image media. Working at the BFI brought to light the limited conversations between academics and those who are actively engaged with producing diverse media (Mains 2003, and more recently, Mahtani 2009). This experience also fed into my next position, as a Lecturer in Human Geography at the University of the West Indies-Mona, in Kingston Jamaica—where I expanded my research on migration and borders by exploring depictions and experiences of transnationalism, mobility and migration in the Jamaican context and more widely in the Americas (Mains 2008a), This research integrated the use of documentary film, drawing on historical film archives held at the BFI (Mains 2007, 2008b). While based in Jamaica I also diversified the forms of media I was exploring, returning to news coverage and widening my analyses to include public art, the latter of which had become a key talking point in popular media discussions about depictions of the body and race in the Caribbean (Mains 2004b).

While living in Jamaica I was struck by the ways in which Caribbean creative writers were narrating stories of migration and the ongoing legacies of slavery. I was also conscious of what seemed to be limited critical discussion of the region more generally within the discipline of (Anglophonic) Geography, particularly in terms of media representations of place, and the problematic application of ideas around "development." Much of this (uneven) development seemed to hinge on tourism, a cause for concern in relation to a lack of public input and/or opaque decision-making processes. Since this period, I have been building on creative writing, media geography, and cultural studies work that critically engages with active learning and what "Caribbeanness" means (for example, the writings of Paul Gilroy, Kamau Brathwaite, Norval (Nadi) Edwards, Olive Senior, Kei Miller, Carolyn Cooper, Rex Nettleford ,Velma Pollard, Paulo Freire, Annie Paul, Mimi Sheller, and Deborah A. Thomas) to examine the interconnected stories of migration, tourism and island media geographies. And although I have crossed the Atlantic (once again!), and am now based in Geography at the University of Dundee, I continue to explore these overarching themes in the context of both Caribbean and Scottish mediated landscapes, as well as other contexts.

My current work expands on the themes outlined above, exploring media representations of tourism developments in the Caribbean and Scotland in relation to the themes of mobility, borders, heritage and security, and as part of a collaborative network of interdisciplinary scholars. This has encompassed an ongoing engagement with the use of media as part of critical pedagogy and participatory research, and I am currently working with a group of Tayside-based creative writers, artists and academics to produce a series of short films exploring diverse relationships with the River Tay and surrounding environs (Mains 2014a, b). In many ways my journey has come full circle, starting from the cine films of my childhood living room to a new process of small scale filmmaking, but with a slightly different twist on the initial process: a new take that also explores the act of recording, filmmaking and editing (as well as viewing), as key components of collaborative storytelling and pedagogical processes. My hope is that these projects, and this edited collection, will enable a greater understanding of the myriad connections between these personal, national and global media narratives of mobility, praxis and place.

Chris Lukinbeal, University of Arizona

In 1991, as an undergraduate at California State University, Hayward, I took a class titled, "The American Landscape." One assignment for that class was to write a paper about a movie. The paper I wrote, "Dick Tracy's Cityscape" (Lukinbeal and Kennedy 1993), helped me combine my mutual interest in landscape and film studies. At the time, I was working as a landscape architect to pay for school. Through this writing exercise and my experience working with suburban landscapes (Lukinbeal and Kennedy 1992), I became absorbed by the mutual ontological underpinning of film and landscape that surfaces when we approach them through questions of representations, perspective, visuality and power. This fascination led to my master's thesis, *A Geography in Film, A Geography of Film* (1995), completed at the same time Aitken and Zonn (1994) released the first book dedicated to film and geography.

Before Aitken and Zonn came on the scene, however, there was already a long and now often forgotten history of media geography in the U.S. This history can be traced back to J. K. Wright's (1947) presidential address to the Association of American Geographers (AAG), "Terrae Incognitae: The Place of the Imagination in Geography." Here, Wright proposed that geographers should investigate "peripheral areas" (subjective studies) that charted the terra incognita of geographical knowledge. To Wright (1947, 10), this terra incognita existed in "books of travel, in magazines and newspapers, in many a page of fiction and poetry, and on many a canvas." Wright coined the term *geosophy* to represent this merging of objective and subjective studies. He further identified *aesthetic geosophy* as an area that would focus on literature and arts.

Taking a similar tack as Wright, Lowenthal (1961) later argued from a humanist perspective that geography includes perception, imagination, and subjectivity. At the time that Lowenthal was writing, humanism (along with Marxism) was a response to the quantitative revolution's exclusion of the human experience as an important area of research. Humanism's focus on the geography of the mind or one's personal geography was elaborated upon by Prince (1961) and Watson (1969). Prince (1961) posited that geographic description must show respect for truth (objectivity), but also inspiration and direction by a creative imagination (subjectivity). Watson (1969, 10) argued that imagination and personal perception are important because,

> Not all geography derives from the earth itself; some of it springs from our idea of the earth. This geography within the mind can at times be the effective geography to which men adjust and thus be more important than the supposedly real geography of the earth. Man has the particular aptitude of being able to live by the notion of reality which may be more real than reality itself.

A complimentary line of thought for humanist geography was provided by J.B. Jackson's focus on vernacular landscape. For Jackson, ordinary and everyday items and landscapes were worthy of study. Like Jackson, Donald Meinig was also interested in pursuing the "ordinary landscape." In his widely read book, *The Interpretation of Ordinary Landscapes*, Meinig (1979, 171) proclaimed that cinema was an "unprecedentedly powerful propaganda medium" that spread an idealized, symbolic image of California suburbia to the world. He went on to suggest that an inventory of landscapes depicted in Hollywood's golden era (1920s–1950s) would allow for ready inferences, and act as a foundation for our understanding of what meanings are being assigned to specific landscapes.

Following the era of humanist landscape studies, Denis Cosgrove's *Social Formation and Symbolic Landscape* (Cosgrove 1984, 13) helped move landscape studies forward by taking a Marxist position and asserting that "landscape is a way of seeing the world," harkening back to Berger's (1972) "Ways of Seeing," a seminal analysis of visual imagery. From here, landscape studies slowly shifted from an examination of the land out there to a threefold focus on the land out there, the way we see the land out there, and the representation of the land out there. This was an important distinction because it allowed for a reexamination of what constituted the "field of study" for geography. Where the Berkeley School privileged material conditions and fieldwork, "new" cultural geography (as this body of work came to be known) emphasized non-material cultural forms and representations (Price and Lewis 1993a b; Cosgrove 1993; Duncan 1993; Jackson 1993). For new cultural geographers in the 1990s media was where fieldwork was done. However, it also became apparent that a purely representational focus by new cultural geographers led to a problematic binary of material versus non-material studies.

It was during this turn toward the non-material that the foremost assay into a purely media geography collection of essays, Burgess and Gold's (1985) *Geography, The Media, and Popular Culture*, appeared. Significantly, in the introduction to this edited collection, the authors outlined an imprecise characterization of media geography as split between a European school and an American school. Though the European school was indeed focused on a Marxist approach that sought to uncover 'hidden agendas' and expose ideology, as Burgess and Gold claimed, the authors wrongly type casted the American school as taking solely a behavioural science approach to media studies that was based on stimulus-response relationships. Though present, this area of research did not nearly encompass the entire spectrum of work being done at the time. Rather, the U.S. geographic work on media was based in humanism, landscape studies, and environmental psychology/transactionalism.

Transactionalism in particular was the primary theory dominating discussions of media geography in the late 1980s and 1990s, especially by those working in environmental perceptions. Transactionalism, according to Aitken (1991, 107), is the study of "person-in-environment contexts as a function of a particular ongoing transaction between persons and environments." If you were doing media geography in the 1990s in the U.S. you were part of the Environmental Perception and Behavioural Geography (EPBG) specialty group in the Association of American

Geographers (AAG), which was split between a science-based behavioural geography and environmental perception. The application of transactionalism to media studies in the 1990s was an obvious choice for me personally, owing to the focus on transactionalism by Christina Kennedy (Zube and Kennedy 1990; Kennedy 1994; Kennedy and Lukinbeal 1997), my master's advisor, Stuart Aitken (Aitken and Bjorklund 1988; Aitken 1991; Aitken and Zonn 1993), my PhD advisor, and Leo Zonn (Zonn 1984, 1985, 1990a), a leader in the development of the film geography sub-discipline. The applications of transactionalism to film geography of the time were many; it helped explain an individual's relationship with mediated landscapes (Zonn 1984, 1985, 1990b; Zube and Kennedy 1990), transactions between film structure and rhythm (Aitken 1991; Aitken and Zonn 1993), and the transactions between a person and their environment in a film's narrative (Kennedy 1994). Similar to phenomenology-based humanist work on landscape, language, and literature (Relph 1976; Tuan 1978, 1979, 1991; Pocock 1981), transactionalism worked by providing a context within which to situate analysis. While its all-inclusive gestalt seemed at the time to foster critical knowledge development, that knowledge development was actually confounded by the theory's reliance on a human-environment metanarrative. Unfortunately for the EPBG group, environmental perception proved to be a gateway drug for the transition from humanism and landscape studies into critical theory and most left EPBG for other more appropriate subfields by the early 2000s.

It was within this trajectory of thought that I began my dissertation research on San Diego's on-location filming (see Lukinbeal 2012b). On-location filming constituted to me a type of thirdspace (Soja 1996) in that it represented the site filmed, the cinematic site in a narrative, and the merging of the two in one place. The problem with this conceptualization, in hindsight, is that it did not explain why one house was chosen for filming over another, nor did it grasp how landscapes are both cultural products (the films themselves) *and* practices (the act of production and consumption). Today, I see the answers behind these cinematic landscapes as lying in an understanding of how the political economy of filmmaking affects the geographies framed in the mise-en-scène. My interest in on-location filming stemmed from a desire to conduct materially focused media geography research, an area that had been ignored by the turn to the non-material that had allowed media geography to take shape. It was apparent that the focus on media by new cultural geography had little to do with an interest in media per se, but was rather a means to explore different socio-spatial phenomena like class, race, gender, or sexuality. Nearly two decades later, Doel and Clarke (2007, 894) would state it thusly: "While much has been written about the geography of film and the geography in film, the geography of film qua film remains largely unexplored."

While cinematic landscapes (Lukinbeal 2005) and on-location filming (Lukinbeal 2006) are central to my research, I remain committed to geographic education research related to media. For me this centers on the issue of geographic media literacy, defined as "the ability to locate, evaluate, effectively use, and produce geographic information," while merging critical thinking and praxis (Lukinbeal 2014, 41). Geographic media literacy positions GIS, cartography, film, television and social media side by side. It incorporates visual literacy, information technology literacy, information literacy, and media literacy. It includes analysis of media as text as well as the production of media (Lukinbeal et al. 2007; Lukinbeal and Craine 2009).

As this collection of essays will show, media geography has evolved from a focus on media as text and textual readings, to a more broad based engagement with media as practice. I feel Doel and Clarke's call for studies on film form is central to my research future. My examination of the cartographic paradox sought to expose the relationship between horizontal (linear perspective) and vertical (orthographic) scopic regimes, which provide an architecture to create a coherent representation of the world (Lukinbeal 2010). Further, these scopic regimes rely on a representative and expressive analogy called scale. Scale is central to understanding how space, place and landscape function in the mise-en-scène (Lukinbeal 2005). Scale provides spatial structure that clarifies and explains, but in so doing it must conceal its alterity, its schizophrenia. In cinema, scale often references distance and proportion. However, the schizophrenia of scale is exhibited through the coexistence of dissimilar and incompatible elements such as unity/fragmentation or coherence/the infinite (Lukinbeal 2012a). Whether through a focus on scopic regimes or scale, the future of media geography requires a focus on form.

Julie Cupples, University of Edinburgh

I became a geographer really by accident, becoming a media geographer resulted from a mixture of circumstances and agency. In the mid-1990s, while living in Newcastle upon Tyne in the UK, I was looking for a taught Masters with a focus on Latin American studies and as a result ended up doing a Masters called International Cultural Change in the Department of Geography at Newcastle University. This course, while giving me the possibility to develop my interests in both Latin American studies and development studies, also provided with me with an in-depth and simultaneous introduction to both human geography and media and cultural studies. One of the three academics delivering the programme was Professor Kevin Robins, whose pioneering scholarship on media, globalization and cultural identities both individually and in collaboration with David Morley (see for example, Morley and Robins 1995), was fundamental to the development of what is now a well-established subdiscipline of human geography. This introduction to Birmingham School cultural studies—to the work of Raymond Williams and Stuart Hall—resonated with me. At this point, I didn't embark on research in media geography, although my exposure to cultural studies meant I could stop feeling guilty about watching Coronation Street and worrying about my son's Playstation use. But I continued to identify with and work with cultural studies perspectives. Consequently, my geography and my media geography have always had a strong cultural studies focus.

I did a PhD in Nicaragua on a gender and development theme, working primarily with Latin American women who were single mothers, political activists and the main income earners in the household. I spent much of my fieldwork in both 1999 and 2001 watching *telenovelas* with my participants and other Nicaraguans, including the Colombian original of what later became the global Ugly Betty phenomenon (*Yo soy Betty la fea*) and came to appreciate their important sense-making properties in the context of these women's everyday lives. I also came to understand the community-building and citizenship properties of media consumption when half of the residents of the rural community of El Hatillo would all squeeze into the home of one of my participants at the same time every day to watch the *telenovela*. This community incidentally showed high levels of communal reciprocity and mutual support after it was hit by Hurricane Mitch (Cupples 2004). After that, a more focused scholarly interest in media began to develop as a result of a number of related events.

In 2001, I published a first piece to deal specifically with the media (Cupples and Harrison 2001) as a result of an important media event in 2000 in Christchurch, the city in which I lived. A prominent Christchurch GP and deputy mayor, Morgan Fahey, had been accused of sexually assaulting a number of his female patients. The story was covered on news and the current affairs show, *20/20*, and produced a decisive anti-feminist backlash from a range of media that asserted Fahey's innocence and cast doubt on the veracity of the women's allegations. A second episode involved a former patient confronting Fahey in his surgery wearing a secret camera and led to intense mediated debates over the ethics of a secret camera. Using broadly Gramscian and cultural studies approaches, we analyzed the struggle over the hegemonic meanings of place and gender as they played out in a range of media texts.

Not long after I joined the permanent faculty at the University of Canterbury, I became a member of the Cultural Studies programme. This was, and remains, the first and only Cultural Studies major to be offered in New Zealand. It was an interdepartmental programme so I was able to interact with scholars in English, American Studies, Art History, Chinese, and Anthropology who had an interest in media. I got to co-teach courses on media culture, supervise dissertations with a media focus and co-organize conferences and other events.

Throughout this period, my research in Central America continued and I became increasingly interested in how media resources were being harnessed by ordinary people to contest suspected electoral fraud (Cupples 2009), to challenge the mainstream media's support of the Central American Free Trade Agreement or CAFTA (Cupples and Larios 2010), and to contest the botched privatization of electricity distribution (Cupples 2011).

But I also got interested in other kinds of media use. With a colleague, we interviewed New Zealand teenagers about their use of text messaging, a project that took me into actor-network theory and questions of posthumanism. We both had teenage children and were fascinated by the cultural dynamics surrounding the use of text messaging, a phenomenon that had teachers and parents in a moral panic at the time (see Thompson and Cupples 2008; Cupples and Thompson 2010; see also Cupples 2014 for reflections on the moral panic).

Over the past decade, I continued to engage with the literature in cultural studies but increasingly also began to focus on the work being done on media by geographers. While this growing body of literature was evidently making substantial contributions to our understandings of the ways in which mediations are situated, how they contribute to the reworking of place meanings, and how they facilitate reconfigurations of scale, I also became aware of some gaps, particularly within political geography and the body of literature known as popular geopolitics. So while there is no doubt, as Adams et al. (2014, 2) write, that British cultural studies provided the "predominant methodological inspirations" for some of the early work in media geography, much media geography seems to proceed without an awareness of the well trodden paths in cultural studies (see Glynn and Cupples 2015). This frustration with what I see as a neglect of cultural studies has also been felt with respect to the large number of undergraduate and graduate dissertations I have supervised and examined over the past decade that deal with the media in one way or another. Many human geography students are interested in media and wish to do some kind of media analysis in their project, but they often do so badly because of quite simplistic understandings of how media work. In particular, I have noticed a tendency to replicate a highly problematic media effects model (see Gauntlett 1998), or to equate the political economy of the media with the cultural politics of the media. It is not just students, there remains a residual dismissiveness of the media, and in particular media consumption, among established human geographers. Without finding out what it is people are doing with what he problematically calls 'weapons of mass distraction' or why they invest their time in watching/surfing/playing, Harvey (2014, 278) dismisses sit-com watching, web surfing, or video game playing as "useless" activities. In other words, geographers need to read more media studies, and they need to embrace media and cultural studies in the way that many media and cultural studies scholars have embraced geography and geographically-attuned forms of analysis (see for example Fiske 1993; Morley 2000, 2006; McCarthy 2001; Couldry and McCarthy 2004; Parks 2005, 2014; Curtin 2007; Berry et al. 2010), as well as the Chapters by Curtin, Glynn, Gurevitch, Parks, Spigel, and Wilson in this volume).

More recently, through a collaborative research project on the geographies of media convergence, my collaborator and I have been trying to think what is specifically geographical about the question of convergence. Taking the dramatic changes in the ways that people know, consume and produce media as a starting point, and given the characteristics of the global historical conjuncture characterized by persistent yet highly contested forms of neoliberalism and securitization, our project attempts to address how the new media environment is able to facilitate democratizing or decolonizing forms of cultural citizenship, particularly for those marginalized by dominant epistemologies. We are trying to take the debates on convergence into new terrains, thinking for example about the convergence between indigenous media and so-called mainstream media and about the political work that is done by entertainment media. We are working closely with the community and indigenous television channels on the Caribbean Coast of Nicaragua (see Glynn and Cupples 2011, for some preliminary findings from this research) and with Māori Television in New Zealand. We are also thinking about the convergent activities that surround humanitarian emergencies and disasters (Cupples and Glynn 2013, 2014) and how ordinary people and marginalized groups make do with the media resources that they have at their disposal, and most importantly the potential of this making do for the creation of forms of belonging and cultural citizenship. Much of this work is focused on the critical geopolitics of contemporary TV drama, including *ER*, *Treme*, *Commander in Chief*, and *Orange is the New Black*.

I am therefore trying to make a decisive contribution to the productive cross-fertilizations between human geography and media studies. There is much important work to be done, and it is hoped that this volume will prove to be a key resource for those of us thinking geographically about media, and thinking about the difference media makes to geographies.

Chapter Organisation

To produce a book that addresses media/mediated geographies, is—literally and metaphorically—(in the words of Usain Bolt) a project to take on the world. It is both an exciting and daunting prospect, and one that we feel has provided us a unique opportunity to bring together diverse, inspiring and rich stories of how geography and media interweave. While we recognise that it is impossible to include every key area of media geography research, we have attempted to encourage as broad and inclusive a discussion as possible—and hope that any remaining gaps provide an opportunity for further research, dialogue and publications. Although several of the chapters illustrate connections and parallel concerns, in order to highlight key debates, formats and specific discussions, we have divided the collection into seven thematic sections. This first section provides an introduction and overview of a range of media formats and the ways in which spatial understandings and ideas have emerged in relation to varied media and geographic content. Section 2 explores more specific contexts of media production drawing on a range of international contexts. This is followed by a third section focusing on emerging research examining geospatial technologies and offers an exciting entryway into these rapidly changing communication processes and spaces. Section 4 continues these conversations with an exploration of new media platforms such as webcams, blogging and satellite radio. The fifth section focuses on pivotal events and their associated spaces to explore the ways in which key moments and activities, such as 9/11 and the Nobel Prize, become emblematic—or contradictory—in relation to how media and geographical knowledge are strategically deployed and challenged. Section 6 explores convergent media, indigenous media representations and dance, and highlights the importance of transnational media practices and diverse media geographies of the Global South. The final, and seventh section, points to innovative and practice-based explorations of the pedagogical nature of media geographies and the possibilities for the further development of engaging future media/mediated geographies. We hope that you enjoy your travels through the varied chapters to follow and have as stimulating, creative and thought provoking journey as we have had while compiling this exploration of *Mediated Geographies and Geographies of Media*.

References

Adams, P. A., Craine, J., & Aitken, S. (2014). Introduction: Geographies of media. In P. Adams, J. Craine, & S. Aitken (Eds.), *The Ashgate research companion to media geography* (pp. 1–14). Farnham: Ashgate.

Aitken, S. (1991). A transactional geography of the image-event: The films of Scottish director, Bill Forsyth. *Transactions. Institute of British Geographers, 16*, 105–118.

Aitken, S. (2014). No tartan, no shortbread—King Creosote portrays a different Scotland. *The Guardian*. http://www.theguardian.com/music/musicblog/2014/jun/23/king-creosote-scotland-film-commonwealth-games. Accessed 21 Nov 2014.

Aitken, S., & Bjorklund, E. (1988). Transactional and transformational theories in behavioral geography. *The Professional Geographer, 40*, 54–64.

Aitken, S., & Zonn, L. (1993). Weir(d) sex: Representation of gender-environment relations in Peter Weir's picnic at hanging rock and Gallipoli. *Environment and Planning D: Society and Space, 11*, 191–212.

Aitken, S., & Zonn, L. (Eds.). (1994). *Place, power, situation and spectacle: A geography of film*. Totowa: Rowan and Littlefield Publishers.

Berger, J. (1972). *Ways of seeing*. London/Harmondsworth: British Broadcasting Corporation/ Penguin.

Berry, C., Soyoung, K., & Spigel, L. (Eds.). (2010). *Electronic elsewheres: Media, technology, and the experience of social space*. Minneapolis: University of Minnesota Press.

Bryden, B., & Bailey, D. (1990). *The ship*. Govan: Harland & Wolff's Former Shipyard. Performance.

Burgess, J., & Gold, J. (Eds.). (1985). *Geography the media and popular culture*. New York: St. Martin's Press.

Cosgrove, D. (1984). *Social formations and symbolic landscapes*. Totawa: Barnes and Noble Books.

Cosgrove, D. (1993). Commentary on "The reinvention of cultural geography" by Price and Lewis. *Annals of the Association of American Geographers, 83*(3), 515–516.

Couldry, N., & McCarthy, A. (2004). *MediaSpace: Place, scale and culture in a media age*. London: Routledge.

Cresswell, T. (1996). *In place/out of place: Geography, ideology, and transgression*. Minneapolis: University of Minnesota Press.

Cupples, J. (2004). Rural development in El Hatillo, Nicaragua: Gender, neoliberalism and environmental risk. *Singapore Journal of Tropical Geography, 25*(3), 343–357.

Cupples, J. (2009). Rethinking electoral geography: Spaces and practices of democracy in Nicaragua. *Transactions of the Institute of British Geographers, 34*(1), 110–124.

Cupples, J. (2011). Shifting networks of power in Nicaragua: Relational materialisms in the consumption of privatized electricity. *Annals of the Association of American Geographers, 101*(4), 939–948.

Cupples, J. (2014). Bodies. In P. Adams, J. Craine, & J. Dittmer (Eds.), *The Ashgate research companion to media geography* (pp. 171–184). Farnham: Ashgate.

Cupples, J., & Glynn, K. (2013). Postdevelopment television? Cultural citizenship and the mediation of Africa in contemporary TV drama. *Annals of the Association of American Geographers, 103*(4), 1003–1021.

Cupples, J., & Glynn, K. (2014). The mediation and remediation of disaster: Hurricanes Katrina and Felix in/and the new media environment. *Antipode: A Radical Journal of Geography, 46*(2), 359–381.

Cupples, J., & Harrison, J. (2001). Disruptive voices and boundaries of respectability in Christchurch, New Zealand gender. *Place and Culture, 8*(2), 189–204.

Cupples, J., & Larios, I. (2010). A functional anarchy: Love, patriotism and resistance to free trade in Costa Rica. *Latin American Perspectives, 37*(6), 93–108.

Cupples, J., & Thompson, L. (2010). Heterotextuality and digital foreplay: Cell phones and the culture of teenage romance. *Feminist Media Studies, 10*(1), 1–17.

Curtin, M. (2007). *Playing to the world's biggest audience: The globalization of Chinese film and TV*. Berkeley: University of California Press.

Doel, M., & Clarke, D. (2007). Afterimages. *Environment and Planning D: Society and Space, 25*, 890–910.

Duncan, J. (1993). On "The reinvention of cultural geography" by Price and Lewis. *Annals of the Association of American Geographers, 83*(3), 517–519.

Elam, K. (1980). *The semiotics of theatre and drama*. London: Methuen.

Escobar, A. (1995). *Encountering development: The making and unmaking of the third world*. Princeton: Princeton University Press.

Fiske, J. (1993). *Power plays, power works*. London: Verso.

Fugard, A., Kani, J., & Ntshona, W. (1976). *Sizwe Banzi is dead*. London: Viking Press.

Gauntlett, D. (1998). Ten things wrong with the "effects model". In R. Dickinson, R. Harindranath, & O. Linné (Eds.), *Approaches to audiences: A reader* (pp. 120–130). London: Arnold.

Glynn, K., & Cupples, J. (2011). Indigenous mediaspace and the production of (trans)locality on Nicaragua's mosquito coast. *Television and New Media, 12*(2), 101–135.

Glynn, K., & Cupples, J. (2015). Negotiating and queering US hegemony in TV drama: Popular geopolitics and cultural studies. *Gender, Place and Culture, 22*(2), 271–287.

GUMU. (1985). *War and peace news*. Milton Keynes: Open University Press.

Harvey, D. (2014). *Seventeen contradictions and the end of capitalism*. London: Profile Books.

Jackson, P. (1993). Berkeley and beyond: Broadening the horizons of cultural geography. *Annals of the Association of American Geographers, 83*(3), 519–520.

Kennedy, C. (1994). The myth of heroism: Man and desert in Lawrence of Arabia. In S. Aitken & L. Zonn (Eds.), *Place, power, situation and spectacle: A geography of film* (pp. 137–160). Totowa: Rowan and Littlefield Publishers.

Kennedy, C., & Lukinbeal, C. (1997). Towards a holistic approach to geographic research on film. *Progress in Human Geography, 21*, 33–50.

Lowenthal, D. (1961). Geography, experience, and imagination: Towards a geographical epistemology. *Annals of the Association of American Geographers, 51*(3), 241–260.

Lukinbeal, C. (2005). Cinematic landscapes. *Journal of Cultural Geography, 23*(1), 3–22.

Lukinbeal, C. (2010). Mobilizing the cartographic paradox: Tracing the aspect of cartography and prospect of cinema. *Digital Thematic Education, 11*(2), 1–32.

Lukinbeal, C. (2012a). "On Location" filming in San Diego County from 1985—2005: How a cinematic landscape is formed through incorporative tasks and represented through mapped inscriptions. *Annals of the Association of American Geographers, 102*(1), 171–190.

Lukinbeal, C. (2012b). Scale: An unstable representational analogy. *Media Field Journal, 4*, 1–13.

Lukinbeal, C. (2014). Geographic media literacy. *Journal of Geography, 113*(2), 41–46.

Lukinbeal, C., & Craine, J. (2009). Geographic media literacy: An introduction. *GeoJournal, 74*, 175–182.

Lukinbeal, C., & Kennedy, C. (1992). Suburban landscapes of the East Bay. *California Geographer, 32*, 77–93.

Lukinbeal, C., & Kennedy, C. (1993). Dick Tracy's cityscape. *Association of Pacific Coast Geographers Yearbook, 55*, 76–96.

Lukinbeal, C., Kennedy, C., Jones, J., Finn, J., Woodward, K., Nelson, D., Grant, Z., Antonopoulos, N., Palos, A., & Atkinson-Palombo, C. (2007). Mediated geographies: Critical pedagogy and geographic education. *Yearbook of the Association of Pacific Coast Geographers, 69*, 31–44.

Mahtani, M. (2009). The racialized geographies of news consumption and production: Contaminated memories and racialized silences. *GeoJournal: An International Journal on Geography, 74*(3), 257–264.

Mains, S. P. (1999). Statutes of liberty: Migration and urbanism in the Borderlands. *Yearbook of the Association of Pacific Coast Geographers, 61*, 42–66.

Mains, S. (2000). An anatomy of race and immigration politics in California. *Journal of Social and Cultural Geography, 1*(2), 143–154.

Mains, S. P. (2002). Representing national identity, borders, and migration: Making news in Southern California. *Hagar: International Social Science Review, 2*, 271–298.
Mains, S. P. (2003). 'The future of multi-ethnic Britain': Media, diversity, and regional identity. In S. Ralph, H. Manchester, & C. Lees (Eds.), *Diversity or anarchy* (pp. 225–232). Luton: University of Luton Press.
Mains, S. P. (2004a). Imagining the border and Southern spaces: Cinematic explorations of race and gender. *GeoJournal, 59*(4), 253–264.
Mains, S. P. (2004b). Monumentally Caribbean: Borders, bodies, and redemptive city spaces. *Small Axe: A Caribbean Journal of Criticism, 16*, 179–198.
Mains, S. P. (2007). Hurricanes in Toronto: Distance, identity, and the Jamaican diaspora. *Focus on Geography, 50*(1), 1–8.
Mains, S. P. (2008a). 'English Fever': Documenting the Caribbean diaspora in the colony. In A. Escher, C. L. Anton, & S. Zimmermann (Eds.), *The geography of cinema – A cinematic world* (pp. 137–153). Stuttgart: Franz Steiner-Verlag.
Mains, S. P. (2008b). Transnational communities, identities, and moving populations. In E. L. Jackiewicz & F. J. Bosco (Eds.), *Placing Latin America: Contemporary themes in human geography* (pp. 205–234). Lanham: Rowman and Littlefield.
Mains, S. P. (2014a). Fieldwork, heritage and engaging landscape texts. *Journal of Geography in Higher Education, 38*(4), 525–545.
Mains, S. P. (2014b). *Ties to the tay: Narrating stories of belonging, community and place.* http://tiestothetay.wordpress.com/. Accessed 4 Dec 2014.
McCarthy, A. (2001). *Ambient television: Visual culture and public space*. Durham: Duke University Press.
McGrath, J. (1974). The cheviot, the stag and the black black oil. In E. Davison (2014) (Ed.), *BFI Screen Online*, http://www.screenonline.org.uk/tv/id/466025/. Last accessed 26 Nov 2014.
Meinig, D. (1979). Symbolic landscapes. In D. W. Meinig (Ed.), *The interpretation of ordinary landscapes* (pp. 164–192). Oxford: Oxford University Press.
Morley, D. (2000). *Home territories: Media, mobility and identity*. London: Routledge.
Morley, D. (2006). *Media, modernity, technology: The geography of the new*. London: Routledge.
Morley, D., & Robins, K. (1995). *Spaces of identity: Global media, electronic landscapes and cultural boundaries*. London: Routledge.
Natter, W., & Jones, J. P., III. (1993a). Pets or meat: Class, ideology, and space in 'Roger and Me'. *Antipode, 25*(2), 140–158.
Natter, W., & Jones, J. P., III. (1993b). Signposts towards a postructuralist geography. In J. P. Jones, W. Natter, & T. R. Schatzki (Eds.), *Postmodern contentions: Epochs, politics, and space* (pp. 165–204). New York: Guilford.
Paddison, R. (1989). Spatial effects or the poll tax: A preliminary analysis. *Public Policy and Administration, 4*(2), 10–21.
Paddison, R. (1993). City marketing, image-reconstruction and urban regeneration. *Urban Studies, 30*, 339–350.
Parks, L. (2005). *Cultures in orbit: Satellites and the televisual*. Durham: Duke University Press.
Parks, L. (2014). *Media spaces and global security: Coverage after 9/11*. New York: Routledge.
Peake, L., & Schein, R. (2000). Racing geography into the new millennium: Studies of 'Race' and North American geographies. *Journal of Social and Cultural Geography, 1*(2), 133–142.
Philo, C. (1998). Reading drumlin: Academic geography and a student geographical magazine. *Progress in Human Geography, 22*(3), 344–367.
Pocock, D. (1981). *Humanistic geography and literature: Essays on the experience of place*. London: Croom Helm.
Price, M., & Lewis, M. (1993a). The reinvention of cultural geography. *Annals of the Association of American Geographers, 83*(1), 1–17.
Price, M., & Lewis, M. (1993b). Reply: On reading cultural geography. *Annals of the Association of American Geographers, 83*(3), 520–522.
Prince, H. (1961). The geographical imagination. *Landscape, 11*(2), 22–25.

Quinn, A. (2014). *Glasgow media group: GMG timeline 1974–2004*. http://www.glasgowmediagroup.org/about-us/timeline. Accessed 24 Nov 2014.

Relph, E. (1976). *Place and placelessness*. London: Pion.

Rogerson, R., Findlay, A., Morris, A., & Paddison, R. (1989). Variations in quality of life in urban Britain. *Cities, 6*(3), 227–233.

Shange, N. (1980). *For colored girls who have considered suicide when the rainbow is Enuf: A Choreopoem*. New York: Bantam.

Sibley, D. (1997). *Geographies of exclusion*. New York: Routledge.

Soja, E. (1996). *Thirdspace: Journeys to Los Angeles and other real-and-imagined places*. Cambridge MA: Blackwell Publishers.

Thompson, L., & Cupples, J. (2008). Seen and not heard: Text messaging and digital sociality. *Social and Cultural Geography, 9*(1), 95–108.

Tuan, Y. (1978). Sign and metaphor. *Annals of the Association of American Geographers, 68*(3), 363–372.

Tuan, Y. (1979). Sight and pictures. *The Geographical Review, 69*(4), 413–422.

Tuan, Y. (1991). Language and the making of place: A narrative-descriptive approach. *Annals of the Association of American Geographers, 81*(3), 684–696.

Watson, W. (1969). The role of illusion in North American geography: A note on the geography of North American settlement. *The Canadian Geographer, 14*(1), 10–27.

Women and Geography Study Group of the IBG. (1984). *Geography and gender: An introduction to feminist geography (explorations in feminism)*. London: Harper Collins.

Wright, J. (1947). Terrae incognitae: The place of the imagination in geography. *Annals of the Association of American Geographers, 15*, 1–15.

Zonn, L. (1984). Landscape depiction and perception: A transactional approach. *Landscape Journal, 3*, 144–150.

Zonn, L. (1985). Images of place: A geography of the media. *Royal Geographical Society of Australia, South Australian Branch, 84*, 35–45.

Zonn, L. (Ed.). (1990a). *Place images in media*. Savage: Rowan and Littlefield Publishers.

Zonn, L. (1990b). Tusayan, the traveler, and the Imax theater: An introduction to place images in media. In L. Zonn (Ed.), *Place images in media* (pp. 1–5). Savage: Rowan and Littlefield Publishers.

Zube, E., & Kennedy, C. (1990). Changing images of the Arizona environment. In L. Zonn (Ed.), *Place images in media* (pp. 183–203). Savage: Rowan and Littlefield Publishers.

Film Geography: A Review and Prospectus

2

Laura Sharp and Chris Lukinbeal

With this paper we review past works that have established film geography as a sub-discipline. The paper is organized around the author-text-reader (ATR) model and pays particular attention to its role in defining the area of study and how it is approached theoretically and methodologically. The textual metaphor from which the ATR model is derived is a signifying practice associated with the cultural production of meaning through various forms of representation. Textual analysis is a hermeneutical method that became hegemonic in film studies beginning in the 1970s following Christian Metz's influential application of semiotics to film, which occurred concomitantly with the establishment of film theory as a serious discipline (c.f. Shiel 2001). The method came to geography later during the "linguistic turn" in the social sciences that did not take full effect until the late 1980s (Lukinbeal and Zimmermann 2008). While the ATR model consists of three modalities, researchers have tended to focus on only one at a time (Dixon et al. 2008). An author-centered approach focuses on the pre-filmic processes of meaning creation. Here, the emphasis is on production, labor, the auteur, the generative process of meaning creation, and the overall economic conditions within the creative industries. A text-centered approach analyzes the construction of meaning within the film's diegesis and mise-en-scène. Reader-centered approaches investigate film as a spectatorial practice, the audience as market, the situatedness of consumption, the ethnography of film audiences, and film exhibition.

The textual metaphor is not the only way to approach film geography. Moreover, though it has enjoyed a period of dominance within the field (Cresswell and Dixon 2002), it is not without its problems. As Dixon et al. (2008) have demonstrated, because the ATR model relies on the temporary stabilization of textual meaning

L. Sharp (✉) • C. Lukinbeal
School of Geography and Development, The University of Arizona, Harvill Building 437B, 85721 Tucson, AZ, USA
e-mail: laurasharp@email.arizona.edu; chris.lukinbeal@arizona.edu

S.P. Mains et al. (eds.), *Mediated Geographies and Geographies of Media*,
DOI 10.1007/978-94-017-9969-0_2

relative to a specified context – the conditions of production, the researcher's own theoretical and intertextual framework, or the time and place of audience viewing and interpreting – the ATR model operates on an essentialist ontology, even as it draws from otherwise poststructuralist theories. A second noted concern undergirding the metaphor is the real/reel binary. This lurking epistemological trap conceives of the reel, onscreen world as a representation of the real, offscreen world. This manner of thinking has significant implications for how geographers approach film. Attempts to overcome the binary have been made by deploying dialectics, simulacra, and haptics. Where dialectics and simulacra maintain the conceptualization of film as text, haptics moves the discussion away from text and optics and onto a reconceptualization of film as an embodied and emotional event. Despite the flaws of the ATR model, because much film geography research up to this point has implicitly or explicitly relied upon it, it remains a useful heuristic for discussing the breadth of film geography thus far (Lukinbeal 2009, 2010). Recently, some welcome avenues away from the text and its attendant binary have begun to appear. Therefore, although we rely on this model as a framework, we also expand upon it to explore the ways that it can be used to accommodate non-textual approaches.

Author-Centered Approach

Often, studies of films from the author modality employ the notion of *auteurship*, the idea that films possess a unique aesthetic and philosophical outlook that is inscribed in the film by the director (the auteur, or author). Film theorist Andrew Sarris (1962) identified three aspects of auteurship: technical competency, aesthetic signature, and the creation of meaning that derives from the relationship between the director and the film content. In this vein, some geographers have been interested in the spatial meanings found in the works of specific directors, including Bill Forsyth (Aitken 1991); Peter Weir (Aitken and Zonn 1993); Gus Van Sant (Lukinbeal and Aitken 1998); Werner Herzog, Carlos Diegues, and John Boorman (Godfrey 1993); Jacques Tati (Marie 2001); independent documentary filmmakers McGuiness, Fugate, and Palos (Dixon 2008); and Sergio Leone (Starrs 1993). These analyses are classically textual in that they use information about the director(s) as a launch pad for their interrogation of the construction of geographic meaning within the text. In contrast to the emphasis placed on the auteur by the author-centered approach via the textual metaphor, non-textual approaches to this modality position it within a broader economic series of productive practices.

The most common non-textual author-centered approach is to focus on the economics of production, which can be sub-divided into political economy and economic geography. Writers outside of geography have taken a Marxist and/or critical approach to examine how power relations within the media industries contribute to the hegemony of globalization (Bagdikian 1992; Miller et al. 2005). In the current compilation Brett Christophers's chapter takes this approach, exploring the geographically distinctive nature of media's political economy. Approaching the television industry from the standpoint of labor, Vicki Mayer (2011) has taken a similarly critical approach by illustrating the mutually-producing relationship between television commodities and the subjectivities of the unseen "below the line" labor (anyone who is not an actor, director, producer, or writer) that makes them possible.

The more common approach to film production within geography, however, is to use economic theories to examine national centers of production. Most notable in this area is the work done by Susan Christopherson (2002, 2006, 2008, 2013; Christopherson and Rightor 2010; Christopherson and Storper 1986, 1989) and Michael Storper (1989, 1997; Storper and Christopherson 1985, 1987), as well as Allen Scott (2005). These researchers have focused on the economic geography of the Hollywood film industry. They have charted the transition of Hollywood from a craft-based industry, through its heyday during the Golden Age when it was underlined by Fordist practices, to its more current configuration under flexible specialization. Flexible specialization is characterized as networks of sub-contracted companies and major conglomerates that are flexible enough to come together around specific projects and then dissolve and re-configure as needed. Further, through sub-contracting, the industry has at its disposal more specialized skills available to hire for specific projects. This reorganization of the industry has created a unique landscape where production is concentrated around specific agglomerations of industry, and yet dispersed by way of location production incentives, tax credits, and below the line talent pools.

Where American geographers have focused primarily on the Hollywood film industry, others have examined the geographies of other global film industries. In this compilation Curtin's chapter explores the historical geography of Chinese cinema through a focus on how the cinema of Hong Kong has been affected by Hong Kong's incorporation into Mainland China. Both Mike Gasher (1995, 2002) and Neil Coe (2000a, b) have detailed the cultural and economic geographies of the film industry of British Columbia, Canada. Where other studies tend to naturalize a national film industry, the case of British Columbia highlights the effects of globalization and the phenomenon of Hollywood's runaway production. Neil Coe does this by focusing on the economic geographies of capital and labor relations, while Mike Gasher is more concerned with the cultural implications of Hollywood's hegemony and British Columbia's control over their own representation. According to Gasher (1995, 234), "it is the definition of a Canadian reality which is at stake in the struggle for control of the mediascape."

Inspired by Gasher's work, Lukinbeal (1998, 2004, 2006, 2012) has emphasized an approach to film geography that combines cultural studies and cultural economy. This approach makes two demands. The first is that we understand the creation of a cultural text or product as ontogenetic, embedded within the ongoing political and economic practices of the industry. The second is that we recognize that hermeneutical analyses of cultural texts necessitate an engagement with the political economy of the production practices that went into the text's creation. Rather than examining the representation of gender relationships, sacrifice, and territoriality during the American Civil War in North Carolina – all themes that could be key topics of interest in a textual analysis of the film *Cold Mountain* – Lukinbeal's (2006) analysis focuses instead on the politics associated with where the film was made and how this impacted the text. To save money, the producers of *Cold Mountain* chose to film in Romania. To save face, they argued that North Carolina did not offer enough historical realism. This statement, which obfuscates the producer's economic motivation, is questionable however, as the book's setting in Asheville, North Carolina was used just a decade earlier to shoot *The Last of the Mohicans*, a story based on the French and Indian War. Amidst calls for a boycott of *Cold Mountain* at the Oscars for its runaway production practices, one cannot but be critical of the state's own practice of doubling for other locations in film and television.

Text-Centered Approach

Textual inquiry assumes that cultural products and practices, such as landscape or film, are systems of signification that can be interpreted if one knows the "language" in which they are written. This approach to film analysis is not exceptional to geography. As Shiel (2001, 3) notes, "Film Studies has been primarily interested in film as *text* [...] and with the *exegesis* of the text according to one or other hermeneutic." He goes on to suggest that this approach is largely due to the fact that the origins of film studies lie in literary analysis. In geography, the rise of the textual metaphor can be traced to the linguistic turn and the adoption of post-structuralist theories in the social sciences. It is important to recognize the significance of the linguistic and discursive nature of geography's adoption of post-structuralist thought, influenced as it was by thinkers such as Foucault, Derrida, and Barthes; it was because of this linguistic-epistemological approach that the textual metaphor became widespread in geography and dominant as a method of research on film in particular (c.f. Natter and Jones 1993a; Cresswell and Dixon 2002; Dixon et al. 2008; Lukinbeal and Zimmermann 2008).

In order for a text to be interpreted it must be understood in relation to some context – of production, for instance – that might include factors such as the cultural era in which it was produced and the personal vision of the auteur. This juxtaposition between text and context leads to the oft-noted issue of the real/reel binary (c.f. Cresswell and Dixon 2002), the belief that film is a representation of reality. Some examples of the binary have been overt, taking it as a research topic in itself. This can be seen in Benton's (1995) "Will the Real/Reel Los Angeles Please Stand Up?" or Horton's essay where he argues, "All landscapes in cinema are 'reel'" (2003, 71). Frequently, however, it is more insidious, appearing as the omission of the researcher's ontological orientation to the object of study, accompanied by slippery language about what the researcher hopes to gain by examining a film. Often, the researcher is interested in a given film because it is taken to be either a particularly astute or a particularly problematic representation of what is understood to be the "real" situation "outside" of the film. An example of this is Klaus Dodds's (2013) paper examining the depiction of the US-Canadian border in the film *Frozen River*. Introducing the film and topic, Dodds writes:

> In marked cinematic contrast to the US-Mexican border, the US-Canadian border, and as *Frozen River exemplifies*, the 'internal borders' of the United States, are rather 'hidden' in comparison. The *actual border*, as the film *vividly portrays*, is not simply one between the Unites States and Canada [. . .] *The border, in this film*, is shown to [be] a complex space of alienation, containment, dispossession, and incorporation. (2013, 2–3, emphasis added)

The remainder of Dodds's article is an interpretation of the film, which acts as foil for Dodds's discussion of his primary concerns: the material, social, and biopolitical nature of borders. Here, while Dodds maintains a distinction between the "actual border" and the border in the film, or between the real and the reel, he nevertheless implicitly treats the film as an unmediated window to the world, a representation of such caliber that it can be used, in effect, as a case study on which to base claims about reality outside of the film. This subtle contradiction is enabled by Dodds's failure to account ontologically for his object of study. Making a similar critique, Glynn and Cupples (2014) have also pointed to the often binarizing analyses of media texts in Dodds's work, as well as to the tendency of Dodds and other popular geopolitics scholars toward conceiving of texts as coherent, self contained systems lacking in attention to the theoretical complexities therein. As Glynn and Cupples show, this demonstrates the need for popular geopolitics to engage with the extensive literature of cultural studies (and vice versa). As our own example suggests, this argument should be extended to include the need for scholars of popular geopolitics to engage with the broader media geography community in order to enrich and strengthen the theoretical conceptualizations of both sub-disciplines.

The real/reel binary provides a simplistic ontology that has overt ramifications for film geography. First, it positions film as a secondary object, a cultural text that functions only to reflect lived conditions. In so doing, it enforces a hierarchy of research in which "true" meaning production comes only from first-hand or real experiences and not from second-hand or mediated experiences. This hierarchy strengthens the normative belief in geography that film is mere entertainment (Gold 1984; Harvey 1990). The second implication of the binary is that it constricts research to the geography *in*, *of*, or *from* film (c.f. Hopkins 1994). Whereas research on the geography *in* film focuses on the production of meaning that occurs within the film text, research on the geography *of* film examines the spatial practices of film production and consumption. The study of geography *from* film looks at how a text can influence geography outside of the text. The relationship of the real/reel binary to the textual metaphor is especially thorny owing to the epistemological and ontological implications of the textual metaphor. The oft-cited dictum, "There is nothing outside of the text" (Derrida 1998, 158) negates a distinction between the real and reel at the level of epistemology, rather than at the level of ontology, by suggesting that there is no way to look beyond our own linguistic-cultural ways of knowing and thus no way to access an ontological reality outside of the text. The result of this ontological evasion is that the focus of research becomes a textual analysis of a film's narrative, with an occasional reference to the mise-en-scène and film form. Thus, while geographers frequently use film as a means to explore concepts of gender, sexuality, race, colonialism, or class, very few have paid attention to film qua film (Doel and Clarke 2007).

Three Approaches to the Real/Reel Binary

One of the first responses to the real/reel was to approach it through dialectics, which is to suggest that social life and representation are mutually producing. In their textual analysis of the Michael Moore documentary *Roger and Me*, Natter and Jones (1993a) investigate the portrayal of economic decline on Flint, Michigan, arguing that Moore's elision of traditional "objective" documentary techniques in favor of an overtly politicized narrative renders the film more authentic. For the authors, it is because of the dialectic nature between representations and their contexts of production and reception that geographers need to take movies such as Moore's seriously. In their words, "the power of representations to intervene in the ongoing reformation of material life should not be underestimated" (1993, 156; c.f. Jones and Natter 1999). Significant to this approach is that film is still seen as "representation," signaling the continued belief that film is attempting to be something other than itself. A second way of getting beyond the real/reel complicates the idea of representation by calling on Baudrillard's notion of simulacra.

According to Baudrillard (1994), in the society of late capitalism the signs and symbols necessary for meaning production have ceased to have any relation to reality and instead are simulacra, a copy without an original. To say that film is simulacra then, is to say that there is no reality other than the film itself, that film is its own reality. While it was David Clarke (1997) who first began to ponder the implications for film geography that a simulacral approach might offer, it has been in Clarke's prolific work with Marcus Doel that there developed a historical account for how the simulacral nature of film came to be. In a series of papers Doel and Clarke (Clarke and Doel 2006, 2007; Doel and Clarke 2007; Doel 2008) document the transition from animated photography of the late nineteenth and early twentieth centuries to cinema. Where anima-photographers were driven to capture and present reality as found, through the adoption of the montage editing technique, cinema became a means to manipulate space and time and thus engineer a reality-effect.

A third approach to the binary has been to treat film as no longer representational, but as a geography in and for itself. This approach eschews the notion of an immobile viewing subject segregated from an object of vision, the film. Bruno (2002) highlights this distinction by juxtaposing the conceptual pairings of *sight*- and *site*-seeing and *voyeur* and *voyageur*, shifting our understanding of cinema from an optical to a haptical affair. Haptics focuses on corporeality and the porous boundary between inside and outside, the skin, thereby repositioning our attention to the sense of touching and being touched (Laine 2006). According to Sharp and Lukinbeal (forthcoming) "This redirection occurs not merely as an add-on to vision, but as an emotional resonance affecting the body." In a similarly tactile approach Craine and Curti (2013) suggest that we treat televisual realities as "bodies among bodies" (2013) and, along with Stuart Aitken, propose that through our affective relations "we become the image" (Craine et al. 2013, 264). The emphasis placed here on viewing as a (bodily) experience further applies to the reader-centered approach.

Reader-Centered Approach

The reader-centered approach takes as axiomatic "the death of the author," the notion that a film text is always re-written as it is read depending on the positionalities of the viewers and their contexts of viewing. Following this, reader-centered studies tend to focus on the multiplicity of meanings derived during film reception and effects these interpretations have, as well as the situated act of consumption. Of the ATR model's three modalities the reader-centered has been the least discussed by geographers. In part, this may be attributed to the theoretical rather than empirical nature of traditional spectatorship research within film studies (Mayne 1993), as well as to the problems of essentialism and relativism that are quickly unearthed within these theories. Essentialism in spectatorship research is when the researcher's interpretation of the audience's perspective is taken to be everyone's viewpoint. Relativism, or the fact that meaning is always dependent on the viewer's positionality, works to expose that meaning has been made to appear natural, rather than contingent.

Some ways forward into audience studies by geographers have been discussed by Jancovich et al. (2003), who divide practices of consumption into four categories: the audience as market, the situatedness of reception, ethnographies of reception, and exhibition. The first category, the audience as market, focuses on audience demographics and tastes, as well as how the audience is conceptualized and targeted by the film industry. Ethnographies of film consumption, Jancovich, Faire, and Stubbings's second category, engages the everyday practices and motivations of viewing. Here, the preference of the viewer is of less interest than the social activity of cinema-going and the opportunities for interaction that it affords. The third category, the situated approach, is archival and intertextual in nature; it explores movie reviews, marketing material, news stories, billboards, and other media that help contextualize the reception of a film by a social group in a given era. The fourth category is the place of the audience, which looks at the history and geography of film exhibition sites. Some venues of interest to exhibition studies have been film festivals (Stinger 2001; Elsaesser 2005; Wong 2011), movie theatres (Zonn, Chap. 9, current volume; Jones 2001; Bruno 2002), and the home (Klinger 2006).

One of the most significant elements of film reception to the exhibition studies approach is the historical geographies and architecture of film viewing venues. Allen (1990) has demonstrated how empirically based historical research on film can break down myths about movie going, for instance the belief that cinema has always been a primarily urban trend. Rather, in the United States during the first decade of cinema's commercialization, 71 % of the population lived in small towns and rural areas. Thus, although urban nickelodeons and vaudeville theatres are often cited as cinema's origins, Allen points out that these origins are as much a small town and rural phenomenon, with film exhibition occurring wherever equipment, space, and interest aligned. Allen's research is only the beginning of a fascinating and under explored historical geography of film exhibition in the United States. This history begins with film's origins in the public spaces of small towns and rural areas, as well as in urban nickelodeons and vaudeville theatres in ethnic enclaves. Pursuing a dream of respectability, theatre owners sought out locations at the edge of shopping and central business districts, giving rise to the movie palace and the transformation of film into an architectural event (Merritt 1979; Hanson 1991). As televisions became readily available reception locations became striated. The movie palaces followed the post-World War II demographic shift to the suburbs and away from the struggling downtowns (Christopherson and Storper 1986). Suburban shopping malls, as they began to appear, were a natural site for housing the increasing size and numbers of screens that theatres maintained (Friedberg 1993). Facing years of neglect, movie palaces across the country fell into decline. Today's cinema theatre landscape is a variegated one, filled with the carcasses of movie palaces, penny-arcade museums, abandoned drive-ins, and monster-plexes hungry for more attendance. While in some cities undergoing gentrification movie palaces are enjoying a rebirth, others have been cut up, spawning a postmodern spectacle: the palace-plex.

The Viewing Subject

The reader-centered approaches cannot be understood without thinking about the people that have populated the festivals, nickelodeons, home-theatres, and roadside attractions of exhibition studies: the viewing subjects. How we conceive of the viewing subject has important ramifications for how we understand and research cinema generally, and the ATR model specifically. The historical trajectory of spectator theory has roots in Marxism, semiotics, and psychoanalysis, all of which came to occupy a prominent position in film studies of the 1970s through *apparatus theory*, as applied by such influential film critics as Christian Metz, Laura Mulvey, and Jean-Louis Comolli. According to this theory it is through the filmmaking apparatus and the mechanics of film construction (camera movement and angle, for instance) that a film's meaning is brought into being. Moreover, because film is an always-flawed attempt at mimesis it is inherently ideological. Through Lacan's mirror stage this imperfect representation of reality constructs the viewing subject in ideology. The understanding of the spectator through apparatus theory is a voyeuristic or one-way model that creates an automatic binary distinction separating the subject (the viewer) from the object (the film). This voyeuristic aspect of the film spectator has two connotations. The first is of an immobile subject tied to the disembodied gaze. The second is sexual and alludes to scopophilia. These questions have led film theorists to consider who is the "ideal" viewing subject. For the influential feminist film theorist Laura Mulvey (1975), Hollywood cinema is founded on the male gaze, where the male character looks, the audience looks, and the female character is looked at; through the male gaze of the camera both male and female audience members are constructed as male. Hollywood's ideal viewing subject is not only male, however, but also white, young, middle class, and an agnostic Christian (c.f. Mayne 1993). As Gledhill (1998) as pointed out, however, the question of the female viewer identity is more complicated than Mulvey and others have proposed in that it must take into account the different sites of negotiation: the text's production, the text itself, and the text's reception.

While psychoanalytic theories remain strong, the more recent uptake of haptical and affective theories has posed significant challenges to the mind/body dualism of the voyeuristic approach to cinema as a primarily visual-cerebral activity. Bruno's (2002) haptical mobilization of the spectator and transformation of voyeur into voyageur (discussed above) has been significant for helping conceptualize film viewing as an emotional and bodily experience. Under the voyeur model the notion of film as representation is upheld by keeping the theoretical focus on optics and sight. Due to cinema's much-heralded visual realism, the focus on optics positions the viewing subject as occupying a reel or real space, producing an indexical relationship between image and reality. Through haptical mobilization the spectator is not chained in Plato's cave, but free to wander (Bruno 1993). Additionally, where optical theories produce a heterological subject, haptics produce an embodied subject whose senses work in cooperation. Our attention thus shifts to corporeal experience of the subject; consciousness is not established through the mind/body dualism, but rather resides in the porosity of the skin (Kirby 1996; Laine 2006). According to Laine (2006, 104), "consciousness has no permanent 'place' anywhere; rather it arises whenever one touches another, in the mutual act of shaping." This haptical voyage of the spectator is, moreover, architectural in character, as architecture transforms film into cinema by providing a house within which the perceptual journey can take place, one that is "topophilically re-collected for public housing and exploration" (Bruno 2002, 50).

Carrying on Lacan's conception of the subject as formed through the gaze, Crang (2002), like Bruno (2002) and Craine et al. (2013) discussed above, also argues against a subject/object dualism, but does so by suggesting that we understand the image and the observer as always-already united, a coproduction wherein one does not exist without the other. For Crang, this is particularly relevant given the ubiquity of mobile cameras in today's hyper-mediated society, a fact that draws our attention to the mobile process of observation and capture and its relation to the screened content. Rather than thinking of capture, image, and reception as discrete moments (à la the ATR model), film becomes an assemblage, an active process of connecting people and things in space and time. By shifting our attention to the self's becoming with the image we are able "to move from a focus on the motion of images swirling around an analytically stationary and embattled subject to a view of the subject in motion and occupying the same terrain as the images" (Crang 2002, 27). In other words, while it has been conceptually useful to break cinema into the three modalities outlined here, continuing to do so may blind us to alternative orientations to film geography and the questions that can and need to be asked.

Conclusion

What is the future of the author-centered approach? While interest in certain directors will always come and go, this approach has not distinguished itself as being significantly different than text-centered approaches. For auteur studies to move forward they will need to situate the author and authorship within a broader milieu of productive practices and/or historically contingent inter-textual relationships. Where economic geography has embraced the importance of the film production industry, much of the political economy research has not had a geographic focus. Further, political economic or Marxist geographers have failed to give any credence to film or the film industry (excepting Christophers). This is perhaps due to a normative belief that film is mere entertainment or, as David Harvey has said (1989, 322), just "a sequence of images upon a depthless screen" that do not have "the power to overturn established ways of seeing or transcend the conflictual conditions of the moment." We find this interesting, considering the emphasis by the Frankfurt School on film and media, as well as in the works of Raymond Williams and Guy Debord.

The most fruitful area of future research for author-centered studies lies at the intersection of cultural economy and cultural studies. Within this there are two different tracts. The first comes out of the approach developed by Gasher and Lukinbeal, which emphasizes a critical engagement of textual meaning through a focus on the economics of location production and the politics of on-location filming. The second emphasizes the underlying power relations, inequalities, and uneven developments of the industry and how these influence meaning production. Meaning production here focuses on how social and economic production practices perpetuate, reify, and naturalize the hegemonic ideologies complicit in their own production.

What is the future of the text-centered approach? As people continue to study film-as-text, issues over the real/reel binary and the infinite deferral of research away from film qua film will continue. A dialectic understanding of the real/reel binary reifies the simplistic ontology of film as mere representation. We therefore see simulacra and haptics as the most productive avenues of future research in the text-centered approach. Simulacra provides a means to discuss representational discourse in relation to film without connoting that film is merely an image. It also helps shift the focus toward geographies of film form and its relation to the diegesis. Rather than segregating the viewer-reader from the text, a haptical approach positions the reader within an ontological understanding of the film viewing experience. A haptical understanding of film moves away from the connotation of subject/object relations, where film is merely a cultural object/product. In a similar way, geography's engagement with performance and non-representational theories seeks to overturn the focus of studying cultural products by switching the focus to cultural practices, affect, emotion, and the body. Particularly exciting paths of future text-centered research are those that combine simulacra and haptics with author or reader centered approaches. Fletchall et al. (2012), for instance, contextualize how the production of Orange County, California, or "the OC," derives from a simulacral palimpsest of media texts. It is through an understanding of the OC as simulacra that reality television shows depicting this area are understood by viewers, rather than through the actual happenings of daily life in Orange County. Further, they argue that through emotional and geographic realism reality show fans engage in a haptical practice of place-making.

What is the future of the reader-centered approach? The most exciting aspects of reader-centered studies are twofold. The first approach in need of further development is the theoretical advances regarding the viewing subject, especially as this relates to haptics and psychoanalysis. These theories are of most interest because they challenge traditional notions of the spatiality of film reception centered on voyeurism, thus allowing for more nuanced understandings. Where voyeurism allowed us to point out certain underlying power issues relating to the images (e.g. Mulvey 1975), it also delimits a terrain that is constricted by the subject/object and mind/body binaries that lead to heterology, an unsustainable theoretical construct. On the other hand, the promise of affect, emotion, and non-representational theories in geography parallels the emphasis in film studies on haptics, the body, and psychoanalysis. Cross-pollination between these fields is much needed and these theories point the way.

The second needed area of research in the reader-centered approach is engagement with historical geographies of exhibition, especially as this relates to spectatorship and how we understand the production of meaning within the diegesis and mise-en-scène. Researchers here have mainly focused on the historical context of viewing within the United States and Britain. Little attention has been paid, however, to comparing the situatedness and spatiality of exhibition across countries. Further, the ATR model assumes a textually-centered spectator. Although poststructural theory has challenged this assumption through concepts of essentialism and relativism, Hanson's (1991) work highlights how, during the 1910s and 1920s, the textually-centered spectator was mutually co-constructed with Hollywood's classical paradigm of narration. The classical paradigm attempted to homogenize meaning across a national scale by eliminating the "empirically variable acts of reception" (Hanson 1990, 55). The development of the Hollywood narrative style would take decades to develop and is not a static construct (Bordwell 2002).

In this paper we have used the ATR model as a heuristic device to discuss past trajectories and future possibilities within film geography. Despite significant drawbacks to this approach, its continued relevance is seen in the work of the many geographers who continue to deploy it, wittingly or not. For this reason, we have found it necessary to provide a clear delineation of how the model has become entrenched in the discipline, as well as how it can be used to move forward. By recognizing the ATR model and its accompanying real/reel binary for what they are – scaffolding that allows researchers to safely and slowly work towards a stronger, more theoretically sound paradigm – it is our hope that we will soon be at a point where we can move beyond this approach to discover new and exciting vistas of research for film geography.

References

Aitken, S. C. (1991). A transactional geography of the image-event: The films of Scottish director, Bill Forsyth. *Transactions, 16*(1), 105–118.

Aitken, S. C., & Zonn, L. E. (1993). Weir (d) sex: Representation of gender-environment relations in Peter Weir's picnic at Hanging Rock and Gallipoli. *Environment and Planning D, 11*(2), 191–191.

Allen, R. C. (1990). From exhibition to reception: Reflections on the audience in film history. *Screen, 31*(4), 347–356.

Bagdikian, B. (1992). *The media monopoly*. Boston: Beacon Press.

Baudrillard, J. (1994). *Simulacra and simulation* (S. F. Glaser, Trans.). Ann Arbor: University of Michigan Press.

Benton, L. M. (1995). Will the real/reel Los Angeles please stand up? *Urban Geography, 16*(2), 144–164. doi:10.2747/0272-3638.16.2.144.

Bordwell, D. (2002). Intensified continuity: Visual style in contemporary American film. *Film Quarterly, 55*(3), 16–28.

Bruno, G. (1993). *Streetwalking on a ruined map: Cultural theory and the city films of Elvira Notari*. Princeton: Princeton University Press.

Bruno, G. (2002). *Atlas of emotion: Journeys in art, architecture, and film*. New York: Verso.

Christopherson, S. (2002). Project work in context: Regulatory change and the new geography of media. *Environment and Planning A, 34*(11), 2003–2015. doi:10.1068/a34182.

Christopherson, S. (2006). Behind the scenes: How transnational firms are constructing a new international division of labor in media work. *Geoforum, 37*(5), 739–751. doi:10.1016/j.geoforum.2006.01.003.

Christopherson, S. (2008). Beyond the self-expressive creative worker: An industry perspective on entertainment media. *Theory, Culture & Society, 25*(7-8), 73–95. doi:10.1177/0263276408097797.

Christopherson, S. (2013). Hollywood in decline? US film and television producers beyond the era of fiscal crisis. *Cambridge Journal of Regions, Economy and Society, 6*(1), 141–157. doi:10.1093/cjres/rss024.

Christopherson, S., & Rightor, N. (2010). The creative economy as "Big Business": Evaluating state strategies to lure filmmakers. *Journal of Planning Education and Research, 29*(3), 336–352.

Christopherson, S., & Storper, M. (1986). The city as studio; the world as back lot: The impact of vertical disintegration on the location of the motion picture industry. *Environment and Planning D: Society and Space, 4*(3), 305–320.

Christopherson, S., & Storper, M. (1989). The effects of flexible specialization on industrial politics and the labor market: The motion picture industry. *Industrial and Labor Relations Review, 42*(3), 331–347.

Clarke, D. B. (1997). *The cinematic city*. New York: Routledge.

Clarke, D. B., & Doel, M. A. (2006). From flatland to vernacular relativity: The genesis of early English landscapes. In M. Lefebvre (Ed.), *Landscape and film*. New York: Routledge.

Clarke, D. B., & Doel, M. A. (2007). Shooting space, tracking time: The city from animated photography to vernacular relativity. *Cultural Geographies, 14*(4), 589–609. doi:10.1177/1474474007082295.

Coe, N. (2000a). The view from out West: Embeddedness, inter-personal relations and the development of an indigenous film industry in Vancouver. *Geoforum, 31*, 391–407.

Coe, N. (2000b). On location: American capital and the local labour market in the Vancouver film industry. *International Journal of Urban and Regional Research, 24*(1), 79–94.

Craine, J., & Hadi Curti, G. (2013). Prologue: Introducing the fight to stay put: Social lessons through media imaginings of urban transformation and change. In G. Hadi Curti, J. Craine, & S. C. Aitken (Eds.), *The fight to stay put: Social lessons through media imaginings of urban transformation and change* (pp. 9–18). Stuttgart: Franz-Steiner Verlag.

Craine, J., Hadi Curti, G., & Aitken, S. C. (2013). Cosmopolitan sex, monstrous violence and networks of blood. In G. Hadi Curti, J. Craine, & S. C. Aitken (Eds.), *The fight to stay put: Social lessons through media imaginings of urban transformation and change* (pp. 249–265). Stuttgart: Franz-Steiner Verlag.

Crang, M. A. (2002). Rethinking the observer: Film, mobility, and the construction of the subject. In T. Cresswell & D. Dixon (Eds.), *Engaging film: Geographies of mobility and identity* (pp. 13–31). Lanham: Rowman & Littlefield.

Cresswell, T., & Dixon, D. (Eds.). (2002). *Engaging film: Geographies of mobility and identity*. Lanham: Rowman & Littlefield.

Derrida, J. (1998). *Of grammatology* (G. C. Spivak, Trans.). Baltimore: Johns Hopkins University Press.

Dixon, D., Zonn, L., & Bascom, J. (2008). Post-ing the cinema: Reassessing analytical stances toward a geography of film. In C. Lukinbeal & S. Zimmermann (Eds.), *The geography of cinema – A cinematic world* (pp. 25–50). Stuttgart: Franz-Steiner Verlag.

Dodds, K. (2013). "I"m still not crossing that': Borders, dispossession, and sovereignty in frozen river (2008). *Geopolitics, 18*(3), 560–583. doi:10.1080/14650045.2012.749243.

Doel, M. A. (2008). From animated photography to film: The formation of vernacular relativity in early English films (1895–1908). In C. Lukinbeal & S. Zimmermann (Eds.), *The geography of cinema – A cinematic world* (pp. 87–100). Stuttgart: Franz-Steiner Verlag.

Doel, M. A., & Clarke, D. B. (2007). Afterimages. *Environment and Planning D: Society and Space, 25*(5), 890–910. doi:10.1068/d436t.

Elsaesser, T. (2005). *European cinema face to face with Hollywood*. Amsterdam: Amsterdam University Press.

Fletchall, A., Lukinbeal, C., & McHugh, K. (2012). *Place, television, and the real Orange County*. Stuttgart: Franz Steiner Verlag.

Friedberg, A. (1993). *Window shopping: Cinema and the postmodern*. Berkeley: University of California Press.

Gasher, M. (1995). The audiovisual locations industry in Canada: Considering British Columbia as Hollywood north. *Canadian Journal of Communication, 20*(2), 231–254.

Gasher, M. (2002). *Hollywood north: The feature film industry in British Columbia*. Vancouver: University of British Columbia Press.

Gledhill, C. (1998 [1988]). Pleasurable negotiations. In J. Storey (Ed.), *Cultural theory and popular culture: A reader*. Essex: Prentice Hall.

Glynn, K., & Cupples, J. (2014). Negotiating and queering US hegemony in TV drama: Popular geopolitics and cultural studies. *Gender, Place, & Culture: A Journal of Feminist Geography*. doi:10.1080/0966369X.2013.855711.

Godfrey, B. J. (1993). Regional depiction in contemporary film. *Geographical Review, 83*(4), 428–440.

Gold, J. (1984). *The city in film: A bibliography. Architecture series: Bibliography*. Monticello: Vance Bibliographies.

Hanson, M. (1990). Adventures of goldilocks: Spectatorship, consumerism and public life. *Camera Obscura, 22*, 51–71.

Hanson, M. (1991). *Babel and Babylon: Spectatorship in American silent film*. Cambridge, MA: Harvard University Press.

Harvey, D. (1990). *The condition of postmodernity: An enquiry into the origins of cultural change*. Malden: Blackwell Publishers Inc.

Hopkins, J. (1994). A Mapping of cinematic places: Icons, ideology, and the power of (Mis)representation. In S. Aitken & L. Zonn (Eds.), *Place, power, situation, and spectacle: A geography of film*. Lanham: Rowman and Littlefield.

Jancovich, M., Faire, L., & Stubbings, S. (2003). *The place of the audience: Cultural geographies of film consumption*. London: British Film Institute.

Jones, J. P., III, & Natter, W. (1999). Space "and" representation. In A. Buttimer, S. D. Brunn, & U. Wardenga (Eds.), *Text and image: Social construction of regional knowledges* (pp. 234–237). Leipzig: Institut für Länderkunde.

Kirby, K. (1996). *Indifferent boundaries: Spatial concepts of human subjectivity*. New York: The Guilford Press.

Klinger, B. (2006). *Beyond the multiplex cinema, new technologies, and the home*. Berkeley: University of California Press.

Laine, T. (2006). Cinema as second skin. *New Review of Film and Television, 4*(2), 93–106.

Lukinbeal, C. (1998). Reel-to-real urban geographies: The top five cinematic cities in North America. *California Geographer, 38*(1), 65–78.

Lukinbeal, C. (2004). The rise of regional film production centers in North America, 1984–1997. *GeoJournal, 59*(4), 307–321.

Lukinbeal, C. (2006). Runaway Hollywood: Cold Mountain, Romania. *Erdkunde, 4*, 337–345.

Lukinbeal, C. (2009). Film. In R. Kitchin & N. Thrift (Eds.), *International encyclopaedia of human geography* (Vol. 4, pp. 125–129). Oxford: Elsevier Press.

Lukinbeal, C. (2010). Film and geography. In B. Warf (Ed.), *Encyclopedia of human geography* (2nd ed.). London: Sage.

Lukinbeal, C. (2012). "On location" filming in San Diego County from 1985–2005: How a cinematic landscape is formed through incorporative tasks and represented through mapped inscriptions. *Annals of the Association of American Geographers, 102*(1), 171–190. doi:10.1080/00045608.2011.583574.

Lukinbeal, C., & Aitken, S. (1998). Sex, violence and the weather: Male hysteria, scale and the fractal geographies of patriarchy. In H. Nast & S. Pile (Eds.), *Places through the body* (pp. 356–380). New York: Routledge.

Lukinbeal, C., & Zimmermann, S. (Eds.). (2008). *The geography of cinema – A cinematic world*. Stuttgart: Franz Steiner Verlag.

Marie, L. (2001). Jacques Tati's *Play Time* as the new Babylon. In M. Shiel & T. Fitzmaurice (Eds.), *Cinema and the city: Film and urban societies in a global context* (pp. 257–269). Oxford: Blackwell.

Mayer, V. (2011). *Below the line: Producers and production studies in the new television economy*. Durham: Duke University Press.

Mayne, J. (1993). *Cinema and spectatorship*. New York: Routledge.

Merritt, R. (1979). Nickelodeon theaters: Building an audience for the movies. *Wide Angle, 1*, 4–9.

Miller, T., Covil, N., McMurrin, J., Maxwell, R., & Wang, T. (2005). *Global Hollywood 2*. London: British Film Institute.

Mulvey, L. (1975). Visual pleasure and narrative cinema. *Screen, 16*(3), 6–18.

Natter, W., & Jones, J. P., III. (1993a). Signposts toward a poststructuralist geography. In J. P. Jones III, W. Natter, & T. R. Schatzki (Eds.), *Postmodern contentions: Epochs, politics, space* (pp. 165–203). New York: Guilford Press.

Natter, W., & Jones, J. P. (1993b). Pets or meat: Class, ideology, and space in Roger & Me. *Antipode, 25*(2), 140–158.

Sarris, A. (1962). Notes on the auteur theory in 1962. *Film Culture,* Winter 1962–1963, *27*, 1–8.

Scott, A. (2005). *On Hollywood: The place, the industry*. Princeton: Princeton University Press.

Shiel, M. (2001). Cinema and the city in history and theory. In M. Shiel & T. Fitzmaurice (Eds.), *Cinema and the city: Film and urban societies in a global context*. Oxford: Blackwell Publishers.

Starrs, P. (1993). The ways of western death: Mor(t)ality and landscape in Cormac McCarthy's novels and Sergio Leone's films. *Wide Angle: A Quarterly Journal of Film History, Theory, Criticism, and Practice, 15*(4), 62–76.

Stinger, J. (2001). Global cities and the international film festival economy. In M. Shiel & T. Fitzmaurice (Eds.), *Cinema and the city: Film and urban societies in global context* (pp. 134–144). Malden: Blackwell Publishers Inc.

Storper, M. (1989). The transition to flexible specialization in the U.S. film industry: External economies, the division of labour, and the crossing of industrial divides. *Cambridge Journal of Economics, 13*(2), 273–305.

Storper, M. (1997). *The regional world: Territorial development in a global economy*. New York: Guilford Press.

Storper, M., & Christopherson, S. (1985). *The changing organization and location of the motion picture industry* (Research report R854). Los Angeles: Graduate School of Architecture and Urban Planning, University of California.

Storper, M., & Christopherson, S. (1987). Flexible specialization and regional industrial agglomerations: The case of the U.S. motion picture industry. *Annals of the Association of American Geographers, 77*(1), 104–117. doi:10.1111/j.1467-8306.1987.tb00148.x.

Wong, C. H. (2011). *Film festivals culture, people, and power on the global screen*. Piscataway: Rutgers University Press.

TV and the Spaces of Everyday Life

3

Lynn Spigel

In 1958 the Radio Corporation of America (RCA) promoted its vision of television's future with an image of a stylishly modern home (Fig. 3.1). Equipped with a "picture frame" flat screen TV mounted on a wall near a huge picture window, the living room was overcome by the postwar dream of TV leisure where views of the outside world (gleaming through the window) were now competing with virtual views on the TV screen. Adding to the attractions of this domestic utopia are a "television control unit" and a mini-fridge on wheels so that the residents are spared the quotidian "challenges" of simply moving around. As the RCA promotional rhetoric suggests, television offers a new and thoroughly modern form of spectacular intimacy where the virtual and the material co-exist, and where the object world is easily manipulated through technical and architectural tricks that allow for (at least the fantasy of) mastery over the environment.

Despite its somewhat antiquated version of the future, today this RCA home is easily recognized as a media space, a space not just full of media, but rather created in part by it.

In this respect, television is part of a longer history of communication and transportation technologies (the train, the airplane, the telegraph, the telephone, radio, cinema, the computer) that have contributed to changes in the way people experience time and space, making the world *seem* both smaller in scale and more readily accessible at the stroke of a keyboard or touch of a switch. Not surprisingly, television, satellites, and media networks have been central to theories of postmodern geography. As David Harvey (1990) argues, the media have contributed to the "time-space compression" that paradoxically creates uneven development within a highly unified global economy composed of homogenous products and manufactured spaces across the advanced capitalist world. With a similar concern

L. Spigel (✉)
School of Communication, Northwestern University, Chicago, IL, USA
e-mail: lspigel@northwestern.edu

S.P. Mains et al. (eds.), *Mediated Geographies and Geographies of Media*,
DOI 10.1007/978-94-017-9969-0_3

Fig. 3.1 Home of the future (RCA 1958)

for spatial homogenization, in his influential analysis of "supermodernity" Marc Augé (1995) claims that rather than a traditional anthropological sense of place, the networked consumer societies of the postindustrial world have constructed a series of "non-places"–supermarkets, air terminals, freeways–places we pass through rather than inhabit. On the one hand, television follows and even precipitates this pattern. Its live transmissions turn "there" into "here" on a daily basis while international franchises like *Survivor* or dubbed reruns like *Friends* circulate in places around the globe, and even on airplanes in between. On the other hand, television is, as David Morley (1991) puts it, both a global and a "sitting room" technology, so that forces of globalization and spatial homogenization must be conceptualized in relation to particular places and everyday experiences of home and homeland.

Here, I consider television's relation to the spatial geographies of everyday life, and in particular I explore the history of TV's impact on the relations between, and social construction of, private and public space. However obvious, it should be said at the outset that public and private are not essential givens but rather historical and geographically specific constructs (so that what is considered appropriate public or private behavior in one place or time may not be in another). So too, television is implemented differently in diverse nations and locations. In the following pages, I mostly explore television in western industrialized contexts. I consider its technological incarnations (as domestic TV, mobile TV, and "jumbo" screens in urban centers) in relation to the increasingly "mediatized" spaces of everyday life. In offering this large-scale overview, I want to consider some common threads, as well as divergent claims, among different approaches to studying TV as a spatial apparatus.

One of the perplexing issues for anyone interested in television is its own status as a "migrant" object across the disciplines of the humanities and social sciences. TV has been a subject for communication theory; history (especially cultural history and the history of technology); visual/textual analysis (that often involves methods from film and literary studies); visual anthropology and ethnography; journalism (particularly with regard to documentary realism); feminist and queer theory; media sociology and institutional analysis; critical theories of simulation and virtuality; critical race theory; phenomenology and philosophical questions of ontology; critical geography and urban studies; and art and architecture history/theory. In the work I present here, methods of textual analysis, discourse analysis, industry analysis, ethnographic and qualitative research, oral history, archival research, and even autobiography are all central to research. Moreover, often approaches overlap, making TV a truly interdisciplinary object. While the subjects are certainly vast and the methods are often quite different, my overview is intended as a map through which to understand connections among different avenues of inquiry into television and the spaces of everyday life. The subjects as I divide them here crystallize around: (1). Television's arrival in homes after WWII and its continued place in domestic space; (2). Television's relation to suburbanization and its privatization of public amusements, especially the theater; (3). Television's centrality to what Raymond Williams (1975) called "mobile privatization" and to related fantasies of virtual travel; (4). Television's aesthetics of liveness and "telepresence"; and (5). Television's shift from a predominantly domestic medium to a mobile technology and "everywhere" cultural form. In dividing my attention across these subjects I hope in to demonstrate the contact points between television and the dynamics of public and private space in media cultures.

Domestic Space and Family Intimacy

Although its installation has been uneven around the globe, in the first two decades after WWII, the new medium posed immediate concerns with regard to the intimate spaces of home, its relation to public spaces, and related issues of gender and generation. Historical research on a range of national contexts (including the US, Britain, Sweden, Italy, West Germany, Russia, Australia, Argentina, and Austria) have detailed the hopes and fears (which circulated in both popular and scholarly venues) about television's effects on family life, gender roles, and domestic space (Spigel 1992; Boddy 2004; Smith 2012; Olofsson 2012; Penati 2013; Perry 2007; Roth-Ey 2007; Darian-Smith and Hamilton 2012; Varela 2005; Bernold and

Ellmeier 1997). While the national responses vary in intensity and orientation, some commonalities –as well as differences–emerge.[1]

In the 1950s commentators often predicted that television would reunite the war-torn families of the previous decade, and in this respect TV was often depicted as a spatial apparatus that brought families closer together. In the US and Britain, people referred to television as an electronic "hearth" linking it to the traditional centers of domestic life (Frith 1983; Tichi 1992; Spigel 1992; Morley 2000; Smith 2012). So, too, as I detail in my book *Make Room for TV*, US advertisements often showed television sets in family circle iconography with mom, dad, and kids huddled around the screen. Sociologists suggested families were indeed using television as a means of keeping families together. A mother from a Southern California study claimed, "Our boy was always watching television [at other people's houses], so we got him a set just to keep him home." A mother from a Georgia study similarly enthused, "We are closer together . . . Don and her boyfriend sit here, instead of going out" (McDonagh et al. 1956, p. 116; Stewart cited in Bogart 1956, p. 100).

To be sure, not everyone was convinced of television's unifying power over family life. Sociologists and popular critics just as often depicted television as a divisive force that would disrupt traditional forms of intimacy and especially the gendered spaces of the home. In the US, women's magazines spoke constantly of the family fights television caused and recommended ways to balance the ideals of family togetherness with the divided interests of individual family members. Everything from room dividers to earplugs served as means by which to carve out television places within common spaces of the home. Women's household labor has historically presented a special dilemma for these twin ideals of unity and division because women were (and still often are) expected to perform chores while still taking part in family leisure time pursuits. Ads for television sets often showed housewives doing both at once – serving snacks, cradling babies, or drying dishes while watching with the family group (Spigel 1992). Even when women were depicted in family scenes, they were often spatially remote from men or children in the room. For example, in the RCA ad with which I began, it is clear that despite their mutual occupation of living space, the men and woman in the room are divided in their interests. The woman appears to be sideways glancing at the clown on screen while the men are involved in conversation, ignoring both her and the TV set. (Nevertheless, one of the men has his finger on the TV remote, apparently still dominating his wife's TV pleasures.)

In addition to its role in articulating spatial arrangements of family unity and division, television is also a symbolic object in the home that communicates a message about the resident. In his study of television's arrival in Japanese homes

[1]Histories of television's installation in homes is a relatively recent phenomenon, and to date there is no single comparative historical study of this on a transnational level. My effort to do so here, therefore, is based on my preliminary attempts to merge some of these studies. I want to thank generous colleagues for either translating or sharing essays, findings, and resources, especially Mirta Varela, Cecile Panati, William Uricchio, and Judith Keilbach.

of the 1960s, Shunya Yoshimi (1999) observes that the color television set (along with the air conditioner and car) functioned as a symbol of what it meant to be a modern family. Historical research on British, US, Australian, Swedish, Italian, West German, and Austrian television similarly finds that the TV set functioned as powerful symbol of the modern family home and/or social mobility in the postwar period (O'Sullivan 1991; Spigel 1992; Darian-Smith and Hamilton 2012; Olofsson 2012; Penati 2013; Perry 2007; Ellmeier 1997). As Kristin Roth-Ey (2007) claims, even in the Soviet Union, "The symbolism of a television set in every Soviet apartment" functioned as "proof of socialism's ability to deliver the good life." She goes on to show that "many foreigners who visited the USSR in the first years after Iosif's Stalin's death were struck by the presence of television technology in a country evidently still struggling to provide the basics of food, clothing and shelter" (pp. 181–2).

That said, in some households television was (and still is) a site of shame rather than an object of conspicuous consumption. In their oral history of Australian homes, Darian and Hamilton (2012) find that despite its almost universal presence by the 1960s, "a minority of middle class families believed television was too popular and too crass" and refused to buy a TV set (p. 41). In the US, as television moved from being a rich man's toy (in the experimental period of the late 1930s and 1940s) to a mass medium (by 1960 almost 90 % of homes had a least one TV), its class status also changed; and in this respect the spaces that TV occupied had to be carefully managed. High-end designers (and even some of the more socially aspirant middle-class home magazines) called TV an eyesore and often advised hiding it from view (for example, behind a wall of more "highbrow" objects like books or paintings). Speaking of television's introduction into Swedish homes in the 1950s, Jennie Oloffson (2012) observes a similar dynamic. While early adapters often displayed their TV sets as a sign of prestige, "social elites in Sweden later on concealed the TV object in the closet, a move that occurred in conjunction with the increasing retail of the TV object to non-elites" (p. 14).

In qualitative studies reflecting trends from the 1980s to the present, researchers have shown that television continues to have a central relation to family life and domestic spaces, both in its symbolic and its practical dimensions. People often use the television set as a site of display for personal expressions of family history and cultural heritage (McCarthy 2000; Morley 2000; Leal 1990; Silverstone 1994; Gauntlett and Hill 1991; O'Sullivan 1991; Olofsson 2012; Ureta 2008). In her ethnographic study of television in Brazil, Ondina Fachel Leal (1990) examines how working-class suburban families make television meaningful within an "entourage" of objects around it (photos, plastic flowers, knickknacks, religious pictures, etc.). Exploring media in Iban society of Malaysian Borneo, John Postill (1998) finds that people place family photographs on top of the television set, which in turn serves a means of symbolically re-connecting the family to deceased relatives or to family members who have migrated away from home.

People also manage TV spaces (and more recently computer and gaming spaces) in ways that fit with household routines, gender roles, and/or taste preferences (Morley 1985; Lull 1990; Burke 2003; Lally 2002; Ureta 2008; Aslinger 2013;

Young 2007). In his path-breaking *Family Television* (1986), which looks at a group of working-class British families, David Morley observes that the power dynamics of gender are intimately related to TV watching (for example, men tended to dominate the remote control). Later in *Home Territories* (2000), Morley reports on his and other studies that show how people use television and other communication technologies to "negotiate difference" in the family (along lines of age and gender especially) and to demarcate space. More generally, David Gauntlet and Annette Hill (1991) observe, "Television… is often a primary determining factor in how British households organize their internal geography," and the majority of their respondents watched TV apart from other family members in what they call "routinized dispersion" (p. 38).

The movement of television from the central spaces of the home to more private rooms (especially bedrooms) is often accompanied by deeply felt cultural practices of social etiquette and boundary marking. In his ethnography of low income families in Santiago, Chile, Sabastian Ureta (2008) finds that the newest and largest TV set is typically located in the family's "ceremonial" central living space where it is used as a symbol of class status and family pride rather than being a functional machine. In fact, some families felt it improper to keep the living room television turned on when guests arrive. Instead, family members watch older and smaller screen TVs in bedrooms. He notes, however, that bedroom TV has its own perceived pitfalls so that, for example, women expressed concerns about its effects on intimacy between couples. Writing about media and "bedroom culture" in the UK, Sonia Livingston (2007) argues that children's and teenagers' use of television and related media in bedrooms is symptomatic of the larger risk society: as adults consider outside spaces more dangerous for young people, and as public alternatives become scarce, parents channel young people's pleasure into the perceived safe space of the home, and the "media-rich bedroom" plays a key role in keeping children inside. Although articulated in relation to personal spaces and mobile platforms, such strategies recall the tactics of the previously cited mothers at the dawn of the TV age, who also used TV to keep children and teenagers home.

Today, particularly for people with broadband and mobile devices, the status of television as the symbolic center of the home is certainly in transition. However, the transitions are marked by ambivalences and paradoxes that suggest the complexity of interpreting trends. For example, in 2010 the Pew Research Center's nationwide Social and Demographic Trends Project reported a sharp decline in the number of people in the US who thought TV was a household necessity—from 64 % in 2006 to 42 % in 2010. And of younger adults, fewer than three-in-ten (29 %) "said they needed a television set." Nevertheless, the study also reported that the perception of need did not actually match up with purchasing behavior. "Even as fewer Americans say they consider the TV set to be a necessity of life, more Americans than ever are stocking up on them. In 2009, the average American home had more television sets than people" (Taylor and Wang 2010, p. 1).

This overabundance of TV correlates with the industry's push toward "personal TV" and related time-shifting and place-shifting technologies through which people can watch media (on numerous platforms) in any room of the house or on mobile technologies away from home. Television's convergence with the Internet allows for social interactions with networks of people outside the home; for example, now many people "tweet" TV as it happens, sharing comments on TV news coverage or responses to fiction shows. That said, even while Net-enhanced "smart television" makes it possible to interact in a more immediate way, such interactions are not entirely revolutionary. Even in the past, people often chatted about television programs on the telephone or at the water cooler (at work) or through fanzines so that private viewing was often turned into occasions for social exchange outside the family context. Rather than an absolute break with the past, today television is a hybrid spatial experience; people encounter and use TV a number of ways and in a variety of places. Even now, despite the push toward personal viewing, manufacturers are still also marketing TV as a collective household ritual through home theater technologies that have been part of the spatial imagination for television since its inception.

Home Theaters and Suburban Towns

The development of television as a domestic device occurred in the context of broader geographical shifts that influenced both its object form and the programs that emanated from its screen. In the US, Australia, and Britain, historians have considered television in the context of suburban expansion and new forms of postwar community, consumerism, and communication (Haralovich 1991; Lipsitz 1992; Spigel 1992; Hartley 1999; Darian-Smith and Hamilton 2012). In his account of television in everyday life, Roger Silverstone (1994) suggests that television is not just related to particular suburban histories, "but is itself suburbanizing "and has been central to the "suburbanization of the public sphere." Television, he claims, "is suburban in expression and reinforcement of the particular balance of isolation and integration, uniformity and variety, global and parochial identifies and cultures, that are, indeed, the hallmark of suburban existence" (pp. 55–57). Silverstone's insights into spatial hybridity and the balance of contradictory social categories are useful ways to think about how television negotiates the spatial conditions of everyday life in the modern world. Nevertheless, television is always also a site-specific spatial practice so that its relation to particular suburbs, cities, or rural towns may well result in important differences. For example, in her study of Italian television, Penati (2013) shows that the communications infrastructure, which was located in cities, made early television a predominantly urban phenomenon. But even in nations that did undergo massive suburbanization, the particular histories of space (both at the level of actual location but also in terms of broader issues of spatial justice) clearly make a difference in the ways in which different populations have experienced television (and newer media).

The spatial dislocations and re-orientations that accompanied television's rise in the US suburbs suggest how television was related, for example, to the history of racist geographies entailed in US migrations. Faced with a severe housing shortage in US cities, people migrated to new mass-produced suburbs that sprang up at enormous speed after WWII. Signaling the centrality of television to these new suburban homes, in 1952 (even before most Americans had a television set) Levittown (one of the first and most famous suburban developments of its time) offered a house with a television set built into its living room wall. But because these new developments were built with government financing that sanctioned racial exclusions (through zoning laws) suburban spaces were notoriously homogeneous, so much so that migration to the suburbs was known as "white flight." The suburban sitcoms that proliferated on early US television, with their all white nuclear families, reified and helped to naturalize the exclusionary geographies of suburban towns.[2] But in a more general sense, as a medium, television provided a kind of "antiseptic electrical space" that filtered out as much as it brought into the home. Programs with titles like *Admiral Broadway Revue* or *TV Dinner Date* offered new suburbanites simulated forms of urban entertainments, devoid of actual social contact with the more heterogeneous crowds in the city.

Perhaps the most blatant expression of television's relation to America's racist geographies came in the words of sociologist Raymond Stewart who claimed that television would be a boon for "Southern Negros who are ... barred from public entertainments" (cited in Bogart, p. 98). Stewart cited one African American man who observed that television and radio allowed him to bypass the degradation experienced in ball parks or theaters that "require that we be segregated and occupy the least desirable seats" (cited in Bogart 1956, p. 98). Rather than seeing television as a symptom of the longer history of segregation that produced this kind of spatial humiliation for African Americans, Stewart saw television a technological cure. In a similar way sociologist David Riesman (1954) claimed that the "television set is exactly the compensation for substandard housing the [slum] family can best appreciate – and in the case of Negroes or poorly dressed people, or the sick, an escape from being embarrassed in public amusement places" (p. 23). This concept of television as compensation for social/spatial/housing inequality was, then, the flip side of television's link to suburban expansion and the white privilege entailed in white flight.

More generally, the rise of television in various regions of the U.S. correlated with a general slump in spectator amusements, especially movie attendance but also in baseball, hockey, theater, and concert admissions. (*Fortune* editors, 1956). Even before the postwar period, the concept of the "home theater" was a dominant metaphor in the popular imagination. In 1912, the mass periodical *The Independent* ran an article titled "The Future Home Theater" in which "talking pictures" were transported through the ether into the home where residents could see "vistas of reality" channeled into their living space (Gilfillan 1912). By the 1950s advertisers

[2]For an excellent analysis of the rise of the suburban family sitcom see Haralovich (1992). Note, however, that television programs, including family sitcoms, also offered more critical perspectives via allegories of suburban alienation, racism, and isolation for women. See for example, Chapter 4 ("Static and Stasis") in Sconce (2004) and Chapter 4 ("From Domestic Space to Outer Space: The 1960s Fantastic Family Sitcom") in Spigel (2001).

referred to television as a “home theater,” “armchair theater,” “family theater,” and so forth. Advertisers promised that the new home theaters would provide women at least an imaginary escape from the isolating role of housewife. Ads showed couples dressed in ballroom gowns and tuxedoes while watching TV in their living rooms, as if they were out for a night on the town. Ironically, sociological studies revealed that women feared television’s isolating effects on their lives, and articles in women’s magazines discussed television as a potential threat to romance that would compete for their husband’s attention (Spigel 1992). As one woman complained, “I would like to go for a drive in the evening, but my husband has been out all day and would prefer to watch a wrestling match on television” (McDonough et al. 1956, p. 119).

More than just a US phenomenon, historical research on television’s innovation in a number of national contexts indicates that, for better or worse, television served as a kind of threshold technology between public and private space that mediated and helped produce new forms of sociality. In her oral history of 1950s Italian television, Penati finds that her interviewees fondly recall the collective viewing rituals that took place in the few homes in rural villages that were equipped with TV. Conversely, US, British, Swedish, and Australian histories demonstrate that collective TV could also induce the opposite reaction: people who were the first on the block to buy a new TV complained about the influx of unwanted guests eager to watch programs (O’Sullivan 1991; Olofsson 2012; Spigel 1992; Darian-Smith and Hamilton 2012). In her research on Argentina, Mirta Varela (2005) shows that while 1950s discourses on television in popular magazines often followed (and even directly translated) the US predications (for example, television was often called a “hearth” and discussed as a family medium), the actual situation was quite different. Public/political meetings, attendance at theaters and sporting events, and other forms of public culture thrived during of this period of political transition. In this regard, she argues, regardless of the similarities at the level of popular discourse, television did not serve as a replacement for public culture to the degree that it did in the US. As these competing historical findings suggest, television’s relation to the spaces of everyday life depends on broader geographies and shared assumptions about the production of social space.

Beyond the practical concerns regarding neighbors and social contact, television’s integration into the home was accompanied by more fantastic fears regarding privacy and surveillance. In the US, commentators often used military metaphors, picturing television as an “invader” of domestic territory. In 1957 *The New York Times* reported that a “clanking army of television sets… has entered the American home” (Reif 1957). Military metaphors of being “blasted,” “detonated,” or “bombarded” by TV (and the “noise” it produced) were common (Spigel 2009). Architects and interior designers also used military metaphors. In 1951, the trade journal *Interiors* ran an article titled “Cyclops,” which observed, “Television attacks the American eye, and the American eye, to our military way of thinking, is something for the designer to worry about” (Allen 1951, p. 62). In his study of West German TV, Joe Perry (2007) finds similar language, citing one prominent critic who called TV “the invader in the living room.”

While hyperbolic, the use of military metaphors had a very real association for people who were familiar with the use of radio and television as reconnaissance and surveillance technologies during wartime. Even while commercial broadcasting was designed for one-way communication in the home (with no capabilities for monitoring residents), people often treated TV as if it were a surveillance medium that could look into living rooms and spy on residents. In addition to the Cyclops analogy, commentators called TV a "prying eye," a "hypnotic eye," and even a "Peeping Tom." In this last configuration, the military trope of invasion was coupled with eroticized accounts of television's voyeuristic gaze. Writing for *Architectural Forum*, one critic worried that "making love in front of in front of television… [would be the same thing as] making love in the same room with an interested hypnotist" ("Television, Its Hypnotic Screen," 1948, p. 119). In her study of Italian TV, Penati (2013) discusses a cartoon from the popular press that shows a naked woman in her bedroom watching a man on TV. Afraid that the actor on screen will see her, she hides behind a dressing screen.[3] In this and other cases, turning the home into a home theater winds up putting the resident in the spotlight. Rather than a spectator, the woman is an object of spectatorship beyond her control.

Despite its disputed value, the home theater remains a dominant fantasy and a powerful marketing ploy for television and related media. Today, the home theater has morphed into gigantic 50+ inch flat screen displays that, at their most extreme, promise residents not just a simulated night out on the town but rather a complete immersive audio-visual environment. Magazines like *Electronic House* and *Home Theater* feature rooms designed to look like silent era movie palaces, sports bars, arcades, and the like. Just as skyline or ocean views drive up the price of the housing market, the views offered by wide screen home theaters have become a kind of virtual real estate. For example, in 2013 homeowners in an upscale community of Naples, Florida built a patio theater with an 80-in. TV screen that competes with a view of the ocean as residents relax in a nearby hot tub. According to *Electronic House*, "While the homeowner prefers the outdoor AV experience on most days, sometimes it rains. For those moments the home also includes an indoor theater with a 100-in. retractable screen" (Clauser 2013). While luxury home theaters are designed for the "viewing elite," average homeowners can find more affordable versions in big box stores or online sites. Recalling the theater date and family circle logic of early TV theaters, Samsung's website tells prospective consumers, "Your home entertainment centre can make a night at home just as much fun as an expensive night out. Get the family together for movies, sporting events, games, and more."[4]

[3]Note that such confusions between material space and media space, and the jokes about this, were not new to television. As Carolyn Marvin (1988) observes, these kind of jokes circulated in the trade journals of electrical engineers, who often specifically poked fun at children, women, and people of color who they depicted as "technical illiterates" who were unable to distinguish material from electrical spaces.

[4]See www.samsung.com/ae/consumer/tv-audio-video/home-theatre. Retrieved June 1, 2010.

As in these cases, the promotion and design of home theaters typically emphasize what Barbara Klinger (2006) calls a "fortress" mentality in which the propertied classes attempt to enclose themselves in the safe space of the home and which corresponds to the rise of gated communities and private security systems. Often home theaters are sold as part of a more general "smart" home automation package that bundles together the twin ideals of middle-class home ownership—privacy and security on the one hand, and luxury and comfort on the other. Smart homes are part of the post-Fordist service economy where residents rely on private sector wireless services to make their homes communicate. (For example, your internet-connected smart fridge can email you to let you know when you run out of tuna or you can use your iPhone to turn on your sprinklers.). In this context, TV is linked to DVRS and interactive streaming video services that make home theaters more insular in their ability to receive media on demand (without going anywhere to get it). Yet at the same time the home becomes more vulnerable to market research forces that, for example, record your media choices and sell this data to other corporations. The home has become a central site for monitoring and motion sensing technologies that present themselves as everyday conveniences but which ultimately hook occupants into elaborate systems of feedback and control (Andrejevic 2007; Allon 2004; Heckman 2008; Spigel 2005). So ironically, while smart home automation is marketed as a form of insular luxury, it is nevertheless also feared as a violation of personal privacy. As the *Wall Street Journal* reports, everything from mundane mechanisms like Internet enabled heaters to uncanny contraptions like "smart" toilets are now a risk of being hacked by people who can make "technology do terrible things." As might be expected, the television set is also a prime target. The *Journal* reports that hackers can "make a new Samsung TV set—which features a camera—watch you" even "when you think the TV is turned off." (Yadron 2013) So, as the new medium of smart TV develops, the old fears of surveillance return.

Mobile Privatization and Virtual Travel

As the case of the hacked TV suggests, even while historically linked to bourgeois ideals of domestic havens, television is also a product of social forces outside the home. Since the late nineteenth century, communication technologies have been intimately connected to the patterns of commerce, community, and mobility that shape the industrial and post-industrial world. Considering the history of telecommunications in these terms, Raymond Williams (1975) coined the term "mobile privatization," a phenomenon he tied to the simultaneous rise of privatized suburban housing and mobile urban industrial centers in the latter decades of the nineteenth century. The advent of telecommunications, Williams argues, offered people the ability to maintain ideals of privacy while providing the mobility required by industrialization, and broadcasting in particular held out the promise of bringing the public world indoors (pp. 26–8). In this respect, while labor-saving technologies (like washing machines or refrigerators) were marketed as private luxuries that reduced the need *physically* to travel to public spaces (like public laundries or markets), media technologies like radio and television offer the ability to *imaginatively* travel to distant locales (and to commune with virtual strangers) while ensconced in the safe space of the home.

The dream of virtual travel has been central to the promotional rhetoric surrounding television. In the 1940s and 1950s, advertisers placed TV sets against exotic backdrops of spectacular nature or cosmopolitan landmarks like the Eiffel Tower or Big Ben (Spigel 1992). Similarly, in Germany, an ad for Shaub-Lorenz television sets promised to turn "your home into a peaceful island of relaxation [with] enticing voyages of exploration of world events" (cited in Perry, p. 586). Even more thrilling in its premise of armchair travel, a 1956 ad for Germany's Philips TV showed a housewife sitting in a modern-styled chair watching TV against a background that displayed a futuristic car driving down a bridge in a city of tomorrow.[5] The ideals of modern mobility and virtual contact with the outside world continue with the global marketing for widescreen TV. In its 2010 advertising campaign, Samsung displayed exotic peacocks, Alaskan huskies, and breathtaking views of the sea emanating from its "Galaxy" flat screen TV.[6] In 1996, the Indian company Videocon marketed its "double wide window" TV that was, according to the ad, "bringing the world to India" (Kumar 1996, p. 80).

As in the Videocon example, the touristic pleasures of television have often been conceptualized within utopian ideals of global connectivity. Even before Marshall McLuhan (1962) predicted that television and satellite technologies would offer a "global village," NBC President Sylvester "Pat" Weaver promised that television would make the "entire world into a small town, instantly available . . . to all." But Weaver's concept of the small town was indeed a very small place, as he went on to describe broadcasters' need to keep all television programming within the "areas of American agreement" and to make television the "shining center of the home" (Weaver 1954). Here as elsewhere, the utopian ideal of global connectivity was tied to hegemonic practices of nation building and logics of center-periphery. At the level of policy (particularly with regard to the history of satellites), this mentality resulted in uneven flows of global communication. But, at the level of everyday practices, Weaver's formulation harked back to earlier cultural fantasies about virtual travel applied to telegraphy, telephony, and radio. As Susan Douglas (1987) and Carolyn Marvin (1988) demonstrate, intellectuals, technical experts, and popular critics of the late nineteenth and early twentieth century variously claimed that long distance telecommunications, whether across the oceans or just across town, offered a way to convene with strangers while avoiding the perceived threat of actual contact with unfamiliar people and contexts.

At the dawn of the television age, popular fantasies of virtual travel often took on specifically colonialist tones. In 1944, even before the massive adaptation of

[5]For the ad see http://www.tvhistory.tv/1950s-Siemens-TV-Ad-Germany.JPG. Retrieved May 1, 2010.

[6]For the ad see http://www.samsung.com/us/2013-smart-tv/#smart-tv-4. Retrieved June 1, 2010.

television in US homes, an ad for Dumont TV sets told future consumers, "You'll be an armchair Columbus!" and "sail with television through vanishing horizons in exciting new worlds."[7] Contemporary lifestyle television programs like *House Hunters International* and globetrotting TV chefs like Anthony Bourdain continue to promote this experience of armchair tourism, often with rhetoric that associates distant locals with exoticism.[8] *Survivor*, for example, turns the traditional travelogue into a parlor game (complete with tribes and a ritual tribunal), literally bringing the exotic back home. But as the tribe scenario suggests, the dream of virtual travel through television also depends on the colonialist fantasy of static populations cut off from (western notions of) mobility and progress.

Considering tele-travel from the point of view of media ontology and phenomenological questions of experience, Paul Virilio (1989) argues that audiovisual media are "the last vehicle," the endgame of modernity's promise of progress through technological manipulations of time and space. Comparing audiovisual media to a Japanese swimming machine that requires no actual destination (just endless laps in place), Virlio claims that television and related media are the ultimate form of "stasis," promoting a "sedentary" culture of "domestic inertia" (p. 109). The logical extension is the automated smart home, "the cadaver-like inertia of the interactive dwelling... whose most important furniture is the seat, the ergonomic armchair..." (p. 119).

While I agree that television and interactive media do often leave you exactly where you are (both physically and mentally), Virilio's argument nevertheless falls into a genre of anti-domestic rhetoric that has historically connected the home to passivity while equating mobility and public space with tropes of activity. As Rita Felski (1999) argues, the seminal philosophies of modernity privileged mobility and the city street (a space historically dominated by men) as a site for action and resistance, often neglecting or else denigrating the home and women's lives there.[9]

[7]For the ad see http://library.duke.edu/digitalcollections/adaccess_TV0445/. Retrieved July 19, 2013.

[8]For analysis of cooking shows along these lines see Strange (1998) and Bell and Hollows (2007). There is also a growing body of literature on television's depictions of places that considers, for example, the racial dynamics of TV cities in dramatic programs like *The Wire* or the identity politics of regional location. See, for example Lipsitz (2011) and Johnson (2008).

[9]Felski also observes that feminist theorists (she names, for example, Betty Friedan) have also often denigrated the home as have classic writers on everyday life such as Henri Lefebvre. Regarding the latter, it is also the case that while I am generally drawing on Lefebvre's (1991) important insights about the social production of space, and while his volumes on everyday life are central to any consideration of the topic, he saw television in negative terms as a source of alienation. Particularly apropos to my discussion here, see his 1958 Introduction to the first volume of *Critique of Everyday Life* where he calls television as "leisure machine" that is part of a more general conversion of "spontaneous" social needs into a form of "social organization" that modifies and directs those needs. In the same passage his also uses the phrase "armchair reading" negatively with regard to "escapist" mass culture items like "travel books" or "reader's digests," and he especially names "images" in "films... which are as far from *real life* as possible." (Lefebvre 1958; reprinted 1991, pp. 32–33, emphasis his).

In its association with the home (and with mass culture[10]), television came to have a key place in this anti-domestic, anti-feminine discourse—so much so that the classic image of a spectator is the "couch potato" (usually a slovenly overweight man half asleep on his sofa or "lazy boy" chair). Symptomatically here, at its misogynist extreme, television has been configured (or explicitly called) an "emasculating" and "feminizing" machine (Spigel 1992).[11] Yet, the home cannot really be reduced to a place of stasis, and the notion of "armchair tourism" does not capture the myriad experiences that television affords.

Comparing studies from a variety of national contexts (from the Australian Outback to own his research project in the UK) Morley (2000) shows how, for example, satellite television encourages people to rethink the boundaries between home, homeland, and elsewhere, and can allow people to reorient themselves in the world.[12] For people living in the Diaspora, television can provide a link between host country and homeland yet can also serve as a flash point for generational conflict as younger people use TV to break with parental/ethnic traditions associated with the homeland (Gillespie 1995; Morley 2000). Television can also provoke the re-imagination or reconfiguration of gender relations in a variety of national and transnational contexts (Mankekar 1999; Gillespie 1995; Ang 1995, 2004; Morley 2000; Abu-Lughod 2004).

With a specific focus on public and private space, Ratiba Hadj-Moussa (2010) shows how the introduction of satellites in Argentina is both symptomatic of but also transformative for the traditional ideological divides between (male) public and (female) private space. Satellite television has encouraged men to migrate from the neighborhood to the home where they dominate the television set and police women's access to French programs that men think will have a bad influence. Nevertheless, women do gain access to the television space and while they often prefer Arabic–language soaps, they also like French programs that promote western sensibilities. And (like the men) they watch global news on the Al-Jazeera network. Here, as elsewhere, satellite TV does not on its own change politics or social relations but rather intervenes in everyday experiences as people rethink and re-arrange entrenched social practices and beliefs.

[10]For a classic discussion of mass culture's association with tropes of femininity and passivity see Andreas Huyssen's chapter "Mass Culture as Woman: Modernism's Other" in Huyssen (1986).

[11]It's also important to point out however, that feminists have also often objected to television and seen it as a source of boredom or an outright tool of patriarchy for women's domestic confinement and role as consumer. Outright attacks on the medium especially ran through feminist writing of the early second wave, such as Betty Friedan's *The Feminine Mystique* (1963) and specifically in a series of articles published in *TV Guide* (Friedan 1964). By the late 1970s, and especially with the rise of feminist film theory and cultural studies, feminists developed a much more varied literature that (while still holding onto a negative critique) also understood television's relation to women's pleasure and to their everyday lives in more diverse and complex ways. For a bibliography see Brunsdon and Spigel (2007).

[12]Morley draws especially on Moore (1997) and Green (1998).

So too, television can serve as an important source of information and fantasy for people who do not want to live in the hetero-normative family spaces that suburban architects have historically designed. In their intriguing attempt to fathom a house as a queer space, Moon et al. (1994) discuss television and "entertainment/information centers" as "points of contact…between the protected realm of home and the world at large, through which unsanctioned information may find its way into the house." "Just like the closet (which they call a space of "transgression" for finding hidden secrets), media centers "are the sites to which the adolescent goes to uncover information about sexuality" (p. 36). Remembering his childhood fondness for the 1960s sitcom *That Girl* (about a fashionable suburban twenty-something single girl who moves to Manhattan with dreams of becoming an actress), architect/theorist Ernest Pascucci (1997) recalls the sitcom's importance to his sense of queer identity. Or as he puts, the program offered the possibility of "cross-identification—these overwhelming urges to be *That Girl*" (p. 52). Accordingly, he claims, despite the popular conception of sitcoms as mere trivial pursuits, the program offered him and other gay men of his generation a profound space in which to imagine lifestyles and identities that were markedly different from those encouraged by his hetero-normative suburban home.

Liveness, Telepresence, and Spatial De-realization

In its capacity to bridge the near and the far, television promotes the experience of "telepresence" –a sense of being on the scene of presentation that is produced through what media scholars typically refer to as TV's aesthetics of "liveness"– its sense of immediacy, simultaneity, and intimacy (Bourdon 2004; Feuer 1983; Scannell 1996; Moore 2004; Boddy 1989). Television executives and producers have historically capitalized on TV's ability to transmit sounds and images live through network feeds by deploying audio-visual techniques designed to make people feel as if they are present at live events. These strategies include: (1). Direct address (when, for example, a news presenter speaks into the camera directly at the viewers); (2). The use of studio audiences or canned laughter (which encourage viewers to feel as if they are participating in a real-time social event); (3). Self-reflexivity (for example, when news programs display TV cameras, making viewers feel they are privy to backstage insider knowledge in the TV studio; (4). The presentation of everyday people (as in game shows or reality TV); (5). Performance conventions and acting styles that create a sense of sincerity and ordinariness; and (6). Continuity editing and sound fidelity, which create a sense of real time and space. These aesthetic features are most apparent in live originated program formats such as news, sports, or special events, but they also have come to define the more general aesthetic features of television, even in recorded fare. Taped daytime soaps, for example, promote a sense of presence and simultaneity by having characters celebrate holidays in real time while filmed or taped sitcoms are famous for their laugh tracks that make viewers believe they are chuckling along with a crowd.

In "Television and Modern Life" Paddy Scannell (1996) examines how broadcasting's aesthetics of liveness produced new ways of addressing citizens in the privacy of their homes, and in the process helped change the nature of public life in Britain. Scannell writes, "It is not just that radio and television compress time and space. They create new possibilities of being: of being in two places at once, or two times at once" (p. 9). The event unfolds where it occurs in material space and in the space in which people watch it on TV. The ability for people to watch a public event via television (he uses the 1953 coronation of Queen Elizabeth II as a case in point) means that the event can be witnessed through all sorts of demeanors (for example, people can watch while drinking in bars or at home in pajamas). The public no longer has to perform the outward show of social etiquette and respect traditionally required of such events; and broadcasters (who know this) are especially careful to orchestrate events in ways that take account of the cameras.

While Scannell focuses on the new form of sociality and publicness that broadcast liveness enabled, liveness has even more often been analyzed for its alienating effects. Along these lines, in one of the best accounts, Margaret Morse (1991) argues that television should be seen in relation to the other dominant spaces of postwar everyday life—the freeway and the mall. This historical conjuncture of spatial arrangements, she claims, has produced new forms of simulated social life based on distraction. So for example, television news programs simulate interpersonal communication through modes of direct address, creating models of talk (but not actual feedback) between the audience and news presenter. Morse connects television to the "nonspace" of freeways that cut through local communities (often destroying them) to produce (for the driver) a sense of space divorced from context. Finally, she analyzes television's simulated universe in relation to the mall – a world in miniature that promotes distracted, disengaged forms of social relations based on consumption rather than community.

Certainly, Morse's picture of spatial distraction is bleak. Although she does draw on Michel de-Certeau's seminal "Walking in the City" (1984) in order to speculate on the possibilities for human agency within the highly controlled nonspaces she investigates, Morse too readily overlooks the social communities, sense of place, and site specific differences that do arise in consumer-oriented mediatized environments – whether this be the TV and new media fan communities that Jenkins (1992) has analyzed; the differences among malls and the uses women shoppers make of them (Morris 1998); or the more unfortunate social relations of road rage that take place on freeways. Nor does Morse consider television's own depictions of place or the investment people have in televised places as exemplified by viewers' pilgrimages to locations where TV shows are shot (Couldry 2000) or people's charged memories of imaginary TV settings (as with Pascucci's memories of *That Girl's* Manhattan locale). Nevertheless, Morse provides a powerful critique of television as a mode of spatial engineering that encourages "spatial de-realization." With television, as with freeways and with malls, you are often not where you think you are.

From Mobile to Everywhere TV

The experience of spatial de-realization has been central to the promotion of mobile television and its predecessor, the portable TV. By the late 1950s, in their attempts to sell people a second TV, advertisers for portable sets promised consumers that TV was not just a window on the world, but also a way to extend one's private life into public spaces. Unlike the 1950s console model, which was typically placed in a central area of the family home, ads for portable receivers often showed people on the move, carrying their tote-able TV sets to beaches, picnics, and even, in one humorous 1967 Sony ad, nudist colonies. Rather than homebodies gathered around the family tube, now spectators were presented as adventurous heroes toting portables on motorcycles or else as liberated women wearing mini-skirts accessorized by mini-TVs that looked like purses. In this respect, the advent of portable media inverted the experience of "mobile privatization" that Williams first associated with the rise of telecommunications. Instead, portable television offered people a fantasy of "privatized mobility" (Spigel 2001). Although market research showed that most people did not even move their portable TVs around the house (no less take it outdoors), the fantasy of privatized mobility had a power of its own.

Today, mobile communications (like PDAs, lap tops, iPhones, iPods, etc.) promise new forms of privatized mobility by allowing people to store and carry their private lives around with them on an ever-expanding array of handheld devices. Media are being relocated to the public sphere so that people increasingly experience being at home while in public. Speaking specifically of the mobile phone, Thomlinson (2001) argues that mobiles should be seen as "technologies of the hearth . . . by which people try to maintain something of the security of cultural location" (p. 17). In her interview-based study of mobile phone use in Morocco, Maya Kriem (2009) found that people who felt increasingly alienated in urban environments used the mobile phone to connect back to home and family. Writing about the mobile phone in Denmark, Toke Hunstrup Christensen (2009) similarly finds that family members who are separated in physical space use the device as a mode of "connected presence."

Considering this from the point of view of public culture, Michael Bull (2004) talks about the disturbing consequences of the "private bubbles" that people inhabit when they use cell phones and mobile sound devices (from the Sony Walkman to the iPod). "As we become more and more immersed in our mobile media sound bubbles of communication, so then those spaces we habitually pass through in our daily lives increasingly lose significance and progressively turn into the 'non-spaces' of daily lives which we try, through the self-same technologies, to transcend" (p. 290). While Bull's concept of "private bubbles" captures the strange new forms of social life brought on by mobile media, his declination narrative regarding public space may well be hasty. Alienation had been the core problematic for critical philosophies of modernity and urban space way before the introduction of mobiles (and Bull draws on this literature in the essay). So in that sense, it seems to me, Bull's argument fits into the genre conventions of a well-established discursive trope that has a powerful rhetorical sway, but which may not really capture the variety of experiences mobiles offer. For example, in their contemporary uses, mobiles have also become instruments for social congregation, helping to organize flash mobs or to capture and disseminate footage of political demonstrations, as in the case of the Arab Spring (Abaza 2013).

So too, even while mobile devices can afford everyday forms of virtual solitude, the spatial experience they help produce is less a complete tuning out of the material environment than a melding of telepresence with physical embodiment as a preferred mode of experiencing place. In other words, rather than total alienation, mobiles allow people to go in and out of virtual and physical worlds, and thereby to control (or at least have a fantasy of control over) the environments of daily life. Just as John Thomlinson (2007) observes how young people often prefer to be in online worlds and actually feel "at home" in them (he points to the case of teenagers sitting side by side in the library but choosing to email back and forth on computers), the spaces of mobile telepresence need not just be about alienation or disconnection. Instead, as Tomlinson argues, we should think about "telepresence ... as a distinctive existential mode of presencing." "Telepresence," he continues, "should not be regarded and evaluated as a shortfall from the 'definitive' existential mode of embodiment" in physical space (Thomlinson 2007, Chapter 5, Section 6, par. 6).

Although Tomlinson is referring to interpersonal communication via the Internet and mobile phones, mobile TV offers similar forms of place-shifting, virtual connectivity, and "presencing." Like early TV before it, mobile TV services capitalize on liveness and telepresence, offering people virtual attendance at exclusive live events like the World Cup. In this sense, even while mobile TV is (like the iPod) a system of digital storage on which you can download TV programs you might also watch at home, mobile services trade on the currency of publicness. Rather than connecting people back home (as in the case of the cell phone), advertisers promote mobile television as a means of disconnecting viewers from it. Making this point crystal clear, the website for "Mobi-TV" (a mobile TV service company) claims that whereas TV had once been "locked in the living room," today it is an "everywhere" technology that offers people "flexible" modalities of use.[13]

In his study of advertising for mobile TV, Max Dawson (2007) shows how early ads for mobile TV services, which were targeted at affluent men aged 18–34, linked this anti-domestic rhetoric to a gendered conception of home. Showing men romping through urban spaces (or else humorously chained to old living-room TV sets and dragging them, like balls and chains, through the streets), the ads promoted mobile TV as "an escape [from] the social and spatial constraints of the home—as well as the feminine connotations of domestic viewing—for more interactive (and presumably masculine) forms of perambulatory leisure." Nevertheless (as with portable television before it), Dawson observes that people actually use the mobile devices most often in the home (p. 233).

[13]See http://www.mobitv.com. Retrieved August 4, 2013.

Mobile television is just one among many devices and services that have turned television from a place-bound medium into a ubiquitous technology and cultural form that people encounter in a range of times and places. In this respect, television is so central to daily routines inside and outside the home that it often fades into the background forming what Anna McCarthy (2000) calls a mode of "ambient" experience. To be sure, since the 1930s and continuing through the twentieth century, television could be found in public places—from factories to shops to bars to classrooms to fairs to museums to hospitals to drive-in churches (Allen 1983; McCarthy 2000; Olsson 2004; Acland 2009; Spigel 2009; Feuqua 2012; Robles 2012; Bernold and Ellmeier 1997). And television's public uses were historically tied to political agendas for gathering citizens *en masse*, as in the case of Nazi Germany (Uricchio 1989) or President Juan Perón's 1951 plan to distribute TV sets to Peronist locales in Argentina where he hoped people would gather for his daily broadcasts (Varela 2014). But today, as it converges with digital technologies, television is part of a more general "screening" of public space.

Paradoxically, as mobile screens get smaller and more portable, the built environment in large cosmopolitan centers is constructed through gigantic screens that flash everything from news updates to sporting events to stock market reports to advertisements to snippets of Viennese opera or Broadway shows. Global cities often invest in giant screens to signify their place on the map of progress, but at the same time (in a more positive sense) urban planners also hope to provide forms of public access to events (as in the case of the "Big Screen" HD TVs set up in cities across the UK for the 2012 London Olympics). Whatever their intended uses, screens can serve as a means of herding the movements and capturing the attention of large populations. Media conglomerates like Viacom, CBS, NBC, and Microsoft are investing in digital "out of home" ads that track consumers along the routes of their daily itineraries. At the supermarket interactive digital screens on shopping carts can advise you what to buy and track your purchases for market research; "Adwalkers" wander the streets wearing interactive digital displays; and large screen digital billboards target individuals via location mapping technology that senses pedestrians' movements and emits "personalized" messages about what movies to see or products to buy (Boddy 2011). Tracing the history of multiscreen environments, architecture historian Beatriz Colomina (2001) suggests we are "enclosed by images" that compete for our attention and leave us in a state of perceptual distraction (markedly different, she thinks, from the "critical distraction" Walter Benjamin (1936) theorized at the dawn of cinema and consumer culture).

Meanwhile, in popular culture, the ubiquity of screens has given rise to a whole genre of media dystopias, and television has historically assumed a privileged role here. To be sure, TV dystopias can be found well before the present-day proliferation of screens. Think, for example, of Fritz Lang's *Metropolis* (1927) with its fascist industrialist controlling workers via his futuristic TV phone; or Charlie Chaplin's *Modern Times* (1936) with its evil boss monitoring Charlie on the factory's large screen bathroom TV; or George Orwell's *1984* (first published in 1949) with its fears of Big Brother. But in more recent decades these TV nightmares are not just about surveillance and mind control per se. Instead, television is now also a rhetorical figure in tales about the total ontological confusion between virtual and physical universes. Perhaps the "urtext" of this genre is Philip K. Dick's science fiction masterpiece *Ubik* (1969) that renders a world of spatial simulation and annihilation through a potent mix of TV advertising jingles, insolent smart homes (that make you pay to open the door), and ghostly half-dead bodies speaking "live" on TV from wireless coffins. Playing with similar tropes of ontological confusion, but presenting this in more accessible stories that maintain (for spectators if not for the heroes) the distinction between simulated and real worlds, popular films from *Poltergeist* (1982) to *The Truman Show* (1998) to *Pleasantville* (1999) revolve around characters trapped in TV. Premièring in 2011, the British television series *Black Mirror* (described by Channel Four as "a twisted parable for the twitter age") offered its share TV nightmares. Season One's "15 Million Merits," for example, follows a hero who lives in a media controlled eco-system where people have to peddle energy-producing bicycles in order to earn "merits" to pay for daily goods, all the while forced to watch mindless TV shows, pornography, and ads flickering on screens everywhere around them. The only chance of escape from this toxic combination of dreary peddling and force-fed TV comes with a chance to star on a reality TV show (and this, as might be expected, results in only more spatial entrapment and misery). Meanwhile, actual reality TV shows like *Big Brother* turn these nightmarish visions into a ludic sport. Even video games have now chosen TV as a source of narrative intrigue. *Persona 4* (Atlus 2008) features a Japanese town in which bodies are found dangling from television antennas. As the game progresses we learn that TVs turn into portals that can pull people into a nightmarish world that eventually kills them. The sheer amount of these TV dystopias (as well as the proliferating tales of computer nightmares in cyberfiction) shows that the media are themselves entirely self-reflexive about their own social and spatial ubiquity. On the one hand, these stories promote the more general anti-TV rhetoric that has for many years admonished TV as a "vast wasteland," "boob tube," or "plug-in drug."[14] On the other hand, they can also provide what Constance Penley (1991) calls a "critical dystopia," a form of negative thinking in science fiction doomsday narratives that nevertheless can serve a productive function for contemplating alternatives.

In a similar sense, the new large screen installations in public spaces might be used to provide productive estrangement and defamilarization from mundane routines. Along these lines, just as with the longer history of happenings, video, performance, and/or installation art, media artists and activists use public screens as sites for artistic interventions (McCarthy 2000; Harbord and Dillon 2013). And just as de Certeau's "Walking in the City" (1994) theorizes the possibility that people can navigate their own routes within the planned grids and dominant maps of modernity, it seems likely that people can articulate their own experiences as they

[14]Itself a spatial metaphor, the term "vast wasteland" (which was coined in 1961 by Newton Minow, the Chair of the Federal Communications Commission) uses the title of T.S. Elliot's poem to encapsulate the ruinous (and over-commercialized) state of US TV in that period. For his full speech see Minow (1961).

move through screened and networked cities. Considering the more positive group dynamics that can result, Cubitt et al. (2008) argue that large urban screens should be conceptualized less as individual modes of subjectivity and distraction than as forms of "transient media" that confront physically mobile people and offer new possibilities for spontaneous social situations. While acknowledging that screens are often carefully planned for the control and management of populations, the authors nevertheless follow the trajectory of Guy DeBord's concept of the "situation" to imagine an "ecology" and "ethics" of screens that might promote dialogues among crowds (and, I presume, less alienating ways of occupying public places).

Locating TV Studies

Whether in popular media or media scholarship, the ubiquity of television in everyday life raises fundamental questions about space and place as we attempt to figure out where we are and how to proceed. If corporations have historically offered domestic utopias (as in RCA's TV future with which I began), television has provoked a much more complex spatial imaginary, and people engage TV through a much more diverse set of spatial practices. Rather than the utopian spaces of corporate futurism or dystopian nightmares of spatial entrapment and annihilation, at its best the critical inquiry into television's spatial geographies has opened up an important agenda for understanding how people live with and through media in everyday life. Whether it is the research on domesticity, privacy, and public space on which I have primarily focused here, or the equally important scholarship on television's relation to nationalism and global flows of culture, the scholarship on television's material histories and geographies demonstrates that TV is never one thing going in one direction (whether that be euphoria or doom). Like all technologies, television affords possibilities that are differently articulated at different times in different locations.

In that sense, the study of television should not just be the study of troubling encounters with non-places and nowheres. Instead, both as a material object and as a transmission medium, television helps to produce the "somewheres" in which daily life takes place. Finding a way to theorize "somewheres" in a mediatized world is, it seems to me, one of the main contributions of television studies. As an audio-visual device, television merges material spaces with virtual "presence," a spatial phenomenon that has, with digital media, become an increasingly dominant way in which space and place are constructed and experienced. Being here and there, home and elsewhere, near and far, is the quintessential TV state of mind.

To understand these hybrid mergers requires something more robust than cultural pessimism about the loss of authentic anthropological places and related grand theories of social decline. Even if television is often rightly taken to task for forms of social injustice, its relation to social life is much more varied than the "negative influence," "spatial annihilation," or "escape" paradigms suggest. Rather than asking just what TV does to some idealized form of a priori "real" space (in the technological determinist sense), the point is to consider how television helps people assemble social worlds. In evoking Bruno Latour's *Reassembling the Social* (2005) and Actor Network Theory, I am suggesting that we think of television as part of a feedback loop of activities and actions among humans and artifacts that help to produce social environments. Media spaces are not "unreal" in the sense of a technological or aesthetic illusion; yet they do allow people to dwell in places accessed and designed through technical means. Rather than replacing anthropological space (as with Augé's theory of non-place), media spaces are both human and non-human, places where we form relationships not just with people but also with things (interfaces, sensors, and voices like Siri's). Thinking about the possibilities that media spaces offer, as well as the human contact they sometimes shut down, should be central to the goals of media ecology. TV scholarship has been crucial to opening up these more thoughtful debates about the increasingly mediatized worlds in which we live.

That said, among television scholars—and media scholars more broadly—there is often disagreement over appropriate methods, subjects, and perspectives. In that sense, while I have presented a synthetic overview, it's important to remember that the scholarship on television comes from numerous domains of social scientific and humanistic research, and even within those broad paradigms there is often conflict about how to approach television (or even if this so called "lowbrow" medium is worth studying in the first place). In my view, there has been an unfortunate implicit assumption among media theorists that small-scale studies—for example, narrative analysis of a TV program, the historical case study, or the empirical analysis of a local practice–are somehow less important than large-scale philosophies of media ontology, the macro-politics of globalization, or blanket statements of cultural judgment. In other words, and perhaps because of its ubiquity in daily life, people have often wanted "Big Bang" theories of television's effects –and in a hurry (or at least in time to sell data to a government think tanks or corporations). Resisting the urge for grand theories does not, however, reduce the use value of our findings. As Morley (2000) argues, we should not mistake scale (the macro-politics of the global vs. the micro-politics of the local) for significance.[15] The "somewhere" of media studies helps to illuminate the contradictions, ambivalences, differences, and often the utter messiness and unpredictability of television in everyday life. Studies of TV and related media will, I hope, follow this road to somewhere, even if the complexity involved means that this will take some time.

References

A room that does everything. *American Home* (May 1951), no author, p. 40.

Abaza, M. (2013). In J. Harbord, C. Berry, & R. Moore (Eds.), *Public space, media space* (Chapter 4, E-book edition). London: Palgrave Macmillan.

Abu-Lughod, L. (2004). *Dramas of nationhood: The politics of television in Egypt*. Chicago: University of Chicago Press.

[15]In making this observation Morley draws upon Massey (1997).

Acland, C. (2009). Curtains, carts and the mobile screen. *Screen, 50*(1), 148–166.

Allen, D. (1951). Cyclops... the nature of a new household pet. *Interiors* CX, *12*(July), 62–63.

Allen, J. (1983). The social matrix of television: Invention in the United States. In E. A. Kaplan (Ed.), *Regarding television: Critical approaches—An anthology* (pp. 109–119). Frederick: University Publications of America.

Allon, F. (2004). An ontology of everyday control: Space, media flows and 'smart' living in the absolute present. In A. McCarthy & N. Couldry (Eds.), *Mediaspace: Place, scale and culture in a media age* (pp. 253–274). London: Routledge.

Andrejevic, M. (2007). *I Spy: Surveillance and power in the interactive age*. Lawrence: University of Kansas Press.

Ang, I. (1995). *Living room wars: Rethinking media audiences*. London: Routledge.

Ang, I. (2004). The cultural intimacy of TV drama. In K. Iwabuchi (Ed.), *Feeling Asian modernities: Transnational consumption of Japanese TV dramas* (pp. 303–310). Hong Kong: Hong Kong University Press.

Aslinger, B. (2013). Make room for the Wii. In J. Hartley & J. Burgess (Eds.), *A companion to new media dynamics* (pp. 209–219). West Sussex: Blackwell.

Augé, M. (1995). *Non-places: An introduction to supermodernity*. London: Verso.

Bausinger, H. (1984). Media, technology and daily life. *Media, Culture, and Society, 6*(4), 343–351.

Bell, D., & Hollows, J. (2007). Mobile homes. *Space and Culture, 10*(1), 22–39.

Benjamin, W. (1936; reprint. 1969). The work of art in the age of mechanical reproduction. In W. Benjamin (Ed.), *Illuminations: Essays and reflections* (H. Arendt, Trans., pp. 217–252). New York; Schocken.

Bernold, M., & Ellmeier, A. (1997). Addressing the public: Television, consumption and the family in Austria in the 1950s and 1960s. In M. Nava, A. Blake, I. MacRury, & B. Richards (Eds.), *Buy this book: Studies in advertising and consumption* (pp. 4551–4881). New York: Routledge. ebook location.

Boddy, W. (1989). *Fifties television: The industry and its critics*. Champaign Urbana: University of Illinois Press. 1990.

Boddy, W. (2004). *New media and the popular imagination: Launching television, radio, and digital media in the United States*. London: Oxford.

Boddy, W. (2011). Is it TV yet? The dislocated screens of television in a mobile digital culture. In J. Bennett & N. Strange (Eds.), *Television as digital media* (pp. 76–104). Durham: Duke University Press.

Bogart, L. (1956). *The age of television: A study of viewing habits and the impact of television on American life*. New York: Frederick Unger.

Bourdon, J. (2004). Live television is still alive. In R. C. Allen & A. Hill (Eds.), *The television studies reader* (pp. 182–195). New York: Routledge.

Brunsdon, C., & Spigel, L. (2008). *Feminist television criticism: A reader*. Berkshire: Open University Press.

Bull, M. (2004). 'To each their own bubble': Mobile spaces of sound in the city. In Couldry & McCarthy (Eds.), *Mediaspace* (pp. 275–293). New York: Routledge.

Christensen, T. H. (2009). Connected presence in distributed family life. *New Media and Society, 11*(3), 433–451.

Clauser, G. (2013). *Stunning outdoor house*. www.electronichouse.com/article/stunning_outdoor_theater_highlights_florida_home) (July 11 online issue) Retrieved July 20, 2013.

Colomina, B. (2001). Enclosed by images: The Eameses' multimedia architecture. *Grey Room, 2*(Winter), 6–29.

Couldry, N. (2000). *The place of media power: Pilgrims and witnesses of the media age*. London: Routledge.

Cubitt, S., Papastergiadis, N., & McQuire, S. (2008). Transient media. *Public, 37*, 76–85.

Darian-Smith, K., & Hamilton, P. (2012). Part of the family: Australian histories of television, migration and memory. In K. Darian-Smith & S. Turnbull (Eds.), *Remembering television: Histories, technologies, memories* (pp. 30–51). Cambridge: Cambridge Scholars Publishing.

Dawson, M. (2007). "Little players, big shows" format, narration, and style on television's new smaller screens. *Convergence, 13*(3), 213–250.
de Certeau, M. (1984; reprinted 1994) *The practice of everyday life* (S. Rendall, Trans.). Berkley: University of California Press.
Felski, R. (1999). The invention of everyday life. *New Formations, 39*, 15–31.
Feuer, J. (1983). The concept of live television: Ontology as ideology. In E. A. Kaplan (Ed.), *Regarding television* (pp. 12–22). Los Angeles: The American Film Institute.
Feuqua, J. (2012). *Prescription TV: Therapeutic discourse in the hospital and at home*. Durham: Duke University Press.
Fortune Editors. (1955). "30 billion for fun" Reprinted (1958). In E. Larrabee & R. Meyerson (Eds.), *Mass leisure* (pp. 161–172). Glencoe: Free Press.
Friedan, B. (1963). *The feminine mystique*. New York: W.W. Norton.
Friedan, B. *Television and the feminine mystique*. February 1–8, pp. 273–275.
Frith, S. (1983). The pleasures of the hearth: The making of BBC light entertainment. In T. Bennett & F. Jameson (Eds.), *Formations of pleasure* (pp. 101–123). London: Routledge and Kegan.
Gauntlett, D., & Hill, A. (1999). *TV living: Television, culture, and everyday life*. London: British Film Institute.
Gilfillan, S. C. (1912, October 17). The future home theater. *The Independent*, 886–891.
Gillespie, M. (1995). *Television, ethnicity, and cultural change*. London: Routledge.
Gray, A. (1992). *Video playtime: The gendering of leisure technology*. London: Comedia.
Green, L. (1998). *Communications and the construction of community: Consuming the remote commercial television service in Western Australia*. Unpublished PhD thesis, Murdoch University.
Hadj-Moussa, R. (2010). The unpredictable and the irreversible: Satellite television in the Algerian public arena. In C. Berry, S. Kim, & L. Spigel (Eds.), *Electronic elsewheres: Media, technology and social space* (pp. 117–136). Minneapolis: University of Minnesota Press.
Haralovich, M. (1992). Sitcoms and suburbs: Positioning the 1950s homemaker. In L. Spigel & D. Mann (Eds.), *Private screenings: Television and the female consumer* (pp. 111–141). Minneapolis: University of Minnesota Press.
Harbord, J., & Dillon, T. (2013). Encountering screen art on the London underground. In C. Berry, J. Harbord, & R. Moore (Eds.), *Public space, media space*, Chapter 8, E-book edition. London: Palgrave Macmillan.
Hartley, J. (1999). *Uses of television*. London: Routledge.
Harvey, D. (1990). *The condition of postmodernity*. Oxford: Blackwell.
Heckman, D. (2008). *A small world: Smart houses and the dream of the perfect day*. Durham: Duke University Press.
Huyssen, A. (1986). *After the great divide: Modernism, mass culture, postmodernism*. Bloomington: Indiana University Press.
Jenkins, H. (1992). *Textual poachers: Television fans and participatory culture*. New York: Routledge.
Jenkins, H. (2006). *Convergence culture: Where old and new media collide*. New York: New York University Press.
Johnson, V. (2008). *Heartland TV: Primetime television and the struggle for US identity*. New York: NYU Press.
Kingler, B. (2006). *Beyond the multiplex: Cinema, new technologies, and the home*. Berkeley: University of California Press.
Kriem, M. S. (2009). "Mobile telephony in morocco: A changing sociality", media. *Culture & Society, 31*(4), 617–632.
Kumar, S. (2005). *Ghandi meets primetime: Globalization and nationalism in Indian television*. Champaign-Urbana: University of Illinois Press.
Latour, B. (2005). *Reassembling the social: An introduction to actor-network-theory*. New York: Oxford.
Leal, O. F. (1990). Popular taste and erudite repertoire: The place and space of television in Brazil. *Cultural Studies, 4*(1), 19–29.

Lefebvre. (1958; reprinted 1991). *Critique of everyday life* (Vol. I, J. Moore, Trans.). London: Verso.
Lefebvre. (1991). *The production of space* (D. Nicholson-Smith, Trans.). Oxford: Blackwell.
Lipsitz, G. (1992). The meaning of memory: Family, class, and ethnicity in early network television programs. In L. Spigel & D. Mann (Eds.), *Private screenings* (pp. 71–108). Minneapolis: University of Minnesota Press.
Lipsitz, G. (2011). *How racism takes place*. Philadelphia: Temple University Press.
Lull, J. (Ed.). (1988). *World families watch television*. Beverly Hills: Sage.
Lull, J. (1990). *Inside family viewing*. London: Comedia.
Mankekar, P. (1999). *Screening culture/viewing politics: An ethnography of television, womanhood, and nation in postcolonial India*. Durham: Duke University Press.
Marvin, C. (1988). *When old technologies were new*. New York: Oxford University Press.
Massey, D. (1991). Flexible sexism. *Environment and Planning D: Society and Space, 9*(1), 31–57.
McCarthy, A. (2000). *Ambient television*. Durham: Duke University Press.
McDonagh, E. C., et al. (1956). Television and the family. *Sociology and Social Research, 40*(4), 117–119.
McLuhan, M. (1962). *The Guttenberg galaxy: The making of typographic man*. Toronto: University of Toronto Press.
McLuhan, M. (1964). *Understanding media: The extensions of man*. New York: McGraw Hill.
Miller, D. (2009). *Stuff*. Cambridge: Polity.
Minow, N. (1961, May 9) "*The vast wasteland," address to the thirty-ninth annual convention of the national association of broadcasters*. Washington, DC.
Modleski, T. (1983). The rhythms of reception: Daytime television and women's work. In Kaplan (Ed.), *Regarding television* (pp. 67–75). Los Angeles: AFI.
Moon, M., Sedgwick, E., Gianni, B., & Weir, S. (1994). Queers on single family space. *Assemblage, 24*, 30–37.
Moore, S. (1997). *Satellite television and everyday life*. Luton: University of Luton Press.
Moore, S. (2004). The doubling of place: Electronic media, time-space arrangements and social relationships. In McCarthy & Couldry (Eds.), *Mediaspace* (pp. 21–36). London: Routledge.
Morley, D. (1985). *Family television: Cultural power and domestic leisure*. London: Routledge.
Morley, D. (1991). Where the global meets the local: Notes from the sitting room. *Screen, 54*(2), 1–15.
Morley, D. (2000). *Home territories: Media, mobility and identity*. London: Routledge.
Morley, D. (2007). *Media, modernity and technology: The geography of the new* (pp. 199–234). London: Routledge.
Morley, D., & Robins, K. (1995). *Spaces of identity: Global media, electronic landscapes, and global media*. London: Routledge.
Morris, M. (1998). Things to do with shopping centers. In M. Morris (Ed.), *Too soon too late: History in popular culture* (pp. 64–92). Bloomington: Indiana University Press.
Morse, M. (1991). An ontology of everyday distraction: The freeway, the mall, and television. In P. Mellencamp (Ed.), *Logics of television: Essays on cultural criticism* (pp. 193–221). Bloomington: Indiana University Press.
Olofsson, J. (2012). Revisiting the TV object: On the site-specific location and objecthood of the Swedish television during its inception. *Television & New Media, 20*(10), 1–6.
Olsson, J. (2004). One commercial week: Television in Sweden prior to public service. In L. Spigel & J. Olsson (Eds.), *Television after TV: Essays on a medium in transition* (pp. 249–269). Durham: Duke University Press.
O'Sullivan, T. (1991). Memories and culture of viewing, 1950–65. In J. Corner (Ed.), *Popular television in Britain: Studies in cultural history* (pp. 159–181). London: BFI.
Pascutti, E. (1997). Intimate televisions. In S. Harris & B. Berke (Eds.), *Architecture of the everyday* (pp. 52–54). Princeton: Princeton Architectural Press.
Penati, C. (2013). *Remembering our first TV set: Personal memories as a source for television audience history*. View 1:3, online at http://journal.euscreen.eu/index.php/view/rt/printerFriendly/jethc026/48. Retrieved 20 July 2013.

Penley, C. (1991). Time travel, primal scene, and the critical dystopia. In C. Penley, L. Lyon, S. Willis, L. Spigel, & J. Bergstrom (Eds.), *Close encounters: Film, feminism, and science fiction* (pp. 63–82). Minneapolis: University of Minnesota Press.

Perry, J. (2007). Healthy for family life: Television, masculinity, and domestic modernity during West Germany's miracle years. *German History, 24*(4), 560–595.

Postill, J. (1998). *Little by little, a national subculture is made: The role of modern media in the institutional transformation of Iban Society*, Unpublished Paper, Department of Anthropology, University College London. Cited in Morley, *Home Territories*.

Reif, R. (1957). Suggestions to cut down home noise. *New York Times, 21*, 30.

Reisman, D. (1954). Recreation and the recreationist. *Marriage and Family Living, 16*(1), 23.

Robles, E. A. (2012). The crystal cathedral: Architecture for mediated congregation. *Public Culture, 24*(3), 577–600.

Roth-Ey, K. (2007). Finding home for television in the USSR, 1950–1970. *Slavic Review, 66*(2), 278–306.

Scannell, P. (1996). *Radio, television and modern life*. Oxford: Blackwell.

Sconce, J. (2004). Static and stasis. In J. Sconce (Ed.), *Haunted media: Electronic presence from telegraphy to television* (Chapter 4). Durham: Duke University Press.

Silverstone, R. (1994). *Television and everyday life*. London: Taylor and Francis.

Smith, K. M. (2012). Domesticating television: Changing attitudes in postwar Britain. *Interiors, 3*(1–2), 23–42.

Spigel, L. (1992). *Make room for TV: Television and the family ideal in postwar America*. Chicago: University of Chicago Press.

Spigel, L. (2001). *Welcome to the dreamhouse: Popular media and postwar suburbs*. Durham: Duke University Press.

Spigel, L. (2005). Designing the smart house: Conspicuous production and posthuman domesticity. *European Journal of Cultural Studies, 8*(4), 403–426.

Spigel, L. (2009). *TV by design: Modern art and the rise of network television*. Chicago: University of Chicago Press.

Spigel, L. (2012). Object lessons for the media home: From storagewalls to invisible design. *Public Culture, 24*(3), 535–576.

Spigel, L., & Dawson, M. (2009). Television and digital media. In M. Halliwell & C. Morley (Eds.), *American thought and culture in the 21st century* (pp. 209–318). Edinburgh: Edinburgh University Press.

Strange, N. (1998). Perform, educate, entertain: Ingredients of the cookery programme genre. In C. Gheraghty & D. Lusted (Eds.), *The television studies book* (pp. 101–112). London: Arnold.

Taylor, P., & Wang, W. (2010). The fading glory of the television and telephone. *Pew Research, Social and Demographic Trends*. Retrieved September 1, 2001, from http://www.pewsocialtrends.org/2010/08/19/the-fading-glory-of-the-television-and-telephone/

Thomlinson, J. (2001). *Instant access: Some cultural implications of 'globalizing' technologie* (University of Copenhagen: Global Media Cultures Working Paper No. 13). Copenhagen: Department of Film & Media Studies, University of Copenhagen.

Thomlinson, J. (2007). *The culture of speed: The coming of immediacy* (Ebook edition). London: Sage.

Tichi, C. (1992). *Electronic hearth: Creating an American television culture*. New York: Oxford University Press.

Ureta, S. (2008). There is one in every home': Finding the place of television in new homes among a low-income population in Santiago, Chile. *International Journal of Cultural Studies, 11*(4), 477–497.

Uricchio, W. (1989). Rituals of reception, patterns of neglect: Nazi television and its historical representation. *Wide Angle, 11*(1), 48–66.

U.S. Census Bureau. (2012, April). Households and families: 2010. *2010 Census briefs*. https://www.census.gov/prod/cen2010/briefs/c2010br-14.pdf

Varela, M. (2005). *La televisión criolla. Desde sus inicios hasta la llegada del hombre a la Luna, 1951–1969*. Buenos Aires: EDHASA.

Varela, M. (2014, February 23) personal email to Lynn Spigel.

Virilio, P. (1989). The last vehicle. In D. Kamper & C. Wulf (Eds.), *Looking back at the end of the world* (D. Antal, Trans.). New York: Semiotext(e).

Weaver, S. L. (1954, January 6). The task ahead: Making TV the 'shining center of the home,' and helping create a new society of adults. *Variety,* 91.

Williams, R. (1975). *Television, technology and cultural form.* New York: Schocken Books.

Yader, D. (2013, July 31). Hackers expose how connected toilets, heaters, and lightbulbs are at risk. *Wall Street Journal.* Retrieved August 10, 2013, from online.wsj.com/.../SB10001424127887323997004578640310932033772.html

Yoshimi, S. (1999). 'Made in Japan': The cultural politics of home electrification in postwar Japan. *Media, Culture & Society, 21*(2), 149–171.

Young, B. M. (2007). The disappearance and reappearance and disappearance of the player in videogame advertising. Authors & Digital Games Research Association (DiGRA), situated play. In *Proceedings of DiGRA 2007 conference*. Online at http://citeseerx.ist.psu.edu/viewdoc/download?doi=10.1.1.190.2888&rep=rep1&type=pdf. Retrieved 14, June, 2013.

Cultural Industries and the (Geographical) Political Economy of the Media

4

Brett Christophers

This chapter provides an overview of the central findings and propositions of a multifaceted literature dealing with the political economy of "the media" and of the cultural industries whose products constitute today's most recognizable media forms: television, radio, music, magazines, the Internet, and so on. While its primary focus is on the *geographical* political economy of the media, its scope is wider and covers political economy more broadly conceived – partly since any specifically geographical issues cannot be understood except in terms of that wider framework, and partly in view of the fact that the political economy of the media is in many respects geographically-distinctive *by its very nature*. The chapter argues that notwithstanding the considerable progress that has been made in elucidating the media's geographical political economy, much remains to be explored, particularly in view of rapid and ongoing technological transformations in the cultural industries and the manifold political-economic implications thereof.

The chapter understands "political economy of the media" in a number of connected ways. To begin with, it follows Peter Golding and Graham Murdock (1997, pp. 11–14) in distinguishing media political economy from media economics. Where the latter treats "the media economy" as a separate sphere of activity, the former emphasizes "the interplay between economic organization and political, cultural and social life"; it actively considers both the economic *and* symbolic dimensions of the media, and it attempts to demonstrate "how different ways of financing and organizing cultural production have traceable consequences for the range of discourses and representations in the public domain and for audiences' access to them."

What becomes obvious when examining these issues, however, is that the media, while displaying their "own" such political economy, are pivotal to contemporary

B. Christophers (✉)
Institute for Urban and Housing Research, Uppsala University, Uppsala, Sweden
e-mail: brett.christophers@ibf.uu.se

S.P. Mains et al. (eds.), *Mediated Geographies and Geographies of Media*,
DOI 10.1007/978-94-017-9969-0_4

political economy much more generally. Hence Robert McChesney's (2003) definition of the political economy of communication as not only the question of how ownership structures, economics and government policies influence media and communication systems, technologies, practices and content, but also how media and communication systems and content impact on wider social relations. While the main focus of this chapter is on the former, therefore, it also touches on key dimensions of the latter.

As we shall see, the political economy of the media and the cultural industries is in many respects idiosyncratic. There are numerous reasons for this. We will encounter several of these as we proceed, but two merit foregrounding at the outset because their relevance will be apparent throughout and because they help to explain the distinctive nature not only of the media's political economy but also of its critique. For, this critique, *unlike* many other strands of scholarly political economy, has been undertaken exclusively in recent decades and, more pertinently, with little direct "assistance" from classical political economy (whether of bourgeois or Marxian variants), even if the latter has decisively shaped its mood and basic analytical parameters.

The first of these two reasons concerns the fact that the media or cultural "commodity" is itself relatively distinctive. The political-economic structures and dynamics that crystallize around it do not exactly mirror the capitalist political economy more widely because the commodity produced and circulated is substantively different. In turn, the way this economy is conceptualized must also be distinctive, even if it leans on "generic" political economy. Hence this chapter follows Nicholas Garnham in his assertion that the "economic nature of the cultural industries can be explained in terms of the general tendencies of commodity production within the capitalist mode of production *as modified by the special characteristics of the cultural commodity*" (1987, p. 30; my emphasis). In short, classic political-economic analysis needs substantial reversioning since it did not – indeed, could not – grapple with the peculiar attributes of commodities that did not exist until the twentieth century.

The second reason ultimately derives from the first. According to Garnham, one of the principal implications of the distinctive nature of the cultural commodity is that the primary locus of power and profit in the cultural industries is found not in the production of such commodities but in their distribution (cf. Christophers 2008). In other words, to come satisfactorily to grips with the political economy of the cultural industries we must focus more on distribution than production, since power and its exercise and effects are what political economy, at heart, is about. And here we confront the second "limit" to the conceptual purchase of classical political economy. As is widely recognized, such political economy, in the hands of Adam Smith, Thomas Malthus and David Ricardo *and* of Karl Marx, was oriented predominantly towards the relations and forces of *production*. Production was theorized as general, and law-like; distribution (or what such economists termed "exchange" or "circulation"; "distribution" was something else entirely) was deemed particularistic, and contingent, and was effectively assumed away (Harvey 2012). The upshot, once again, is that the purchase of classical political economy is significantly circumscribed where the political economy specifically of the media is concerned.

As such, over the past three or four decades, political economists of the media have had to start, to some extent, from scratch. They have learned *from* classical political economy – and primarily, it would be fair to say, from Marx – in the sense that it is recognized that, as Ronald Bettig (1996, p. 3) observes, "the logic of the [cultural] commodity must be situated within the larger context of the logic of capital." But they have had to both creatively adapt the insights of such literature to the cultural industries *and* add substantially to them. In what follows, I sketch out these adaptations and additions in the form of ten linked sets of propositions.

Geographies of Production (Proposition 1)

Classical political economy, and Marx's political economy in particular, was heavy on time and the temporal dimensions of the productive process but light on space and the *geographies* of production. One of the core findings of research into the cultural industries in recent decades, however, is that such industries are characterized by a marked spatial clustering. Producers of cultural and media products, in other words, tend to locate close to one another in dense agglomerations. Not only that, but such agglomerations are overwhelmingly *urban* agglomerations, found in – and often lending considerable cultural visibility to – major cities such as Los Angeles, London, Paris, Tokyo and Milan (Scott 2000).

It is true, of course, that the urban clustering of productive activity has been identified in many other industry sectors, too; the phenomenon is not exclusive to the media environment. Yet it does appear to be especially pronounced here. And while there is considerable debate over why this might be the case, researchers typically appeal to distinctive features of the cultural commodity such as the particular innovation processes involved in its development. Mark Lorenzen and Lars Frederiksen (2008), for example, argue that cultural industries benefit from both localization economies (positive externalities arising from co-location with similar firms) and urbanization economies (externalities related to co-location with other clusters and to the broader benefits of urban location) since they rely upon three different types of innovation (termed "variety," "novelty" and "radical" innovation).

But such analysis of media production clustering, while a mainstay of economic-geographical analysis since the mid-1980s and the seminal contributions of Susan Christopherson and Michael Storper (Christopherson and Storper 1986; Storper and Christopherson 1987), does not usually warrant the label of "political economy." More often it takes the form of a "weak" economics in the sense that it is modeled on the insights into clustering originally explicated by Alfred Marshall in the late nineteenth century, but generally lacks the formal, quantitative model*ing* required by the "strong" neoclassical economic tradition that Marshall did much to help institute. Indeed, where such economic-geographical work on the cultural industries begins to approximate more to *political* economy, as in the work of Neil Coe and Jennifer Johns (2004), it tends to extend inquiry – as the title of their contribution suggests – "Beyond production clusters." It has done so, in the first place, by way of a dramatic jumping of scales.

From the Local to the Global (Proposition 2)

If analysts of cultural industry clusters have pointed to the peculiarity of cultural-industry products in terms of the innovation processes that spawn them, the political economist Garnham (1987) was minded to look elsewhere. In his view, cultural commodities have two key characteristics. We will consider only the first of these here; the second surfaces in the following section.

The essence of the first key attribute of cultural commodities, argues Garnham (ibid., p. 30) is that "in general the costs of reproduction are marginal in relation to the costs of production." Once one version of a film, television program or piece of recorded music has been created (often at extremely high cost, of course), it costs relatively little – and often, in the digital world, effectively nothing – to produce another. Because of this cost structure, it follows that "the marginal returns from each extra sale tend to grow" (ibid.): it will be vastly more profitable to sell a $10 compact disc to a million people than to one person because while revenues increase proportionately with each sale, costs clearly do not; as such, cumulative profitability increases, as Garnham observes, with each extra sale. And precisely because of the enhanced profitability associated with sales maximization, there is, in the cultural industries, "a very marked drive towards expanding . . . audiences" (ibid.).

Such audience expansion can occur along several axes (the targeting of new, untapped audience demographics, for instance) but perhaps the critical and most fertile one is singularly geographical: growing audiences by selling to the maximum possible number of international markets, and ideally to a global audience. The immaterial costs of reproduction, Garnham therefore shows, have historically tended to lead in the cultural industries to a profound globalizing impetus.

This globalizing impetus has been most frequently observed and most heavily researched in relation to music and *especially* in relation to film and television. The literature on such globalization can be helpfully – although, obviously, not entirely cleanly – segmented into two phases and two corresponding political-economic constellations. From the late 1960s through to the late 1980s, research focused largely on US-led media globalization, which was typically seen as a forceful, unbending political-economic force that trampled on "local" cultures in local markets and thus befitted the label of "cultural imperialism" (see especially Schiller 1969, 1976). Jeremy Tunstall's *The Media Are American* (1977) captured the overall gist of this analysis.

Since the beginning of the 1990s, however, research in this field has become markedly more diverse and textured. For one thing, the applicability of the label of "imperialism" to US media expansion was subjected to close critical scrutiny (e.g. Tomlinson 1991), with scholars suggesting that such a concept, inter alia, rendered inert and thus patronized active audiences around the world who interpreted and evaluated content in highly diverse and culturally-specific ways (e.g. Ang 1985; Liebes and Katz 1990). For another, researchers began to show that while Hollywood remained tremendously powerful amidst the circuits of international media capital, it was not *only* the US's cultural industries that harbored and realized globalizing ambitions. The rise of Bollywood, the international popularity of Latin American *telenovelas* and the emergence of television-formats development-and-export hothouses in the Netherlands and the UK have been just some of the more prominent phenomena to prompt a shift in scholarly discourse away from US cultural imperialism and towards "media globalization." But while the political economy of this new scheme of things looks quite different – less monolithic, less centrist, more competitive, and more open to notions of "active" consumers – the underlying conjuncture of low marginal costs with an associated expansionist impulse remains as pertinent as ever.

Creative Labor, Monopoly Rent and Public Goods (Proposition 3)

Capitalism, writes Bernard Miège (1989, p. 29), tends "to push cultural production towards the dominant model, namely that of material production." And yet, as we shall begin to explore now, the capitalization and commodification *of* cultural production is riddled with contradictions. Miège (ibid.) helpfully highlights one of the more obvious: precisely "in order to be standardized," he argues, "cultural products must continue to be marked by the stamp of the unique, of genius." They must look, and feel, distinctive. One implication of this, in turn, is that as Miège rightly points out, "the star system and the industrial organization of Hollywood are indissolubly linked." But there are two other implications with much wider political-economic significance.

First, something curious and vitally important occurs in the realm of labor. Again, here is Miège (ibid., p. 30): "The penetration of capitalism into the cultural sphere is thus not, unlike what happens in other sectors (commerce, leisure), necessarily accompanied by the generalization of wage labor and the steady disappearance of the independent cultural worker. The old system of royalties is maintained and even tends to grow more complex." In other words, "creative" laborers – writers, directors, producers and so forth – play a critical role, and are able to secure a portion of surplus value in the shape of royalties (cf. Garnham 2005). At the same time, however, the bulk of the labor force *does* comprise wage laborers; hence Mike Wayne (2003, p. 18) depicts a cultural-industry workforce bifurcated into intellectual or creative labor on the one hand and more manual or technical labor on the other. And as Miège (1989, p. 29) insists, *only* the former (a minority) "are in a position to claim a part of the surplus value produced." For the majority, the notion of creative autonomy "is a pure façade: it allows them to be paid at a rate markedly lower than the value of their labor power."

The second crucial implication of cultural commodities being "marked by the stamp of the unique" is that their producers are nominally able to extract from their distribution what political economists term *monopoly rents*. David Harvey (2002, p. 94) explains: "Monopoly rent arises because social actors can realize an enhanced income stream over an extended time by virtue of their exclusive control over some directly or indirectly tradable item which is in some crucial respects unique and non-replicable." For Harvey (ibid., p. 93), "those products designated as 'cultural'" constitute a quintessential example of such items.

But herein lies a profound contradiction or tension, one hinted at by Miège – in his assertion that cultural products must be rendered "unique" *in order to be standardized* – and fleshed out by Harvey (ibid., p. 95): "while uniqueness and particularity are crucial to the definition of 'special qualities', the requirement of tradability means that no item can be so unique or so special as to be entirely outside the monetary calculus." An item must be commodified *to some degree* to enable monetization. But this process rests on a knife-edge; if it goes too far, the very source of monopoly rent is extinguished. "The bland homogeneity that goes with pure commodification," notes Harvey (ibid., p. 96), "erases monopoly advantages. Cultural products become no different from commodities in general."

All of which brings us to the second "special characteristic" of the cultural commodity identified by Garnham: the fact that "the cultural commodity is not destroyed in the process of consumption" (1987, p. 30). Another way of expressing this is to say that the cultural commodity is a "public good" since it is *nonrival in consumption*: consumption by one person does not limit or detract from consumption by another person. The significance of this, meanwhile, is that not only do nonrivalry and the easy, cheap replicability of commodified products not, as Harvey (2002, p. 95) observes, "provide a basis for monopoly rent." They also, more materially, endanger the formation of meaningful economic markets per se. The fact that cultural products are nonrival in consumption means, says Garnham (1987, p. 30), that "it has been difficult to establish the scarcity on which price is based."

What solutions have the cultural industries formulated to resolve this conundrum? What have been their strategies, in Garnham's words, for "artificially limiting access in order to create scarcity" (ibid.)? More pointedly, perhaps, "how to assemble monopoly powers" (Harvey 2002, p. 98)? Three main strategies have, historically, been pursued. We now consider these in turn, in propositions four (the first), five and six (the second) and seven (the third).

Consolidation and Centralization (Proposition 4)

Harvey is quite clear about the first and in some ways most transparent of the strategies used to create scarcity around products to which none "naturally" adheres and, in the process, to enable the extraction of monopoly rents: namely, to "centralize capital in mega-corporations" (ibid.). If the requirement to trade and commodify "the cultural" – and the nonrival status of such commodities – makes monopoly rents elusive, then the simplest way to realize the latter is to *monopolize*.

Dominating cultural commodification institutionally and operationally helps protect the monopoly rents that commodification *itself* threatens to erode.

Not surprisingly, there is a large literature detailing the nature and extent of concentration and consolidation in the cultural industries at both individual national levels and, increasingly, internationally. In the case of the former, the US represents an especially stark example. When Ben Bagdikian published the first edition of *The Media Monopoly* in 1983, he reported that 50 major corporations controlled the national media; but as he revised his text through five successive new editions, the number was gradually whittled down, first to 29 (by 1987), then 23 (1990), 14 (1992), 10 (1997) and finally 6 (2000). At the international scale, Robert McChesney (2008) has similarly chronicled the pace and scale of industry rationalization. Not only has the number of major market participants been reduced through multiple large-scale cross-border mergers, but complex webs of cross-ownership make the pool of pertinent shareholders smaller still, leading McChesney (ibid., pp. 320–1) to conclude that in some respects "the global media market more closely resembles a cartel than it does the competitive marketplace found in economics textbooks."

One particularly interesting and important dimension of this consolidation trend can be brought into focus by returning momentarily to Garnham. We have identified already what he sees as the two special characteristics of cultural commodities – their trivial marginal costs of production and their nonrivalry in consumption – together with some of the respective implications thereof. But it is the *combination* of these two characteristics that is, for Garnham, particularly salient. It is the *combination* of "the drive to audience maximization" with "the need to create artificial scarcity by controlling access" that makes "*cultural distribution, not cultural production*" the "*key locus of power and profit*" (Garnham 1987, p. 31; original emphasis) because both are ultimately distributive (i.e. exchange-oriented) issues.

The relevance of this insight at this juncture in our argument is that because it is in distribution that power is most consequential and that the profitability of cultural commodification is essentially settled (one way or another), it is in distribution that "we typically find the highest levels of capital intensity, ownership concentration and multi-nationalization" (ibid., p. 32). If one is going to seek to monopolize, better to do it in distribution than in production, where the levers of political-economic control – to determine whether the cultural economy makes money or not – simply are not available. All of this is borne out in historical-geographical practice. If we look around the world at such realms as book publishing, video game publishing, television and film, and recorded music, one finds production sectors in which, to be sure, major firms participate, but where large numbers of small, independent, single-territory producers *also* manage to thrive. The same is not true, however, of distribution, where the field is dominated by a small number of consolidated and frequently transnational conglomerates. The creation of economic scarcity, with a view to securing monopoly rents, has been served by monopolization not of production but of distribution.

Copyrighting Scarcity (Proposition 5)

Yet as Garnham was also aware, centralization of power in the form of monopolistic or oligopolistic control of distribution has not alone been sufficient to create the scarcity required of the cultural economy. Even a single, globally-dominant distributor of all cultural products would struggle to generate profits without some way of dealing *technically* and *legally* – as well as economically – with the problem of nonrivalry. What good would selling to one person do if their (nonrival) consumption precipitated "free" consumption by others? While the advent of Internet-based file-sharing technologies has made this more of a live question than ever before, the question has never been an idle one for the cultural industries.

Garnham, interestingly, does not discuss this particular aspect of those industries' solution to the problem of market formation. Harvey, however, does, and he correctly points to the significance of the central legal institution of copyright. Copyright is, quite literally, *the right to copy*, a definition and legal privilege that clearly takes on particular force in the context of an economy of nominally nonrival consumption. (What worth is a right to copy, by way of counterpoint, when consumption of a good by one person intrinsically rules out consumption by another?) It is copyright that enables the cultural industries to make many of their outputs – in theory at least – *rivalrous* (to use the hideous economics terminology). As consumers, we are all aware of the copyright warnings on compact discs and digital versatile discs (DVD). We typically skim these without a moment's thought, but as Harvey (2002, p. 98) observes it is copyright, "[p]atents and so-called 'intellectual property rights'," embedded in the "international commercial laws that regulate all global trade," that enable the corporate world to "secure ever more firmly the monopoly rights of private property" (c.f. Christophers 2016, ch. 3).

Bettig (1996), among others (e.g. Gillespie 2007), has contributed to the writing of a broad and fascinating political-economic history of the work of copyright in enabling the monetization of cultural products. Notably, and not surprisingly, it is a history indelibly interwoven with the history of consolidation and monopolization touched upon in the previous section; strategies to enact scarcity have never operated in isolation from one another. Hence Bettig shows how "new" media technologies – he was writing in the mid-1990s, note – such as cable television and video recorders (VCR) ultimately came to be thoroughly integrated into the prevailing, Hollywood-dominated market structure despite being introduced into the marketplace before a set of copyright laws had been implemented, and despite appearing to harbor the potential fundamentally to disrupt existing markets and market power relations. His narrative is one in which the emphasis is placed on oligopolistic power allied to the role of the state in "recognizing, conferring, and protecting intellectual property rights" (Bettig 1996, p. 3).

Of course, the rapid development and uptake of new digital and networked technologies over the past 10–15 years have put the viability – and hence political economy – of cultural copyright under strain and in the spotlight as never before. Moreover, opponents of the cultural industries' attempts to shore up legacy copyright systems in the face of peer-to-peer file-sharing and related threats include not only large numbers of consumers but also notable influential academics (e.g. Lessig 2004). Hence, the spread of "gift economies" and the effective death of "copyright capitalism" have long been predicted (e.g. Leyshon 2003). But while space does not permit here of a detailed account of actual developments, it is increasingly clear that even in the case of products where the Internet has demonstrably revolutionized distribution practices (e.g. recorded music), the core political-economic trinity of oligopolistic corporate power, legal protections and monopoly rent extraction remains far from broken (Rogers 2013).

Territorializing Scarcity (Proposition 6)

In his analysis of the cultural industries' historical taming of threatening technologies such as cable television and VCRs, Bettig explicitly considers the geographical dimensions of the pertinent political-economic dynamics. Notably, he demonstrates that as Hollywood fought to maintain its iron grip on global film and television markets in the 1980s, it repeatedly "found itself engaged in foreign policy making alongside the U.S. government" (1996, p. 189), primarily to compel foreign governments to enforce the intellectual property rights of copyright owners. Which begs the question: what is, and has been, the geography *of* copyright in the international context?

This is a vital question to answer because the cultural industries' copyright scaffolding – and thus, we can suggest, the political economy of media in general – *has* always had a distinctive and economically-material spatial architecture. This, at the most basic level, has comprised a territorialization of copyright at the national scale: rights owners, whether it be in the recorded music sector, in book publishing, or in television and film, license the rights to distribute their products strictly on a territory-by-territory basis.

Why have they done this? This is not an altogether straightforward matter. Partly it has always been the most pragmatic approach: control over local distribution platforms (e.g. broadcast television, packaged media wholesale, etc.) has generally been nationally-circumscribed, so it has simply made practical sense to parcel up rights accordingly. At a deeper level, there is a rationale of *control* (Christophers 2014b): segmentation of the world into a series of discrete national markets allows the rights owner to tailor the details of local release to each local market (differentiating by price, by timing of release, and even by exact content) in such a way as to maximize overall revenues. At heart, however, we come back to the question of scarcity and market formation: competitive pressures look different in different national markets and buyers in those different markets can display widely varying abilities to pay. What scarcity "looks" like varies considerably from place to place, and so therefore must the supply-and-demand-oriented strategies for configuring it.

There is one especially critical point to note about the territorialization of scarcity and of cultural-products markets. Specifically, this industry strategy has come under considerable pressure in the last two decades from distribution technologies that transparently threaten to render national economic boundaries obsolete. First satellite television and then the Internet – with the latter extending the scope of the threat from television alone to essentially all media products with digital adaptations – have opened up unprecedented technical possibilities for *direct* cross-border distribution of cultural products. How, in such a world, could a strictly territorialized system of rights allocation and exploitation be maintained? Here is not the place to posit an answer. The issues are, necessarily, complex, embracing at once the cultural, economic, technical, political and – not least – legal spheres. Suffice to say that the cultural industries are grappling with those issues on a daily basis and that *despite* the manifold threats, territorialization remains the dominant approach (Christophers 2009, pp. 143–60; 2014a, b).

The Attention Economy (Proposition 7)

In many respects the third and final solution to the problem of manufacturing scarcity is the most interesting since it has effectively entailed abandoning the attempt, through circumscribing access, to make the cultural product *itself* scarce, and exploiting instead an *alternative* – albeit connected – locus of scarcity. That locus is found, ironically, in the stratum constituted by the consumers *of* cultural products, and its mining makes for a fascinating and crucial political and cultural economy.

We are referring, of course, to advertising, where the retailers of stubbornly nonrival cultural products such as, most notably, broadcast television, frequently eschew charging consumers to watch programming and instead charge advertisers for the privilege of access to the consumer and his or her scarce attention. In this respect, advertising is, at heart, an attention economy (Davenport and Beck 2001), one in which the audience is constituted as the commodity for sale and where, as Garnham (1987, p. 31) observes, "the cultural software merely acts as a free lunch." Free-to-air, advertising-funded broadcast television is perhaps the prototypical example of such an economy.

Political economists of the media have written extensively on the nature of the attention economy in the media sector. Among the most influential, early interventions was that of Dallas Smythe (e.g. 1981), who pointed out that classical Marxism had never dealt with the demand-management function of advertising and the mass media because it did not exist until the 1920s at the earliest. His answer to the question of "what is the principal product of the commercial mass media in twentieth century capitalism?" was original and powerful: *audience power*. It is audience power, he argued, that is commodified insofar as it is produced, sold, purchased and consumed. And the term "power" was, for Smythe, vital, since he insisted that the audience *is* working – even if unwittingly – *for* capital. Smythe posited that this was unpaid work. Sut Jhally (1991), deepening and extending Smythe's insights, agreed that audiences "work" for the cultural industries in consuming advertising and hence creating symbolic meaning for advertisers and profit for the media, but theorized entertainment as the "wage" they receive for doing so.

Again, however, new technologies have posed troubling questions for the manufacturers and retailers of scarcity. If the power and profitability of commercial broadcasters is predicated on their ability to aggregate audience "eyeballs" and sell them to advertisers, what do technologies such as digital video recorders – giving viewers the capacity simply to skip commercials – mean for the attendant political-economic configurations? Once more, such issues are in the process of being worked out (and researched); and once more, the industry, through innovations such as product placement and advertiser-funded programming, is proving itself highly adaptable.

Smythe's conceptualization of advertising specifically in terms of capitalist *demand management*, meanwhile, is particularly pertinent. For, what he demonstrated is that advertising – and the media more generally – is doubly implicated in the political economy of capitalism. The cultural industries, as we have seen, represent one part *of* the capitalist economy, and dance broadly to the tune of its political-economic laws, even if certain steps are different in view of the special characteristics of the cultural commodity. But, as the example of advertising shows, the media support and "lubricate" capitalist circulation in a much wider sense. We therefore cannot understand the political economy of capitalism per se without explicating the role of the cultural industries *within* it. In our three closing propositions, we shall reflect briefly on different aspects of that role.

The *Culture* Industry (Proposition 8)

Let us first return momentarily to Miège (1989), who usefully distinguishes between three different "modes of insertion of cultural labour into the general process of production" (p. 25). They are, firstly, non-capitalist production of cultural products (which we will touch upon in proposition number 10); secondly, capitalist cultural production (which has been our focus so far); and thirdly, the "integration of cultural products, normally in the form of non-material performances, into the process of circulation within the framework of the realization of value" (ibid.)

This last is our focus here, but what exactly is Miège getting at? For political economists, the "realization" of value means converting the outputs of production into money through sale in the marketplace. Thus it means, in Smythe's terms, consumer demand; and Miège, like Smythe, is pointing to the critical role of cultural products in general and advertising in particular in helping value to be realized and thus in helping capital – envisioned as perpetual circulation or flow – literally to *survive*. Miège (ibid., p. 36) calls this "the promotion of commerce by culture."

For some writers, this is much more than simply a matter of the mobilization of the media to sell *particular* products. It is about the media being used to sell products – commodities, commodification, capitalism – *in general*. This was the argument advanced by Horkheimer and Adorno in the famous essay on "The Culture Industry" in *Dialectic of Enlightenment* (1972[1944]). How could we understand the fact that workers were so docile – and seemingly, in most cases, contented – despite often-challenging economic circumstances? "The Culture Industry" was the answer: the mass media rendered mass society submissive by *selling* capitalism's merits.

Horkheimer and Adorno's argument was resolutely political-economic but it was also profoundly psychological (and, needless to say, highly controversial). "The Culture Industry" helped keep people "happy" because it convinced them that happiness was a function of commodity acquisition, and thus of the capitalist society that facilitated this. Others, writing more recently, but equally convinced of the role of the media in reinforcing and thus reproducing the political economy of capitalism, have emphasized other dimensions of the media's power. Herman and Chomsky's *Manufacturing Consent* (1988) was a famous example – arguing that the mass media simply reflect and entrench the capitalist interests of the capitalist institutions that own them – but, focusing on the news media, it was arguably not directly pertinent to the cultural industries. McChesney's *The Political Economy of Media* (2008), however, is; its particular polemic alleges the power of the corporate media to constrain the democratic public sphere and, in the process, to cement capitalist political economy.

Political Economies of Time and Space (Proposition 9)

One of the most influential writers on the political economy of the media in the past two decades has been Vincent Mosco. In his *The Political Economy of Communication* (2009), Mosco theorizes three concepts or what he calls "entry points" for the analysis of the political economy of communications media. One is "commodification" – the process of converting use values into exchange values, which much of this chapter has been concerned with to this point. The second is "structuration" – a concept borrowed from Anthony Giddens and intended to curtail a perceived tendency in political economy to focus on the "structures" of the material relations of production at the expense of the "action" or "agency" of individuals. The third, and the one we are interested in here, is what Mosco calls "spatialization."

By "spatialization," Mosco (ibid., p. 14) refers to the pivotal role of communications media historically in enabling the "overcoming [of] the constraints of space in social life." Communications media of all stripes, he argues, have progressively contributed to making the world a smaller place by annihilating space *with* time: reducing the time it takes for geographical distances to be bridged physically and, no less importantly, experientially. The great Canadian political economist Harold Innis had suggested much the same thing in identifying the "spatial bias" of modern communications media – media that were "space-binding" in the sense that they distributed information or entertainment to many people over increasingly long distances (see Carey 1988).

The primary emphasis of the literature on the spatialization associated with modern mass media and communications technologies has been on its cultural ramifications. Books like Arjun Appadurai's *Modernity at Large* (1997) are emblematic; so too is Marshall McLuhan's famous trope of the "global village." But the spatialization evoked by Mosco clearly has political-economic implications as well (hence its centrality to a book *on* political economy), and these extend well beyond the role of the media in facilitating – through spatialization – the globalization sought *by* the media (proposition two). Consider, for instance, the arguments of David Harvey in *The Condition of Postmodernity* (1989). In relation to communications media Harvey coined the phrase "time-space compression" to capture precisely the changing experience of space and time discussed by Mosco. And he argued, significantly, that time-space compression was intimately implicated in a formative shift in Western capitalist political economy from the Fordist system of production to the "flexible accumulation" characterizing the period since the early 1970s.

The Role of the State (Proposition 10)

To this point our focus has been exclusively on media mediated by the framework of capitalist political economy – which is to say, on profit-oriented cultural *industries*. And with good reason: the vast majority of mass-market cultural output of all forms is today produced and distributed by commercial organizations. But recall Miège's three categories of cultural production, the first of which was *non-capitalist* production of cultural products. His category includes, of course, all manner of small-scale cultural activities that occur outside – or at least, nominally outside – the market economy and its profit calculus. Here, however, where our interest is in industrial-*scale* cultural production, the one form of pertinent large-scale non-capitalist activity is that which is funded by the state.

What role or roles, in the West, has the state historically assumed in relation to large-scale media production? Far and away the most significant such role has been the chartering and funding of public-service radio and television broadcasting services. Such broadcasting has a long and (in many eyes) distinguished heritage, dating back to the founding of the British Broadcasting Corporation (BBC) in the UK in the inter-war period. The primary rationale for public-service broadcasting (PSB) has long been that private-sector media organizations tend not to satisfy in isolation all public informational needs: a case of "market failure." As such, news and educational material have always been PSB staples, typically alongside entertainment, programming designed to promote national culture and identity, and programming geared towards minorities and various special interest groups. PSB organizations tend to be funded by the state indirectly – through, for example, license fees – rather than directly out of general taxation, although many are increasingly reliant on commercial sources of revenue as the pressures on state finances tighten.

So far as the political economy of PSB is concerned, we can point briefly to three main issues of interest. First, this public-sector cultural industry is, not surprisingly, characterized by a *different* political economy to commercial media (different motives, different power relations, different economics), although many would argue that the differences have radically diminished as commercial revenue sources become more material and as private-sector management regimes have been widely imported in repeated efficiency drives. Second, PSB not only bears its own political economy but impacts directly on the political economy of private-sector media. Witness the persistent complaints of commercial media groups in countries with a strong PSB presence that the latter "unfairly" distorts market competition; such distortion may in reality be less marked than the complainants allege, but it is indisputable that *some* effect on commercial political economy is exercised. Third, lastly, and most significantly of all, it would be hard to dispute that PSB affects a country's political economy in the very broadest sense. It is frequently observed, for instance, that countries with strong PSB traditions – the Scandinavian countries, for example – tend to be more centrist or left-leaning than those that do not (the US being a good example of the latter). Again, however, the exact nature of this relationship is unclear. Indeed it is, in some respects, a classic chicken-and-egg question. Does a progressive political economy make space for an enduring PSB presence, or does the latter buttress and sustain the former? Ultimately, we do not know.

Conclusion

In this short overview of the cultural industries and of the (geographical) political economy of the media, I have developed a series of interlocking propositions about how researchers have sought to describe and account for that political economy. Hopefully, two general, underlying themes will have become clear, and I will conclude by drawing them out more directly (cf. Christophers 2012).

First, there is the question – to which the final three of our ten propositions sought to speak – about the media and their implication in political economy in the broadest, non-sector-specific sense. Both within geography specifically and indeed in critical social science more generally, political economists have, with notable exceptions, been oddly reluctant to engage with the materiality of the media. Given the brief account above of how important the media and cultural industries *are* to contemporary capitalist political economy, I hope to have shown that this analytical marginality is unsupportable and unsustainable.

Second, there is a narrower disciplinary question. As we have seen, the political economy of the media is, in so many different respects, deeply and fascinatingly geographical. Yet scarcely any economic geographers of the media and cultural industries have examined these political-economic structures and dynamics (which is not the same as saying that geographical political economists have largely neglected the media, note). Instead, we have seen, for the past three decades, a painfully narrow preoccupation with media production clusters – a literature from which, moreover, questions of *political* economy are also typically evacuated. This, too, seems unsustainable, and one therefore hopes that the tentative steps taken here help precipitate a much longer geographical journey.

References

Ang, I. (1985). *Watching Dallas: Soap opera and the melodramatic imagination*. New York: Routledge.
Appadurai, A. (1997). *Modernity at large: Cultural dimensions of globalization*. Minneapolis: University of Minnesota Press.
Bagdikian, B. (2000). *The media monopoly* (6th ed.). Boston: Beacon.
Bettig, R. (1996). *Copyrighting culture: The political economy of intellectual property*. Boulder: Westview.
Carey, J. (1988). Space, time and communications: A tribute to Harold Innis. In his *Communications as culture: essays on media and society* (pp. 142–172). New York: Routledge.
Christophers, B. (2008). Television's power relations in the transition to digital: The case of the UK. *Television & New Media, 9*, 239–257.
Christophers, B. (2009). *Envisioning media power: On capital and geographies of television*. Lanham: Lexington.
Christophers, B. (2012). Markets, the media, and the state of contemporary economic geography. *Dialogues in Human Geography, 2*, 243–245.
Christophers, B. (2014a). Spaces of media capital. In P. Adams, J. Craine, & J. Dittmer (Eds.), *Ashgate research companion on geographies of media* (pp. 363–376). Aldershot: Ashgate.
Christophers, B. (2014b). The territorial fix: Price, power and profit in the geographies of markets. *Progress in Human Geography, 38*, 754–770.
Christophers, B. (2016). *The Great Leveler: Capitalism and Competition in the Court of Law*. Cambridge, MA: Harvard University Press.
Christopherson, S., & Storper, M. (1986). The city as studio; the world as backlot: The impact of vertical disintegration on the location of the motion picture industry. *Environment and Planning D, 4*, 305–320.
Coe, N., & Johns, J. (2004). Beyond production clusters: Towards a critical political economy of networks in the film and television industries. In D. Power & A. Scott (Eds.), *Cultural industries and the production of culture* (pp. 188–204). London: Routledge.
Davenport, T., & Beck, J. (2001). *The attention economy: Understanding the new currency of business*. Boston: Harvard Business School Press.
Garnham, N. (1987). Concepts of culture: Public policy and the cultural industries. *Cultural Studies, 1*, 23–37.
Garnham, N. (2005). From cultural to creative industries: An analysis of the implications of the "creative industries" approach to arts and media policy making in the UK. *International Journal of Cultural Policy, 11*, 15–30.
Gillespie, T. (2007). *Wired shut: Copyright and the shape of digital culture*. Cambridge, MA: MIT Press.
Golding, P., & Murdock, G. (1997). Culture, communications, and political economy. In Golding & Murdock (Eds.), *The political economy of the media*. Cheltenham: Edward Elgar Publishing.
Harvey, D. (1989). *The condition of postmodernity*. Oxford: Blackwell.
Harvey, D. (2002). The art of rent: Globalisation, monopoly and the commodification of culture. *Socialist Register, 38*, 93–110.
Harvey, D. (2012). History versus theory: A commentary on Marx's method in *Capital. Historical Materialism, 20*(2), 3–38.
Herman, E., & Chomsky, N. (1988). *Manufacturing consent: The political economy of the mass media*. New York: Pantheon.
Horkheimer, M., & Adorno, T. (1972 [1944]). *Dialectic of enlightenment*. New York: Herder and Herder.
Jhally, S. (1991). *The codes of advertising: Fetishism and the political economy of meaning in the consumer society*. New York: Routledge.
Lessig, L. (2004). *Free culture: How big media uses technology and the law to lock down culture and control creativity*. New York: Penguin.

Leyshon, A. (2003). Scary monsters? Software formats, peer-to-peer networks, and the spectre of the gift. *Environment and Planning D, 21*, 533–558.

Liebes, T., & Katz, E. (1990). *The export of meaning: Cross-cultural readings of Dallas*. New York: Oxford University Press.

Lorenzen, M., & Frederiksen, L. (2008). Why do cultural industries cluster? Localization, urbanization, products and projects. In P. Cooke & R. Lazzeretti (Eds.), *Creative cities, cultural clusters, and local economic development* (pp. 155–179). Cheltenham: Edward Elgar.

McChesney, R. (2003). Making a molehill out of a mountain: The sad state of political economy in U.S. media studies. In A. Calabrese & C. Sparks (Eds.), *Toward a political economy of culture: Capitalism and communication in the twenty-first century* (pp. 41–64). Lanham: Rowman & Littlefield.

McChesney, R. (2008). *The political economy of media: Enduring issues, emerging dilemmas*. New York: Monthly Review Press.

Miège, B. (1989). *The capitalization of cultural production*. New York: International General.

Mosco, V. (2009). *The political economy of communication: Rethinking and renewal* (2nd ed.). London: Sage.

Rogers, J. (2013). *The death and life of the music industry in the digital age*. London: Continuum Publishing.

Schiller, H. (1969). *Mass communications and American empire*. New York: A. M. Kelley.

Schiller, H. (1976). *Communications and cultural domination*. New York: M. E. Sharpe.

Scott, A. (2000). *The cultural economy of cities: Essays on the geography of image-producing industries*. London: Sage.

Smythe, D. (1981). On the audience commodity and its work. In his *Dependency road: Communications, capitalism, consciousness and Canada* (pp. 22–51). Norwood: Ablex.

Storper, M., & Christopherson, S. (1987). Flexible specialization and regional industrial agglomerations: The case of the US motion-picture industry. *Annals of the Association of American Geographers, 77*, 260–282.

Tomlinson, J. (1991). *Cultural Imperialism: A Critical Introduction*. Baltimore: Johns Hopkins University Press.

Tunstall, J. (1977). *The media are American*. London: Constable.

Wayne, M. (2003). *Marxism and media studies: Key concepts and contemporary trends*. London: Pluto Press.

A Brief History of Mediated, Sensational and Virtual Geographies

5

Stuart C. Aitken and James W. Craine

> *"In a world that really has been turned on its head, truth is a moment of falsehood" (Guy Debord 1994, 15).*

Media geography is a convoluted and complex field of study – running the gamut of maps, movies, internet and social media – that tries to make sense of how we communicate our geographical imaginations. With this chapter, we tap into our own influences and discipline to map a brief trajectory of the field from the very specific focus of our interests in mediated, sensational, affective, digital and virtual geographies. Ours is just one story of the field, perhaps appropriately emanating from Southern California where we both live and where, arguably, the most powerful newspaper, television, film and internet media have held sway this last century. Our substantive interests are with film and digital media. Given this interest, we do not deal directly with newspaper, books, theatre or television, but their contexts are clearly reflected in our discussion. Nor do we follow strictly linear or historical trajectories to describe the field. Rather, we bounce between eras so as to highlight simultaneously our areas of interest and larger discursive tensions. Elsewhere, one of us argues that understanding disciplinary engagements and debates is an appropriate way of appreciating how we come to know the world because it maintains the tensions that propel creativities; although it may be

S.C. Aitken (✉)
Department of Geography, San Diego State University, San Diego, CA, USA
e-mail: saitken@mail.sdsu.edu

J.W. Craine
Department of Geography, California State University, Northridge, CA, USA
e-mail: james.w.craine@csun.edu

S.P. Mains et al. (eds.), *Mediated Geographies and Geographies of Media*,
DOI 10.1007/978-94-017-9969-0_5

pedagogically prudent to think of paradigm changes and the linear development of thought, in actuality at any one time there are a plethora of discursive tensions at play (Aitken and Valentine 2014).

In what follows, we begin with some early media geographies as moments of falsehood, which in many ways set the stage for critical appraisals of representations in Europe through the Birmingham school and in the US through the Santa Cruz school. We then show how some of the first writing by contemporary media geographers brought these two schools of thought together while at the same time bringing geography and media studies closer together as disciplines. This is followed by a discussion of the development of critical spatial theorizing and its influence on media scholarship. We then consider digital media, and show how critical spatial theorizing opens this complex web of virtuality before returning to the embodied, affective material conditions of lived experience. We end with a consideration of space as a mediation, and what that means for contemporary cartographies and spatialities.

Truth as a Moment of Falsehood

The extent to which mediated geographies – as a way to make an impression, to represent ourselves in the world, or to create sensation and spectacle – resonates with the ideas of Guy Debord (1994), it may be argued that it is a way to produce truth as a moment of falsehood. What Debord meant by this is that many (perhaps all) representational mediations between us and the world (and especially maps, news and other seemingly straightforward narratives) comprise spectacles and falsehoods in the guise of truth. His ideas inspired work during the height of postmodernism in the 1980s and 1990s, with its focus on the ways representations skimmed the surfaces of life's depths, and the ways everything is simultaneously true and self-evident on the one hand, and false and superficial on the other. The idea of creating truth as a moment of falsehood is particularly pertinent for looking at the beginnings of contemporary media geographies.

A fashionable understanding of mediated geographies as a specific part of the discipline and national consciousness began in 1888 when the US's *National Geographic Magazine* first hit the news-stands as a non-technical monthly (Rothenberg 2007). The falsehood of photographic realism, benign exploration, and the representation of others spoke truth to the ways a muscular male heterosexual science mediated the world and presented it as a global and cultural *fait accompli*. In a similar vein, and over 60 years later in the UK, *The Geographical Magazine* published a series of articles in the 1950s in collaboration with Dr. Robert Manvell, Director of the British Film Academy, which emphasized the national character and factual basis of film-making. The series of articles outlined the ways film dealt effectively with the culture, customs and behaviors of the everyday lives of people in the country portrayed, and focused on the realism of the medium as its most significant contribution to geography (Aitken 1994). Manvell (1956, 420) described the enterprise as global in reach and a "visual network ... capable of projecting the indigenous portraiture of mankind through the motion picture."

The idea of any veracity in these kinds of images is unavoidably subjective and, in addition, there arises the potential for the misuse of the concept of "objectivity" when it is applied to photographic realism or documentary film-making came into sharp relief when California's (specifically UC Santa Cruz's) 'crisis of representation' school began speaking of the situated knowledge and biases in representation (Clifford 1986; Clifford and Marcus 1986; Haraway 1991). This work helped propel some early writing on media geographies, especially in the US. In the UK, geographers were heavily influenced by the Marxist orientation of Birmingham's Center for Cultural Studies under the directorship of Stuart Hall, which in turn was indebted to Walter Benjamin, Theodor Adorno and the Frankfurt School. The two schools of thought that began in Santa Cruz and Birmingham merge to create a forceful foundation which focuses on concern about what cultural texts and media actually do, and the ways critical theories help unravel a complex set of political orientations. They also provided an important, although not singular, foundation for the last several decades of critical media geographies.[1]

Critical Theories and Complex Political Orientations

Media geographies gained significant academic footing when UK scholars Jackie Burgess and John Gold published *Geography, the Media and Popular Culture* (1985) and US scholar Leo Zonn published *Place Images in Media: Portrayal, Experience and Meaning* (1990). These two edited collections brought together a growing interest in joining together the fields of geography and media studies around the issues of truths and falsehoods that were emerging from the Santa Cruz and Birmingham schools. In addition, although these two edited volumes did not establish a connection to the colonialism of *The National Geographic's* and *The Geographical Magazine's* mediated geographies, they nonetheless brought together the two schools of thought emanating from Europe and the US.

Zonn's *Place Images in Media* was focused on productions of meaning in terms of the portrayers and creators of place images. His collection of essays is also heavily influenced by the phenomenological and perceptual ideas that permeated geography in the 1970s and 1980s, with a focus on the ways places are represented back to us. A little later Zonn co-edited the first book on film geographies that attempted to bring together the work of film studies theorists like Christian Metz, Stephen Heath, Jean-Louis-Baudry, and Thierry Kunstel with emerging critical theories of space and place (Aitken and Zonn 1994). With this book, geographers witnessed one of the first coherent integrations of cognitive, critical and affective perspectives. At around the same time, Natter and Jones (1993) published a

[1]Other areas of interest included social and cognitive psychology, semiotics and, latterly, neuro-science.

trenchant critique of the 'reality effect' of newspaper reporting and documentary filmmaking, which highlighted performed narrative conventions as the outcome of social, cultural and political mediations. Behind this work, and pushing a media savvy troop of young geographers to aspire to new giddy heights, were the spatial and social theories of David Harvey, Doreen Massey, Michael Dear and Ed Soja. Caught in the rational objectivity of the quantitative revolution where the most coherent spatial theories – like Waldo Tobler's pronouncement that things close to other things have the most influence on those things – seemed banal and self-evident, these young academics were inspired by a mostly Marxist-derived set of social theories about the ways media and culture intertwine in sometimes insidious and sometimes subversive ways with a capitalist-industrial-military complex.

In his acclaimed *The Condition of Postmodernity,* Harvey (1989) used *Blade Runner,* amongst other films, to argue for the ways movies elaborate what was referred to at the time as our postmodern condition and, in particular, the conflicts that contemporary capitalism wrought on time scales and spatial resolutions. Unwilling to cede too much power to film, Harvey sees *Blade Runner* as a parable in which postmodern conflicts are set in a context of flexible accumulation and time-space compression. Doreen Massey (1998) is more circumspect about the power of media, commenting on its global reach in a thoughtful essay on filming in the Yucatan and the media contexts of the spatial construction of youth cultures. Massey's (2005) theoretical influences show up forcefully in her later writing, and include Bergson, de Certeau and Laclau. For de Certeau (1984), the importance of spatial stories, with all the nuances of narratives through movement and surprise, is lost to science as "the writing of the world" and to map-making as a practice that attempts to control space. In order to disrupt this mediated spatial coherency and rekindle the dynamism and differentiation of life, Massey (2005, 26) argues that we need to get beyond de Certeau's (and Laclau's) "equation of representation and spatialization." What is required for the opening up of the political, argues Massey (2005, 25), is simultaneously a space of freedom and memory (from Bergson 1910), dislocation (from Laclau 1990) and surprise (from de Certeau 1988).

Ed Soja and Michael Dear are even more sacrosanct about the spatial powers of media. Soja's (in addition to Harvey) work on reasserting space into critical social theory is particularly influenced by Fredric Jameson, who in *The Political Unconscious* (1981) and *The Geopolitical Aesthetic* (1992) not only uses films as parables for conditions of post-modernity but argues convincingly about larger global/capitalist conspiracies that are mediated by the subtle and enduring influences of images and moving pictures. Although Dear is a long advocate of postmodern theory and the postmodern city, his focus on media is much more direct than Soja. In *The Postmodern Urban Condition* he reworks urban theory through film, architecture and filmspace and from the sidewalk to cyberspace. Also influenced by Jameson and Lefebvre, but trying to move beyond the work of Soja to insinuate not only how spaces are produced through representations, Dear (2000) creates an urban and policy agenda for the twenty-first century that simultaneously incorporates local art and global politics.

Virtual and Digital Effects

In terms of how mediated geographies are constructed, more recent geographic engagement with digital media and the production of meaning have facilitated a commitment to virtual affectivity. This suggests a change away from notions of truth and falsehood, the separation of which becomes even more blurred in virtual domains, to a more critical understanding of what media technologies actually do. Some media geographers have built upon semiotics and psychoanalytical theory while others have rejected these more structured philosophical perspectives in order to open up media studies to a fuller engagement with rapid technological advances that continually redefine the modalities of practice – the sciences of optics, the intricacies of visual technology, and the hardware and software of seeing, all components of how the twenty-first century media representations are constructed and circulated (see Johnston 1999) for the role of technogenesis in the production of meaning). Importantly and critically, questions arise regarding the affects that circulate around and through our technological modalities and engagements with the world.

The continued challenge of media geographies is to expand its boundaries to take into account the spatialities of other forms of media such as music, advertising, television, theatre, radio and the cartographic representations generated by GIS mapping programs. General media theory has begun to explore how the relationships between the city, the body and space are mediated in an attempt to unpack the complex relations that produce representations of landscapes. Today it is often literature in the field of communications studies that leads the way in the study of the spatial representations of mediated geographies and the interdisciplinary nature of this field has proved to be rife with ideas for geographers. Jenkins' (2006) work is an example of this dynamic, providing an interesting gateway into the intersection of media and space, especially in terms of how traditional forms of media are blending with newer, emerging media to create new, hybrid representations of mediated space and place. Hansen's (2003, 2006) much more in-depth discussion of the development of media theory and its application to new forms of digital media, with his focus on bodies as code, is relevant to the way geographers approach the concept of mediated geographies. More recently, Parikka (2013) edited a special volume of *Theory, Culture & Society* that brings together the application of cultural techniques to the field of Media Studies. The essays further the growing volume of research on mediated geographies by providing greater ontological insight into the interaction of humans and media that pushes the creation and representation of meaning across space.

Previously, Baudrillard's 1983 and 1987 discussions of *simulacra* and postmodern space – where the screen replaces reality directly – place visual media theory as a communicating agent that moves away from bodiliness to a disincorporated subjectivity that is actively directed toward metaphoric spaces. Baudrillard's theories on the hyperreality of space and on telematic culture can be linked to Gibson's (1984, 2005) invention of the literary term *cyberspace* that is also indicative of realities that exist outside normative cultural experiential spaces. At this point it is interesting to go back a little and revisit the way geography is mediated through the imaginary of science fiction, again recalling the tremendous influence of *Blade Runner* on the initial move into media geography. Scott Bukatman (1993) discusses science fiction literature and film (especially the late 1980s cyberpunk phenomenon) very specifically in relationship to cinematic urban landscapes, especially how they change through time in the various subgenres of the science fiction film. Bukatman further establishes the *digital* domains of our culture as sites of contestation thus rendering them as useful sources of new information, particularly in view of the way digital technologies are used in today's world to provide mediated representations of space and place.

Media geography today is interested in how people interact with these digital virtualized environments, and specifically how geographers can rethink the relationship between human viewer and digital information. One could argue that, based on current research, geography is in the midst of a new twist in visual theory. If we look at visual culture in its totality, we have seen this happen before – visual critics like Walter Benjamin confronted the full impact of visual training by examining its effects (i.e. his speculation about tactility and distraction in film) and then exploring how visual culture might be reworked through the application of cinematic principles. We (as geographers, in a critical societal sense) are beginning to understand more clearly the damaging effects of new technologies on social life and have begun to ask what might these technologies render geographies on the other side of their (often) capitalist deployment. We have begun to see that perhaps there is another side to the culture of digital immersive experience and, thus, there is indeed a cultural politics that does not leave it to our current theories of visual geography to control every aspect of our digital engagement with geographical space. There is a need for geographers to make more of virtual immersion, to make something else of their mediated digital engagements by reimagining how these engagements have the potential to work outside of current levels of geovisual interaction and geovisual discourse.

Material Affects

In the postmodernist era and now, even past that (if indeed such things can be labeled), there are certain aspects of mediated space that must be understood to the degree that all information has spatial qualities and those qualities can then become a mediative digital narrative that itself serves as a spatial representation that, in turn, creates meaning. There is, in some quarters of media geography and communication theory, a joining of Deleuzian discourse (and the inability to embody 'information') with an engagement that embraces visual theories exploring the affectiveness of mediated space. Massey (1992, 2005) advocated that space is relational, that it is not static and that relationships themselves create and define space and time. Space is thus socially constituted thereby setting the 'stage' for human affective engagement with mediated space, especially in a digital form. Massey further continues (2005, 11):

> Imagining space as always in process, as never a closed system, resonates with an increasingly vocal insistence within political discourses on the genuine openness of the future. It is an insistence founded in an attempt to escape the inexorability which so frequently characterizes the grand narratives related by modernity.

The grand narratives of modernity, in Massey's opinion (1992, 76), accord with the viewpoint of classical, Newtonian, physics. In classical physics, both space and time exist in their own right, as do objects. Space is a passive area, the setting for objects and their interaction. Objects, in turn, exist prior to their interactions and affect each other through force-fields. The observer, similarly, is detached from the observed world. In modern physics on the other hand, the identity of things is *constituted through* interaction.

This modernist form of engagement does not consider *how* it is possible (either individually or collectively) to perceive, embody, experience and thereby become an active participant in mediated geographies. In opening the door to these possibilities, Gilles Deleuze inspired a new generation of media geographers.

With its focus on emotion and affect, Deleuze's (1986, 1989) two-volume study of the cinema provides a central paradigm for the development of media geography theory – particularly the introduction of affection into the discourse – in that the image-event finds perfect instantiation in the cinema through the use of an *interval* (a 'cut' between shots, or the 'movement-image' in *Cinema 1* and the 'time-image' in *Cinema 2*) that introduces a gap between the action and the reaction (Hansen 2006). In *Cinema 1,* Deleuze characterizes the internal/external processes interwoven through image-events as movement-images that comprise simultaneously a *perception-image* that moves us from indistinguished knowledge at the periphery of our universe to a central subject position, and an *action-image* that is about our perception of things at the center of our universe and grasping the 'virtual action' of those things. Concurrently, there is the *affection-image* that "surges in the center of indetermination" between our perceptions and our actions (Deleuze 1986, 65). This is an affect that alludes to the "motion part of emotion that sloshes back and forth between perception and action" (Aitken 2006, 494). In *Cinema 2*, as Hansen (2006) explains, the function of the cut or time-image, for Deleuze, is perfectly homologous with that of the body as a center of indetermination: it is the process by which the body isolates certain aspects of images to generate perception. Thus, for Hansen and others embracing post-Deleuzian media theory, affection – the capacity of the body to experience itself as more than itself, is a specific permutation of the movement-image. Affection, as a post-phenomenological modality of bodily life, gives way to affection as a concrete type of image that is defined exclusively by the protracted interruption of the sensorimotor circuit, the interruption, that is, of the *form* of the movement-image. This, for Deleuze, carries out the progressive disembodying of the center of indetermination. The 'time-image' from *Cinema 2* does not subordinate time to movement in space – it presents time directly and can thus be understood to divorce perception entirely from human embodiment.

In the years following the publication of Deleuze's *Cinema*(s), media theorist Pierre Lévy (1997, 1998) theorized mediated space as a form of affective geovisualization, a view also advocated by Johnston (Johnston 1999) who discussed what he felt were the limitations of Deleuzian theories relating to the movement-image and the time-image, particular in their relations to newer digital forms of media representations that were unknown to Deleuze. Lévy's media theory thus takes Deleuze's formal understanding of cinematic framing into a contemporary neuroscience – there is now much more than a passive correlate of linkages between images. It is now understood that the body has a creative capacity that can form mediated digital space and mediated spatial representations. This is accomplished by becoming *virtual* according to Lévy: we can now make the connection between movement and sensation to the point that the slightest, most literal displacement invokes a qualitative difference –motion thus triggers affection as an active modality of bodily action. Space becomes mediated and our interactions with digital representations of place also become affectively mediated. Lévy privileges the computational power that lies behind digital and virtual technologies, and by so doing offers an opportunity to more fully comprehend geographic data coded as an array of iconic images and representations positioned within digital and virtual space, thus giving us a more comprehensive understanding of mediated geographies.

Producing Media; Mediating Spaces

Craine (2009) explores how mediated geographies work between actual places, television, and the internet by referencing the material landscapes of Los Angeles, the visualized landscapes of the Fox Network television show *The Shield*, and the virtual landscapes of its web presence (http://theshieldrap.proboards45.com). Media geography theory surmises that the mental maps that compose the themes located within geographic space are mediated internally by the systems of signs, symbols and signals people have previously internalized through the experiential negotiation of constructed landscapes. In the hyperreality of our visualized media culture and experience, these maps take on a different register in virtual space – like all symbols they bear a mediated connection to real-world phenomena. Therefore, by engaging these mediated representations of Los Angeles, we remake space and time by specifying spatial locations and temporal identities from the perspective of a transformed and mediated experiential self. In this manner, we find human agency and freedom within the mediated spaces created by the virtual signification of actual place and time. Whether virtual space is real or not, our experience of these mediated spaces, as Lévy (1998) insists, is a 'real' experience.

In the technocratic world of today's spatial representations, this engagement provides the biological self with the means to compensate for the loss of the corporeal in the virtual environment – we (again, the individual or the society) can now simply transform that environment into a space more susceptible to human control. Thus the virtual digital environment becomes a fundamental part of a mediated human experience – there is a literal *projection* of the human into virtual space thereby allowing for the construction of a spatial simulacrum of the previously invisible circulation of information through this simultaneous grounding and dislocating of bodily experience. With this in mind, the discussion of mediated geographies has now moved into the specific modalities of the virtual (i.e. GIS, gaming, social media) utilizing a wide array of theory to more concretely place the mediated virtual into geographic discourse. Hansen (2006) and Aitken and Craine (2006), among others, have used this form of virtual affectivity to facilitate a new line of flight through the realm of mediated geographies.

In this discussion of mediated geographies, we relate how visual tropes are very much mediated cultural constructions and can, therefore, be analyzed to uncover the specific meanings contained within their particular virtual spatialities. The methods of engagement with visual and with mediated space – how visual information can be engaged and analyzed to provide insights into the construction of identity – is very much a product of our technogenetic society. In exploring representations of the 'real world' – whether through traditional maps or mediated digital representations – visual methodologies inform us of how human beings interact with and view their surroundings and the elements within their virtually constructed environment. More recent forms of mediation can now consider how representations make use of a fuller array of human senses, including audio and haptic (technologies that become commonplace in the twentieth century) besides the visual to contrive that affectivity. Baudrillard called these new cartographies *telematic culture*, while science fiction writers (cf Gibson 1984) termed them *cyberspace*. Nonetheless, as with traditional maps there exists the pervasive recognition that a virtual and decentered spatiality exists parallel to, but outside of, the geographic (or any disciplinary) topology of experiential reality. Vivian Sobchack (1990, 56) states "... personal computers form an encompassing electronic system whose various forms 'interface' to constitute an alternative and absolute world that uniquely incorporates the spectator/user in a spatially decentered, weakly temporalized and quasi-disembodied state." Thus, our virtual experience has effectively superseded our human experience of *temporality*. Perhaps this new virtual-time represents the final abstraction of time and its complete separation from human experience, but we doubt it.

With this in mind, we can see the concept of mediated geographies in two ways: as a mode of entry to the virtual and the decentered spatialities of our culture; and, secondly, as a liberation *from* the human body – a virtualization operating through the very medium of embodied affectivity – that provides media geographers with a much more in-depth understanding of how technology works to create a mediated identity through the processes of affective visualization.

Continuing Spatialities and Cartographies

There are new ways to interpret how we understand these spatialities and cartographies. In this context the productions of postmodernism become surprisingly sensible: the decentered forms, which respond to the loss of Fredric Jameson's 'readable cartography' and the displacement of lived space (however, see (Dixon and Zonn 2006) for a critique of Jameson's late capitalism argument in cinema); or, as Scott Bukatman (1993) believes, the withdrawal into an empty historicism, a simulacrum of the past, a period (at least in geography) in which metadata has usurped historical analysis and in which gigabits of RAM have usurped memory. Particularly important to our discussion is that, today, mediated geographies represent the rise of virtual simulation as the prevalent form that configures our understanding of the social structure and construction of identity: as we *experience* digital information in a visual form these mediated digital experiences can now be transformed into realized and experienced geospatial information, rife with meaning.

We've come a long way from the late nineteenth century imperial representations of the *National Geographic Magazine*. Indeed, new media critic Lev Manovich (2001) proclaims the obsolescence of the image in its traditional sense: since the digital image culminates the transition from an indexical basis (film) to a sequential scanning (any digital device that delivers information), it substitutes for the image proper a processural realization of information in time that appears as a traditional image only for contingent reasons. Manovich asks why we still consider a hybrid conception of the image as an analog surface and a digital infrastructure. He further questions why, given the disjunction between surface appearance and materiality, we continue to associate a given set of numerical coordinates of information with a visually perceivable form. As interface or instrument, the image does not comprise a representation of a pre-existent and independent reality, but rather as a *mediated* geography that intervenes in the production of the 'real': we are now all active users who actively *go into* this geography. We have now reached a point where geographers have correlated mediated geographies with actualized experience through the production of embodied sensation. The sensory apparatus in all of us is technically extended and put into direct interface with a mediated 'space' or 'world' created entirely out of data, a fully dynamic space that may well be realized to its fullest extent.

We have come a long way from the imperial and colonial images of the late nineteenth century, but are we any clearer about Debord's admonition to consider the ways our society of spectacle produces truth as a moment of falsehood? Given that *National Geographic* remains a best-selling magazine, Deleuze would argue that our underlying desires might well remain the same. The daunting task of media geographers is to continue to unpack these desires and to give, interrogate and challenge their meanings.

References

Aitken, S. C. (1994). I'd rather watch the movie than read the book. *Journal of Geography in Higher Education, 18*, 291–308.

Aitken, S. C. (2006). Leading men to violence and creating spaces for their emotions. *Gender, Place and Culture, 13*(5), 491–507.

Aitken, S. C., & Craine, J. (2006, February 7). Affective geovisualizations. *Directions: A Magazine for GIS Professionals*. 4(1). www.directionsmag.com/article.php. Article reprinted as a 1,700 excerpt in M. Dodge, R. Kitchin, & C. Perkins (Editors, 2011), *The map reader: Theories of mapping practice and cartographic representation* (pp. 278–81). The Atrium, Southern Gate: Wiley.

Aitken, S. C., & Valentine, G. (2014). Ways of knowing and ways of doing geographic research. In S. C. Aitken & G. Valentine (Eds.), *Approaches to human geography: Philosophies, theories, people and practices*. London: Sage.
Aitken, S. C., & Zonn, L. E. (Eds.). (1994). *Place, power, situation and spectacle*. Lanham: Rowman & Littlefield.
Baudrillard, J. (1983). *Simulations* (P. Foss, P. Patton, & P. Beitchman, Trans.). Semiotext(e) Foreign Agents. New York: Semiotext(e).
Baudrillard, J. (1987). *Forget Foucault* (N. Dufresne, Trans.). Semiotext(e) Foreign Agents. New York: Semiotext(e).
Bergson, H. (1910). *Time and free will*. London: George Allen & Unwin.
Bukatman, S. (1993). *Terminal identity: The virtual subject in postmodern science fiction*. Durham: Duke University Press.
Burgess, J., & Gold, J. R. (Eds.). (1985). *Geography, the media and popular culture*. New York: St. Martin's Press.
Clifford, J. (1986). Introduction: Partial truths. In J. Clifford & G. E. Marcus (Eds.), *Writing culture: The poetics and politics of ethnography*. Berkeley: University of California Press.
Clifford, J., & Marcus, G. E. (Eds.). (1986). *Writing culture: The poetics and politics of ethnography*. Berkeley: University of California Press.
Craine, J. (2009). Virtualizing Los Angeles: Pierre Levy, The Shield, and http://theshieldrap.proboards45.com. *GeoJournal, 74*(3), 235–243.
de Certeau, M. (1984). *The practice of everyday life*. Berkeley: The University of California Press.
de Certeau, M. (1988), *The writing of history* (T. Conley, Trans.). New York: Columbia University Press.
Dear, M. (2000). *The postmodern urban condition*. Oxford: Blackwell.
Debord, G. (1994). *The society of the spectacle* (D. Nicholson-Smith, Trans.). New York: Zone Books.
Deleuze, G. (1986). *Cinema 1: The movement-image* (H. Tomlinson & B. Habberjam, Trans.). Minneapolis: University of Minnesota Press.
Deleuze, G. (1989). *Cinema 2: The time-image* (H. Tomlinson & R. Galeta, Trans.). Minneapolis: University of Minnesota Press.
Dixon, D., & Zonn, L. (2006). Confronting the geopolitical aesthetic: Fredric Jameson, *The Perfumed Nightmare* and the perilous place of third cinema. In M. Power & A. Crampton (Eds.), *Cinema and popular geo-politics*. New York: Routledge.
Gibson, W. (1984). *Neuromancer* (Ace Science Fiction). New York: Ace.
Gibson, W. (2005). *Pattern recognition*. New York: Berkley.
Hansen, M. B. N. (2003). *New philosophy for new media*. Cambridge, MA: MIT Press.
Hansen, M. B. N. (2006). *Bodies in code: Interfaces with digital media*. London: Routledge.
Haraway, D. (1991). *Simians, cyborgs, and women: The reinvention of nature*. London: Routledge.
Harvey, D. (1989). *The condition of postmodernity*. Cambridge: Basil Blackwell.
Jameson, F. (1981). *The political unconscious: Narrative as a socially symbolic act*. Ithica: Cornell University Press.
Jameson, F. (1992). *The geopolitical aesthetic: Cinema and space in the world system*. Bloomington: Indiana University Press.
Jenkins, H. (2006). *Convergence culture: Where old and new media collide*. New York: New York University Press.
Johnston, J. (1999). Machinic vision. *Critical Inquiry, 26*(1), 27–48.
Laclau, E. (1990). *New reflections on the revolution of our time*. London: Verso.
Lévy, P. (1997). *Collective intelligence: Mankind's emerging world in cyberspace* (R. Bononno, Trans.). New York: Plenum Trade.
Lévy, P. (1998). *Becoming virtual: Reality in the digital age* (R. Bononno, Trans.). New York: Plenum Trade.
Manovich, L. (2001). *The language of the new media*. Cambridge, MA: MIT Press.
Manvell, R. (1956). Geography and the documentary film. *The Geographical Magazine, 29*, 417–422.

Massey, D. (1992). Politics and space/time. *New Left Review, 196*, 65–84.
Massey, D. (1998). The spatial construction of youth cultures. In T. Skelton & G. Valentine (Eds.), *Cool places: Geographies of youth cultures*. New York: Routledge.
Massey, D. (2005). *For space*. London: Sage.
Natter, W., & Jones, J. P., III. (1993). Signposts toward a poststructuralist geography. In J. P. Jones III, W. Natter, & T. Schatzki (Eds.), *Postmodern contentions: Epochs, politics, space*. New York: Guilford Press.
Parikka, J. (2013). Afterword: Cultural techniques and media studies. *Special Issue: Cultural Techniques. Theory, Culture & Society, 30*(6), 147–159.
Rothenberg, T. (2007). *Presenting America's world: Strategies of innocence in the national geographic, 1888–1945*. Aldershot: Ashgate.
Sobchack, V. (1990). The scene of the screen: Towards a phenomenology of cinematic and electronic presence. *Post-Script, 10*, 50–59.
Zonn, L. (Ed.). (1990). *Place images in media: Portrayal, experience and meaning*. Savage: Rowman and Littlefield.

Section II

Media Production and Place

Chinese Cinema Cities: From the Margins to the Middle Kingdom

6

Michael Curtin

Writing about screen media and Chinese identities is a contentious and complicated task. In some ways it's similar to writing about "European media," an endeavor that would seem odd to most readers given the diverse terrain traversed by these media, a terrain on which societies have struggled for centuries to sustain distinctive voices in numerous languages. Although political regimes have at times brought Europeans together through military might and strategic alliances, their successes have been comparatively fleeting. Europe today continues to struggle with the process of unification and remains differentiated along lines that were solidified by the Westphalian system of states and the waves of nationalism that washed across European societies during the nineteenth century.

By comparison, China over the course of two millennia experienced enduring dynasties, each anchored by an imperial court that was the center of politics and civilization. Despite a diversity of languages and cultures, imperial elites shared a written script, a sophisticated administrative infrastructure, and an officially sanctioned set of cultural practices. They celebrated the emperor's vast domain as the "middle kingdom," outside of which laid the lands of barbaric others.[1] Even while modern nation-states were emerging across Europe, China's final imperial dynasty soldiered on through the end of the nineteenth century, despite its doddering ineptness in both domestic and foreign affairs. To the very end, imperial elites saw

[1] Even today, the Chinese characters for middle kingdom (中国) are used to represent the nation of China.

M. Curtin (✉)
Department of Film and Media Studies, University of California, Santa Barbara, CA, USA
e-mail: mcurtin@filmandmedia.ucsb.edu

S.P. Mains et al. (eds.), *Mediated Geographies and Geographies of Media*,
DOI 10.1007/978-94-017-9969-0_6

contact with the West as fraught with the risk of cultural contamination and nowhere did that appear more evident than the treaty ports of the east coast, among them Shanghai, Hong Kong, and Macau.

Interestingly, as nationalism first took hold in China, it was crucially dependent on the media infrastructure of these colonial enclaves, which both nourished an air of innovation and global interconnection. Telegraph facilities and diasporic newspapers in Hong Kong and Shanghai helped to knit together the spatially extended networks of an emerging nationalist movement (Zhou 2006). Meanwhile cinema delivered intriguing glimpses of western societies and fashions; and as local filmmakers developed their own productive capacity, their movies became a site of popular deliberation over Chinese traditions, modernization, and social relations (Huang and Xiao 2011; Zhang 2004, 2010). Then as now, media were a means of cultural experimentation and an object of intense scrutiny, sparking passionate debates about identity, authenticity, and contamination. Thus it was at the very margins of an ancient imperial system that China's modern national consciousness first emerged, and cinema played an especially important role.

Film technology first arrived in Shanghai and Hong Kong around the turn of the twentieth century and by the 1920s both cities had thriving exhibition venues and an emerging creative community. Moviemakers soon extended their reach to prosperous overseas Chinese communities in Southeast Asia and to Chinatowns around the world. Although influential, movies played mostly to urban audiences along the major inland trading routes and throughout the network of port cities inhabited by the Chinese diaspora. Civil war on the mainland and the Japanese invasion during the 1930s pushed the commercial industry to expand southward and by the end of World War II the two largest Chinese movie companies were based in Singapore. The establishment of the Republic of China on Taiwan and the surging prosperity of Hong Kong encouraged a remapping of the cinematic landscape with the latter becoming so prolific that it was often referred to as "Hollywood East." Popular commercial cinema was from the outset both mobile and marginal, thriving not at the center of political power but rather on the fringes of the middle kingdom.

In 1997, with the end of the British colonial era and the return of Hong Kong to Chinese sovereignty, the movie industry struggled increasingly under the thumb of a national government that sought to shift the center of cultural power and popularity back to Beijing. As we shall see, the complex history and spatial trajectory of Chinese cinema help to explain the recent media policies of the People's Republic of China, a regime that now holds sway over the world's largest population of Internet users, the second largest television advertising market, and the second largest theatrical market. Given the rapidly growing scale of mainland media, one might expect twenty-first century popular culture to be increasingly dominated by Chinese interests and audiences. Yet despite the government's explicit desire to exercise cultural leadership at home and soft power abroad, state-sanctioned media have proven remarkably ineffective at either task.

This essay focuses on struggles between the Chinese state and commercial film industry, considering the spatial and power dynamics of their relationship in the past, present, and future. It explains why the Chinese film industry has shifted its geographical center of gravity over time and raises intriguing questions about logics that inform the spatial deployment of media institutions, what I refer to as "media capital," a concept that at directs attention to the conditions under which cultural resources concentrate in particular cities (Curtin 2004, 2011, in progress). The first section of the essay explains the concept of media capital and the principles by which it operates. The second describes the institutional characteristics and enabling conditions that transform a city into a media capital. And the third section returns to the Chinese case example, explaining the rise and demise of Hong Kong as a media capital and exploring the Chinese state's efforts to reassert the cultural leadership of Beijing. Although the state has been arguably successful in some respects, the essay concludes by explaining why, under current conditions, Beijing is unlikely to become a media capital and unlikely to extend its cultural influence globally or even regionally.

Understanding Media Capital

Media capital refers to a set of interacting processes that can best be understood by directing attention to three principles that have played a structuring role in screen industries since the early twentieth century. They include: (1) the logic of accumulation, (2) trajectories of creative migration, and (3) contours of socio-cultural variation.

The logic of accumulation is not unique to media industries, since all capitalist enterprises exhibit innately dynamic and expansionist tendencies. As David Harvey (2001: 237–266) points out, most firms seek efficiencies through the concentration of productive resources and through the expansion of markets so as to fully utilize their productive capacity and realize the greatest possible return. These tendencies are most explicitly revealed during periodic downturns in the business cycle when enterprises are compelled to intensify production and/or extensify distribution in order to survive. Such moments of crisis call for a "spatial fix," says Harvey, as capital must on the one hand concentrate and integrate sites of production so as to reduce the amount of time and resources expended in manufacture and on the other hand it must increase the speed of distribution in order to reduce the time it takes to bring distant locales into the orbit of its operations. These *centripetal* tendencies in the sphere of production and *centrifugal* tendencies in distribution were observed by Karl Marx more than a century earlier when he incisively explained that capital must "annihilate space with time" if it is to overcome barriers to accumulation (Marx 1973: 539). As applied to contemporary media, this insight suggests that even though a film or TV company may be founded with the aim of serving particular national cultures or local markets, it must over time re-deploy its creative resources and reshape its terrain of operations if it is to survive competition and enhance profitability.[2] Implicit in this logic of accumulation is the contributing influence

[2]Monopoly rents are an exception, but as shown in this essay, monopoly rents have proven less tenable in an era of changing technologies and increasing transborder flows.

of the "managerial revolution" that accompanied the rise of industrial capitalism (Chandler 1977). Indeed, it was the intersection of capitalist accumulation with the reflexive knowledge systems of the Enlightenment that engendered the transition from mercantile to industrial capitalism. Capitalism became more than a mode of accumulation, it also became a disposition towards surveillance and adaptation, as it continually refined and integrated manufacturing and marketing processes, achieving efficiencies through the concentration of productive resources and the extension of delivery systems (Giddens 1990).

The second principle of media capital emphasizes trajectories of creative migration. Audiovisual industries are especially reliant on creative labor as a core resource due to the recurring demand for new prototypes (i.e., feature films or television programs). Yet the marriage of art and commerce is always an uneasy one, especially in large institutional settings, and therefore the media business involves placing substantial wagers on forms of labor that are difficult to manage. As Aksoy and Robins (1992: 12) observe, "Whether the output will be a hit or a miss cannot be prejudged. However, the golden rule in the film business is that if you do not have creative talent to start with, then there is no business to talk about at all, no hits or misses." In fact, attracting and managing talent is one of the most difficult challenges that screen producers confront. At the level of the firm this involves offering attractive compensation and favorable working conditions, but at a broader level it also requires maintaining access to reservoirs of specialized labor that replenish themselves on a regular basis, which is why media companies tend to cluster in particular cities.[3]

Geographer Allen J. Scott contends that manufacturers of *cultural* goods tend to locate where subcontractors and skilled laborers form dense transactional networks. Besides apparent cost efficiencies, Scott points to the mutual learning effects that stem from a clustering of interrelated producers. Whether through informal learning (such as sharing ideas and techniques while collaborating on a particular project) or via more formal transfers of knowledge (craft schools, trade associations, and awards ceremonies) clustering enhances product quality and fuels innovation. "Place-based communities such as these are not just foci of cultural labor in the narrow sense," observes Scott (2000: 33), "but also are active hubs of social reproduction in which crucial cultural competencies are maintained and circulated."

This centripetal migration of labor encourages path dependent evolution, such that chance events or innovations may spark the appearance of a creative cluster, but industrial development depends on a spiral of growth fueled by the ongoing migration of talent in pursuit of professional opportunities. Locales that fail to make an early start in such industries are subject to "lock-out," since it is difficult to disrupt the dynamics of agglomeration, even with massive infusions of capital or

[3]Although it does not address media industries specifically, an extensive literature discusses the impact of human capital on the clustering of business firms in particular locations (Jacobs 1984; Porter 1998; Florida 2005).

government subsidies. The only way a new cluster might arise is if a dominant media capital were to falter or if a new cluster were to offer an appreciably distinctive product line.

Despite the productive power and structural advantages of media capitals, the symbolic content of media products attenuates their geographical reach. That is, the cultural distance between say Chinese filmmakers and Turkish or Indian audiences introduces the prospect that the meaningfulness and therefore the value of certain products may be undermined at the moment of consumption or use. Although the centripetal logics of accumulation and of creative migration help us identify concentrations of media capital, the centrifugal patterns of distribution are much more complicated, especially when products rub up against counterparts in distant cultural domains that are often served, even if minimally, by competing media capitals.

Cities such as Cairo, Mumbai, Hollywood, and Hong Kong lie across significant cultural divides from each other, which helps to explain why producers in these cities have been able to sustain distinctive product lines and survive the onslaught of distant competitors. These media capitals are furthermore supported by intervening factors that modify and complicate the spatial tendencies outlined above. Consequently, the third principle of media capital focuses on contours of socio-cultural variation, demonstrating that national and local institutions have been and remain significant actors in the global cultural economy.

During cinema's early years of industrial formation, market forces and talent migrations fostered the growth of powerful producers such as Hollywood, but as of the 1920s governments around the world reacted to Hollywood's growing influence by developing policies to limit imports and foster local media production. Attempts to develop local filmmaking institutions often proved difficult, but many countries were nevertheless successful at promoting radio and later television (most of them public service broadcasters) that produced popular shows and attracted substantial audiences. Broadcasting seemed an especially appropriate medium for intervention, since many of its cultural and technological characteristics helped to insulate national systems from foreign competition. The ensuing parade of broadcast news and entertainment punctuated daily household routines, interlacing public and private spheres, thereby situating national culture in the everyday world of its audiences (Scannell 1991; Silverstone 1994; Morley 2000).

It should also be pointed out that state institutions were not the only actors to organize and exploit the contours of socio-cultural variation. Media enterprises have for decades taken advantage of social and cultural differences in their production and distribution practices, especially by employing narratives and creative talent that resonate with the cultural dispositions of audiences within their spheres of influence. They furthermore made use of social networks and insider information to secure market advantages, and they invoked ethnic and national pride in their promotional campaigns. Contours of socio-cultural variation have provided and continue to provide opportunities to carve out market niches that are beyond the reach of powerful but culturally distant competitors.

Overall, media capital is a concept that at once acknowledges the *spatial* logics of capital, creativity, culture, and polity without privileging one among them. Just as the logic of capital provides a fundamental structuring influence, so too do forces of socio-cultural variation shape the diverse contexts in which media are made and consumed. As a concept, media capital encourages dynamic and historicized accounts that delineate the operations of capital and the migrations of talent, while at the same time directing attention to socio-cultural forces and contingencies that can engender alternative discourses, practices, and spatialities. Such an approach furthermore aims to address the supposed tension between political economy and cultural studies scholarship—and between the media imperialism and global studies approaches—by showing how insights from each of these schools can productively be brought to bear on the study of film and television.

Characteristics of Media Capitals

Media *capitals* are powerful geographic centers that tap human, creative, and financial resources within their spheres of circulation in order to fashion products that serve the distinctive needs of their audiences. Their success is dependent on their ability to monitor audience preferences, tap the popular imagination, and operationalize resources within their cultural domain. A media capital's preeminence is therefore relational: its bounty flows outward while in turn it gathers and exploits the very best human and cultural resources within its sphere of circulation. Its preeminence is dynamic and contingent, for it is subject to competition from other cities that aspire to capital status. Dubai, for example, is self-consciously attempting to challenge the leadership of Beirut within the sphere of Arab satellite television and Miami has recently arisen as a transnational competitor to Mexico City. Thus, the concept of media capital encourages a spatial examination of the shifting contours of accumulation and dissemination, which both shape and are shaped by the imaginary worlds of audiences. Such research seeks to understand why some locales become centers of media activity and to discern their relations to other locales. Media capitals emerge out of a complex play of historical forces and are therefore contingently produced within a crucible of transnational competition. Cities as diverse as Hollywood, Mumbai, and Lagos operate as media capitals within their respective spheres of circulation. Although qualitatively different in many respects, cities that become media capitals exhibit a shared set of characteristics with respect to institutional structure, creative capacity, and political autonomy.

Institutionally, media capitals tend to flourish where companies show a resolute fixation on the tastes and desires of audiences. In order to cater to such tastes, they adopt and adapt cultural influences from near and far, resulting in hybrid aesthetics. Such eclecticism and volatility are moderated by star and genre systems of production and promotion that help to make texts intelligible and marketable to diverse audiences. The bottom line for successful firms is always popularity and profitability. Although often criticized for pandering to the lowest common denominator, commercial film and TV studios are relentlessly innovative, as they avidly pursue the shifting nuances of fashion and pleasure. In the early stages of development, a media capital may be characterized by small businesses with an opportunistic outlook, many of them chasing the latest trend with abandon, churning out products on shoe-string budgets and releasing them into the market with little promotion or strategic calculation. As media capitals mature, however, firms begin formalize their institutional practices and in most cases they begin to integrate production, distribution, and exhibition within large corporations. Profitability is derived from structured creativity that feeds expansive (and expanding) distribution systems. Marketing considerations become woven into the conceptual stages of project development and financing. Media capitals therefore emerge where regimes of capital accumulation are purposefully articulated to the protean logics of popular taste. The mercantilist opportunism of an emerging filmmaking community gives way to industrialized modes of production and distribution.

Just as importantly, media capital tends to thrive in cities that foster creative endeavor, making them attractive destinations for aspiring talent. The research literature on industrial clustering shows that creative laborers tend to migrate to places where they can land jobs that allow them to learn from peers and mentors, as well as from training programs that are sponsored by resident craft organizations (Porter 1998). Job mobility and intra-industry exchanges further facilitate the dissemination of skills, knowledge, and innovations. Thus a culture of mutual learning becomes institutionalized, helping to foster the reproduction and enhancement of creative labor (Scott 2000). Workers are also inclined to gravitate to places that are renowned for cultural openness and diversity (Florida 2005). It's remarkable, for example, that the most successful media capitals are usually port cities with long histories of transcultural engagement (Hesse 2010; Jacobs et al. 2010).

It's furthermore noteworthy that national political capitals rarely emerge as media capitals, largely because modern governments seem incapable of resisting the temptation to tamper with media institutions.[4] Consequently, media capitals tend to flourish at arm's length from the centers of state power, favoring cities that are in many cases are disdained by political and cultural elites (e.g., Los Angeles, Hong Kong, and Mumbai). Successful media enterprises tend to resist censorship and clientelism, and are suspicious of the state's tendency to promote an official and usually ossified version of culture. Instead, commercial media enterprises absorb and refashion indigenous and traditional cultural resources while also incorporating foreign innovations that may offer advantages in the market. They do this even though such appropriations tend to invite criticism from state officials and high-culture critics. The resulting mélange is emblematic of the contradictory pressures

[4]London, the national capital of the United Kingdom, is an exception, largely because of the residual advantages of empire that made it such an important maritime and financial center. Its importance as a center of media activity has been perpetuated largely because it has exploited its access to the wealthy global Anglophone market and because the state has exercised restraint in its oversight of creative institutions.

engendered by global modernity, at once dynamic and seemingly capricious yet also shrewdly strategic. The choice of location is no less calculated: media capital tends to accumulate in cities that are relatively stable, quite simply because entrepreneurs will only invest in studio construction and distribution infrastructure where they can operate without significant interference over extended periods of time.

Contested Capital, Contested Identities

The commercial Chinese movie industry was, during its prime, a fundamentally transnational medium. Expansive and mobile, it emerged in the 1920s in Shanghai and Hong Kong, and soon expanded into export markets in Southeast Asia. During by the 1930s, the mainland movie market was beleaguered by war and revolution, so the center of Chinese commercial cinema shifted south to Singapore, only to be buffeted yet again by waves of nationalist fervor on the Malay Peninsula during the 1950s. The industry then relocated to Hong Kong, where it matured and flourished, serving local audiences but also fashioning products with an eye to overseas markets (Fu 2003, 2008; Uhde and Uhde 2000; Zhang 2004). Chinese movie executives pursued opportunities wherever they arose and the industry was therefore proto-global in orientation, even if its products were not ubiquitous worldwide. It was anchored moreover by a resident creative community that tapped talent and resources from near and far, making Hong Kong the central node in the intricate circuits of Chinese popular culture.

Although transnational in orientation, it was also quite local in many respects. In the latter decades of the twentieth century, during the heyday of Hong Kong cinema, filmmakers shot most of their productions on the streets of the city and consciously fashioned their movies for local fans. Hong Kong's film culture was then renowned for midnight premieres, where cast and crew would mingle among the moviegoers, taking the pulse of the audience and sometimes adapting the final cut of the film accordingly (Teo 1997; Bordwell 2010). Movies were made for locals and their response was considered a rough indicator of potential success in overseas markets such as Malaysia, Singapore, and Taiwan. The creative community made its home in a colonial city, among a population that had largely migrated from elsewhere and was then in the process of developing a distinctly indigenous but also cosmopolitan identity. Moviemaking was a local business with a translocal sensibility (Zhang 2010). Aspiring Chinese talent moved to Hong Kong from many parts of Asia—and even as far afield as Europe and North America—seeing the city as the most promising place to build a career. Movie executives similarly saw it as the best place to secure financing, recruit labor, and launch projects.

The movie business operated outside the reach of national politics, sheltered by the benign neglect of the British colonial regime. Producers cobbled together feature films in a freewheeling fashion and at a ferocious pace, turning out popular products, occasional gems, and a good deal of rubbish. Nevertheless the tempo, scale, and diversity of production helped to foster a flexible ensemble of film companies that provided job opportunities to thousands of professionals as well as training for those that aspired to join the industry. Hong Kong became a magnet for talent from near and far, and became an incubator for creative experimentation (Curtin 2007). It was home to Tsui Hark, Maggie Cheung, and Leonard Ho. Home to Ann Hui, Peter Chan, and Michelle Yeoh. Home to Peter Pau, Wong Kar-wai, and Christopher Doyle. It was also home to a vibrant ensemble of newspapers, music labels, and broadcasting stations.

In 1997, the People's Republic of China reclaimed Hong Kong after more than a century of British rule. The terms of transfer provided a 50-year transition in which the city would operate as a relatively autonomous Special Administrative Region, but it was clear from the beginning that Beijing intended to exert its authority and many believed that government scrutiny of the media industries would increase. This posed a problem for Hong Kong film companies that were accustomed to producing satirical and ribald comedies, as well as fantasy, horror, and crime stories. The city's creative class grew nervous as the deadline for transition approached, for the very genres that had proven most prosperous were likely to become targets of censors and propaganda officials. Consequently, many producers, directors, and actors began to explore job options abroad and even those that remained in place quietly began moving resources and families overseas in case of an official crackdown (Chan 2009). The industry also entered into a cycle of hyperproduction, spewing out as many movies a as possible, hoping to maximize profits before the fateful moment of transition. This flooded the market with low-grade products that alienated loyal audiences both at home and abroad. Hong Kong's reputation suffered tremendously as a result, most tragically with its audiences, who by the late 1990s had grown accustomed to cultural alternatives from Tokyo, Seoul, Europe, and Hollywood that were readily available at movie multiplexes, on video, and over the Internet. No longer willing to risk the expense of a theater ticket for a Hong Kong feature film, consumers bought (or downloaded) pirated Chinese movie videos that sold for only a fraction of retail price (Wang 2003). As audiences turned a cold shoulder to the industry, so too did media professionals in other parts of Asia. Distributors stopped buying, producers stopped collaborating, and directors declined to take Hong Kong talent onto their projects. In the decade following the handover, the industry's transnational network of audiences, distributors, and creative talent slowly dissolved (Curtin 2007; Chan et al. 2010; Bordwell 2010).

In retrospect, anxieties about the handover to Chinese sovereignty were somewhat exaggerated and the industry was therefore ill-served by the opportunistic mentality that prevailed throughout much of the 1990s and into the new century. In fact, the industry suffered less from censorship than it did from a fear of censorship that fueled a self-destructive cycle of hyperproduction. The Beijing leadership therefore didn't need to dip its hands directly into the messy mechanics of content regulation. Instead it kept its distance and withheld assistance during a time when the Hong Kong industry was under tremendous stress. Interestingly, Chinese state enterprises hatched numerous joint-ventures with Hollywood partners while snubbing the Hong Kong industry, whose films were treated as foreign imports for seven long and turbulent years after the city's return to Chinese sovereignty. The PRC government essentially starved the industry at a moment of crisis and only opened the door to the mainland market slowly after it was sure it had the upper hand in its relationship with "Hollywood East."

As the irreverent and innovative qualities of Hong Kong media products diminished, export revenues declined and producers were confronted with two options: focus on the tiny domestic market of the SAR itself or enter into projects (usually coproductions) with mainland media partners.[5] The former would entail significant downsizing while the latter would require feature films that were fashioned as much for PRC censors as audiences. The Beijing government furthermore sent signals that it would brook no challenges to the supremacy of state institutions such as China Film and CCTV. If Hong Kong firms were to participate in the rapidly growing mainland media economy, they would do so within parameters established by the Communist Party (Yeh and Davis 2008; Yeh 2010; Davis 2010).

Today, Hong Kong is but one node in a geographically dispersed circuit of deal-making and creative endeavor that is increasingly driven by the exigencies of the mainland market. Filmmakers must be attentive to government officials who explicitly make use of import policies, subsidies, and regulations to shape movie messages and to nurture the development of large national enterprises that they hope will someday compete with Hollywood counterparts. They favor big movies with big stars. Themes and dialogue are cautious, even at times stilted, but the production values are growing more competitive with global standards and much of this has to do with the skills and insights that Hong Kong talent bring to these coproductions. Indeed, the leading box office performers during the first decade of the 2000s were Hong Kong-PRC blockbusters. Most were historical dramas (e.g., *Hero* 2002, *Red Cliff* 2008, *The Warlords* 2008), which were safe with censors because they displaced controversial issues onto a distant past and were furthermore acceptable to officials because they promoted the image of China as a grand and ultimately united civilization with a long and distinguished history (Wang 2009; Zhao 2010).

Audiences in East Asia outside the PRC seem to sense the caution and calculation behind these efforts, and many moviegoers consequently opted for Hollywood products, which are arguably no less cautious or calculated. The difference is that Hollywood filmmaking is periodically rejuvenated by sleeper films and independent features. It also has a film rating system that makes it possible for filmmakers to target particular segments of the audience and to explore mature themes and offbeat topics. These structural mechanisms have allowed innovative projects—such as *Juno* (2007), *Slumdog Millionaire* (2008), and *The Hurt Locker* (2009)—to break through the institutional inertia and insider dealings of the industry. As currently constituted, the mainland movie industry has few such mechanisms. Instead there is a yawning gap between state-sanctioned feature film extravaganzas (all of them G-

[5]Recently, a third option has begun to present itself. The terms of the Closer Economic Partnership Agreement between the PRC and Hong Kong has made it possible for filmmakers to target a provincial Guangdong market (Pang 2010). It is still too soon to know whether this possibility will provide greater autonomy to the Hong Kong industry.

rated) and sadly undernourished mid-range and independent movies (Zhang 2010; Song 2010). Chinese independent films are micro-budgeted projects that are either destined for the international festival circuit or they are opportunistic features that are produced largely for the satellite television market. The former are seen as unprofitable art cinema that rarely make it into theaters while the latter tend to be "main melody" films that are subsidized by the state and conform to ideological guidelines that favor uplifting characters and pro-social themes (Song 2010).[6]

Television likewise suffers from various institutional constraints, so that mainland China—which has by far the world's largest national television audience—remains a net importer of programming (Keane 2010). Low-cost genres (talk, reality, and variety) flourish, but few are innovative and those that are find themselves quickly besieged by imitative competitors (Keane et al. 2007). Drama and comedy—signature genres of the world's most successful television enterprises—remain underdeveloped, largely because of the same caution and calculation that prevails in the movie business. Besides the constraints on content, mainland television enterprises also suffer from structural limitations. Shanghai and Guangzhou media have exploded in size and Hunan provincial television has proven itself to be a shrewd innovator, but most TV companies are run by provincial or municipal units of government that are eager to maintain their authority and ownership status. This makes it difficult for companies to merge and makes it difficult to shake out the weakest performers (Diao 2008). Provincial and municipal TV enterprises are hampered as well by regulations that favor the state-sanctioned national champion, China Central Television (Zhu 2012). Provincial and municipal telecasters are furthermore discouraged from building overseas distribution channels, a privilege that largely belongs to Beijing-based institutions that nest snugly under the wing of the state, where they are closely monitored for content and tone.

If today there is a geographic center to Chinese media, it is within the Communist Party offices in Beijing, not because the party micromanages the day-to-day operations of television and film enterprises but rather because it systematically doles out favors and franchises to those that acknowledge its supremacy. The party leadership is quite successful at keeping a leash on domestic players and at exploiting joint venture partners from overseas.[7] The PRC government has cagily manipulated both Western and Hong Kong movie companies to serve its own

[6]In 2013, several popular micro-budgeted films (*Lost in Thailand*, *So Young*, and *Tiny Times*) bucked the system, proving massively popular with audiences due to their off-beat and blatantly commercial qualities, but they quickly became the subject of official condemnation, leaving many to wonder if one of the periodic government campaigns against "cultural pollution" might be in the offing.

[7]For example, both News Corporation and Warner Bros. have experienced very uneven success after decades of effort devoted to the mainland market. Many Western media executives express outright frustration with the manipulative practices of Chinese officials, arguing that India is a better bet for foreign investors (Frater 2008). Most prominently, News Corporation decided to sell its ownership stake in Star TV after close to twenty years of assiduous attention to the mainland market, literally throwing in the towel and redeploying its resources (Young 2010).

ambitions, which are to build a movie infrastructure that will ultimately be popular with national audiences and competitive with Hollywood, both at home and abroad. To the extent that it has succeeded, it is largely because China's moviegoing public is expanding at a breath-taking pace, with box office receipts of $2.7 billion in 2012, making it the second largest theatrical market in the world (China Screen 2013). Television is undergoing a similar growth trajectory with CCTV announcing that its 2010 annual advertising auction drew $1.9 billion in revenue, rising more than 15 % over the preceding year (Coonan 2010b); by 2013 that figure had doubled.

Yet despite these impressive figures, mainland media have little influence abroad, casting barely a ripple through media markets in Korea, Taiwan or Japan, let alone Europe or the United States. This is due largely to the fact that PRC cinema has, since its inception, been an instrument of the state, a bridge between the Communist Party and the people. Since the 1980s, the government has reorganized and marketized the national economy so that media institutions now operate in a more decentralized fashion and they pursue audiences as they might pursue media consumers, but their overriding mission is to serve the party and therefore media ownership remains squarely in the hands of the state (Zhu 2003; Diao 2008). This system of control is fairly obvious to viewers in the mainland who commonly seek alternatives via the Internet and the DVD black market. Young people especially rely on Internet viewing, employing a host of strategies to circumvent the "Great Firewall" in order to acquire products that could never find their way into cinemas or onto the airwaves (Barboza 2010; Chua and Iwabuchi 2008). As for overseas markets, audiences seem occasionally interested in historical dramas that emanate from the mainland, but their tastes are quite diverse and they have access to a great range of media products and services. Mainland movies have therefore performed modestly overseas and television exports have proven to be of little interest to audiences in Hong Kong, Taipei, and Singapore. As for projecting soft power even further afield in Asia, state media products have enjoyed little success in Tokyo, Seoul, or Bangkok.

Conclusion

Beijing's apparent success at controlling its domestic film industry runs counter to what many critics and researchers see as the unrelenting global expansion of Western media conglomerates. It raises the prospect of a new center of cultural power based in the national capital under the watchful eye of the state, suggesting that under certain conditions, state regimes may indeed be able to assert their cultural influence domestically, and may furthermore be able to tame the power of nearby competitors in cities such as Hong Kong. Yet regime's apparent triumph also undermines its explicit soft-power ambitions. For Beijing is unlikely to become a global media capital so long as it is remains the seat of national government. This is because media capital flourishes at cultural crossroads, not at the centers of political power. Beijing may build and manage a vast domestic media infrastructure, but it is likely to struggle in its efforts to influence popular culture beyond its borders. Quite simply, media capitals tend to prosper at a distance from state power—Mumbai vs. New Delhi, Lagos vs. Abuja, Miami vs. Mexico City. Although London might be seen as an exception, that's largely because its creative industries tend to be resolutely capitalistic and are insulated from state pressures by a common law tradition that sets limits on the exercise of state power. Even the national public broadcaster, BBC, has a long tradition of creative independence and insulation from the political imperatives of the ruling regime. If Chinese media can someday achieve this level of relative autonomy, Beijing might indeed emerge as a transnational media capital, but until that time its cluster of media institutions is perhaps more appropriately seen as a center of *official* rather than *commercial* media, as an instrument by which the ruling regime reaches outward in its quest to impose an monological vision of China and Chineseness.

As for Hong Kong, the (soft) nationalization of its film industry helps to explain why its status as a media capital is declining. Once known for its rambunctious, reflexive, and visceral cinema, the city's creative community has shriveled and those who remain have capitulated to a system that is built around the cautious, calculated blockbuster feature film that will appease state censors, party officials, and major financial backers. Audiences matter, but not the way that they used to matter when the Hong Kong industry was in its prime, and not the way they continue to matter to filmmakers in Los Angeles or Mumbai. Chinese commercial cinema has by comparison turned inward and one therefore wonders where (and if) a new center of gravity will emerge? Will it remain a national industry nestled in Beijing under the watchful eye of the state or might a transnational media capital reemerge in one of the seaport cities along China's coast? For if China is truly to assert its cultural soft power, it is less likely to emanate from a state apparatus based in Beijing than it is from a popular culture industry located at a cosmopolitan crossroads.

References

Aksoy, A., & Robins, K. (1992). Hollywood for the 21st century: Global competition for critical mass in image markets. *Cambridge Journal of Economics, 16*(1), 1–22.

Barboza, D. (2010, April 18). For Chinese, web is the way to entertainment, *The New York Times*, p. B1.

Bordwell, D. (2010). *Planet Hong Kong: Popular cinema and the art of entertainment* (2nd ed.). http://www.davidbordwell.net/books/planethongkong.php

Chan, P. interview (2009, April 2). A discussion with producer/director Peter Ho-San Chan on global trends in Chinese-language movie production. *Taiwan Cinema* website, Government Information Office, Republic of China (Taiwan). http://www.taiwancinema.com/ct.asp?xItem=58252&ctNode=124&mp=2

Chan, J. M., Fung, Y. H., & Ng, C. H. (2010). *Policies for the sustainable development of the Hong Kong film industry*. Hong Kong: The Chinese University of Hong Kong Press.

Chandler, A. (1977). *The visible hand: The managerial revolution in American business*. Cambridge: Belknap.

China box office round-up 2012: China becomes world's second biggest market. (2013, January 21). *Screen Daily*. http://www.screendaily.com/china-box-office-round-up-2012-china-becomes-worlds-second-biggest-market/5050843.article

Chua, B., & Iwabuchi, K. (2008). *East Asian pop culture: Analysing the Korean wave*. Hong Kong: Hong Kong University Press.

Coonan, C. (2010b, November 8). Ads auction up 15.5% for CCTV. *Variety*. http://www.variety.com/article/VR1118027181

Curtin, M. (2004). Media capitals: Cultural geographies of global TV. In J. Olsson & L. Spigel (Eds.), *Television after TV: Essays on a medium in transition* (pp. 270–302). Durham: Duke University Press.

Curtin, M. (2007). *Playing to the world's biggest audience: The globalization of Chinese film and TV*. Berkeley: University of California Press.

Curtin, M. (2011). Global media capital and local media policy. In J. Wasko, G. Murdock, & H. Sousa (Eds.), *Handbook of political economy of communication*. Malden: Blackwell.

Curtin, M. (in progress). *Media capital: The cultural geography of globalization*. Manuscript.

Davis, D. W. (2010). Market and marketization in the China film business. *Cinema Journal, 49*(3), 121–125.

Diao, M. M. (2008). *Research into Chinese television development: Television industrialisation in China*. Dissertation, Macquarie University, Sydney.

Florida, R. (2005). *Cities and the creative class*. New York: Routledge.

Fu, P. (2003). *Between Shanghai and Hong Kong: The politics of Chinese cinemas*. Stanford: Stanford University Press.

Fu, P. (Ed.). (2008). *China forever: The Shaw Brothers and diasporic cinema*. Urbana: University of Illinois Press.

Frater, P. (2008, August 1). Hollywood weighs China yin-yang. *Variety*. http://www.variety.com/article/VR1117989939?refCatId=2520&query=yin+yang

Giddens, A. (1990). *The consequences of modernity*. Stanford: Stanford University Press.

Harvey, D. (2001). *Spaces of capital: Towards a critical geography* (pp. 237–266). New York: Routledge.

Hesse, M. (2010). Cities, material flows and the geography of spatial interaction: Urban places in the system of chains. *Global Networks, 10*(1), 75–91.

Huang, X., & Xiao, Z. (2011). *Shadow magic* and the early history of cinema in China. In S. H. Lim & J. Ward (Eds.), *The Chinese cinema book* (pp. 47–55). New York: BFI Palgrave.

Jacobs, J. (1984). *Cities and the wealth of nations*. New York: Random House.

Jacobs, W., Ducruet, C., & De Langen, P. (2010). Integrating world cities into production networks: The case of port cities. *Global Networks, 10*(1), 92–113.

Keane, M. (2010). Keeping up with the neighbors: China's soft power ambitions. *Cinema Journal, 49*(3), 130–135.

Keane, M., Fung, A. Y. H., & Moran, A. (2007). *New television, globalisation, and the east Asian cultural imagination*. Hong Kong: Hong Kong University Press.

Marx, K. (1973). *Grundrisse: Foundations of the critique of political economy* (p. 539). New York: Vintage.

Morley, D. (2000). *Home territories: Media, mobility, and identity*. New York: Routledge.

Pang, L. (2010). Hong Kong cinema as a dialect cinema? *Cinema Journal, 40*(3), 140–143.

Porter, M. (1998). Clusters and the new economics of competition. *Harvard Business Review, 76*, 77–90.

Scannell, P. (1991). *A social history of British broadcasting*. Cambridge: Blackwell.

Scott, A. J. (2000). *The cultural economy of cities*. Thousand Oaks: Sage.

Silverstone, R. (1994). *Television and everyday life*. New York: Routledge.

Song, T. (2010). *Independent cinema in the Chinese film industry*. Dissertation, Queensland University of Technology, Queensland.

Teo, S. (1997). *Hong Kong cinema: The extra dimensions*. London: BFI.

Uhde, J., & Uhde, Y. N. (2000). *Latent images: Film in Singapore*. Singapore: Oxford University Press.

Wang, S. (2003). *Framing piracy: Globalization and film distribution in greater China* (p. 2003). Lanham: Rowman & Littlefield.

Wang, T. (2009). Understanding local reception of globalized cultural products in the context of the international cultural economy: A case study on the reception of *Hero* and *Daggers* in China. *International Journal of Cultural Studies, 12*, 299–318.

Yeh, E. Y. (2010). The deferral of pan-Asian: Critical appraisal of film marketization in China. In M. Curtin & H. Shah (Eds.), *Reorienting global communication: Indian and Chinese media beyond borders*. Urbana: University of Illinois Press.

Yeh, E. Y., & Davis, D. W. (2008). Re-nationalizing China's film industry: Case study on the china film group and film marketization. *Journal of Chinese Cinemas, 2*(2), 37–51.

Young, D. (2010, August 9). News corp sells controlling stake in china tv channels. *Reuters*. http://www.reuters.com/article/idUSTRE67810L20100809

Zhang, Y. (2004). *Chinese national cinema*. New York: Routledge.

Zhang, Y. (2010). Transnationalism and translocality in Chinese cinema. *Cinema Journal, 49*(3), 135–139.

Zhao, Y. (2010). Whose hero? The 'spirit' and 'structure' of the made-in-China blockbuster. In M. Curtin & H. Shah (Eds.), *Reorienting global communication: Indian and Chinese media beyond borders* (pp. 161–182). Urbana: University of Illinois Press.

Zhou, Y. (2006). *Historicizing online politics: Telegraphy, the internet, and political participation in China*. Stanford: Stanford University Press.

Zhu, Y. (2003). *Chinese cinema during the era of reform: The ingenuity of the system*. Westport: Praeger.

Zhu, Y. (2012). *Two billion eyes: The story of China central television*. New York: The New Press.

US Television Travels Abroad: Global TV and the Formatting Trend

7

Ann Fletchall

From a global perspective, the US, or "Hollywood," has long dominated the entertainment industry. As early as 1939, an estimated 65 % of films shown worldwide had a US origin and today, Hollywood blockbusters routinely gross just as much or more from international distribution as they do from domestic receipts (Miller et al. 2005). Television, often thought to be the domain of mostly national audiences, is increasingly becoming a global phenomenon as well. A recent development along these lines is that of "formatting;" it has become quite common for the format, or concept, of a popular television program, to be sold overseas and remade for local audiences. It is this formatting trend that is the primary focus of this chapter. Formatting raises several intriguing issues, including the sustained US dominance of the medium, whether formats promote or counter cultural homogenization, and questions about the transferability of the unique sense of place that characterizes certain television series. It is important to first establish the historic hegemony of US television in the global marketplace, as well as to provide a brief history of formats and a description of the top players in this industry, before exploring how these questions might be answered.

The Ubiquity of US Television

As usual, at the 54th Annual Monte Carlo Television Festival, American fare took top honors in the audience awards category. CBS's *NCIS* was named the most-watched show in the world in 2013 with more than 57 million viewers, and ABC's *Modern Family* was the most popular comedy (Festival de Télévision

A. Fletchall (✉)
Department of History, Political Science and Geography, Mississippi University for Women, Columbus, MS, USA
e-mail: amfletchall@muw.edu

S.P. Mains et al. (eds.), *Mediated Geographies and Geographies of Media*,
DOI 10.1007/978-94-017-9969-0_7

2014). As this example suggests, US television shows have long dominated the international market. Although exact figures are difficult to pin down, it is estimated that US-based media conglomerates receive approximately 75 % of television export revenues worldwide (Hoskins et al. 1997; Havens 2006; Esser 2010). TV programming rights held by the US were valued at 60 % of the world's $10 billion total in 2006 (McMurria 2009), and the US generates an estimated 70 % of television programming worldwide, as measured by hours produced (Christophers 2009). Fictional programming from the US (including both TV series and films) made up nearly 70 % of total imported content on European screens in 2000 and reached nearly 80 % in some European markets in 2001 (Miller et al. 2005).

There are several explanations for the US dominance of the television trade, as put forth by Hoskins et al. (1997). First, as television markets and programming decisions are usually defined on a national scale, the advantages of producing programming for the US audience are many. The US is one of the largest domestic markets in the world because the majority of viewers are English-speaking. The US is also a very affluent country, by world standards. Large, consumerist audiences like these attract advertiser dollars, and thus more money is available to finance productions. Thus, US television productions have much larger budgets and higher production values than do those in other countries. With the exception of the many globally successful *telenovelas* produced in Latin America, networks in poorer countries and smaller "geolinguistic regions" (Sinclair et al. 1996, p. 12) simply do not have the financial resources to bankroll elaborate productions and cannot produce enough volume to fill their schedules. Furthermore, as described by Christophers (2009), because television programs cost very little to *re*produce, US distributors tend to charge low licensing fees in smaller markets, making it much more affordable for those networks to purchase US imports rather than to produce original programming. US audiences have not had to suffer dubbing and subtitles, unlike so many in the world.

However, US television abroad suffers from what has been termed "cultural discount" (Hoskins et al. 1997, p. 32). Quite often, programming produced within and for a particular cultural group will be less popular in other places as viewers might find several aspects of the show or film difficult to identify with; cultural discount applies particularly to situation comedy and less so to the action drama genre (Hoskins et al. 1997). Moran (1998) provides an illustration of cultural discount using the internationally popular game show *Wheel of Fortune*:

> [the imported] version will always be "foreign": American-English will not be the language of the national population and the program may have to be subtitled or dubbed; the version will draw on cultural knowledges and abilities most available to Americans; prizes will be in the form of goods and services deemed desirable by Americans but not necessarily by other nations' populations.

Due to cultural discount, it has been proven that viewers in many countries generally prefer local or same-language programming to imports, although Hollywood films are also very popular (Sinclair et al. 1996; Havens 2006; Sakr 2007). For example, in his study of TV scheduling in Hungary, Havens (2006) found that networks were likely to schedule Hollywood movies during primetime,

Latin American *telenovelas* in the mornings, burn US series acquired as a part of larger entertainment packages overnight, and save the precious few locally-produced programs to fill out the primetime schedule. In their empirical study of Western European programming in the 1990s, De Bens and de Smaele (2001), found a similar "bipolarization" of the primetime lineup, consisting of both American movies and nationally-produced series. As in Hungary, fictional series from the US were generally found to be shown outside of primetime.

Global Formats

As a result of cultural discount, the trend has shifted away from the purchase of complete US series to the purchase of "formats," whereby the rights to make a domestic version of a popular foreign program are acquired. One industry executive describes the format as akin to the crust of a pie: "the crust is the same from week to week, but the filling changes" (Moran 1998, p. 13). As Moran (2008) explains, "what is marketed and distributed is not a completed programme but rather a body of knowledge and accompanying resources that will help in the remaking of a programme." In short, a format is a "program concept" (Oren and Shahaf 2012, p. 2). Formatting has now become a $5 billion industry (National 2012).

A formatting agreement usually covers the rights to use logos, set designs, and in the case of fictional series, scripts (Havens 2006). Often a consultant from the original production is included in the agreement, and also a "bible," a document containing detailed information on budgets, set designs, casting procedures, and other production knowledge deemed necessary to produce the show (Moran 1998; Chalaby 2011). Thus, in addition to the concept, "formats constitute a transfer of expertise" (Chalaby 2011, p. 295). This excerpt, taken from Baltruschat's (2009) personal communication with a line producer on *Canadian Idol*, provides an example of such a formatting arrangement:

> During pre-production, Insight [a Canadian production company] producers . . . went to Los Angeles to attend *American Idol* auditions and to "figure out how it works and to get some pointers on the whole operation." The producers also watched the entire seasons of *American Idol* and *Pop Idol* to become familiar with the program and its production matrix. Throughout the first season of *Canadian Idol*, FremantleMedia remained in close contact with its Canadian co-producers, "because it is in their [FremantleMedia's] interest that the show does well in Canada and adheres to the format." Consultations via telephone and face-to-face meetings ensured that the format maintained its recognizability while being adapted to Canada's television culture and viewer sensibilities. FremantleMedia, therefore, checked the rough cuts and premiere tapes of *Canadian Idol* to make sure that production standards were met.

As this example illustrates, it is of utmost importance to the format distributor (FremantleMedia, in this case) that the brand not be compromised by a substandard production.

Format adaptation is not a new phenomenon. In the early days of television, many radio programs were adapted for the small screen (Moran 1998; Chalaby 2012). The world's first television format was the British version of the US show *What's My Line*, which debuted in the UK in 1951 (Chalaby 2011). The first format imported to the US was the now-classic *All in the Family*, adapted from the British sitcom *Till Death Us Do Part* (Chalaby 2011). Game shows, near the bottom of the television hierarchy, were long the core of the formatting industry and were mostly exported by the US, such as *Jeopardy* and *Wheel of Fortune* (Chalaby 2011).

There are several factors which explain the recent formatting boom. New communications technologies such as satellites and digital transmission, as well as deregulation and privatization of the television industry in many countries, have led to a proliferation of new channels (Sakr 2007; Moran 2008). For example, between 1989 and 2002, channel offerings in Western Europe alone increased from 40 to 1,500 (McMurria 2009). With new channels come audience fragmentation and increased competition, resulting in a need for both more programming and more appealing programming (Moran 1998). Formats filmed in the local language, with some cultural adjustments made, have been found to be more appealing and popular than a traditional import that has been dubbed or subtitled (Moran 1998; Havens 2006). In his study of Hungarian programming, Havens (2006) found that the Hungarian version of *Big Brother* was much more popular than the American feature film that once occupied its prime timeslot. According to a German network executive, European audiences grew "tired of the formulaic American situation comedies and dramas" and "required more local production that was relevant to them in their local languages and tastes" (Pfanner 2006, para. 6–7). Furthermore, in countries that set a quota on television imports, local productions of global formats are not considered such and are often more politically acceptable (Moran 1998, 2009; Waisbord 2004; Oren and Shahaf 2012). Additionally, Esser's (2010) study of programming in the US found that the Writer's Guild Strike of 2007–2008 drove many networks to acquire formats as a strategy to mitigate the lack of new scripts for its dramatic and comedy series. Indeed, the number of internationally traded formats increased from 259 in 2002–2004 to 445 in 2006–2008 (Esser 2010).

The reality TV boom of the late 1990s/early 2000s further explains the formatting phenomenon. This period saw the introduction of many new types of programs (such as talent competitions with dramatic elimination procedures based on audience voting) that have proven quite compelling to viewers, and thus very profitable to license for overseas adaptation (Chalaby 2011). Four such game-changing reality shows debuted between 1997 and 2001: *Who Wants to Be a Millionaire*?, *Survivor*, *Big Brother*, and *Pop Idol* (eventually *American Idol* and other versions). Dubbed "super-formats" by Bazalgette (2005), these shows quickly became extremely popular as formats. According to Moran (2009, p. 115), "*Big Brother* has become the most watched program in world television history." Twenty new *Big Brother* series are made in the world each year, seen in more than 60 countries (Endemol 2013). As of 2009, there were 199 iterations of Britain's *Pop Idol*, reaching 460 million viewers worldwide (FreemantleMedia 2009), and more than 50 "local format versions" of *Survivor* are now produced (Survivor Format 2009, para. 5). *Who Wants to Be a Millionaire*? is currently broadcast in 118 countries (What We Do 2012).

Adding to the attractiveness of the super-formats and similar reality fare is their cost effectiveness. According to Havens (2006), formats are typically sold at 80 % of the cost of finished, or "canned" programs, and therefore present a less expensive alternative to the production of original programming.[1] While still highly rated, formats are generally less expensive to produce than an original drama or sitcom series, due to lack of need for highly-paid actors, script writers and several elaborate sets. According to data compiled by the Format Recognition and Protection Association (FRAPA), non-scripted talent shows cost between $1 million and $1.6 million per episode to produce, while dramatic series range from $1 million to $2.5 million per episode (FRAPA Report 2009). Accordingly, reality shows (including talent competitions and other "factual" programming) were by far the most-formatted genre with 10,757 episodes exported in 2008 compared to 3,188 episodes of scripted shows (Chalaby 2011, citing FRAPA Report 2009). Even in the US, a market that historically has not imported television content, Esser (2010) found that in the 2007–2008 television season, formats accounted for approximately 33 % of hours broadcast on the major networks (ABC, CBS, NBC, FOX, and the CW) during primetime.[2]

The US is a leading player in format exports; however, the US is not nearly as dominant internationally in the format market as it is in finished programs or in film. For example, during the period 2006–2008, the US exported a total of 159 program formats while the UK exported 275. The US imported a significant number of shows during this period as well, 116 (FRAPA Report 2009); this number is exceptional in a market that had once been nearly impenetrable to imports (Waisbord 2004). Many Americans would be surprised to learn that of the four "super-formats," none originated in the US (See Table 7.1). It should be noted, however, that once the "super-formats" saw success in the US market, their popularity as global formats increased greatly (Chalaby 2011, 2012).

Although not the leading exporter of formats, Hollywood still holds a great deal of leverage in licensing arrangements. As Christophers (2009) explains, imported formats are usually produced by US companies, preventing format exporters from making too much profit from their program concepts through show production, thus perpetuating the paradigm of US control over much of the global television industry.

The International Distributors

While it is beyond the scope of this chapter to untangle the web of horizontally-integrated media conglomerates that dominate the global entertainment industry, several of the biggest players in the formatting industry are not US-based companies, and they are selling a great deal of content that was not originally produced in

[1]The term "canned" programming derives from the time before digital transmission when film was shipped in cans (Moran 2009).

[2]This figure includes formats originating in the US, such as *Law & Order* and *The Bachelor*.

Table 7.1 Origin of popular unscripted formats

Program	Country of origin
American Idol (Pop Idol)	UK
America's Got Talent	USA
America's Next Top Model	USA
Big Brother	The Netherlands
Dancing with the Stars	UK
Deal or No Deal	UK
Extreme Makeover: Home Edition	USA
I'm a Celebrity – Get Me Out of Here	UK
Project Runway	USA
So You Think You Can Dance	USA
Survivor	Sweden
The Apprentice	USA
The Bachelor	USA
The Biggest Loser	USA
The X Factor	USA
Undercover Boss	UK
Who Want to Be a Millionaire?	UK
Wife Swap	UK
Wipeout	USA

Source: FRAPA Report (2009)

the US. This section explores the format and finished program catalogues offered by three leading players in the industry: Endemol, FremantleMedia, and Sony Pictures Television (SPT). Generally, large distributors such as these will acquire the rights to a program (if it isn't one of their own productions), sell it to networks around the world, and will often assist in the production of local versions of these programs (Moran 1998).

Endemol, based in the Netherlands, is, as stated on its website, "the largest independent television and digital production company in the world" (Who We Are 2013), and is also the largest "format specialist" in the world (Moran and Malbon 2006). Endemol offers more than 2,000 formats "with over 50 new formats commissioned around the world each year, across genres including entertainment, reality TV, game shows, animation, comedy and drama" (Formats 2013, para. 1). Endemol's most successful format is *Big Brother*, but it is also responsible for the distribution of other hit shows, including *Deal or No Deal*, *Extreme Makeover: Home Edition*, and *Fear Factor*.

FremantleMedia is a UK-based company that has expanded through several international acquisitions. From its acquisition of All American Television, the company owns the rights to many classic game shows (*Family Feud* and *The Price is Right*), as well as the global mega-hit, *Baywatch* (Moran and Malbon 2006). The crown jewel of FremantleMedia's format library is the *Idols* series, but the company also distributes *The Apprentice*, *The X Factor*, and (*America's*) *Got Talent* (International Distribution 2009).

Sony Pictures Television (SPT) is the distribution arm of a more typical Hollywood studio, Sony Pictures. Headquartered in Culver City, California, Sony Pictures is the archetypical contemporary media conglomerate. Owned by the Japanese Sony Corporation, Sony Pictures acquired Columbia Pictures in 1989, part and parcel of the significant trend of mergers and acquisitions in the cultural industries (Jin 2012; Sony Corp 2013). According to Havens (2006), among the Hollywood majors, SPT was one of the leaders in the "local-language production" trend, developing production deals in Europe in the 1990s: "By 1999, [SPT] was producing more programming abroad than in the domestic market, including local versions of sitcoms such as *Bewitched, I Dream of Jeanie* and *Who's the Boss* in Brazil, Germany, Italy, Spain and the UK . . . " (Havens 2006, p. 33).

An exploration of program catalogues available on each company's website revealed several trends. First of all, these catalogues reiterated the popularity of reality, or unscripted, programming as formats. Endemol, the world's leading formatter, advertises 41 unscripted formats in its MIPCOM 2012 catalogue (MIPCOM is a major international television trade show), but only 24 scripted program formats (excluding soap operas and *telenovelas*) (Formats: MIPCOM 2012; Scripted Programs 2012). SPT's unscripted format offerings total 73 (from its "game shows," "factual," and "factual entertainment" categories), while only 19 dramas and comedies are featured in its "scripted formats" catalogue (Sony Pictures Television International Formats Distribution 2013).

Of the unscripted formats included in Endemol's catalogue, only eight out of 41 programs (20 %) originated in the US (Formats: MIPCOM 2012).[3] Of Endemol's scripted program formats, none were from the US (Scripted Programs 2012). Similarly, the vast majority of SPT's unscripted offerings are not originally American shows (only ten could be definitely identified as originating in the US). However, in terms of scripted programs, all but two featured in SPT's catalogue were from the US. SPT is the format distributor of several classic sitcoms that have found success abroad, including *Everybody Loves Raymond, The King of Queens, The Nanny*, and *Who's the Boss* (Sony Pictures Television International Formats: Factual 2013).

On the other hand, two-thirds of the finished reality programs (non-formats) that Endemol advertises are from the US (Reality: MIPCOM 2012). Finished programs would be broadcast as is, in English, or subtitled or dubbed for a different language market. FremantleMedia did not provide a catalogue of formats on their website, but does showcase 43 finished programs for international distribution (FME Screenings 2013). Of these, roughly half were from the US. These results further demonstrate that finished US programs are indeed preferred for export, while the market for unscripted series is not nearly so dominated by US programming.

It is also instructive to peruse the program descriptions contained in these catalogues to get a sense of how these shows are marketed and clues to their cross-cultural appeal. Endemol, for example, promotes a reality format titled *Your Face Sounds Familiar*:

[3]The country of origin of each show was provided in the catalogue in most cases; in other cases, the country of origin was determined by an internet search.

> *Your Face Sounds Familiar* is the only talent competition that pushes a group of celebrities to take on a new identity as an iconic music performer. Every week, the same group of celebrities find out who they have to impersonate by simply pressing a buzzer. A giant screen displays multiple images of different singers in rapid succession. The minute the celebrity hits the buzzer, the images freeze and the selection is made. Whether it's an older or younger person, or even someone of the opposite sex, it doesn't matter – vocal coaches, choreographers, make-up artists and stylists are ready to prep them for their big performance. A panel of judges and competing celebrities award points based on various categories, including singing, style and believability... Toward the end, only the four celebrities who accumulated the most points throughout the show qualify for the series final. After their very last performance, the celebrity with the most points receives the top cash prize for a charity of their choice (Formats: MIPCOM 2012).

This show has not yet come to the US, but has been produced in Chile, China, Italy, Lithuania, Portugal, Spain, Turkey, and the Ukraine.

SPT promotes the American syndicated talk show, *The Dr. Oz Show*, with the following description:

> Feeling good is contagious... The Dr. Oz Show has become a coast-to-coast ratings phenomenon. This fast-moving magazine talk show is full of audience participation and viewer interaction with its highly qualified medical host and team of specialist advisers serving up daily Plans and Challenges for ordinary people. At the heart of the show is the quest to help viewers live a longer, happier and healthier life. It's playable as a stripped daily or major weekly event and is designed to appeal equally to male and female demographics. The high-quality content that underpins the format's year of success in the U.S. is available in a package that can be customized to work in any territory (Sony Pictures Television International Formats: Factual 2013).

Finally, SPT's *Everybody Loves Raymond* pitch reads as follows:

> Raymond has it all: a fed up wife, overbearing parents, and an older brother with lifelong jealousy... Ray Barone, a sportswriter, lives with his wife, Debra, a 5-year-old daughter, Ally, and a set of two-year-old twin boys, Michael and Geoffrey. On their own, the family is fairly normal. Complicating matters, though, is the fact that Ray's mom (Marie), dad (Frank), and elder brother (Robert), a 40-year-old policeman still living at home, live next door and keep butting into Ray's life (Sony Pictures Television: Scripted 2013).

Described in these ways, the format's wide appeal is made apparent. No matter the nationality, it would seem that many can relate to a hectic family life and meddling relatives or could benefit from regular health and lifestyle tips. Or, are the values expressed in these examples a form of Western cultural imperialism to be imposed on audiences around the world?

Homogenization or Localization?

As an industry executive adroitly describes a successful format, "the key to most of these things is to have the kind of idea that works locally for everyone" (Navarro 2012, p. 25). As implied by this paradoxical description, global formats occupy an ambiguous place at the intersection of the global and the local. On the surface, and judging by the immense popularity of super-formats such as the *Idol* franchise, it is easy to position global formats as yet another driver of cultural homogenization amid globalization. As the UK, the US, and several other European countries dominate the trade (with Japan being the non-Western exception) (Moran and Keane 2006), along with these popular formats, might we also be exporting our Western/American brand of competitive self-interest to other parts of the world?

In support of this rather dim view of popular formats, Brennan (2012) explains that television in general, and formats in particular, are best-suited to communicating extrinsic, tangible rewards rather than inner happiness or improved self-esteem. Obviously, it is much more visually entertaining to reward someone with one million dollars than simply the pride gained from winning! Among the popular formatted programs included in his study, Brennan (2012) finds that the overwhelming majority are competition-based formats that reward individual success and scheming over collective strength and self-improvement. Baltruschat (2009) agrees, adding that "even in programs where alliances between contestants are encouraged, one individual's selfish pursuit is rewarded in the end." In fact, the Japanese version of *Survivor* famously failed because it focused on inner conflict and friendships rather than the single-minded pursuit of winning which characterizes the successful US adaptation (McMurria 2009). Similarly, Galander (2008) warns that game show formats such as *Wheel of Fortune* and *Who Wants to Be a Millionaire*? promote consumerism, risk-taking, and gambling to Muslim audiences, at odds with the teachings of the religion.

Furthermore, some indigenized versions of formats remain very Western in character. The *Idol* franchise provides several examples. A great many songs by US artists have been featured on *South African Idol*, which leaned heavily toward the country's minority white demographic, and on *West African Idol*, contestants were dismissed for speaking heavily accented English (Ndlela 2012). As Ndlela (2012, p. 254) describes, "The search for the *West African Idol* could thus be construed as a search for the most Americanized aspirant."

Another key argument which positions formats as a vehicle of cultural homogenization is the increasingly shared sensibilities of television executives around the world (Waisbord 2004; Havens 2006; Oren 2012). Most of those making programming decisions, regardless of their nationality, have been educated at the same trade shows, have followed the same Hollywood/Western trends, and "increasingly share similar concepts and attitudes about 'what works' and 'what doesn't in commercial television" (Waisbord 2004, p. 364). These executives, therefore, tend to buy the same formats and put similar types of programs on the air. Given such examples, and the astounding popularity of several global formats, it would not be inaccurate to say that the world is increasingly watching the same basic television content.

Appadurai (1996) has successfully argued that the indigenization of cultural forms, from cricket to cinema, is an inevitable feature of the globalization process, and as such, we must also consider formats as a way for local cultures to be expressed within an industry that has historically been open only to those with the biggest production budgets (i.e. Hollywood). There is a definite flexibility to most formats, and it would be wrong to describe the process as a mechanical reproduction, as Moran (1998) explains. A plethora of local tweaks to global formats can be cited: The version of *Extreme Makeover*: *Home Edition* that plays in Brazil, *Lar Dolce Lar*, focuses more on drama between the locals, whereas the US version is primarily a "feel-good" show (Navarro 2012). While the US version of *Wife Swap* concludes with a message of personal transformation and self-improvement, according to Sharp (2012), in a play to *machismo*, the Chilean edition has been adapted to eliminate the "rules change" portion of the program where the family must abide by the household rules of the visiting wife. As Sharp (2012) explains, "the emphasis is on the household patriarch judging the housekeeping of his new wife." In adapting the Colombian *telenovela Yo Soy Betty, la Fea* (*Ugly Betty*) for US audiences, the program had to become a weekly, rather than daily program, resulting in significant changes to the story arc. Also, the original character of Betty's mother, a woman "who stoically absorbs all of life's adversities" was replaced by a single father, a figure drawn as less passive and thus more relatable to US audiences (Rohter 2007, para. 19). Finally, producers of the upcoming Colombian version of *Breaking Bad*, *Metastasis*, have revealed that meth cooks will be performed in a beat-up school bus, rather than an RV, a vehicle which played a pivotal role in the first season of the series (Roxborough 2013a). Often, these national changes are so effective that audiences would not guess that the show was not an original format, as was the case with the Polish adaptation of *The Honeymooners*, a classic 1950s American sitcom (Moran 2009).

Closer analyses of specific local adaptations position formats as very much indigenized, rather than homogenized, cultural products. One such example can be found in Hetsroni and Tukachinsky's (2003) comparative study of *Who Wants to Be a Millionaire*? Their analysis of the game show's quiz questions found significant differences between the versions from the US, Russia, and Saudi Arabia. Questions in the Saudi Arabian and Russian versions most frequently reflected a theme of national identity, a value promoted in those countries, while the US version fielded the highest share of popular culture questions, mirroring the characteristic American fixation with such. Even each show's title reflects cultural attitudes toward game show winnings:

> ... in Saudi Arabia it is "Who Will *Earn* a Million," because Islam prohibits gambling or prizes that do not demand investment. The show's name in Russia, "Oh, Lucky – You are a Millionaire," reflects that the first prize ... is unseemly high for the hard-working Russian (Hetsroni and Takachinsky 2003, p. 168).

Game shows and reality programming are considered relatively simple formats to export, but sitcoms are notoriously more difficult to adapt due to cultural differences in comedic preferences. Therefore, Beeden and de Bruin's (2010) comparison of the surprisingly successful US adaptation of the sitcom *The Office* to the British original presents an illuminating case study. Beeden and de Bruin (2010) observed that significant changes were made to the series to make the US version distinctly American. Issues of race, rather than class, provided the humor in some episodes, and there was a more hopeful aura to the US version. Beeden and de Bruin

(2010) conclude that "successful format adaptations appear to be those that are an interpretation of the original rather than simply a copy."

As documented in the film *Exporting Raymond* (2010), adapting the *Everybody Loves Raymond* format to Russian audiences was a challenging task. Prior to *Raymond*, US sitcoms such as *The Nanny* and *Married... With Children* had become hits in Russia, and their success is explained by Teeter (2010, para. 6,8) in that they are "caricature-fests" whose "over-the-top" brand of comedy is not difficult for foreign audiences to translate. *Raymond's* humor, however, is more situational and "grounded in reality" (Czarnecki and Rosenthal 2010). *Exporting Raymond* depicts many differences of opinion between the show's American creator (Phil Rosenthal) and the Russian production team regarding humor, casting, script adaptations, and even the "manliness" of the lead character, prompting Rosenthal, to comment toward the end of the process, "I get it. The show is for them. They have to make it their own" (Czarnecki and Rosenthal 2010). The series, called *The Voronins*, emerged as a hit after its 2009 premiere.

These examples provide evidence that when examined closely, successful formats do allow (and must, to be successful) for the expression of distinct and unique cultural values. Even the *Idol* format provides relevant examples. New Zealand's version makes a point of featuring non-white contestants to ensure that the program serves a nation-building agenda (de Bruin 2012), and *Afrikaans Idol* in South Africa is a vehicle for language and cultural preservation (Ndlela 2012). The most successful formats are, in fact, likely to be unrecognizable as such to their audiences. As Christophers (2009) explains in the case of *Who Wants to be a Millionaire*?:

> ... this success has clearly been *enabled*, if not caused, by the fact that viewers have no reason to believe that these are *not* wholly American shows. Certainly, there is anecdotal evidence that many American viewers believed *Millionaire*... to be home-grown. Its UK origins were not widely publicized, and in both form (the quiz show) and objective (wealth) it seemed an archetypal American product.

Christophers (2009) suggests that formats work because a format is merely a "program skeleton" that "comes stripped of the trappings of place, allowing it to be invested instead with the specificity of the place where it is being imported and reproduced." When formats are viewed in this way, when foreign producers are given freedom to diverge from the original, when formats are *interpretations* rather than reproductions (cf. Beeden and de Bruin 2010), it is clear that they have the potential to become a truly local product, in a cultural sense. The remarkable success of *Arab Idol* is another striking example, and is explained by a network executive in this way:

> It is true that these programs are a blend of western-made shows, but we must remember it has proven to be successful in 127 countries. *Ultimately, the success is owed to Arabic culture*... These shows provide the Arabic public with the songs of Umm Kalthoum, *the poetry of* Ahmed Shawqi, and the voice of Abdel Halim Hafez—*more so than any programs*" (Haddad 2013, para. 5; emphasis added).

Another executive adds: "It is quite simply one of the main sources of the great Arab dream" (Haddad 2013, para. 6).

Active Audiences and the Question of Place

To determine whether global formats are to be accurately positioned as local products, we must, as do contestants on *Millionaire*, "ask the audience." In-depth comparative audience studies are needed to determine whether viewers enjoy different versions of these formats in different ways and for different reasons. After all, what matters most in addressing the effects of such programming is not the content itself, but how viewers interpret, react to, and take pleasure from the content. As Liebes and Katz (1993) suggest:

> To understand the messages perceived by viewers of a television program, one cannot be satisfied with abstract generalizations derived from content analysis, however sophisticated. The actual interaction between the program and its viewers must be studied.

As such, we may take our cue from their research into the astounding global popularity of *Dallas*. As a finished imported program (not a format), *Dallas* was successful in at least 90 countries, but failed in only a few (Japan, Brazil, and Peru) (Ang 1985; Liebes and Katz 1993; Sinclair et al. 1996). Liebes and Katz (1993) found that the universality of the show's themes allow many viewers to relate the show to their own realities, but that the show's themes are interpreted and negotiated in significantly different ways by viewers in several different cultures. Conclude Liebes and Katz (1993), "Such programs may beam a homogenous message to the global village, but our study argues that there is pluralism in the decoding." Given these findings, it would greatly benefit our understanding of global formats to uncover the processes of indigenization that occur through audience agency – not only how the content of indigenous versions diverges, but also how these variations may be differently decoded by active viewers within and across cultures.

For geographers in particular, another question arises when considering the adaptability of shows whose appeal and affect is very much tied to a specific setting. It has been argued by Fletchall et al. (2012) that television shows, by using effective production techniques, have the power to communicate a strong sense of place. This sense of place can pack a powerful emotional resonance, or affect, and viewers of such shows actively participate in a process of televisual place-making. Such televisual places, created via viewer engagement with a series over time (several years if the program is successful), can be rendered very meaningful in and of themselves (Fletchall et al. 2012). Therefore, when a place-centered program is formatted, do the elements needed to create a meaningful televisual place get translated too? If so, how? And, will viewers decode these shows in ways similar to their original US audiences?

The reception of one upcoming adaptation, in particular, may allow us an opportunity to address these questions: a Turkish adaptation of the popular US dramatic series, *The OC*. When it aired in the US, as posited by Fletchall et al. (2012), the show's Orange County, California setting played an outsized role. By fashioning televisual Orange County as a wealthy, sunny wonderland, it made the previously little-known Orange County, California a household name. Through a combination of sweeping scenery shots, evocative music, and of course, a soap opera-like narrative to tug at audience emotions, *The OC*, as a televisual place, became a focus of meaning to many viewers (Fletchall et al. 2012). Like the original, the Turkish version will "center on a group whose lives are forever changed when an outsider arrives at their affluent ocean-side residence" (Goldberg 2013, para. 2), but it remains to be seen whether this adaptation will produce a comparable place-centric affect amongst its Turkish audience. In other words, can *The OC* be successful without the OC? Will this localized version portray its setting in such a way that allows viewers to make meaningful connections with the place created in these shows? Will not some affective quality of place be lost in the translation? Formatted versions of such place-based shows warrant analysis in order to determine whether international formats can, along with the basic structure of a series, successfully transfer the basis for viewers' deeper engagements with televisual place.

Conclusion

Curiously, it should be noted, finished scripted series from the US are making something of a comeback over formats in some European markets. According to Arango (2008), the number of hours of US programming on major networks in Europe in 2000 was approximately 214,000 h, and this figure had increased to 266,000 in 2006. Arango (2008) provides a quote from the chief executive of the RTL group, a major player in European TV: "Let's say, at the beginning of the decade, more or less all over Europe you saw on the big channels almost no U.S. series on prime time… Now, all over Europe you have a lot of American series in prime time." In fact, a brief look at RTL's [a major network in Germany] weekly lineup reveals that Tuesday nights are devoted to US shows dubbed in German: *CSI*: *Vegas*, *Bones*, *Dallas* (2012), and *Monk* air consecutively during primetime. On Thursday nights, two episodes of *CSI*: *Miami* play back-to-back, followed by an episode of the US version of *Law & Order*. Sunday night features a Hollywood movie (*PS I Love You* at the date of writing), once a common anchor of TV schedules in Europe (Havens 2006). There are a few noticeable formats on the RTL schedule as well, such as *Wir Wird Millionär?*, *Der Bachelor*, and *Deutschland Sucht den Superstar* (German *Idol*), but finished US shows are inarguably very strongly present (TV-Programm 2013).

What explains this apparent reversal? First of all, US television is currently experiencing something of a renaissance, led by cable channels such as HBO and Showtime (and most recently AMC and Netflix), which are not bound by traditional network rules and sensibilities. The result has been shows that are riskier, edgier, more violent, that do not have to appeal to the lowest common denominator of US viewership. As cable series have pushed the envelope, major networks tend to follow suit and all programming benefits as the bar of quality is raised. It would seem that Europeans appreciate these changes as well. As stated by an executive at a UK network whose niche programming is US shows, "We are seeing bright, intelligent and beautifully made drama coming out of America. In the U.K., many people abhor the politics of the U.S. but eagerly embrace the culture" (Pfanner 2006, para. 16).

At the same time, viewers may be tiring of standardized reality formats, and voicing a preference for original concepts, regardless of their national origin. In the US, ratings for *American Idol's* season 13 finale reached a new low, garnering only ten million viewers, far fewer than its 2006 peak of 36 million (Hinckley 2013; O'Connell 2014), and *Idol* ratings have declined worldwide (Roxborough 2013b). However, judging from the success of newer talent formats like *The Voice*, the recent buzz surrounding sales of the Israeli format *Rising Star* (Roxborough 2013c), and the continued presence of this genre toward the top of the ratings in many countries, the formatting phenomenon has not yet come close to reaching its conclusion.

References

Ang, I. (1985). *Watching Dallas*. London: Methuen.

Appadurai, A. (1996). *Modernity at large: Cultural dimensions of globalization*. Minneapolis: University of Minnesota Press.

Arango, T. (2008). World falls for American media, even as it sours on America. *The New York Times*. Retrieved from http://www.nytimes.com/2008/12/01/business/media/01soft.html?_r=1&

Baltruschat, D. (2009). Reality TV formats: The case of Canadian Idol. *Canadian Journal of Communication, 34*(1), 41–59.

Bazalgette, P. (2005). *Billion dollar game: How three men risked it all and changed the face of television*. London: Time Warner.

Beeden, A., & de Bruin, J. (2010). The office: Articulations of national identity in television format adaptation. *Television & New Media, 11*(1), 3–19.

Brennan, E. (2012). A political economy of formatted pleasures. In T. Oren & S. Shahaf (Eds.), *Global television formats: Understanding television across borders* (pp. 72–89). London: Routledge.

Chalaby, J. K. (2011). The making of an entertainment revolution: How the TV format trade became a global industry. *European Journal of Communication, 26*(4), 293–309.

Chalaby, J. K. (2012). At the origin of a global industry: The TV format trade as an Anglo-American invention. *Media, Culture & Society, 34*(1), 36–52.

Christophers, B. (2009). *Envisioning media power: On capital and geographies of television*. Lanham: Rowman & Littlefield.

Czarnecki, J. (Producer) & Rosenthal, P. (Producer and Director). (2010). *Exporting Raymond* [Motion Picture]. New York: Samuel Goldwyn Films.

De Bens, E., & de Smaele, H. (2001). The inflow of American television fiction on European broadcasting channels revisited. *European Journal of Communication, 16*(1), 51–76.

de Bruin, J. (2012). *NZ Idol*: Nation building through format adaptation. In T. Oren & S. Shahaf (Eds.), *Global television formats: Understanding television across borders* (pp. 223–241). London: Routledge.

Endemol lands regional market deals for Indian Big Brother. (2013, March 1). *Endemol Distribution*. Retrieved from http://www.endemoldistribution.com/content/readnews.aspx?id=1739

Esser, A. (2010). Television formats: Primetime staple, global market. *Popular Communication, 8*(4), 273–292.

Fletchall, A., Lukinbeal, C., & McHugh, K. (2012). *Place, television and the real Orange County*. Stuttgart: Franz Steiner Verlag.

FME Screenings: Home. (2013). *FremantleMedia Enterprises*. Retrieved from http://www.fmescreenings.com/

Formats. (2013). *Endemol UK Ltd.*http://www.endemol.com/what/formats

Formats: MIPCOM 2012. (2012). *Endemol Distribution.* Retrieved from http://endemoldistribution.com/uploads/documents/ES_MIPCOM2012_Formats_Catalogue.pdf

FRAPA. (2009). *The FRAPA Report 2009: TV formats to the world.* Cologne: FRAPA.

Galander, M. M. (2008). Global communication and cultural desensitisation: Repackaging Western values for non-Western markets. *Intellectual Discourse, 16*(1), 1–19.

Goldberg, L. (2013). 'The OC' gets Turkish adaptation. *The Hollywood Reporter.* Retrieved from http://www.hollywoodreporter.com/live-feed/oc-gets-turkish-adaptation-630619

Haddad, V. (2013). MBC: 'Arab Idol' finale watched by 100 million. *Asharq Al-Awsat.* Retrieved from http://www.aawsat.net/2013/07/article55308009

Havens, T. (2006). *Global television marketplace.* London: BFI Publishing.

Hetsroni, A., & Takachinsky, R. H. (2003). 'Who Wants to Be a Millionaire' in America, Russia, and Saudi Arabia: A celebration of differences or a unified global culture? *The Communication Review, 6*(2), 165–178.

Hinckley, D. (2013). 'American Idol' finale ratings plummet, losing seven million viewers from last year's finale. *New York Daily News.* Retrieved from http://www.nydailynews.com/entertainment/tv-movies/american-idol-finale-ratings-plummet-article-1.1347150

Hoskins, C., McFadyen, S., & Finn, A. (1997). *Global television and film: An introduction to the economics of the business.* Oxford: Clarendon.

International Distribution. (2009). *FremantleMedia.* Retrieved from http://www.fremantlemedia.com/enterprises/International_Distribution.aspx

Jin, D. Y. (2012). Transforming the global film industries: Horizontal integration and vertical concentration amid neoliberal globalization. *The International Communication Gazette, 74*(5), 405–422.

Liebes, T., & Katz, E. (1993). *The export of meaning: Cross-cultural readings of Dallas.* Cambridge: Polity Press.

McMurria, J. (2009). Global TV realities: International markets, geopolitics, and the transcultural contexts of reality TV. In S. Murray & L. Ouellette (Eds.), *Reality TV: Remaking television culture* (pp. 179–202). New York: New York University Press.

Miller, T., Govil, N., McMurria, J., Maxwell, R., & Wang, T. (2005). *Global Hollywood* 2. London: British Film Institute.

Moran, A. (1998). *Copycat TV: Globalisation, program formats and cultural identity.* Luton: University of Luton Press.

Moran, A. (2008). Makeover on the move: Global television and programme formats. *Continuum: Journal of Media & Cultural Studies, 22*(4), 459–469.

Moran, A. (2009). Global franchising, local customizing: The cultural economy of TV program formats. *Continuum: Journal of Media & Cultural Studies, 23*(2), 115–125.

Moran, A., & Keane, M. (2006). Cultural power in international TV format markets. *Continuum: Journal of Media & Cultural Studies, 20*(1), 71–86.

Moran, A., & Malbon, J. (2006). *Understanding the global TV format.* Bristol: Intellect Books.

National Geographic Channel CEO David Lyle receives FRAPA/C21 Gold Medal Award. (2012). *FRAPA.* Retrieved from http://www.frapa.org/2012/04/04/national-geographic-channel-ceo-david-lyle-receives-frapac21-gold-medal-award

Navarro, V. (2012). More than copycat television: Format adaptation as performance. In T. Oren & S. Shahaf (Eds.), *Global television formats: Understanding television across borders*(pp. 23–38). London: Routledge.

Ndlela, M. N. (2012). Global television formats in Africa: Localizing *Idol.* In T. Oren & S. Shahaf (Eds.), *Global television formats: Understanding television across borders*(pp. 242–259). London: Routledge.

O'Connell, M. (2014). 'American Idol' wraps soft season, still tops 'Survivor,' but 'Modern Family' wins. *The Hollywood Reporter.* Retrieved from http://www.hollywoodreporter.com/live-feed/tv-ratings-american-idol-wraps-706481

Oren, T. (2012). Reiterational texts and global imagination: Television strikes back. In T. Oren & S. Shahaf (Eds.), *Global television formats: Understanding television across borders*(pp. 366–381). London: Routledge.

Oren, T., & Shahaf, S. (2012). Introduction: Television formats – A global framework for TV studies. In T. Oren & S. Shahaf (Eds.), *Global television formats: Understanding television across borders* (pp. 1–20). London: Routledge.

Pfanner, E. (2006). US TV fare roars back onto foreign prime time. *The New York Times*. Retrieved from http://www.nytimes.com/2006/10/15/business/worldbusiness/15iht-tv16.3160937.html?pagewanted=all

Reality: MIPCOM 2012. (2012). *Endemol Distribution*. Retrieved from http://endemoldistribution.com/uploads/documents/EWD_MipcomRealityCatalogue2012.pdf

Rohter, L. (2007). How 'Ugly Betty' changed on the flight from Bogotá. *The New York Times*. Retrieved from http://www.nytimes.com/2007/01/07/weekinreview/07rohter.html?pagewanted=all

Roxborough, S. (2013a). Meet Walter Blanco: 'Breaking Bad' gets Spanish-language version. *The Hollywood Reporter*. Retrieved from http://www.hollywoodreporter.com/news/breaking-bad-remake-metastasis-meet-640039

Roxborough, S. (2013b). RTL's CEO on the trouble with 'American Idol' and why she loves Downton Abbey (Q&A). *The Hollywood Reporter*. Retrieved November 23, 2013, from http://www.hollywoodreporter.com/news/rtls-ceo-american-idol-trouble-656725

Roxborough, S. (2013c). Watch out 'Voice': Israel's 'Rising Star' coming to global TVs. *The Hollywood Reporter*. Retrieved from http://www.hollywoodreporter.com/news/israels-rising-star-coming-global-661506

Sakr, N. (2007). *Arab television today*. London: I.B. Tauris.

Scripted Programs: MIPTV 2012. (2012). *Endemol Distribution*. Retrieved from http://endemoldistribution.com/uploads/documents/ES_Scripted_Catalogue.pdf

Sharp, S. (2012). Global franchising, gender, and genre: The case of domestic reality television. In T. Oren & S. Shahaf (Eds.), *Global television formats: Understanding television across borders* (pp. 346–365). London: Routledge.

Sinclair, J., Jacka, E., & Cunningham, S. (1996). Peripheral vision. In J. Sinclair, E. Jacka, & S. Cunningham (Eds.), *New patterns in global television: Peripheral vision* (pp. 1–32). Oxford: Oxford University Press.

Sony Corp. Info. (2013). *Sony Corporation*. Retrieved from http://www.sony.net/SonyInfo/CorporateInfo/History/SonyHistory/2-22.html#block5

Sony Pictures Television International Formats Distribution. (2013). *Sony Pictures Television*. Retrieved from http://www.sonypicturestelevision.com/landings/spt/intlformats

Sony Pictures Television International Formats: Factual Entertainment. (2013). *Sony Pictures Television*. Retrieved from http://portal.sliderocket.com/SPT/Factual-Ent-Formats-2012

Sony Pictures Television: Scripted Formats. (2013). *Sony Pictures Television*. Retrieved from http://portal.sliderocket.com/SPT/Scripted_Formats_Brochure

Survivor Format. (2009). *Castaway Television*. Retrieved from http://www.castawaytelevision.com/survivor_format/

Teeter, M. (2010). Everybody loves Kostya here. The Moscow News. Retrieved from http://themoscownews.com/critic/20110926/189070598.html

The Festival de Télévision de Monte-Carlo and Eurodata TV Worldwide Announce the Winners of the 9th International TV Audience Awards. (2014, June 11). *Médiamétrie*. Retrieved from http://www.mediametrie.com/eurodatatv/communiques/the-festival-de-television-de-monte-carlo-and-eurodata-tv-worldwide-announce-the-winners-of-the-9th-international-tv-audience-awards.php?id=1081

TV-Programm. (2013). *RTL*. Retrieved from http://www.rtl.de/fernsehprogramm. Last accessed 13 Mar 2013.

Waisbord, S. (2004). McTV: Understanding the global popularity of television formats. *Television & New Media, 5*(4), 359–383.

What We Do. (2012). *Victory Television*. Retrieved from http://www.victorytelevision.tv/shows/

Who We Are. (2013). *Endemol UK Ltd*. Retrieved from http://www.endemol.com/who-we-are/

Geographies of the News

8

Mike Gasher

Introduction: Journalism as a Practice of Cartography

This chapter considers journalism as a spatial practice and posits journalists as cartographers, as map-makers, as symbolic workers who forge geographies of news. That is, journalists, through their news coverage, map the "news world," selecting from a constellation of current affairs which events, issues, peoples and places warrant their audiences' attention. The first map journalists sketch is the map of the community their news organization proposes to serve. This community is a construction, comprising the news organization's principal audience and advertising markets, the spatial parameters of its distribution network, and the political, economic, social and cultural institutions within that space. This "community of journalism" (Nord 2001) or "imagined community" (Anderson 1989) becomes the place the news organization seeks to cover with its editorial and advertising content and the vantage point from which it aims to view the rest of the world. Whether we

The author is the principal investigator in the Geography of News Project, based in the Department of Journalism at Concordia University in Montreal. The project has received funding from Concordia University (2000–2003), le Fonds québécois de recherche sur la société et la culture (2002–2005) and the Social Sciences and Humanities Research Council of Canada (2004–2007, 2008–2011, 2011–2014).

M. Gasher (✉)
Department of Journalism, Concordia University in Montreal, Montreal, QC, Canada
e-mail: gashmj8590@yahoo.ca

S.P. Mains et al. (eds.), *Mediated Geographies and Geographies of Media*,
DOI 10.1007/978-94-017-9969-0_8

are talking about a community weekly newspaper, a metropolitan daily, a local radio station, or a national television network, news organizations construct a 'here' and an 'us'.[1]

Like all forms of communication, journalism is an interlocutory act, and what is considered newsworthy is defined by what is deemed by journalists to be important, relevant, interesting and current to a particular, implied audience (see Nielsen 2008, 2009). A state election in New South Wales does not have the same news value in Sydney, Nova Scotia as it does in Sydney, Australia. Thus, through the news packages they compile, journalists sketch out the boundaries of their community and make assertions about its core values, record its debates over shifting values, identify the key components of its political, economic and cultural infrastructure, describe its constituents, position this place with respect to its neighbours, highlight other regions with which its constituents have important political, economic and cultural ties, and relegate to the margins great swaths of the rest of the world. This map-making exercise produces centres and margins, peoples and places within the news world's purview and, of course, other peoples and places beyond that news world's boundaries. Not everyone makes it on the news map.

This conceptualization of the relationship between journalism and news audiences takes its cue from the historical work of Benedict Anderson (1989), who documented the role of media – initially eighteenth-century novels and newspapers – in creating "imagined communities" (p. 30). Anderson argues that "all communities larger than primordial villages of face-to-face contact (and perhaps even these) are imagined" (p. 15). Novels produced a "sociological landscape" perceived by "omniscient readers" who made the connections between settings and characters (pp. 30–31). Newspaper stories were brought together on the page by two factors: their "calendrical coincidence" and "the relationship between the newspaper ... and the market" (pp. 37–38). This market consists of two elements: a product or service and a geographic locale where potential buyers can be reached (see Picard 1989, p. 19). If the production of the newspaper – the inclusion, juxtaposition and framing of stories – is informed by who and where journalists imagine their audience to be, Anderson describes its consumption as a "mass ceremony: the almost precisely simultaneous consumption ('imagining') of the newspaper-as-fiction" (p. 39). "At the same time, the newspaper reader, observing exact replicas of

[1]Digitization, of course, expands the range and increases the velocity of all news coverage and thus our understanding of how and where news stories circulate. This is putting an end to national communication systems (see Carey 1998) and may lead to the "cosmopolitan vision" of Uhlrich Beck (2008) and the "global journalism" imagined by Peter Berglez (2013). Digitization is clearly having some effect on how journalists imagine their audience, but news organizations have been very slow to exploit this potential. Considerable empirical research in the field of news-flow studies points to a news world that remains highly circumscribed, reinforcing the conventional news value of proximity – physical, cultural or emotional closeness – as a strongly determinant factor in producing news organizations' news worlds (Gasher 2007, 2009; Gasher and Klein 2008; Gasher and Gabriele 2004; Wu 2000, 2003). While select news organizations are operating internationally and the internet makes almost all news media available globally, news stories still tend to speak from a particular geographical/political/economic/cultural perspective.

his own paper being consumed by his subway, barbershop, or residential neighbours, is continually reassured that the imagined world is visibly rooted in everyday life" (pp. 39–40).

Numerous scholars have echoed Anderson's observation about the media's role in producing audience communities or publics. David Paul Nord (2001), in a study of the historical relationship between two Chicago newspapers and their readers, insists: "Communities are built, maintained, and wrecked in communication" (p. 2). He describes the Chicago *Daily News* of the late nineteenth century as the first "thoroughly urban" daily, "the first to articulate a vision of public community." Nord defines public community as "a kind of association founded upon communitarian notions of interdependence and identity, of sentiment and sympathy, yet powered by formal organizations and activist governments and guided by the new agencies of mass communication" (pp. 108–109). These newspapers

> provided their audience with a limited, organized, common frame of reference, so that diverse city dwellers could communicate with each other – communicate in the sense that they could think about the same things at the same time and share a vision of social reality. These newspapers saw in the fragmenting forces of urbanization the germ of public community (p. 111).

Communication theorist John Hartley (2008) has described journalism as "the most important textual system in the world," given its daily assertion of objective truths, its production of audiences as publics, and its symbiotic relationship with society's central political, economic and social institutions (p. 312). Journalism does more than merely describe or report on current events; it is a constructive practice, contributing to the formation of communities, publics and audiences. These aggregations do not simply precede the media through which they are represented, but media, in fact, play an important role in their constitution and in their identity. As Herbert Gans (2004) argues, journalists "help impose unity on what is otherwise a congeries of individuals and groups acting inside a set of geographical and political boundaries" (p. 298). In a related vein, journalism scholar James Carey perceives journalism as "worldmaking" and his theory of communication is rooted in the idea that "a medium implies and constitutes a world" (cited in Rosen 1997, p. 196). The news has a "positioning effect" on audiences, Carey argues. "We first produce the world by symbolic work and then take up residence in the world we have produced" (Carey 1989, p. 30).

This production of the news world is governed in part, of course, by the business plans, the marketing strategies and the technical capacities of news organizations working within available distribution networks. News organizations, that is, target specific markets, markets with both geographic and demographic parameters. The news package they produce will present a particular rendering of the world, produced specifically for its target market. The audience, though, is always more than a market. Even if the permeability of digital networks may change this, news audiences are imagined to share a physical locale and a corresponding civic identity, as members of a particular polity whose affairs occupy a central place in news coverage.

But the "worldmaking" Carey describes is also governed discursively, and that is the principal concern here. News is defined, as noted above, by the pertinence of stories to an implied audience, and news judgement is the very subjective exercise of selecting which news items warrant coverage, how much and, most importantly for our purposes, what kind. What we think of as the news is a compilation of information-rich stories that its producers assert is what matters most to its audience at any given time. In addressing a particular audience community, then, the news plays a constitutive role in defining and demarcating the place that audience community occupies, and in situating that place in the news world.[2]

The geography of news, then, is the representational space that news organizations construct, the vantage point from which they report. This space is the predominant coverage area the news organization stakes out for itself, a space which its reportage depicts, describes, defines and positions, the perspective it adopts, the starting point for its "worldmaking." All news organizations mold their own geographies. Whether or not the boundaries of this coverage area were simple to draw and maintain in some bygone age when population centres were more clearly defined, separated from one another and largely self-contained, such mapping is a far more complicated prospect in the mobile and inter-connected world of today. Political boundaries do not neatly coincide with economic or cultural or even physical boundaries, and the constituent members of any one political community have ties to other communities constituted by social, economic, cultural, religious, racial and/or ethnic dimensions. What matters, what is relevant to people – what is *newsworthy* – is not confined to their immediate territorial domain. This means that news maps are constructed through business plans, marketing campaigns and, on a daily basis, the editorial judgements of editors and reporters.

Comparing News Reporting to Map-Making

News reporting and map-making are comparable activities. They do not mirror the world, but instead produce texts; they are representational practices, through which complex and multi-dimensional actualities are rendered discrete, with temporal and spatial borders. Journalists and cartographers construct these texts through symbolic systems; journalists primarily use words, but also employ illustrations, graphics and sometimes even maps, whereas cartographers resort primarily to graphics supported by various kinds of labels (e.g., titles, keys, captions). Both practices share an appeal to objectivity, relying on the fact-finding conventions and verification methods of their respective professions. Both journalism and cartography, in sum, orient their readers to the world, at a time when so much of the world their readers inhabit is experienced in symbolic form through media. As the cultural theorist Tony Bennett puts it: "[T]he power which the media derive from their reality-defining capability is

[2]Audience members have agency; they use media content for their own purposes and can respond to media messages in numerous ways, including critically and oppositionally.

attributable to the service they perform in making us the indirect witnesses to events of which we have no first-hand knowledge or experience" (1996, p. 296).

Because it is not possible for journalism or cartography to *re-produce* the world, or even specific elements of the world, they are activities by which the real world is *re-presented*, or depicted, described, highlighted, made available to us in textual form. There are two basic steps to this process: the selection of elements to represent and the transformation of those elements from their materiality into some form of language (Fowler 2007, p. 2). In other words, the textual practices of journalism and cartography involve a translation of the material features of actuality into words, illustrations, symbols, diagrams, etc.

Translation, as always, entails transformation. At the most fundamental level, both journalism and cartography project a spherical, three-dimensional, interconnected and intensely dynamic material world into partial pictures of some select aspect of actuality, taken from the perspective of their producer, for that producer's particular purpose. Even the most skilled, conscientious and ethical reporter or cartographer must make a series of judgement calls about how best to render a complex and chaotic material world into text and make it comprehensible for the audience.

News stories and maps, then, are necessarily selective and reductive. The very purpose of news articles and maps is, in fact, to highlight some aspect of the material world at the expense of other aspects, to focus our attention on a particular subject matter. They frame rather than mirror. A newspaper story, for example, may be only 800 words in length, occupying part of a larger newspaper page. Those 800 words constitute the frame within which the news event must be described and explained. Because not even the most gifted and insightful journalist can address every conceivable aspect of a news event in 800 words, the reporter must decide which are the most newsworthy aspects of this event and what else must be excluded. This will result in a particular depiction of the news event. Similarly, no map can feature every possible element of the material world, so particular aspects – population centres, major roads, mountains, subway stops, etc. – must be highlighted and other features left out. The geographer Geoff King (1996) argues:

> There is and can be no such thing as a purely objective map, one that simply reproduces a pre-existing reality. Choices always have to be made about what to represent and what to leave out. It is here that cartographic meaning is created. To be included on the map is to be granted the status of reality or importance. To be left off is to be denied (p. 18).

News stories and maps render complex and multi-faceted features and activities of the material world into simplified portrayals, using their own forms of shorthand.

Because story-telling and map-making involve such decisions, they are constructive activities. News stories and maps are constructed by people according to the rules and conventions of their respective fields, according to the editorial decisions of their producers, according to the audience they hope to reach, according to the immediate purpose the story or the map is intended to serve, and according to their individual abilities and proclivities (Wood 1992, 2010). This may seem obvious, but the point is to draw attention to the distinction between actuality and representations of it, to deconstruct the textual practices of journalism and cartography, "to break the assumed link between reality and representation" in the words of the historical geographer J.B. Harley (2001, p. 152).

Both journalism and cartography remain highly subjective practices, even if journalists and map-makers subscribe to a certain, even if naïve, notion of objectivity – seeking to represent the world accurately, independently and devoid of bias – and observe specific codes of ethics. Both practices require their producers to make choices. The use of language – any language – is a signifying practice, in that we assign particular words or symbols to stand for elements in the real world, and we always have a choice about which words to use. The acquisition of language is precisely this process of learning to associate particular signifiers – words, symbols, signs, gestures, sounds – with corresponding signifieds – the things, ideas or actions the signifiers conjure.

Because language is always value-laden, these representational choices produce meaning, forging our "mental maps" of the world. The cultural theorist Stuart Hall (2013) defines representation as "the production of the meaning of the concepts in our minds through language" (p. 3). Linguistics scholar Roger Fowler (2007) emphasizes that language establishes relationships and categories that are not *natural*, "but which represent the interests, values and behaviours of human communities." He writes: "Language and other codes, most importantly language, have a cognitive role: they provide an organized mental representation of our experience" (p. 3).

All utterances of language have both denotative (literal) and connotative (associative) meanings (see Barthes 1968). With the experience of seeing and hearing these various utterances, we learn what they mean, both denotatively and connotatively. And because utterances have no solitary and fixed meaning, each of us may understand them in slightly different ways. When, for example, we hear in the news of a demonstration taking place, we share the understanding at a denotative level that a group of people has assembled for some kind of protest. At a connotative level of meaning, however, that demonstration can take on a different hue, prompted by the language of the news report and/or by our own ideological leanings. The word 'demonstration' has meaning, then, but not always the same meaning. Fowler (2007) insists that "the very notion of 'representation' carries within it the qualification of representation *from a specific ideological point of view*" (p. 66).

Even conventional practices of journalism and cartography are value-laden. At the most basic level, to produce a news story or a map is to assert that the subject of the story or the map is worthy of an audience's attention, it is newsworthy or noteworthy. Further, both journalism and cartography are inherently ethnocentric. They speak to the audience, whether consciously or unconsciously, from a particular perspective or, quite literally, point of view. Lots of events occur in the world each day, but news assigns its subject matter gravity and relevance, a particular meaning to a specified news audience. Why is that event newsworthy to an imagined us? Sometimes the answer to this question is obvious and would prompt considerable consensus about its importance and its meaning. Often, though, this question is more contentious. Similarly, maps do not include every feature of the material world, but include some and exclude others, put some elements at the centre and some at the margins, some at the top and some at the bottom. Maps commonly adopt the perspective of their imagined audience, and come to naturalize this perspective through repetition (e.g., putting north at the top of the map). Maps create hierarchies, privileging those features at the centre and at the top of the map where the eye is drawn, marginalizing, literally, those features at the left, right and bottom edges of its frame. The blank spaces on a map tell us there is nothing there, or at least nothing of importance.

This notion of construction goes one important step further. Not only are our texts constructions in and of themselves, but those texts in turn construct the world for us as audiences. The practice of representation is central to what the sociologists Peter Berger and Thomas Luckmann (1967) mean by "the social construction of reality." We occupy a material world, but our knowledge of it, our perception of it, our access to it, is socially constructed. Even what we come to think of as our direct experience of the world is shaped by our beliefs, values, prejudices, interests, expectations, experiences, history as we have come to understand it, what we perceive as normal and not normal, proper and improper. Morley and Robins (1995) maintain: "We are all largely dependent on the media for our images of non-local people, places and events, and the further the 'event' from our own direct experience, the more we depend on media images for the totality of our knowledge" (p. 133). Berger and Luckmann argue that "knowledge must always be knowledge from a certain position" (p. 10). Speaking to this point, Fowler (2007) argues: "A socially constructed model of the world is projected on to the objects of perception and cognition, so that essentially the things we see and think about are constructed according to a scheme of values, not entities directly perceived" (p. 92).

As noted above, the production of any text is an interlocutory act; a text is not simply put out there, but is proffered to an audience, intended to address an imagined group of readers with a particular purpose in mind: to inform, to convince, to instruct, to assist, to provoke. It is an invitation to people to adopt the subject position of reader, to form themselves into a readership community with the aim of consuming the text. These textual renderings of the world contain numerous rhetorical devices, which both constitute their specific form of audience address and situate audience members vis-à-vis their particular subject matter.

Journalism and cartography are unavoidably rhetorical, gaining much of their rhetorical power from their claims to objectivity. Journalists and map-makers make choices about theme (what the story/map is about), scale (how to bound the story/map), labelling (what to name, how to name), detail (what to include, what to exclude) and relations (how this story/map is connected to other stories/maps). Both news stories and maps affirm existence and significance, signalling that some things, some people, some places are worthy of our notice, and others are not.

Mediating Community

News is presented as a compilation of stories, and as with all stories, news stories have a setting in a specific time and place, a clearly-identified cast of characters, a narrative trajectory pulling these ingredients together and a vantage point from which the story is told. As *news* stories, though, their factuality lends them a certain authority and they have a necessary requirement to make audiences care. By extension, the people, places and events that don't make it into the news are, by definition, rendered unimportant, uninteresting, irrelevant, not worth the audiences' notice.

The communities that news organizations forge have both external and internal boundary markers demarcating spheres of activity and interest. Journalists do not provide blanket coverage of their community, because not everything that happens within that community could be considered news. Instead they cover what they deem to be its most newsworthy people (political and business leaders, first responders, athletes, artists), places (city hall, courts, schools, businesses, clubs, commercial districts), and events (meetings, press conferences, shows, festivals, demonstrations, crimes, accidents). Even though journalists like to think they simply provide a mirror-like reflection of their community to news audiences, they in fact, and necessarily, exercise news judgement to highlight those aspects determined to be of most interest and most import to audiences. Over time, consistent patterns in journalists' representation of their community and its relationship to the surrounding world come to give definition to community, to give this place and its people a particular identity, to explain its links to the rest of the world. The question to be considered here is how that highlighting works, how journalists pick and choose what is newsworthy, what patterns of inclusion and exclusion their editorial decisions produce, and how this matters.

The boundaries of any news organization's coverage area are shaped by a number of factors, but in ways particular to the community being served and particular to the goals – journalistic and economic – of the news organization. It is here that I want to make a distinction between news geographies and markets. While, clearly, both audience and advertising markets are key factors in shaping coverage areas, the economics of news production and distribution does not explain everything. News coverage extends beyond the immediate territory of the advertising and readership markets. Audiences are not interested only in what occurs within their own communities, but in current events elsewhere as well. How journalists determine the news value of such peoples and places cannot be reduced to a simple commercial calculation, but speaks as well to journalists' image of themselves as information-providers representing – and thus interpreting – the needs and interests of the public they seek to serve.

My interest in the geography of news is prompted by three contemporary factors which further complicate any news organization's occupation of a defined social space. The first of these is globalization, understood here not simply as an economic phenomenon, but in the fuller sense of the intensified global circulation of people, goods and services, ideas, investment capital, symbols, weather patterns, environmental degradation and disease. We may dwell in specific places, but the lives we live are more globally inter-connected than in any previous historical period. Globalization has thus dramatically broadened our horizons.[3] Ulrich Beck (2008), in fact, argues that globalization means we need to move beyond our "national outlook" and adopt a "cosmopolitan vision," replacing "the either/or logic with the both/and logic of inclusive differentiation" (pp. 4–5).

With that, it can be argued that our news interests, too, have become more extroverted; what matters to us, what is relevant to us, is not confined to our immediate locale. The boundaries of our working and social worlds have expanded, governing where and how we work and who we work for, where we travel, who are neighbours are, where our friends and family are, where we shop, what we shop for and how we spend our leisure time. Foreign people, foreign places, foreign languages, foreign wars, foreign news are not so foreign any more.

The second, inter-related factor is the digitization of news and its circulation via globally-connected networks of computers. Whether we get our news from web sites, social media, e-readers, satellite television, satellite radio and/or from the newspaper delivered to our doorstep, digitization goes hand in hand with globalization, enabling the rapid exchange of every kind of communication between people, businesses, governments and other types of organization, no matter their location. Digitization has changed how journalists produce and disseminate news and how we consume news.

Digitization means we are no longer beholden to our local newspaper, radio or television station. If we still get at least some of our local news from these traditional sources, it has become commonplace for news audiences to venture farther afield for their news diet. News audiences also seek out portals and blogs that either have no explicit geographical boundaries or determine their boundaries based on topic specialization: the sports world, the arts world, the business world. In every case, the reader/listener/viewer is oriented to the world in a particular way.

The third factor is commercialization, or hyper-commercialization, the increasing tendency by the owners of news organizations to perceive journalism as simply another branch of commercial enterprise, to privilege the profit motive by defining news as a commodity and audiences as markets. If journalism, from the time of the earliest newspapers, has been organized predominantly as some form of commodity production, it has been characterized as well by a strong public-service ethos. In this sense, its business role was for a long time compatible with its sociopolitical role; a viable business could be built by providing the public with an array of news, information and analysis that would include the kind of information citizens of a healthy democracy require. Newspapers, magazines and, later, radio and television news and current-affairs programming included in their packages hard and soft

[3]Globalization, of course, is an intensely uneven phenomenon, expanding the world for some, increasing the isolation of others (see Bauman 1998; Massey 1991).

news, news from the political, business, sports and arts worlds, news of serious import and news to amuse and entertain, thus serving a range of publics.

Beginning in the second half of the nineteenth century, and increasingly during the twentieth, news production shifted from small business to big business, first with the development of regional and national newspaper chains and radio and television networks, then with the creation of cross-media enterprises, and finally with the absorption of these companies within larger, omnibus and multi-national corporations. These media properties are increasingly required to generate profits, completing the transition of news to commodity and of audience to market. As Dallas Smythe (1977, 1994) was first to note, the product of commercial media is not its content, but an audience to be sold to advertisers. This affects what kind of news we get, how stories are told, and who journalists seek to address (and who, thereby, is excluded from the audience), shaping news geographies to accommodate commercial interest; serving the public interest has become serving what interests the public (see Hamilton 2004).

News geographies are in constant flux. The news media have altered their spatial parameters throughout the history of journalism, in periods when social, cultural, political, economic and technological changes encouraged and enabled such expansion. The circulation networks of newspapers, for example, grew with the introduction of train service, automobility, satellite printing and, in digital form, the internet. Local radio and television stations became part of regional and national networks before broadcasting ultimately went global via satellite and internet distribution. If our news consumption habits were for most of the twentieth century governed by the physical circulation of hard-copy newspapers and magazines and by over-the-air broadcast signals, circulation areas that corresponded roughly to political boundaries, today our news consumption is governed by whatever interests us. Journalists are thus saddled with an increased responsibility for understanding, satisfying, stimulating and shaping audience interests within newly and variously configured communities.

Journalism scholars have identified a number of news values, criteria by which journalists determine the newsworthiness of a particular person, place, institution or event. These determinants include: timeliness (events that are immediate or recent); impact (events that affect many people); prominence (events involving well-known people, places or institutions); conflict (events pitting two sides against one another); peculiarity (events that deviate from the everyday); currency (long-simmering events that re-emerge as objects of attention); and proximity (events that are geographically, culturally, or "emotionally close" to the audience) (Mencher 2000, pp. 68–76).

In the current news environment, the news value of proximity has taken on added significance. The intensified mobility of our globalized society brings us physically, culturally and, presumably, emotionally closer to the peoples and places of the world, whether we feel at times excited and at other times threatened by this redrawing of our frontiers. Journalists, whether they work for international news agencies or local media, play a key role in sketching and patrolling these boundaries, determining on our behalf what peoples, places and events we feel close to, and the bases and nature of that proximity. Such news judgement provides an outline of the news world, revealing a range of assumptions on the part of journalists about what matters, and doesn't matter, to their community, as well as about *who* matters and doesn't matter (see Gasher and Gabriele 2004).

The Maps of the News

News organizations produce at least four kinds of maps. As discussed above, the first of these representational spaces is the micro-scale map of 'here,' the place, the community – the 'us' – the news organization intends to serve. This map is drawn gradually over time as the community's key institutions and political, economic and cultural leaders are identified, its constituents described, its history told, its central beliefs and values revealed, and the boundaries which distinguish 'here' from 'there' and 'us' from 'them' come to be drawn. This sense of community is reinforced by the advertisements featured and by the regular consumption by audiences of its news package. Content supply and audience demand constitute a symbiosis.

The second kind of map produced by news organizations is the macro-scale map of the world as seen from 'here.' This map, too, is drawn over time, by the inclusion and exclusion of news stories from various parts of the world, and from cities and regions beyond the borders of our community. This map of the news world only loosely corresponds to the map of the material world because news organizations cover some places much more intensely – e.g., close neighbours, political and military allies, trading partners, tourist destinations – than other places, and leave some off the map entirely (see Wu 2000, 2003). It is the map of the peoples and places we hear about regularly, a map of our perceived and actual connections.

The third kind of map is drawn by extended coverage of one particular event or one particular people, and corresponds to an inset map, a more detailed highlight contained within the larger, overall map. It is usually rendered over a much shorter period of time, and situates 'us' in relation to 'them', 'here' in relation to 'there.' These maps can be produced by news stories from within our own community – a specific district of the city that becomes the subject of dispute – or from farther afield – e.g., a natural disaster in a remote part of the world. They provide us with important information about how these stories are newsworthy, how they connect us to the people, places and events they describe, on what basis the news value of proximity is asserted or explained – why we should care.

Finally, each news story draws its own map, occupies its own territory, draws its own connections, creates its own temporal-spatial setting. That setting may be confined to one locale with no external links to anywhere or anyone else, or its narrative may bring into the picture people and places in other cities or regions or countries, from the present or the past. How the story is set says something about its singularity or, conversely, its universality.

Why News Geographies Matter

We rely heavily on media of all kinds for our access to, and understanding of, the world. Over time, the representational conventions of books, films, TV programs, news reports and maps become our cognitive conventions, such that the distinctions between actuality and representation can be forgotten, that we take their renderings of the world as the way the world actually is. Harley (2001) writes: "Far from holding up a simple mirror of nature that is true or false, maps redescribe the world – like any other document – in terms of relations of power and of cultural practices, preferences, and priorities" (p. 35).

The news situates us in the world, it assigns meanings to people, places, events and institutions, it focuses our attention on some news events and newsmakers and consistently excludes others. This is significant because how we understand the world and our place in it informs how we act in the world. Representations of the world connect us to, or distance us from, other peoples and places. They offer definitions of people, places and events, and they provide us with a menu of viable actions we can take.

These issues are important because journalists perform a validating function through their news judgement. To cover an event, to consider it *newsworthy*, is to define that event as important and relevant, not only to those who may be directly affected, but also to people who may have a more tenuous connection to the event. Beyond informing, this is a matter of granting human dignity, about allowing peoples to share in each other's joy or sorrow, about paying attention to the welfare of other communities. It is also about making connections between peoples, explaining in concrete terms why we should care, why these events matter.

Related to this is the point that journalists, through their reportage, give meaning to community, identifying who belongs, what the community's boundaries are – physically and culturally (its goals, ideals, values, notions of proper conduct) – and how this community is connected to, or cut off from, others. This is a complicated yet critical task in heterogeneous – multicultural, multi-faith, multilingual, multi-ethnic and multiracial – societies, as any number of news stories about school dress codes or immigration laws can attest. Here, journalism is implicated in providing answers to the questions 'Who are we?' and, correspondingly, 'Who are we not?'

Journalists, too, implicate their audiences in news events in a number of ways. The manner in which an event is defined can lead to a range of responses, or no response at all. If, for example, a hurricane is defined exclusively as a natural disaster, the response may be confined to disaster relief. If, however, the damages caused by that hurricane can be related to political negligence or incompetence or poverty, which affect the way storm damages are apportioned, other responses become available. News coverage can mobilize individuals, governments, resources and/or particular solutions to problems, depending on how those problems are framed. A lack of news coverage, similarly, renders events unimportant – at least to us.

Drawing attention to the journalistic, political, economic, social, cultural and technological determinants that shape news geographies calls into question Marshall McLuhan's (1964) iconic and happy global village. Research to date suggests that in spite of what would seem to be the centripetal forces of globalization and digitization, the maps journalists draw are exacerbating rather then eradicating the distinction between the news world and the material world (Kariel and Rosenvall 1995; Chang 1998; Manthorpe 1998; Wu 2000, 2003; Gasher and Gabriele 2004; Gasher and Klein 2008; Gasher 2007, 2009). This has serious implications for how we perceive our world, how we understand our connections to other peoples and places, as well as for how we understand and identify with our own immediate communities, our understanding of who we are, our sense of, and the bases for, belonging.

References

Anderson, B. (1989). *Imagined communities: Reflections on the origin and spread of nationalism*. London: Verso.

Barthes, R. (1968). *The elements of semiology* (A. Lavers & C. Smith, Trans.). New York: Hill and Wang.

Bauman, Z. (1998). *Globalization: The human consequences*. New York: Columbia University Press.

Beck, U. (2008). *Cosmopolitan vision* (C. Cronin, Trans.). Cambridge: Polity Press.

Bennett, T. (1996). Media, 'reality', signification. In M. Gurevitch, T. Bennett, J. Curran, & J. Woollacott (Eds.), *Culture, society and the media* (pp. 287–308). London: Routledge.

Berger, P. L., & Luckmann, T. (1967). *The social construction of reality: A treatise in the sociology of knowledge*. Garden City: Anchor Books.

Berglez, P. (2013). *Global journalism: Theory and practice*. New York: Peter Lang.

Carey, J. W. (1989). *Communication and culture: Essays on media and society*. Boston: Unwin and Hyman.

Carey, J. W. (1998). The internet and the end of the national communication system: Uncertain predictions of an uncertain future. *Journalism and Mass Communication Quarterly, 75*(1), 28–34.

Chang, T.-K. (1998). All countries not created equal to be news: World system and international communication. *Communication Research, 25*(5), 528–563.

Fowler, R. (2007). *Language in the news: Discourse and ideology in the press*. London: Routledge.

Gans, H. (2004). *Deciding what's news: A study of CBS evening news, NBC nightly news, newsweek, and time*. Evanston: Northwestern University Press.

Gasher, M. (2007). The view from here: A news-flow study of the on-line editions of Canada's national newspapers. *Journalism Studies, 8*(2), 299–319.

Gasher, M. (2009). Mapping the on-line news world: A news-flow study of three U.S. Dailies. *Aether: The Journal of Media Geography, IV*(X), 102–116.

Gasher, M., & Gabriele, S. (2004). Increasing circulation? A comparative news-flow study of the Montreal Gazette's hard-copy and on-line editions. *Journalism Studies, 5*(3), 311–323.

Gasher, M., & Klein, R. (2008). Mapping the geography of on-line news. *Canadian Journal of Communication, 33*(2), 193–211.

Hall, S. (2013). The work of representation. In S. Hall, J. Evans, & S. Nixon (Eds.), *Representation* (2nd ed.). London: Sage Publications.

Hamilton, J. T. (2004). *All the news that's fit to sell: How the market transforms information into news*. Princeton: Princeton University Press.

Harley, J. B. (2001). In P. Laxton (Ed.), *The new nature of maps: Essays in the history of cartography*. Baltimore: The Johns Hopkins University Press.

Hartley, J. (2008). Journalism and popular culture. In K. Wahl-Jorgensen & T. Hanitzsch (Eds.), *The handbook of journalism studies* (pp. 310–325). New York: Routledge.

Kariel, H. G., & Rosenvall, L. A. (1995). *Places in the news: A study of news flows*. Ottawa: Carleton University Press.

King, G. (1996). *Mapping reality: An exploration of cultural cartographies*. New York: St. Martin's Press.

Manthorpe, J. (1998). The decline of foreign news in Canada's media. In D. Logan (Ed.), *Journalism in the new millennium* (pp. 131–140). Vancouver: Sing Tao School of Journalism.

Massey, D. (1991). A global sense of place. *Marxism Today, 35*, 24–29.

McLuhan, M. (1964). *Understanding media: The extensions of man*. New York: McGraw-Hill.

Mencher. (2000). *News reporting and writing* (8th ed.). New York: McGraw-Hill.

Morley, D., & Robins, K. (1995). *Spaces of identity: Global media, electronic landscapes and cultural boundaries*. London: Routledge.

Nielsen, G. M. (2008). Conditional hospitality: Framing poverty in the Montreal Press. *Canadian Journal of Communication, 33*(4), 605–621.

Nielsen, G. M. (2009). Framing immigration in the New York times. *Aether: The Journal of Media Geography, IV*(X), 37–57.

Nord, D.P. (2001). *Communities of journalism: A history of American newspapers and their readers*. Urbana: University of Illinois Press.

Picard, R. G. (1989). *Media economics: Concepts and issues*. Newbury Park: Sage Publications.

Rosen, J. (1997). Introduction/'We'll Have That Conversation': Journalism and democracy in the thought of James W. Carey. In E. S. Munson & C. A. Warren (Eds.), *James Carey: A critical reader* (pp. 191–206). Minneapolis: University of Minnesota Press.

Smythe, D. (1977). Communications: Blindspot of western Marxism. *Canadian Journal of Political and Social Theory, 1*, 1–27.

Smythe, D. (1994). In T. Guback (Ed.), *Counterclockwise: Perspectives on communication*. Boulder: Westview Press.

Wood, D. (1992). *The power of maps*. New York: The Guilford Press.

Wood, D. (2010). *Rethinking the power of maps*. New York: The Guilford Press.

Wu, H. D. (2000). Systemic determinants of international news coverage: A comparison of 38 countries. *Journal of Communication, 50*(2), 110–130.

Wu, H. D. (2003). Homogeneity around the world? Comparing the systemic determinants of international news flow between developed and developing countries. *Gazette, 65*(1), 9–24.

9 Remember the Alamo: A Place of Cinematic Experience

Leo Zonn

My grandfather was a vaudevillian who was often part of a theater troupe. He eventually made the transition to film where he had bit parts with a few speaking lines to a series of larger roles in 'B' Westerns. Among his reminiscences of the early days of theater are the stories of the many ways the audiences became involved with the event on stage from heckling to answering questions to cheering the inevitable outcome or punch line, and in fact my grandparents were married on the stage right after a play in which they had roles, with the audience present. It was 1918. Audience participation in early Twentieth Century stage and then film in the U.S. was common, but with modern cinema came a demand for the privacy of experiential space, in essence a cocoon. The sense of community in the U.S. movie theater today is based primarily on the fact that *you* and *they* are *here* to see the same film, but that is where the sharing stops. We wish to remain anonymous and private when we go to the movies and we expect others to do the same, while our engagement with the screen tends to be passive, at least outwardly.

An alternative to this model arose in the 1950s and 1960s when movies could be watched at home on network television and the trend increased dramatically with the introduction and rapid adoption of the VCR in the U.S. home in the mid-to-late 1970s, which was followed by the DVD, streaming cinema, and the iPad mini as examples of the digital revolution. The audience could watch the film within their own space, enjoying the luxuries of food, drink, and the right to openly participate with the film while being removed from any larger community; who cares if your next door neighbor is watching the same film? But the motion picture theater is hardly passé, never disappeared, and now functions as a complement to the home theater and other cinematic sites. Contrary to many predictions, the public theater remains an important place of cinematic exhibition, although the anonymous and

L. Zonn (✉)
Department of Geography and the Environment, University of Texas at Austin, Austin, TX, USA
e-mail: zonn@austin.utexas.edu

S.P. Mains et al. (eds.), *Mediated Geographies and Geographies of Media*,
DOI 10.1007/978-94-017-9969-0_9

passive viewer remains the expected standard within that site. But in an age of increased anomie not everyone wants to be removed from one another, and the filmgoer is no exception. Cinema-going to a select community means more than the experience of sitting in front of a screen in a near vegetative state for 90 min more, then getting up, stretching ones legs, and leaving with no acknowledgement of those who just had the same encounter.

A few theater entrepreneurs have responded to this potential market by creating a model of public film experience that includes food and alcohol delivered to each seat, which has necessarily included the physical restructuring of facilities to accommodate the service. This concept—known as the First Run Cinema Eatery (FRCE)—has slowly emerged throughout the country but still remains somewhat of a novelty when compared to the standard walk-in theater. Some of these theaters have taken the experience a step further by adding filmic programs and events. The result is a unique site of cinematic exhibition, which in essence is a place comprised of the theater, the films, a program that contextualizes and frames the films, and a discriminating audience. A nationally-recognized leader in this place construction is the Alamo Drafthouse of Austin, Texas.

The focus of this chapter is twofold. First, it is upon the Alamo and the ways in which its novel constructions of these particular programs are presented to a self-selected community that seeks experiences that are not provided by standard home or public theater settings. I conclude that this desire to participate in a more community-based cinematic experience reflects an ironic discursive mode as envisioned by Linda Hutcheon (1994) in which the audience seeks a momentary identity that distinguishes itself for even the brief moment of a theater visit. Second, the chapter presents a call for further study of sites of cinematic exhibition as mediated geographies, and as such is part of a larger project in which the many-faceted natures of the cinematic experience are explored through a geographic lens, especially within place-based notions of sites of exhibition.

Framework

I begin by assuming a broad-based cinematic network comprised of elements that are connected by the flows of people, ideas, and things. The elements that comprise the heart of the network—technology, the film and its narrative, the many people who construct the film, the cinema industry, society/culture, sites of exhibition, and audience—should be seen as primary, but not sole, points of reference. The pieces of the larger puzzle and the resulting linkages and flows that connect them can be described as simultaneously being influences and subjects, then, without the constraints of determinism. Here I draw on the cinematic work of Dixon and Zonn (2004), who consider the writings of Bruno Latour (1997) and Actor Network Theory (ANT) only in a broad and relational sense, and so they are interested in:

> ...how people and things are placed in relation to one another, such that issues of inclusion and exclusion, hegemony and dependence, remain at the forefront of analysis. All phenomena in such a network... whether they be human or nonhuman, can be considered

> powerful in the sense that they operate as part of a collective to allow for a particular event to occur or entity to perform; in this sense all such phenomena are of explanatory significance. (p. 246)

The nature of engagement among and between puzzle pieces varies dramatically by scale and yet all are ultimately related to one another. Imagine the role of the cinematographer in the larger cinematic system, as one of many examples. This person and associated contributions can be envisioned within the larger network on the one hand, and as a discrete network with its own but never totally independent set of connected elements on the other. In the same way, the study of cinematic sites of exhibition includes the site itself as the initial and even primary consideration, but it must be contextualized within the frames of many influencing elements found at a variety of scales.

What, then, are examples of these sites of exhibition, these cinematic places? They may be the home theater, the walk-in, the drive-in, the mobile cinema that graces the rural and small-town landscapes from Scotland to India to Poland, the horse-drawn traveling picture show of early twentieth century Australia, the sheet on a wall in Lagos where 25 cents brings you Nollywood, the side of a building with boats floating nearby as in *Cinema Paradiso*, the desktop computer, and the DVD player being watched by a child on a flight from Los Angeles to New York (Corbett 2001; Klinger 2006; Zonn 2007). Each of these many possibilities can be described as a unique pattern, a distinctive scrambling of puzzle pieces that collectively constitute a cinematic place and that can be envisioned within a set of interconnections at a variety of scales. Focus here is upon a nexus of filmic elements and flows that help define the cinematic experience associated with a special place, the Alamo Drafthouse.

The study of cinema has long been dominated by a focus on the metaphor of film as text. Borrowed from literary traditions and cultural studies, textual analysis has spawned an extensive range of theory and forms of critique that enjoy a rich and intriguing history. Some of this work centers on a more direct audience-content engagement, usually under the rubric and nomenclature of 'spectatorship', and includes an array of valuable contributions focused on gendered and racial gazes, intertextuality, cultural politics, and issues of psychoanalysis, as primary examples. But this work is not of immediate value to this chapter, primarily because of its ultimate focus on the film's text. I turn instead to two smaller sets of literature that are concerned more directly with the site-specific film experience.

One group of studies is concerned with the intentions of going to the movie theater, addressing the question 'why do people go to the movies?' Jancovich et al. (2003) begin their fine work by considering "film consumption as an activity" (p. 3), while Corbett (1998) says "... we know relatively little about a central component of the filmic experience: movie watching as a social act" (p. 34). He speaks to different meanings within a community of "going to the cinema" and "how movie audiences historically have used the act of movie-watching in their everyday lives, how symbolically important the act was within their lives, or how it has contributed to forming, maintaining, and transforming their interpersonal relationships" (p. 34).

The excellent set of U.K. based studies by Hubbard (2002, 2003a, b) explore the barriers and trends of film-going in a variety of settings.

This research on film-going, much of it ethnographic, provides valuable insight into the social and cultural natures of the audience, but it usually gives less direct attention to the venue. This gap is covered to a great extent by the 'exhibition of film' literature where the concern shifts to the inclusion of site-specific technological and aesthetic frames as they engage audience experience. Given the interest in technological evolution, these works are often historical. Examples include the reader by Hark (2002), a cultural geography of old theaters by Marling (2001) that notes "architecture of escapism" (p. 12) as a cinematic force, and the fine book by Jones (2003) that addresses a range of architectural history and preservation issues, economic circumstances, cultural memory, and the natures of the audience as she describes the restoration of old movie houses of the South. Other notable works of this substantive literature set include Douglas Gomery's *Shared Pleasures* (1992), which is a history of American cinema in terms of the theater, the audience, technology, and prevailing economies and social trends, and the contributions by Robert Allen (1979, 1980, 1990), who noted that "we tend to talk of films being 'screened' as though the only thing going on in a movie theater were light being bounced off a reflective surface" (1990, p. 352). His work has focused on sites and experiences that range from the Nickelodeon to the theater of 1920s America, where:

> ... many viewers were not particularly interested in what was playing. They were attracted to the theater by the theater itself, with its sometimes bizarre architectural and design allusions to exotic cultures, its capacious public spaces, its air conditioning in the summer, and its auditorium, which may have been decorated to resemble the exterior of a Moorish palace at night—complete with the heavenly dome and twinkling stars. Regardless of what feature the theater chain had secured from the distributor that week, there was sure to be a newsreel, a comedy short, a programme of music by pit orchestra or on the mighty Wurlitzer, and in many theaters elaborate stage shows. (1990, pp. 352–353)

I turn now to a more contemporary site of exhibition called the Alamo Drafthouse. Consideration of the Alamo and its cinematic experiences must certainly assume a vast range of social and technological issues unique to the early Twenty First Century, but there will be some intriguing similarities between the Alamo model and the noted theaters of the silent era that distinguish both from the normative model of the contemporary U.S. movie theater. I approach the Alamo's distinctiveness, in terms of the noted patterns of 'people, ideas, and things' and suggest a means of engaging the subject matter that could prove fruitful for further studies.

Constructing the Alamo

The Alamo Drafthouse creates a unique experience for the filmgoer in physical and programmatic terms. All theaters provide special stadium seating with a sloping floor toward the screen, extra leg room, a table that fronts every viewer, a menu

Fig. 9.1 Author (on *far right*) waiting for 'ticket' at Alamo Drafthouse Slaughter Lane (Photo by Misha Zonn)

for food and alcohol that can be ordered and served during the film, black clad servers who try to be as inconspicuous as possible, and in some places a free-ranging and eclectic entryway and interior design with displays that vary according to the film events. The programs include full-length films, shorts, cartoons, and events that range from the mainstream to the *avant-garde* to the provocative to the bizarre. The theaters that define Austin's Alamo have collectively become a distinctive, well-known and even revered feature of the city's cultural landscape, and over the last few years the concept has been spread by an image, reputation, and a newly defined corporation to other points in Texas and beyond. The Alamo is certainly not the first of its kind, but its unique combination of qualities and its large-scale prospects make it an intriguing model for the industry. As early as 2005, The Alamo made the front cover of *Entertainment Weekly* with an accompanying article that called the theater 'movie-geek heaven' where "One of America's most fanatically unique moviegoing experiences' could be found" (Cruz and Kirschling 2005) (Fig. 9.1).

The origins of the Alamo are straightforward. The story begins with Tim and Kerrie League, who graduated from Rice University in the early 1990s and moved to Bakersfield, California because of his internship with Shell Oil. In 1994, the husband and wife, both just 24, joined with a financial partner to buy a small, old movie theater, the Tejon, and opened it as a movie art house, showing cult films, classics, and foreign films. The low budget nature of the theater required them to manage all aspects of the daily operations, from scheduling films and running the projector, to taking tickets and providing concessions. The experience eventually proved to be invaluable, but a variety of circumstances forced the Leagues to leave the enterprise and in 1996 they headed back to Texas, choosing Austin as the place to start over.

Only a year later, in 1997, they opened a theater in downtown Austin within a lively restaurant and entertainment district, and once again relied on the eclectic movie art house repertoire, but this time they added food and alcohol that was served to each seat upon demand and a variety of unique events. The University of Texas and its 50,000 students is relatively close, and Austin is a progressive city with a highly energetic entertainment scene, especially film and music. The Alamo was an instant success. They opened a second theater (Alamo Village) in 2001, a third (Alamo Lake Creek) in 2003, and a fourth (Alamo South Lamar) in the quirky and trendy South Austin in 2005. In 2007 they moved the original Alamo several blocks away to the tourist and music oriented Sixth Street district, where they renovated the old Ritz Theatre, calling the new venue the Alamo Drafthouse The Ritz, which has two screens and remains the quintessential Alamo. The Alamo South Lamar closed for renovations in February of 2013 and reopened in 2015.

The Leagues had expressed their dislike of operating a large scale business, and so they sold their franchise rights in 2004, keeping only the four initial theaters which they intended to operate themselves. The buyout group, Alamo Drafthouse Cinemas, Ltd., expanded the franchise with new theaters in several venues in Texas and one in Virginia. Tim League returned in 2010 as CEO of a corporation that combined the two temporarily separate units and has taken a strong leadership role in the company's further expansion. There are now four theaters in Austin, with a fifth under renovation, and there are two in Houston, three in San Antonio, one each in Colorado, Missouri, and Virginia. Four more are planned or are under construction in Texas, with others in northern Virginia, California, Michigan and three in New York City (Brooklyn, Yonkers, Manhattan).

Programming the Alamo

All Alamo theaters except the Ritz include current and mainstream cinema, classic, independent and international films, and the staging of filmic and even sports events, including English Premiere League soccer. Mainstream films clearly predominate at these theaters, although all have many or even most of the unconventional programs discussed below. The Ritz relies exclusively on alternative programs and remains the philosophical heart of the Alamo Drafthouse enterprise and concept.

The cinematic experiences of the alternative programming described here are likely to reflect a greater sense of audience engagement than with the mainstream case, if nothing else because the viewers are apt to be more narrowly defined in social, political, and sometimes demographic terms, at least partially because of the select nature of the films. Imagine the audience and experiences for *The Battle of*

Algiers, where the norms of behavior may still apply, although the communal bonds may be stronger. Then proceed to the bizarre 'Spike and Mike's Twisted Festival of Animation' and the outlandish 'Live Freaky, Die Freaky!', a punk rock musical of Charles Manson's crimes, where the audiences are likely to tend toward a greater homogeneity of community and barriers of non-engagement may begin to crumble.

An example of the next phase away from a normative non-engagement is probably the classic *Rocky Horror Picture Show*, which is shown on occasion at the Alamo and has been entertaining audiences across America since 1975 by enticing and evoking specific and collective audience responses at important junctures of the film. In some ways, this particular filmic experience is a prototype for many of the creative programmatic events at the contemporary Alamo, whereby audience, engagement, spectacle, and community become increasingly integrated. As suggested in the earlier metaphor of the puzzle, the piece known as the film becomes less dominant as the center of the total experience. This chapter now turns to several popular examples of these events at the Alamo, each with a unique, interactive, and community-based sensibility. Most of the descriptions rely on the promotional information provided by the Alamo websites.

Sing-Along

These popular events encourage the audience to sing with a musical or full-length collection of videos, while associated interactive activities and props are often included. The Alamo defines the Sing-Along events as:

> . . . full-on theatrical dance parties: taking the best parts of going to a concert and mixing it together with the best parts of going to a movie theater to create something completely new (and overwhelmingly awesome). You never feel out of place because, even if you're not on your feet or dancing, they lovingly project all the lyrics on the screen so you can wail along with a theater full of people who love these songs just as much as you do, even if you don't know every single word. http://drafthouse.com/series/action_pack/

One of the more popular examples has been the 'Love Bites: the 1980s Power Ballads Sing-Along Extravaganza', where you can join the audience by:

> . . . singing and pumping your fist in the air to a soul shaking collection of both pro-love and love sucks power ballad music video hits from the '80s and '90s . . . We'll hold lighters in the air and sway, we'll pound our fists at the sky in defiance of those who would dare not love us, and we'll do it all with teased hair and animal print tights on. It's the perfect dinner date or post-dinner date or even no date activity! http://drafthouse.com/movies/action_pack_love_bites_sing-along/austin

Other popular Sing-Along examples have included *Grease*, *Buffy the Vampire Slayer*, *Moulin Rouge*, the 'Justin Timberlake Sing-Along', and the 'Queen Sing-Along'.

Quote-Along

These events call for full audience participation, by taking:

> . . . your favorite movies and add in karaoke-like subtitles for the best lines so that everyone can yell out, "I know you are, but what am I?" right on cue with Pee Wee . . . And so even though the Alamo is usually very strict about kicking out anyone who talks during the movie, at The Action Pack's Quote-Along series we're more likely to kick you out if you *don't* talk during the film. http://drafthouse.com/series/quote_alongs/houston

One of the more popular examples is *The Big Lebowski*, with its direct community appeal:

> We hope all of you Little Lebowski Urban Achievers will come out and enjoy yourselves, so you can die with a smile on your face, without feelin' like the good Lord gypped you . . . Enjoy discounted, $5 White Russians during the show!! http://drafthouse.com/movies/biglebowskiquotealong/austin

Other popular examples are *Elf*, *Goonies*, *The Princess Bride*, *Robin Hood*: *Men in Tights*, and the classic *Monty Python and the Holy Grail* (Fig. 9.2).

Fig. 9.2 Author, a White Russian, thinking about White Russians. Alamo Drafthouse Slaughter Lane (Photo by Misha Zonn)

Heckle-Vision

These events bring a previously unused technology into the film experience equation as yet another small but significant audience piece of the puzzle whereby the spectators rely directly on their cell phones as a means of engagement. In Heckle-Vision, the Alamo chooses:

> …some hilariously horrible movies (but movies we can't help but love anyway) and through the magic of MuVChat technology we let you text your heckles silently from your seat as they then magically appear up on the screen alongside the main action. ANYTHING goes in this all out assault on crappy cinema and you can bask in the glory of your textual genius or recoup from a whiff as you prepare your next text. When you don't have anything to contribute to the heckle conversation you can sit back and enjoy other peoples' snarky comments. So charge up those cell phones and prepare your wit because it's heckling time! http://drafthouse.com/series/hecklevision/austin

A popular example has been "From the minds of the Spice Girls (seriously: they got a writing credit for the idea)…*Spice World* had every excuse to be a bad movie—except it wasn't. No wait, it was just a terrible, awful film. But it's so bad, it's INSANELY ENJOYABLE! Which is why it creates the perfect recipe for….Hecklevision." http://drafthouse.com/movies/hecklevision_spice_world/austin. Other examples include *Indiana Jones and the Kingdom of the Crystal Skull*, *Jingle All the Way*, *Burlesque*, *Johnny Mnemonic*, and *Anaconda*.

Food and Film Events

These very popular events take advantage of films in which food consumption is integrated with the narrative by providing relevant cuisine that is served as the audience watches the screen. The chefs are well-trained and so the price is often not cheap, usually more than $50. A very popular example is the annual Lord of the Rings Trilogy Feast (now held exclusively at the Austin Ritz) in which:

> we snuggle up together in the theater and brave our way through the ENTIRE extended-cut versions of the LORD OF THE RINGS TRILOGY—nearly 12 hours—all while devouring food and imbibing drink inspired by Tolkien's Middle Earth! Each of the seven courses is served during Hobbit meal times (which is pretty much ALL the time)! When we reach the end and the King (spoiler alert!) ascends to his rightful throne, you too will be ready to return to the Shire and be greeted as a hero by your loved ones who didn't think you would make it all the way…and then pass out in a blissful food- and film-induced coma. http://drafthouse.com/movies/lord_of_the_rings_trilogy_hobbit_feast

The cost is around $100, more if wine is included. Other popular Feasts focus on *Gone with the Wind*, *Like Water for Chocolate*, *Oh Brother Where Art Thou*, and *Lawrence of Arabia*.

The Master Pancake Theater

Held every weekend in Austin's Alamo Ritz, this event has been popular for over a decade, and now has its own Facebook page. Originally called the Sinus Show, it is comprised of several comedians who sit in the front row with their microphones and give running commentary on popular but critically panned films and so it "is part of the Alamo tradition of pairing bad movies with live comedy (and beer)! It's smart, fast, and bust-a-gut funny." Recent examples include *The Hunger Games*, *Twilight*, *Spider-Man*, and *Robocop*. In addition there is the:

> 'Choose Your Own Pancake'...where you, the viewer, get to pick a movie for Master Pancake to mock, and then we make it up right there on the spot! Just bring the DVD of your choice, Master Pancake will narrow it down to 10 favorites, then the audience will vote on which of those they want to see mocked. Be prepared to give a short speech in defense of your movie. The audience has final say... CYOP is always a fun night, with lots of audience participation... http://drafthouse.com/movies/choose_your_own_pancake/Austin

Videoke

The Videoke experience is "Kinda like Karaoke, but with acting," and relies on the individual's contribution to what is shown on the screen and with subsequent public interactions. The participant brings a clip from his/her favorite movie, and using subtitles on the monitor in front of them, acts or sings a scene. Dialogue may include material straight from the film or it may be original scripts, which range from the serious to the provocative to the absurd. The quality of participant ranges from local professionals to novices just having fun and, needless to say, alcohol is often involved. There are judges, cheering, booing, a final round, and prizes for the winner.

Open Screen Night

Closely related to the Videoke is the case of open screen night, where the audience actually determines the tenor of the film-going experience by bringing to the screen a film (VHS or dvd) of choice, with the only major rule being that it can last no longer than 8 min, usually with a minimum of two. The Alamo description of this event says:

> Inspired by open mic nights at your local coffee house, Alamo programmer Henri Mazza decided to put together the best bring-your-own-video night that's going on anywhere in the country. The concept is simple: we will show anything you bring with you. Anything at all. That's the whole point, see. You bring in a tape or a DVD, and then we all watch it together on the big screen. You can bring in an old Lego animation you made when you were 8, your thesis film from college, or a church youth group training video that you bought off of Ebay. Anything you'd like to show an audience... We have a rowdy audience, too, so generally after the two minutes is up, they'll boo and yell out for the gonger to do their bidding. The

final call, however, rests in the hands of the person with the mallot. That person can gong a movie that everyone loves if he so chooses, or he can let a movie run that everyone's booing. It's all up to the gonger. http://drafthouse.com/series/open_screen_night/austin

Celebrity Guest

In this case, the Alamo brings directors, stars and other filmmakers to the screening of one of their films. The guests conduct Q&A sessions and on occasion "lambaste the audience and perform impromptu song and dance numbers. In any case, we feel this greatly enhances the film experience, and gives everyone who attends an intimate look behind the scenes of the movies we love." Recent examples include the awarding of the Bad Ass Hall of Fame Award to Pam Grier at the opening of the Colorado Alamo, the inclusion of the actors who played children Mike TeeVee and Veruca Salt for the 'Ultimate Willy Wonka Party', and Crispin Glover, who performed "a one-hour dramatic narration of eight different profusely illustrated books. The images from the books are projected behind Mr. Glover during his performance." Quentin Tarantino has hosted his own film fest several times at the Austin theaters, while Austinite Robert Rodriguez has been a guest several times, including his appearance to celebrate the 20th anniversary of *El Mariachi*, his directorial debut.

Rolling Roadshow

Although this event does not take place at theaters it deserves brief mention here, because it is a moment when the filmic experience extends from the theater itself to other sites. In 2002 the theater created the Rolling Roadshow Tour whereby films are shown in surroundings that resemble those of the particular film being shown and are often presented at the actual locations where they were made. This event has been extremely popular for a decade and involves a concept of the filmic experience that includes a new technology set and venue that is beyond the scope of this paper but that deserves research attention in the near future.

Remembering the Alamo, Ironically

Allow me to return to the framework of Latour. This view is of a large puzzle—an assemblage—in which the parts are highly dynamic, interchangeable, and engaged, while the pieces may be human or even objects that have been constructed by individuals or larger communities. The larger shape in this case is the cinematic venue—a site of exhibition as the literature often calls it—that is a place certainly deserving of attention from researchers interested in mediated geographies, because it lies at the heart of the cinematic experience. The character and role of each part of the assemblage is likely to vary by design and each can readily shift its character by the moment, thus providing differential experiences over short periods of time. The primary example here, of course, is the introduction of simple but unique programs and advertising aimed at select audiences.

So, in broad descriptive strokes, we can begin with the building itself and its interior design, including the walls, seats, tables, lighting, and acoustics, and then can proceed to the technical elements of the site, such as the sound system, the quality of projection and the screen itself. Then there are the programs, food and drink being served to the table, and there is the audience, beginning with you as an individual at that moment, the immediate audience that comprises your filmic neighborhood, and then there are yet broader stories of the nature of the film industry, trends in popular culture, politics, and more, which harkens to the multi-scalar nature of the assemblage. And finally, of course, there is the film.

My broader contextualization of the filmic process is not meant to minimize the nature of the film itself, but to help understand that a total focus on the film and/or its spectators can easily become a reductionist argument. Stepping back to contextualize film into a mediated venue-based experience enhances the understanding of where film engages American culture in both cause and effect. This does not mean, however, that in a broader view the complete experience becomes a site or excuse for a textual analysis of its own, which would allow the same pitfalls of the current text-based studies of cinematic geography that have provided an impressive array of intriguing but somewhat idiosyncratic works that have resulted in limited results as a collective sensibility.

Geographers have produced substantive works, in fact, that extend well beyond a primary focus on the textual image, beginning with economic geographies of cinematic construction (Lukinbeal 2002, 2004, 2006; Scott 2005) and more integrative theoretical and methodological analyses (Lukinbeal 2012), but these works remain the exception. Our emphases tend to be on the text and often the visual, although Craine (2007) stresses this shortcoming when he says that we should not "…prioritize the visual over the aural or any other sort of sensory-based form of media" (p. 149) and Andrew (2002) emphasizes the need for a more eclectic frame when he notes that "neither the producer, the text, the apparatus, nor the viewer is stable enough to hold us within a universal theory of the film experience" (p. 162). I do not suggest that taking a contextualized view of cinematic sites of exhibition in the form described here will provide immediate answers to overarching questions that have been asked and even those that should, but I do think it opens the study of film as a mediated geography to broader possibilities.

I should note that this essay began with several a priori assumptions about theater behavior that were clearly U.S.-centric. It may seem anecdotal, but my experiences of being a member of a crowded theater watching a Bollywood tale from the balcony in Dehradun, India, *Indiana Jones and the Temple of Doom* in Penang, Malaysia, and *Hero* in Singapore, suggest that expected and tolerated behaviors vary dramatically beyond our somewhat limited realm. This story remains, then, one of normative behaviors and those that 'push the envelope' at the Alamo Drafthouse. This chapter is about a constructed and mediated geography in which, for a brief moment, the nature of the public sphere is shifted ever so slightly by the producer's design. Hansen (1991) writes that "... the emergence of cinema spectatorship is profoundly intertwined with the transformation of the public sphere" (p. 2), and so this seemingly modest shift of a few parts of the assemblage produces a considerably different experiential sense.

I recently attended the showing of Monsters University within the standard closed space of an Alamo Theater (http://drafthouse.com/austin/slaughter_lane) with a group of 'tweens for my granddaughter's birthday, and weeks later I sat in what may have been the exact same room for the aforementioned Big Lebowski Quote Along. The nature of the public space and its audience shifted dramatically in these two cases, certainly as a function of the film but also by the nature of the program. In the first, the audience was quiet and reserved, while in the second there was a communal participation and realized camaraderie, which certainly included a higher rate of alcohol consumption! The experiences were very different, while the room, lights, technology, and composite surroundings remained the same.

Hansen (1991) refers to a "... a collective, public form of reception shaped in the context of older traditions of performance and modes of exhibition" (p. 3), and uses the example of the entertainment factor in the showing of a film of the Corbett-Fitzsimmons fight in 1897 in New York City that included running commentary and intermittent vaudeville acts in addition to the fight itself (1990, 1991). The works of Allen (1979, 1980, 1990) and Gomery (1992) cited above, are among a large and fascinating literature set that describes the experiences of these multi-modal sites of filmic exhibition, and, along with Hansen emphasize the nature of the audience's composition. The programmatic nature of much of the Alamo's itinerary is more than vaguely reminiscent of these settings, if nothing else because the film in itself is no longer the total focus and the audience has a different form of engagement than what we normally experience. The technology has changed dramatically of course, but the seemingly small shift in the assemblage—program, food, perhaps drinks, commentary, and audience participation—provide a unique experience and call for a distinct audience, or clientele.

And so, ironically, I return to my grandfather for one final thought about site and audience. I wonder how he—a fascinating, eclectic, and open-minded character—would have enjoyed these actions of the Alamo that are bent on integrating the audience, site, and show into an experiential place in which a community becomes a feature of the performance; indeed, I believe that in his eyes it would be a 'show'. In this case, *you* and *they* are certainly *here* in an assemblage of 'people, ideas, and things', but instead of the normative experience of anomie, there is a demonstrated and self-conscious need for contact, even if it so often revolves around the seemingly absurd. But just who is this community? To paraphrase a student of mine who was speaking tongue in cheek but with some truth, the Alamo is adept at commodifying the 'hipster', among several vaguely-defined communities. Is the need for some form of social contact the sole reason for joining this club of vaguely like-minded film viewers, or is there something more?

Perhaps a part of the answer lies within the nature of irony, which Linda Hutcheon (1994) calls "the mode of the unsaid, the unheard, the unseen" (p. 9), this "strange mode of discourse where you say something you don't actually mean and expect people to understand not only what you actually do mean but also your attitude toward it" (p. 2). An important presumption of irony as seen by Hutcheon is that it should be viewed as a discursive strategy in which the intentions and attribution of meaning are embedded into communications within a community that help to reinforce the existence of that group, especially in distinction from others, and that should be considered "not as an isolated trope to be analyzed by formalist means but as a political issue, in the broadest sense of the word" (p. 2). To say the use of irony in the case of the programming and events of the Alamo is an expression of alterity or even subversion may be a stretch, but it is very possible that the Alamo has produced, and a select community—'hipster nation' as my student suggests—has bought, a product that at least partially reflects a common need to, quite simply, be different than the mainstream. If this is true, the content of much of the event material could be far less important that the act of presenting it. Regardless, it is clear that Tim and Kerri League have recognized a distinctive public need, and in incorporating it with other creative features of the film-going experience, have created a unique place that can be as engaging as my grandfather's vaudeville stage.

References

Allen, R. (1979). Motion picture exhibition in Manhattan, 1906–1912: Beyond the nickelodeon. *Cinema Journal, 18*(2), 2–15.

Allen, R. (1980). *Vaudeville and film, 1895–1915*. New York: Arno.

Allen, R. (1990). From exhibition to reception: Reflections on the audience in film history. *Screen, 31*(4), 347–356.

Andrew, D. (2002). Film and society: Public rituals and private space. In I. R. Hark (Ed.), *Exhibition, the film reader* (pp. 161–172). London/New York: Routledge.

Corbett, K. (1998). Empty seats: The missing history of movie-watching. *Journal of Film and Video, 50*(4), 34–48.

Corbett, K. (2001). The big picture: Theatrical moviegoing, digital television, and beyond the substitution effect. *Cinema Journal, 40*(2), 12–34.

Craine, J. (2007). The medium has a new message: Media and critical geography. *ACME: An International E-Journal for Critical Geographies, 6*(2), 147–152.

Cruz, G., & Kirschling, G. (2005, August 5). 10 theaters doing it right: We tell you about the movie houses that make watching films a dream. *Entertainment Weekly*. http://www.ew.com/ew/article/0,,1090060,00.html

Dixon, D., & Zonn, L. (2004). Film networks and the place(s) of technology. In S. Brunn, S. Cutter, & J. W. Harrington (Eds.), *Geography and technology* (pp. 243–266). Dordrecht: Kluwer Academic Publishers.

Gomery, D. (1992). *Shared pleasures: A history of movie exhibition in America*. London: British Film Institute.

Hansen, M. (1990). Adventures of Goldilocks: Spectatorship, consumerism, and public life. *Camera Obscura, 22*, 51–71.

Hansen, M. (1991). *Babel and Babylon: Spectatorship in American silent film*. Cambridge, MA: Harvard University Press.

Hark, I. R. (Ed.). (2002). *Exhibition: The film reader*. London: Routledge.

Hubbard, P. (2002). Screen-shifting: Consumption, 'riskless risks', and the changing geographies of cinema. *Environment and Planning A, 34*, 1239–1258.

Hubbard, P. (2003a). Fear and loathing at the multiplex: Everyday anxiety in the post- industrial city. *Capital & Class, 27*(2), 51–75.

Hubbard, P. (2003b). A good night out? Multiplex cinemas as sites of embodied leisure. *Leisure Studies, 22*(3), 255–272.

Hutcheon, L. (1994). *Irony's edge: The theory and politics of irony*. London: Routledge.

Jancovich, M., Faire, L., & Stubbings, S. (2003). *The place of audience: Cultural geographies of film consumption*. London: British Film Institute.

Jones, J. (2003). *The Southern movie palace: Rise, fall, and resurrection*. Gainesville: University of Florida Press.

Klinger, B. (2006). *Beyond the multiplex: Cinema, new technologies, and the home*. Berkeley: University of California Press.

Latour, B. (1997). *On actor-network theory: A few clarifications*. Keele: STOT Resources, Center for Social Theory and Technology, Keele University.

Lukinbeal, C. (2002). Teaching historical geographies of American film production. *Journal of Geography, 101*, 250–260.

Lukinbeal, C. (2004). The rise of regional film production centers in North America, 1984–1997. *GeoJournal, 59*(4), 307–321.

Lukinbeal, C. (2006). Runaway Hollywood: Cold Mountain Romania. *Erdkunde, 60*(4), 337–345.

Lukinbeal, C. (2012). 'On location' filming in San Diego County from 1985–2005: How a cinematic landscape is formed through incorporative tasks and represented through mapped inscriptions. *Annals of the Association of American Geographers, 102*(1), 171–190.

Marling, K. A. (2001). The cultural geography of the American movie palace. In P. C. Adams, S. Hoelscher, & K. E. Till (Eds.), *Textures of place: Exploring humanist geographies* (pp. 8–23). Minneapolis: University of Minnesota Press.

Scott, A. (2005). *On Hollywood: The place, the industry*. Princeton: Princeton University Press.

Zonn, L. (2007). Going to the movies: The filmic site as geographic endeavor. *Aether: The Journal of Media Geography, 1*, 63–67.

Section III

Transforming Geospatial Technologies and Media Cartographies

10 Vertical Mediation: Geospatial Imagery and the US Wars in Afghanistan and Iraq

Lisa Parks

Immediately before and after the 9/11 attacks the United States' National Reconnaissance Office (NRO) launched three new satellites from California's Vandenberg Air Force base, planned long before 9/11, sending up a Lacrosse radar imaging satellite on August 17, 2001, a signal detection satellite on September 9, and a K-11 satellite known as USA-116 on October 5 (Yugoslavia... 2006). In addition, two new commercial remote sensing satellites—Ikonos owned by SpaceImaging, which later became GeoEye, and QuickBird owned by Digital Globe—were launched in 1999 and 2001 respectively. These state and commercial satellite projects, planned long before the 9/11 attacks, have been key elements of US global reconnaissance in the context of the war on terror. To support the US invasion of Iraq in March 2003, the NRO reportedly had six spy satellites flying over the country each hour (Yugoslavia... 2006). Between 2001 and 2013, the NRO launched an estimated 24 more satellites into orbit, with several more waiting in the wings, and US private remote sensing companies have launched at least 6 more earth imaging satellites.

While aerial reconnaissance long precedes the historical conjuncture of the war on terror and the age of the satellite, extending back centuries, as Caren Kaplan shows in her crucial work on the history of ballooning (2013), to include more recent events like the Cuban Missile Crisis of 1962, the use of vertical vantage points has arguably intensified over the past decade as geospatial images have been mobilized to fight a war imagined as global and perpetual—an "everywhere war," as Derek Gregory (2011) has called it, and a "forever war" as Dexter Filkins (2009) describes it. The contemporary use of geospatial images by US state and corporate entities is part of a broader aero-orbital assemblage that also includes the power to commandeer the airwaves, to regulate activity in and out of the air through airport security, and to use unmanned aerial vehicles or "drones" to monitor, target, and

L. Parks (✉)
Department of Film and Media Studies, UC Santa Barbara, Santa Barbara, CA, USA
e-mail: parks@filmandmedia.ucsb.edu

S.P. Mains et al. (eds.), *Mediated Geographies and Geographies of Media*,
DOI 10.1007/978-94-017-9969-0_10

destroy sites and people on the earth's surface. Such aero-orbital maneuvers are interwoven with what Eyal Weizman (2002) calls "the politics of verticality," which involves, among other things, the assertion of control over the airspace above a territory as part of an effort to regulate and control what happens on the ground beneath it. Though Weizman's concept emerges from a detailed analysis of Israel's attempts to "control the air" over Palestine, similar strategies have informed US practices of targeted killing and broader struggles for aero-orbital domination since 9/11 (Hajjar 2013).

What further distinguishes the current historical conjuncture is the shifting institutional terrain of aerial and satellite imagery. During the past decade, the restructuring of government agencies, the emergence of new remote sensing companies, and the proliferation of digital technologies combined have increased consumer access to high-resolution aerial and satellite images. Now referred to as the *geospatial industry*, this sector includes a host of federal agencies and private firms that participate, often collaboratively, in the production, distribution and interpretation of aerial and satellite imagery. Historically, strategic US aerial and satellite reconnaissance and image intelligence activities occurred under the federal umbrella of the National Reconnaissance Office and the National Imagery and Mapping Agency (NIMA). In 2003 a federal bill authorized NIMA to change its name to the National Geospatial-Intelligence Agency (NGA). The NGA, whose motto is "Know the Earth . . . Show the Way . . . .Understand the World," employs 16,000 people and most of whom work in one of the largest federal buildings in Washington DC (Office of the NGA Historian 2011). In 2013 the NRO's annual budget was $10.3 billion while the NGA's was $4.9 billion (Andrews and Lindeman 2013). Combined, the two budgets exceed that of the Central Intelligence Agency's. Alongside these federal agencies, two privately owned remote sensing satellite operators, Digital Globe and GeoEye, which initially surfaced during the 1990s and merged in 2012, have become major players in the geospatial sector, selling high-resolution satellite imagery on the international market and enjoying a steady flow of major US government contracts. In 2012 Digital Globe's revenue was $421.4 million, up 24 % from 2011 (Digital Globe 2013). And in 2011 GeoEye's brought in $356.4 million, an 8 % increase from 2010 (GeoEye 12 Mar 2012). Most of their revenue comes from US government contracts. In 2010 both companies were awarded 10-year contracts from the NGA to develop the "Enhanced View" program, which amounts to a combined total of $7.35 billion if all options are exercised (Hubler 10 Aug 2010). Finally, information giant, Google, whose revenue hit a record $50 billion in 2012, began emerging as another key player in the geospatial sector when the company purchased the digital mapping company, Keyhole, Inc. in 2004 (Google 2013). Keyhole's 3-D interactive mapping interface became the basis for Google Earth, which by 2011 had more than one billion downloads (Shaer 6 Oct 2011). Keyhole had been backed by the CIA private venture firm Intel-Q and was named after the Keyhole reconnaissance satellite program.

The government restructuring of aerial and satellite reconnaissance, the growth of private remote sensing, and the increasing availability and use of geospatial images via Google Earth form the backdrop of issues to be explored in this chapter. The close collaboration of federal agencies such as the NRO and NGA with private firms such as Digital Globe, GeoEye, and Google Earth interventions makes it increasingly challenging to differentiate state, military and civilian activities in relation to geospatial images and the actions that they are used to mobilize and rationalize. Private companies such as Digital Globe and Google wield a growing amount of political and economic clout as their products support the US Defense Department in various strategic initiatives and worlding maneuvers. Though these private firms are not typically thought of as media companies, they have become integral to what James Der Derian (2001) calls the military-industrial-media-entertainment network. Extending Der Derian's work, Caren Kaplan traces the military histories of satellite-based geolocation systems such as GPS and GIS, and argues that systems designed to support "surgical strikes" in the Persian Gulf War have become part of broader consumer culture that now transposes subjects with targets and "habituates citizen/consumers to a continual state of war understood as virtual engagement" (2006, 705).

Building on this work, this chapter explores how geospatial images have functioned as part of militarized media culture in the midst of the US wars in Afghanistan and Iraq and considers what is at stake when major state decisions such as whether, where, and how to fight war pivot around the capacity to detect and display light and heat patterns on the earth's surface. Geospatial images are not benign abstractions; they are used to catalyze geopolitical agendas, rationalize military interventions, and develop postwar futures. As the co-editors of *Observant States* suggest, in these images "the logic of geopolitical reason is ... inseparable from its visual representation" (MacDonald et al. 2010, 7–8). At the same time, however, as Laura Kurgan (2013) and others have suggested, the unique vantage point of the overhead view might enable ways of thinking about earthly matters from oblique political angles. To explore these issues further, this chapter examines a series of geospatial images as vertical mediations of the US wars in Afghanistan and Iraq. By *mediation* I am referring not only to the technical process of transforming material phenomena into framed images, but also to a multitude of imprints, traces, or residues left in the air, on the ground, or in the water by acts of war. As Kember and Zylinska explain, mediation is more than representation; it "can be seen as another term for 'life'—for being-in and emerging-with the world" (2012, 23). Approaching the geospatial image as a *vertical mediation* involves exploring the stretch of space between the earth's surface and aero-orbital platforms as part of vital processes, life worlds, and ways of life. It involves explicating the kinds of capacities and forces the geospatial image is used to demonstrate, enact, or mobilize, while remaining attentive to its limits and constraints and the unpredictable reversals of power it may be implicated within as well. In short, it involves treating the geospatial image not as a static frame of image data, but as part of biopolitical processes, as part of processes of "... becoming, of bringing-forth and creation" (Kember and Zylinska 2012, 22).

In an effort to approach geospatial images in such a way, the chapter opens by describing what I refer to as the "microphysics of geospatial imagery"—the technical and power-laden processes by which electromagnetic radiation traveling through the atmosphere becomes geospatial imagery. Then, placing these technical processes in a particular historical context, I explore how US commercial or declassified geospatial images have been used to expose "enemy hideouts," spotlight alleged weapons of mass destruction, pinpoint bombed infrastructure sites, prospect for natural resources, and monitor reconstruction projects. In the process, I argue that geospatial images have been used not only to *represent* or *reveal* conditions on earth, but also to stage, enact, and bring about material transformations on the earth's surface, and I close with a discussion of the vertical remediation of Afghanistan and Iraq—the process of transforming the territories of sovereign nation-states into geospatial data that becomes the intellectual property of the US government or corporations so that it can be stored, shared, acted upon, or traded in the global digital economy and used in postwar reconstruction initiatives.

The Microphysics of Geospatial Imagery

Though aerial and satellite images involve different technologies, organizations, and companies, when they circulate in media culture they are often used interchangeably and without source information, which impedes the production of public literacies around their use. US State or Defense Department officials who release these images as declassified intelligence usually offer little, if any, detail about the satellite, aircraft or sensing instruments that acquired the image-data. Captions or taglines in the press often reiterate officials' statements or press releases and provide little information about the provenance of the image and sometimes exclude the date of its acquisition. Furthermore, it is challenging for most viewers to distinguish a satellite image from an aerial image since both look down on the earth's surface from slightly oblique angles and rely on similar sensing instruments and imaging software. Compounding this confusion is the fact that government agencies, military units, the press, and private companies refer to aerial and satellite images in different ways. When the US State or Defense Department uses aerial or satellite images they are called *reconnaissance* images, and historically have also been referred to as PHOTOINT, IMAGEINT and, more recently, GEOINT. When geographers or earth scientists use these images they are *remote sensing* or *geospatial images*. Those who want to convey that a satellite or an aircraft acquired the image-data might refer to them as a *satellite image* or an *aerial image*, or, more vaguely, as *overhead images*. The integration of aerial and satellite imaging, interactive mapping software, graphic design, and computer networking within the NGA and Google Earth has resulted in a shift away from platform-specific terminology (satellite reconnaissance) and toward the more integrative concept of *geospatial imagery or intelligence* or GEOINT. The discussions in this chapter are largely focused on declassified or commercial satellite images released or used by US officials and agencies, but I have decided to refer to them generally as *geospatial images*. How are such images generated?

Contemporary remote sensing satellites are equipped with instruments that can "sense" visible light and other frequencies of electromagnetic radiation reflected off of or emanating from objects or surfaces on Earth. As Jody Berland (2009) suggests, remote sensing augments human perception by making phenomena visible that would not otherwise be perceived. A geospatial image in the visible light register reveals surfaces and objects on the ground as well as the sunlight or artificial light reflecting off of them, which is what makes them visible. A geospatial image in the infrared register shows surfaces and objects on the ground as well as the infrared or thermal radiation they emanate. Infrared radiation has longer wavelengths than visible light and is imperceptible to the naked eye. Infrared geospatial images can show the relative temperature of objects and surfaces on or below the ground and are also used to increase in-the-dark visibility because they reveal the contours of surfaces and objects based on their thermal radiation rather than their reflection of visible light. As a result, infrared images are often used to track and target heat-bearing objects such as energy plants, communication transmitters, moving vehicles, weapons, or bodies.

As remote sensing and spy satellites move through low earth orbits, they pass over and scan particular areas of the earth's surface, turning those areas into swaths or scene footprints. Commercial satellites such as Ikonos, QuickBird, and KH-11 and KH-12 spy satellites, all used in the wars in Afghanistan and Iraq, carry multi-spectral sensors that detect radiation across various parts of the electromagnetic spectrum, from visible light to infrared to radio waves. Multi-spectral sensing is geared toward the production of efficient and information-rich geospatial imagery since data across multiple frequencies of the spectrum can be simultaneously collected during one satellite pass. For instance, multi-spectral sensors on QuickBird collect data in the blue, green, red, and near infrared bands, which can be used to generate images with resolutions ranging from 60 cm to 2.4 m (Wikipedia, Quickbird 2014b). Quickbird has a storage capacity of 128 Gb, which is equivalent to approximately 57 single area images. A single image represents an area of 18 by 18 km and Quickbird can revisit a site every 1–3.5 days. Quickbird's replacement satellites, WorldView-1 and 2, were financed in part by the National Geospatial Intelligence Agency and launched in 2007 and 2009 respectively. Worldview-1 can gather 750,000 km^2 of 0.5 m resolution imagery per day (Wikipedia, Worldview-1 2013). Another commercial satellite, GeoEye-1, funded by the NGA and Google, was launched in 2008 and can acquire images with 16 in. ground resolution (Wikipedia, GeoEye-1 2014a). The NRO increasingly relies on commercial geospatial images because they are not subject to the same classification rules as NRO-operated satellites and thus can be shared more readily with allies (Wesigerber 23 oct 2013).

Once a satellite gathers data, it is temporarily stored, encrypted and transmitted back to earth where it is archived in databases. To generate a satellite image, a user must extract data from the database and calibrate it using radiance and geolocation software so that it can be rendered as a grid or raster made up of pixels. For each pixel in the image there are multiple channels of information that can be activated, depending on how many frequencies of the spectrum data was collected in. These channels or registers can be turned on or off during image processing to support what the user seeks to convey or highlight in the image. These images can appear in black and white or color and can be used comparatively or in a layered manner so that satellite or aerial images of the same site can be contrasted, analyzed or composited. When declassified geospatial images are publicly released and discussed by US officials they often appear in black and white and are inscribed with various graphics that are designed to guide interpretation.

The geospatial image is one of the most technologized kinds of images since its production is based on so many layers of machine automation—remote sensing, data encryption, signal transmission, data storage, and image processing. As a rendering of detected electromagnetic radiation, the geospatial image is a computational image: its view has been scanned by electronic sensors rather than seen through a looking glass. While it is somewhat in vogue in media studies to celebrate the machinic aspects of computational images (Kittler 1999, Kittler and Enns 2009; Ernst and Purikka 2012), there is a tendency to overlook the myriad forms of human labor that support their production. To think of the geospatial image as purely machinic or computational, I would argue, ignores its complex materialisms—the scattering of materials, labor, energy, affect, and discourse that undergirds its production, circulation and signification. Despite the computational status of the geospatial image, humans participate in the design and manufacture of satellites and sensing instruments and extract natural resources to make them. They monitor interfaces in earth stations to track these objects from afar and determine which parts of the earth they scan. They navigate software menus and make decisions in the process of rendering and interpreting image data. They use geospatial images to advance scientific arguments, make business speculations, and carry out military assaults. And humans and non-humans across the planet are profoundly impacted by such uses. Since the geospatial image is as reliant upon humans as it is upon machines it makes more sense to approach it as part of a techno-social formation, actor network, or human-machine assemblage than it does to posit it as an autonomous technical form. The geospatial image is made not only by remotely controlled aerial and orbital machines; it is arguably the product of janitors who ensured the clean room was "clean," communications specialists who track satellites, and mechanics who fueled rockets before take-off.

Part of a techno-social formation, the geospatial image is, like other images, imbricated within what Michel Foucault calls the "microphysics of power" (1995, 26)—the strategies, tactics, techniques and concrete functionings of power. The transformation of electromagnetic radiation into data, image, and discourse brings it within the realm of power and enables it to affect and become part of—to mediate, in a most vital way—human and non-human relations, territories, and actions on and beyond the earth. As Foucault insists, power can move "through progressively finer channels, gaining access to individuals themselves, to their bodies, their gestures and all their daily actions" (1980, 152). Like multi-spectral scanning, power is mobilized across multiple "bandwidths" and generates higher "resolutions." It sets out to make everything and everybody visible. Though the geospatial image can be understood as participating in the quest for what Foucault calls "power through transparency" (1980, 153) or "subjection by illumination," (1980, 154) its unique qualities–namely, its capacity to detect the presence of human *and non-human phenomena* such as radiation, landscapes, vegetation, animals, and objects compels a recognition of its potential to activate imaginings of difference, estrangement, and Otherness.

The aesthetic qualities of geospatial imagery at once make us to strain to see the human in its abstraction and demand object-oriented modes of engagement as it always renders phenomena that are non-human. To embolden the analysis of power and geospatial imagery, we might turn to recent formulations by Jane Bennett and Graham Harman, who develop critical theories and philosophies to account for the presence and dynamism of non-human entities, objects, and matter. Bennett sets out "to highlight what is typically cast in the shadow: the material agency or effectivity of nonhuman or not-quite-human things" (2010, ix). In the process of elaborating a theory of "vibrant matter," she argues, "the image of dead or thoroughly instrumentalized matter feeds human hubris and our earth-destroying fantasies of conquest and consumption. It does so by preventing us from detecting (seeing, hearing, smelling, tasting, feeling) a fuller range of nonhuman powers circulating around and within human bodies. These material powers, which can aid or destroy, enrich or disable, ennoble or degrade us, in any case call for our attentiveness or even 'respect' ...." (2010, ix). Significantly, she asks, "How would political responses to public problems change were we to take seriously the vitality of (nonhuman) bodies? By 'vitality' I mean the capacity of things—edibles, commodities, storms, metals—not only to impede or block the will and designs of humans but also to act as quasi agents or forces with trajectories, propensities, or tendencies of their own" (2010, viii).

Relatedly, Graham Harman insists we must account for objects "that are neither physical nor even real," explaining, "Along with diamonds, rope, and neutrons, objects may include armies, monsters, square circles, and leagues of real and fictitious nations. All such objects must be accounted for by ontology, not merely denounced or reduced to despicable nullities" (2011, 5). For Harman "Objects are units that both display and conceal a multitude of traits" (2011, 7). Emphasizing the idea that the ontology of objects exceeds the visible he sets out to describe "how objects relate to their own visible and invisible qualities, to each other, and to our own minds—all in a single metaphysics" (2011, 7). Beyond this, Harman embraces the complexity of objects and suggests they have realities that are distinct from human subjects or consciousness. He writes, "Objects need not be natural, simple, or indestructible. Instead, objects will be defined only by their autonomous reality. They must be autonomous in two separate directions: emerging as something over and above their pieces, while also partly withholding themselves from relations with other entities" (2011, 18). Ultimately, Harman offers a philosophy that refuses reductionism, and insists "... the world in itself is made of realities withdrawing from all conscious access" (2011, 38).

Taking the ideas of Bennett and Graham into consideration would entail not only treating satellites and their images as vibrant matter or complex objects, but also recognizing that objects have visible and invisible qualities, the capacity to become something beyond themselves, and to have relations to other objects. Satellite images, for instance, always mediate a multitude of objects, non-human bodies, vital things—trees, buildings, vehicles, lands, plants, rock, snow, insects, minerals, roads, etc. Such objects are often latent or dormant in the visual field, perceptible but unintelligible, present but not seen, locatable but without a position in a story or discourse. We might call this the *surplus matter of geospatial imagery*—the earthly stuff that is detected by sensors and turned into imagery yet appears as background or peripheral and hence is not immediately registered as significant or of interest. The geospatial image becomes part of the microphysics of power not only by mediating sites and objects of interest but also by *overlooking* myriad other material forms—both in the sense of passing over and in the sense of abstracting and minimizing their presence and/or significance. The geospatial image's broad inventory of surplus objects and phenomena beckons the viewer to recognize the complex materialisms that constitute this view. Building on the ideas of Harman and Bennett, I want to explore how the surplus matter of the geospatial image might complicate, destabilize, or obstruct its strategic/militaristic deployment. Put another way, lurking within every geospatial image's registry of spectral radiation is a story to be plumbed about vibrant matter.

To summarize, then, on the one hand, I am critiquing the Berlin school of media's investment in the non-human and non-discursive aspect of machines, leading to an erasure or suppression of human labor, energy, and affect and a fixation on internal function, design, and specification—or diagrammatic operations—and, on the other hand, I am embracing elements of object-oriented ontologies in an attempt to develop a critical analysis of wartime uses of geospatial imagery that recognizes the presence, traits, qualities and potentials of a broader repertoire of human and non-human actors, organic and inorganic materials, visible and invisible things. To explicate this mode of analysis we might consider how strategic US uses of geospatial images can be complicated by the traits of objects included in or inferred by the views. For instance, in the weeks after 9/11, US news media circulated several Digital Globe satellite images allegedly revealing "enemy hideouts" in the Darunta Lake region of Afghanistan (Globalsecurity.org 2011b). These images use squares and arrows to identify particular sites as potential targets, while flattening the stature of the mountains nearby. As geological objects with massive scale, solidity and vertical depth, these mountains are challenging to navigate on the ground or from the air, and there have been countless stories about US troops' inability to physically maneuver through such treacherous terrain in Afghanistan, which became a rationale for drone warfare. Excavating and raising questions about the surplus matter in such images–in this case, the mountains–can complicate the ways in which graphics are used to overdetermine geospatial images as targets, and bring Other matter, whether visible or inferred, to the surface. Approaching the geospatial image in such a way not only acknowledges the potential of the mountains to obstruct a strategy of precision targeting, but also brings other objects and matter into the discussion of war and vertical mediation.

To delineate another example of this mode of analysis, we can turn to geospatial images of bombed communication infrastructure sites in Afghanistan and Iraq. Declassified images released by the US Defense Department not only show before and after scenes of targeted sites, but also infer the presence of invisible frequencies of the electromagnetic spectrum, which carry military and civilian communication through vertical domains (Richelson 2003; Globalsecurity.org 2011a). Like mountains, such frequencies can be used to thwart US military assaults, hence the forceful annihilation of transmission facilities as evidenced in these views. The US attacks on communication facilities may have disrupted access to and use of the airwaves by Iraqi and Taliban military units from these sites, but they did not destroy the spectrum itself. In an era in which the US military audaciously claims to have the power to see, know, and destroy everything, it is worth adopting a diffractive position in relation to such geospatial images, and using them (perhaps paradoxically) to acknowledge that some things cannot simply be seen and destroyed.

In making this point, I do not mean at all to diminish the horrific and systematic military violence that US troops have perpetrated on people and things in Afghanistan and Iraq during the past decade. Rather, I am seeking to formulate a posthumanist critique of geospatial images that acknowledges the humans and non-humans that are constitutive of and connected to these views and their uses and to highlight the wide array of organic and inorganic objects, materialities, or phenomena that are part of any war theater and thus are potentially impacted by geospatial imaging and aerial assaults. Even though the US has been able to visualize and attack sites throughout Afghanistan and Iraq from above, insurgents in both countries have used ground tactics to challenge US vertical hegemony, reinforcing the reality that it technologized power is not total. Developing critical dispositions and literacies in relation to geospatial images seems all the more urgent given their ongoing use in military campaigns, their integration within everyday media culture, and the billions of US taxpayer dollars spent on generating them. And while I want to explore different ways of critically engaging with geospatial images, I also want to consider how the commercialization of geospatial imaging and its integration within media culture is articulated with broader hegemonic efforts to reorder and remediate life in Afghanistan and Iraq.

Remediating Afghanistan and Iraq

Google Earth emerged in 2005 as a digital platform offering unfettered access to geospatial views of the planet to anyone with Internet access and a computer. In a matter of years, Afghanistan and Iraq went from being strictly regulated visual domains to ones open for anyone in the world to see as Google Earth circulated geospatial images of these war torn countries as part of a new form of commercial media culture. As geospatial images have become more widely available through the Internet, strategic practices historically associated with intelligence agencies such as the NRO and NGA have been normalized as part of everyday civilian life, creating a culture of open-source GEOINT on demand. In this sense, Google Earth not only serves as another media site for circulating geospatial imagery; it is also a material enactment and manifestation of a set of technological, economic, and political relations between federal intelligence agencies and the commercial geospatial sector. Though its composited interface is made of geospatial image data gathered and licensed by satellite and aircraft operators around the world, its digital architecture—that is, the capacity to arrange and display the world in this way—is the intellectual property of Google and the result of a history of coordinated federal and corporate financial, technological, and political transactions.

The Google Earth interface turns this model of GEOINT on demand into a privatized consumer media experience that is accessible as a download on computers and smart phones. The interface is made of composited, publicly and privately sourced aerial and satellite imagery that is periodically updated. Datasets turned into graphic displays known as "layers" can be formatted as kmz or kml files and dropped into Google Earth so they can be superimposed on the geospatial interface. Not only is geospatial imagery in Google Earth highly processed and composited, it is covered up with icons, shading, and text as layers are activated. This results in a version of geospatial imagery that is so heavily inscribed with graphics that the initial image data recedes and becomes little more than a background for the inscription of iconography, reducing the potential for geospatial literacy.

For instance, when the Digital Globe layer is activated, color-coded square lines and DG brand icons appear in the visual field. The color-coded square lines, called "scene footprints," function as traces of a satellite's pass over a specific part of the earth. When composited, they form a historical record of satellite image data acquisitions during a certain time period, as well as reveal a slice of Digital Globe's inventory. The Google Earth interfaces of Afghanistan and Iraq provide Digital Globe satellite coverage from 2002 to 2010. Clicking on a DG icon opens a frame with data about the image including the acquisition date, cloud cover, and an environmental quality rating. If the user clicks on "preview," she enters a meta-browser featuring the single satellite image captioned with information about how to purchase it or others from Digital Globe.

The US exercised "shutter control" to restrict access to satellite imagery of Afghanistan for 3 months in 2001, but after 2005 allowed Google, Digital Globe and GeoEye (previously, Space Imaging) to conduct an international business that turned the territories of Afghanistan and Iraq (as well as those of other countries) into digital real estate—intellectual property produced, owned and distributed by US corporations. Just as the leveling of communication infrastructure provided opportunities for US contractors to restructure and rebuild Afghan and Iraqi broadcast and telecom systems, Google Earth's vertical mediation of Afghanistan and Iraq is designed to boost the business potentials and profits of US companies as the geospatial image has been used both to stage the eradication of Taliban and Hussein systems and as a platform upon which to imagine, design and map new ones. In this way, the geospatial image is implicated in the material restructuring and remediation of nation-states. As described in the examples below, aero-orbital platforms are not only technologies of observation, but of inscription.

Mediascaping

Google Earth has been used, for instance, to map Iraq's newly privatized media sector. A layer called "Iraq Media Mapping" released by the Open Source Center visualizes a plethora of new commercial television and radio networks that sprouted up in Iraq after the fall of Saddam Hussein's regime and during the US occupation. The layer specifies the names of TV and radio stations, the locations of their headquarters and transmitters, and potential audience size, and uses color-coded shading to indicate the stations' coverage zones. While the layer provides a useful overview of the media industry within Iraq, many of the key stakeholders of Iraq's media system exist beyond the country's borders. Several of the new Iraqi TV networks, for instance, were developed through US Defense Department contracts or are owned by wealthy Iraqis in exile. Al Iraqiya, for instance, an Iraqi news and entertainment TV network, was founded as Iraqi Media Network in 2003 and developed by US defense contractor Science Applications International Corporation for an initial no-bid contract of $82.3 million, which was supplemented by hundreds of millions of dollars (Goldstein 2008; Auster 2004). Another TV network, Al Sharqiya, was launched in 2004 and is owned by Iraqi media tycoon Saad Al-Bazzaz who lives between London and Dubai (Arango 28 April 2013). And TV channel, Al Baghdadiya, which emerged in 2005, is owned by Cairo-based Iraqi businessman, Awn Al Khashlouk. This "free" Google Earth interface articulates the re-mediation of Iraq through multiple layers of media privatization, using privately owned satellite imagery accessible through a privately owned web-platform to map largely privatized use of Iraqi airwaves. In this digital world, Iraqis who live in Iraq have minimal ownership or control over their airwaves, geospatial views of their territory, or their lands (Fig. 10.1).

Resource Speculation

In addition to mapping the restructuring of Iraq's broadcast sector, US geospatial imagery has also been to identify and develop other forms of material value and exchange in and around sites that have been bombed or destroyed. The US Geological Survey has used Digital Globe satellite images and geographic information systems more powerful than Google Earth such as ArcGIS to scout natural resources in Afghanistan and Iraq and share this information with interested parties and investors (USGS undated-b; USGS undated-c). Geospatial images have been used to pinpoint Afghanistan's coal, oil and natural gas, mineral, and hydrologic assets (Chirico 2006), and function as treasure maps or invitations for foreign corporate development and extraction. As the USGS explains, geospatial data sets are "vital to short-term and long-range planning regarding management of these resources, as well as for identifying potential new resources that may attract foreign investment and create employment opportunities for Afghans" (USGS undated-c). The organization even offers geospatial "information packages" as

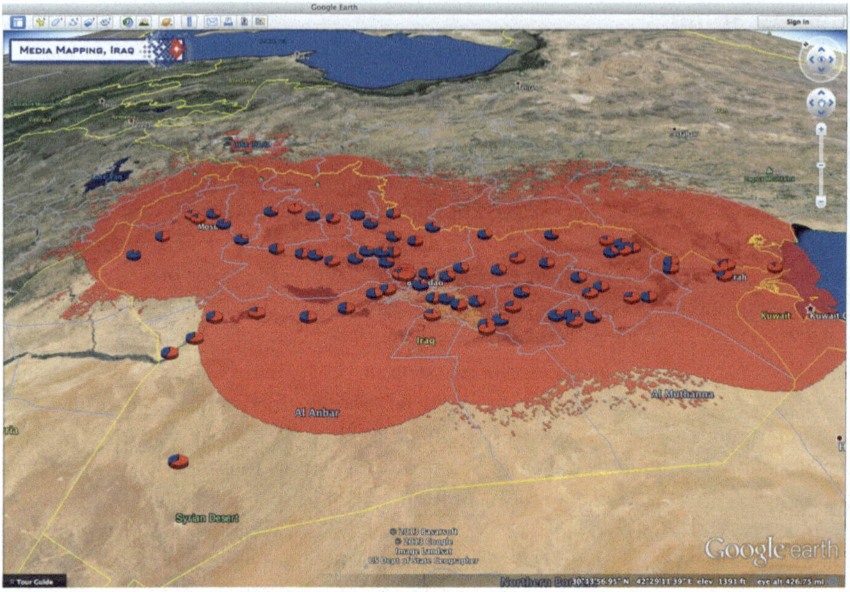

Fig. 10.1 The media mapping Iraq layer in Google Earth demonstrates the mediascaping of Iraq after the 2003 war (Source of layer: Open Source Data Center)

well as an interactive "Afghanistan Oil and Natural Gas Viewer" for potential energy developers, using geospatial images to encourage the tapping of Afghani resources underground (USGS undated-a). Here the geospatial image becomes part of extractive geo-economic strategies that resemble colonial-era practices. What differs, however is the way in which the widespread availability of such images through the web and Google Earth has the effect of normalizing these views as ethical.

Monitoring Reconstruction

At the same time that geospatial systems are being used for mediascaping and resource prospecting, organizations such as USAID have adopted Google Earth to track and verify the flow of reconstruction funding, which is sometimes parachuted into remote regions. Given the corrupt handling of US reconstruction funds in Afghanistan (billions is siphoned out of Afghanistan each year), Mercy Corps to send Afghanis to contentious or remote regions with GPS cameras to photograph people performing work leading to the fulfillment of US contracts. The photos are uploaded into a "Google Earth-style program" so that Mercy Corps "can track projects and their participants" (Hodge 2010). The system, reportedly, allows Mercy Corps to "extend its reach" and work in areas where "it's too insecure to work, or too remote" (Hodge 2010). Here, Google Earth and geospatial imagery are used to remotely monitor financial flows and reconstruction processes—to ensure that projects on the ground conform to externally developed plans and visions.

Predictive Analytics

Finally, geospatial images have been used to monitor insurgents' movements in Afghanistan and Iraq and formulate predictive analytics designed to mitigate improvised explosive device (IED) attacks. In 2009 the Open Source Data Center released a detailed report (including a Google Earth layer) about insurgent incidents that occurred in Afghanistan and the FATA region of Pakistan between 2004 and 2008, offering a "hot spot" analysis designed to "provide valuable information for those responsible for operations in the region" (Open Source Data Center 2009). The Google Earth layer enables users to view the data in different ways, according to incident density, deaths, wounded, kidnappings, and perpetrators. Using data from the National Counterterrorism Center's Worldwide Incidents Tracking System, the report mapped a total of 4,129 incidents and projected their locations onto a Landsat image. During the 4-year period, 64 % of the incidents were allegedly committed by the Taliban, 42 % were caused by IEDs, and 36 % were armed attacks. Like weather forecasters using geospatial imagery to predict the intensity of a storm, Open Source Data Center analysts assess past data to predict the location of future terrorist attacks coding high probability areas in sienna and low probability areas in yellow. As Mark Andrejevic suggests, the tendency of such projects "is to portray predictive analytics as a crystal ball whose view of the future becomes clearer with every new piece of data about the present—as if at the very point when we can capture the entirety of the present in a database, the future will simultaneously be pinned down" (2013, 32).

Such vertical mediations of Afghanistan and Iraq reinforce the temporal and spatial logics of the "forever war," or "everywhere war" by commandeering aero-orbital platforms and digital technologies, historically subsidized by US taxpayers, to turn the world's territories into privately owned and strategically mobilized digital intellectual property that can be used both to stage military assaults and remodel Others' territories. This re-mediation of Afghanistan and Iraq appropriates geospatial imagery to demonstrate and enact the privatization and globalization of mediascapes, the prospecting and speculation of natural resources, and the monitoring and policing of local communities (Fig. 10.2).

Conclusion

Given these enactments and inscriptions of power, the geospatial image can be understood as a signature site for studying vertical mediation and the war on terror. I have approached the geospatial image, on the one hand, as a site for a critical and diffractive object-oriented analysis designed to complicate strategic looking, and, on the other hand, as a site that makes hegemonic use of the vertical field

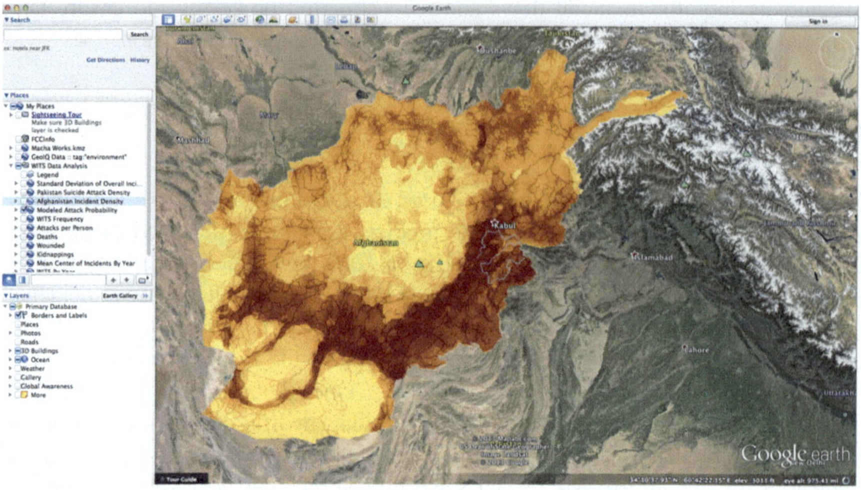

Fig. 10.2 This Google Earth layer predicts the likelihood of future terrorist attacks along the Afghanistan and Pakistan border coding high probability areas in sienna and low probability areas in *yellow* (Source of layer: Open Source Data Center)

intelligible and exposes efforts to remediate the landscapes and lifeworlds of untold and unnamed Others. Much more than remote views, theses vertical mediations are implicated in the material restructuring of life on earth. They are part of broader regimes of targeting, striking, and eradicating, which leave holes and eliminate people such that new governments, social orders, and built environments must be imagined, developed and put in place. At the same time, by detecting human and non-human, organic and inorganic, earthly and circumterrestrial objects and phenomena, geospatial images reveal that war always exceeds the frame, as Rey Chow (2006) and Judith Butler (2010) powerfully remind us, and is inscribed as traces, residues, layers, chemical compounds in the air, the earth's crust and water, as well as in the flesh and minds of bodies. Geospatial images help us to think about the ways war remediates life itself.

It is crucial that hegemonic uses of geospatial images and their institutional underbellies be excavated in the process of critically engaging with them. At the same time, the fact that state and military organizations use geospatial images strategically does not mean we have to inherit and idly adopt those ways of looking at and using them. In fact, one might argue that the increased circulation of geospatial images has created a crucial turning point in visual culture in which we have to struggle to maintain the right to interpret itself, especially given that so many geospatial images come into circulation as always already read—that is, with dense layerings of graphics, icons, and arrows inscribed in the view, which regulate acts of interpretation and sense-making. True GEOINT—geospatial intelligence—is not housed in the NRO or the NGA, but would involve fostering geospatial imaging literacies among citizens so they could engage with such images on different terms and from multiple vantage points and begin to question and fracture the now normalized process of waging war and conducting foreign policy by sensing radiation emanating from the surface of the earth.

What I am ultimately suggesting is a need to further situate the geospatial image within critical dialogues on media and democracy and media conglomeration. The geospatial image is not only helping to shape perceptions and worldviews; it is also being used to re-model life on earth. Given the high stakes of geospatial imaging, it is vital that scholars from an array of disciplines interrogate the effects of its development and continuing use. For instance, to what extent are Iraqi or Afghani citizens able to access and use Google Earth's "free" platform to support their agendas and interests? What policies and regulations have been applied to Google Earth and Digital Globe as they turn the planet into a proprietary digital archive and platform? Why are there so few humanitarian geospatial projects (like Satellite Sentinel) and so many militaristic ones? While it might seem on the surface that the geospatial image lies beyond the purview of media and communication studies, I hope I have shown that it has a generative potential to extend research in new directions.

Acknowledgement I would like to thank members of the Militarization Research Focus Group at UC Davis and the Scholars Program in Communication and Culture at the Annenberg School of Communication at the University of Pennsylvania for feedback and support while I worked on versions of this essay, especially Caren Kaplan and Barbie Zelizer and two anonymous readers. I also thank Julie Cupples and Kevin Glynn for their intellectual support and collegiality and for inspiring me with their work.

References

Andrejevic, M. (2013). *Infoglut: How too much information is changing the way we think and know*. London: Routledge.

Andrews, W., & Lindeman, T. (2013). The black budget. *The Washington Post*. http://www.washingtonpost.com/wp-srv/special/national/black-budget/. Accessed 10 Sept 2013.

Arango, T. (2013). Iraq revokes licenses of Al Jazeera and 9 other TV channels. *The New York Times*. http://www.nytimes.com/2013/04/29/world/middleeast/iraq-suspends-al-jazeera-and-other-tv-channels.html?_r=0. Accessed 15 Jan 2014.

Auster, B. B. (2004). Broadcast blues. *U.S. News and World Report*. http://www.corpwatch.org/article.php?id=7884. Accessed 20 July 2014.

Bennett, J. (2010). *Vibrant matter: A political ecology of things*. Durham: Duke University Press.

Berland, J. (2009). *North of empire: Essays on the cultural technologies of space*. Durham: Duke University Press.

Butler, J. (2010). *Frames of war: When is life grievable?* London: Verso.

Chirico, P. (2006). *Natural resource assessments in Afghanistan supported by high resolution digital elevation modeling and multi-spectral image analysis*. Presented at Civil Commercial Imagery Evaluation Workshop. March 14–16, Laurel, MD. http://calval.cr.usgs.gov/JACIE_files/JACIE06/Files/213Chiri.pdf. Accessed 5 Sept 2013.

Chow, R. (2006). *The age of the world target*. Durham: Duke University Press.

Der Derian, J. (2001). *Virtuous war: Mapping the military-industrial-media-entertainment network*. New York: Basic Books.

Digital Globe. (2013). *Digital globe reports fourth quarter and full year 2012 results*. Digital Globe Press release. Available at http://media.digitalglobe.com/press-releases/digitalglobe-reports-fourth-quarter-and-full-year--nyse-dgi-990509. Accessed 1 Nov 2013.

Ernst, W., & Purikka, J. (Trans.). (2012). *Digital memory and the archive*. Minneapolis: University of Minnesota Press.

Filkins, D. (2009). *The forever war*. New York: Vintage.

Foucault, M. (1995). *Discipline and punish: The birth of the prison*. New York: Vintage.

Foucault, M. (1980). *Power/knowledge*. New York: Vintage.

GeoEye. (2012). GeoEye reports fourth quarter and fiscal year 2011 earnings results. *PR Newswire*. http://www.prnewswire.com/news-releases/geoeye-reports-fourth-quarter-and-fiscal-year-2011-earnings-results-142380305.html. Accessed 1 Nov 2013.

Globalsecurity.org. (2011a). Afghanistan – October 23, 2001. http://www.globalsecurity.org/intell/library/imint/kabul_radcom-station-nw-post.htm. Accessed 21 July 2014.

Globalsecurity.org. (2011b). *Afghanistan-Darunta Camp Complex*. http://www.globalsecurity.org/military/world/afghanistan/darunta.htm. Accessed 25 Aug 2013.

Goldstein, C. S. (2008). A strategic failure: American information control policy in occupied Iraq. *Military Review*. http://www.army.mil/professionalWriting/volumes/volume6/april_2008/4_08_1_pf.html. Accessed 20 July 2014.

Google. (2013). Google, Inc. announces fourth quarter and fiscal year 2012 results. *Google Website*. http://investor.google.com/earnings/2012/Q4_google_earnings.html. Accessed 1 Nov 2013.

Gregory, D. (2011). The everywhere war. *The Geographical Journal, 177*(3), 238–250.

Hajjar, L. (2013). Lawfare and targeted killing: Developments in the Israeli and US contexts. *Jadaliyya*. http://www.jadaliyya.com/pages/index/4049/lawfare-and-targeted-killing_developments-in-the-I. Accessed 15 May 2013.

Harman, G. (2011). *The quadruple object*. London: Zero Books.

Hodge, N. (2010, February 25). Using Google Earth and GPS to track Afghanistan cash. *Wired*. http://www.wired.com/dangerroom/2010/02/using-laptops-cameras-and-gps-to-track-afghanistan-cash/. Accessed 4 Sept 2013.

Hubler, D. (2010). GeoEye, Digital Globe gain $7.3b in NGA contracts. *Washington Technology*. http://washingtontechnology.com/articles/2010/08/10/geoeye-digitalglobe-nga-contracts.aspx. Accessed 21 Aug 2013.

Kaplan, C. (2006). Precision targets: GPS and the militarization of consumer identity. *American Quarterly, 58*(3), 693–714.

Kaplan, C. (2013). The balloon prospect: Aerostatic observation and the emergence of militarized aeromobility. In P. Adey et al. (Eds.), *From above: War, violence and verticality* (pp. 19–40). London: Hurst.

Kember, S., & Zylinska, J. (2012). *Life after new media: Mediation as a vital process*. Cambridge, MA: MIT Press.

Kittler, F. (1999). *Gramophone, film, typewriter*. Palo Alto: Stanford University Press.

Kittler, F., & Enns, A. (Trans.). (2009). *Optical media*. Cambridge: Polity Press.

Kurgan, L. (2013). *Close up at a distance: Mapping, technology, and politics*. New York: Zone Books.

MacDonald, F., Hughes, R., & Dodds, K. (2010). *Observant states: Geopolitics and visual culture*. London: I. B. Tauris.

Office of the NGA Historian. (2011). *The advent of the National Geospatial Intelligence Agency*. https://www1.nga.mil/About/Documents/04a_History.pdf. Accessed 1 Nov 2013.

Open Source Data Center. (2009). *Afghanistan: Geospatial analysis reveals patterns in terrorist incidents 2004–2008*. http://www.fas.org/irp/dni/osc/afghan-geospat.pdf. Accessed 4 Sept 2013.

Richelson, J. (2003). *Eyes on Saddam: US overhead imagery of Iraq* (National Security Archive Electronic Briefing Book No. 88). http://www2.gwu.edu/~nsarchiv/NSAEBB/NSAEBB88/. Accessed 21 July 2014.

Shaer, M. (2011). Google Earth passes 1 billion download mark. *The Christian Science Monitor*. http://www.csmonitor.com/Innovation/Horizons/2011/1006/Google-Earth-passes-1-billion-download-mark. Accessed 1 Nov 2013.

USGS. (undated-a). *USGS Afghanistan oil and natural gas viewer*. http://afghanistan.cr.usgs.gov/flexviewer/. Accessed 5 Sept 2013.

USGS. (undated-b). *USGS Iraq project: USGS activities in support of economic development and stabilization in the natural resource sectors of Iraq*. http://iraq.cr.usgs.gov/. Accessed 5 Sept 2013.

USGS (undated-c). *USGS projects in Afghanistan: A brief history*. USGS Projects in Afghanistan website. http://afghanistan.cr.usgs.gov/background. Accessed 5 Sept 2013.

Weizman, E. (2002). Control in the air. *Open Democracy*. http://www.opendemocracy.net/ecology-politicsverticality/article_810.jsp. Accessed 20 Sept 2013.

Wesigerber, M. (2013). DigitalGlobe: Lift US restrictions on commercial satellite imagery. *DefenseNews*. http://www.defensenews.com/article/20131023/DEFREG02/310230020/DigitalGlobe-Lift-US-Restrictions-Commercial-Satellite-Imagery. Accessed 21 July 2014.

Wikipedia. (2014a). *GeoEye-1*. http://en.wikipedia.org/wiki/GeoEye-1. Accessed 21 July 2014.

Wikipedia. (2014b). *Quickbird*. http://en.wikipedia.org/wiki/QuickBird. Accessed 21 July 2014.

WorldView-1. (2013). *Wikipedia*. http://en.wikipedia.org/wiki/WorldView-1. Accessed 1 Nov 2013.

Yugoslavia, Afghanistan, Iraq: The Satellite Wars. (2006). *Space Today*. http://www.spacetoday.org/Satellites/YugoWarSats.html#afghanistan. Accessed 21 Aug 2013.

The Digital Globe and New Mediations of the Environment

11

Leon Gurevitch

On December 7th, 1972 at 18,000 miles above the earth and 5 h after launch, the crew of Apollo 17 took a photograph of the earth. Apollo 17 was the last manned mission to the moon and so the image (assigned by NASA the innocuous title 'AS17-148-22727', but later reassigned 'The Blue Marble') was not by any means the first to be taken of the earth.[1] It was, however, one of the few to show the earth fully illuminated with no terminator line and as such, made newspaper front pages upon its release. This photograph is not only interesting for the often-cited fact that it is the most 'reproduced image in history'[2] but also for the quantity and range of popular and academic comment that it has attracted.

Before the photograph even emerged from among thousands taken by NASA, the possibility of a 'whole earth photograph' was narrated in a number of ways. As Denis Cosgrove (1994: 281) has detailed, both the United States and the Soviet Union were keenly aware of the powerful propaganda potential inherent in the universalist iconography of a whole earth photograph. Even before the image was taken, commentators such as Fred Hoyle and Stuart Brand made claims of paradigm shifting cultural impacts of such a picture. Hoyle predicted in 1948 that 'Once a photograph of the Earth, taken from the outside, is available... a new idea as powerful as any in history will be let loose' (Redfern 2003: 1). Since its emergence in the 1970s, the discursive interpretations of *The Blue Marble* and its cultural significance have grown. The image is said to have contributed to the rise of James

[1]For more detail on the many 'first' photographs of the earth from space see Robert Poole's account of the US/Soviet space race in the book *Earthrise*: *How Man First Saw the Earth*.

[2]Both Al Gore and Wikipedia make this claim though it is not clear how this can be verified or what is its significance even if it could be.

L. Gurevitch (✉)
School of Design, Victoria University of Wellington, Wellington, Aotearoa/New Zealand
e-mail: Leon.Gurevitch@vuw.ac.nz

S.P. Mains et al. (eds.), *Mediated Geographies and Geographies of Media*,
DOI 10.1007/978-94-017-9969-0_11

Lovelock's Gaia theory[3] and with it a paradigm shift in earth sciences (Lovelock 1995, 2006), the commodification of nature within industrial culture (Haraway 1997); the materialisation of Marshal McLuhan's Global Village (Roberts 1998); the formation of a 'panhumanity' in which the global social is united by a shared human nature; a shared global culture (Franklin et al. 2000: 30) and a transformation in economic theory – based as it previously was on the notion of infinite resources (Poole 2008: 158).

Crucially, central to such discourses has been the claim that images of the earth from space alerted the public and professional spheres alike to the concept that the planet constitutes a closed, fragile and interconnected environmental system. Following on from this was the observation, now claimed by Castels (2010: 177) to be a fundamental lynch pin of twentieth century environmentalism, that the earth's ecosystems operated beyond the confines of nation state boundaries. Echoing this, successive groups of astronauts returned home from space citing the lack of identifiable national boundaries and the fragility of the earth's ecosystems as one of the most profound revelations to spring from beholding the earth in its entirety. This has led Frank White to describe their experiences, and those of humanity (once such images started to filter back) to be the result of what he calls the "overview effect" (White 1998).

Despite the great quantity of public and academic commentary that these images attracted, they have now been superseded as the primary source of representation both for the planet and for its environment. While whole earth images are not in any danger of disappearing from contemporary media circulation, the earth image that has increasingly found favour over the past decade across multiple media formats has been that of the digital earth. And though there are many alternative digital earth programs, it has been Google Earth that has gained popular and widespread public attention in years since it emerged in 2005. Its arrival, though notable across a range of public media from the personal computer to the evening television news, has still not received the quantity of academic analysis that it warrants.

From a social, cultural and representational point of view, Google Earth is a very different object to the spatially unified, visually solidified, indexical photography that preceded it. Immediate differences between whole earth photography and the digital globe are obvious: the first is a representation of the actual earth, frozen in time from the instant that it was taken, the second is a dynamic, interactive model, around which is draped a mosaic/patchwork of constantly changing, digitally captured satellite images. Beyond these immediate differences, however, are sets of more subtle distinctions that, once we start looking, become almost overwhelming in their quantity. Perhaps the most significant and overarching distinction lies in its function as a program: subject to the laws and principles of the coded cultural form.

[3]Robert Poole argues this most forcefully, saying that the origins of James Lovelock's Gaia Theory can be approached through the viewfinder of the Apollo missions 'Blue Marble' photographs and quoting Lovelock himself who as stated that, "When I first saw Gaia in my mind I felt as must as astronaut have done as he stood on the Moon, gazing at our home, the Earth."

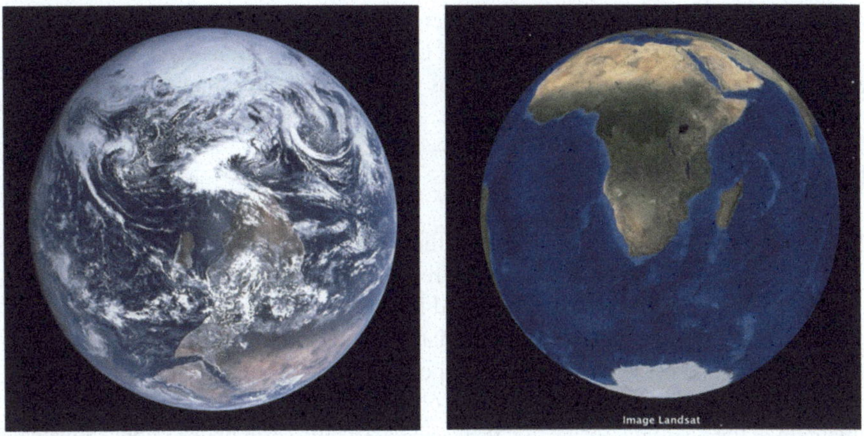

Fig. 11.1 *Left*: NASA's 'AS17-148-22727' photograph later renamed 'The Blue Marble'. *Right*: A screenshot of Google Earth 7.1.2.2041, Build date 2013

Where, for instance, the Blue Marble image presented an ecologically interrelated biosphere, unconstrained by anthropogenic factors such as political borders, Google Earth's start-up skin strips the earth of its cloud and weather systems, reinstating the primacy of the nation state in bright yellow lines, operating as a polar opposite to its predecessors with their one world message. Rather than presenting a fragile, interconnected ecosystem under threat from twentieth century industrial civilisation, Google Earth offers a profoundly twenty first century alternative. The Google Earth viewer does not see one interconnected ecosystem from afar, they drill down through a plurality of programmed 'ecosystems' to reveal layers of information, simulation and scale that constitutes a new environmental conception (Fig. 11.1).

In this chapter I will consider two interrelated areas of analysis most in need of immediate consideration following the transition to a composite digital globe as a primary reference point for the earth. In the first area I will ask, what is the digital globe? More specifically, what is Google Earth, how did it come into being, why is it the predominant representation of our world, and will it continue to be so? The second area of analysis concerns the changes that have taken place in environmental representation in recent years with the advent of the digital globe. Chiefly, what are the current and future implications of the digital globe and how does this feed into education and mediation of the environment?

All of these questions are ultimately interrelated to a broader assertion that I will make here, as I have made elsewhere: that the replacement of the whole earth genre of photography with digital earth programs marks a profound shift in the way that we are able to envision the environment, not only because the latter is an interactive form of digital media, but because it is also a cultural form governed by the twenty first century logic of the industrially engineered, computational new media object. From this, it is a short leap to go from beholding an industrially engineered and endlessly reprogrammable representation of the earth object to regarding the earth itself as an industrially engineerable object in its own right. Before we address this claim in full however, we must first return to our initial question. What is the digital globe and how does it differ from previous whole earth representations?

What Is the Digital Globe?

The first point to make when considering the notion of the digital globe is that while it is undoubtedly a new media form, as a remediated digital object it is reminiscent of many of the media forms that preceded it: it is cartographic, it is topographic, it is photographic, it is panoptical, it is a data-visualisation, it is mathematical, it is algorithmic, it is representational. The important difference between it and previous forms lies in the fact that it can be all of these things at the same time. As an amalgam of the analogue forms that it succeeded (whole earth photographs, physical globes, cartographic data-sets) the digital globe brings with it many of the social, cultural, political, economic and scientific characteristics found in its predecessors. Indeed, this fact is one of the factors that presents scholarship of the digital globe with its first challenge: when it comes to the digital globe the potential avenues of study are literally vast. One is faced with the wealth of scholarship already conducted in cartographic construction (Hale 2007; Woodward 2007), cartographic technology (Brotton 1999), cartographic sociology (Morse 2007) philosophy and cartography (Watts 2007), politics and cartography (Kagan and Schmidt 2007) globes and model-making (Dekker 2007) and more from prehistory to the present. To say that all of this requires a nuanced understanding of the differences between the respective analogue forms that fed into the digital globe is an understatement.

Perhaps a better route to take, then, is one that seeks out those aspects of the digital globe that are specifically unique in the contemporary context. A good start here is, as Elle Dekker has already argued in quite a different context, to recognise the distinctive differences between notions of "the globe" from those of "cartography". As Dekker asserts, 3D globes were differentiated from 2D cartography at their inception during the renaissance by the fact that they stood for more than the three dimensional materialisation of cartographic knowledge. Understanding globes, she argues, requires that they be "considered as (mechanical) representations that facilitate a spatial understanding of things, concepts, conditions, processes, or events in the human world" (Dekker 2007: 136). What is true of the development of the physical globe during the renaissance is doubly so for the digital globe in the contemporary context. At the 6th International Symposium on the Digital Earth in Beijing in 2009, the declaration was made that the:

> Digital Earth is an integral part of other advanced technologies including: earth observation, geo-information systems, global positioning systems, communication networks, sensor webs, electromagnetic identifiers, virtual reality, grid computation, etc. It is seen as a global strategic contributor to scientific and technological developments, and will be a catalyst in finding solutions to international scientific and societal issues. (2009 Beijing Declaration on Digital Earth)

Among the "societal issues" (reminiscent of Dekker's description of the "things, concepts, conditions, processes, or events in the human world" of the fifteenth century and likely similar in nature) the symposium identified, "natural resource depletion, food and water insecurity, energy shortages, environmental degradation, natural disasters response, population explosion, and, in particular, global climate change."

Intriguingly, Dekker argues early globes can be regarded in many ways as some of the first computers, claiming that, in their capacity to allow users to work out the relationship between local times at different places, early renaissance globes "served as an analogue computer" (Dekker 2007: 150). In this sense, the digital globe can be seen as a geo-political tool many hundreds of years in the making. The most important facet of Dekker's claim is that it highlights a relationship between the representation of the earth as an object and notions of nature as mechanically predictable. Through the mechanical representation of daylight hours, seasons and time zones, the earth and the natural world were integrally linked from the earliest stages of the production of globes. It also tellingly demonstrates that notions of the globe as mathematically computational, underpinned the cultural logic of these objects from early in their history: a feature only accelerated in our contemporary context as digital globes become subject to the laws of the computationally coded computer program.

Where the digital globe differs from Dekker's analogue predecessors is the scale and scope. In the Beijing Declaration we have many concepts in play simultaneously: the coincidence of modern communications networks with software and computer visualisation; earth systems surveillance, monitoring, relay and rendering as data; cartography, weather systems eco-systems constructed, as both architectural and computational forms to name a few. What unites these multiple inputs and consequent outcomes of the earth represented as a digital globe is that of industrialisation: the industrial collection of data, the industrial scale of the data collected, the computer automated and therefore industrialised representation of such data, the industrial programmes required to build and launch surveillance and data gathering satellites and to keep them in orbit.

This brings us to the first central point about modern digital globes like Google Earth; that they are fundamentally industrial objects. At a visual level this goes beyond the mechanically reproduced "photography" of the whole earth genre and permeates the structure of the digital globe as a computer generated, mathematically specific model facilitated by the automated rendering of three-dimensional space.

Beyond the purely visual however, the digital globe also now represents the visualisation of a far wider industrial project to *gather and represent data about the world*. In other words, Google Earth is an exercise in industrial data-visualisation. Such an understanding of the digital earth should not be seen as necessarily precluding it from also functioning as an ecological object. On the contrary, we could argue that it is precisely because it is an industrial object that it is also an ecological one. Google Earth is a panoptical project that combines military industrial data-gathering technology at the same time as it represents a profound new example of the cultural production of "ecology". In a typically prescient observation from 1972, Marshal McLuhan observed the intimate relationship between the military industrial capacities of satellite technology, emergent information based ecology and the earth as a mediated "artefact":

> Perhaps the largest conceivable revolution in information occurred on October 17, 1957, when Sputnik created a new environment for the planet. For the first time the natural world was completely enclosed in a man-made container. At the moment that the earth went inside this new artefact, Nature ended and Ecology was born. "Ecological" thinking became inevitable as soon as the planet moved up into the status of a work of art. (McLuhan 1962: 49)

McLuhan's description here highlighted the relationship between panoptical observation of the earth and an acute associated awareness of its status as an ecological entity described in the opening of this chapter. But McLuhan also drew a fascinating parallel between the "information revolution" inherent in the whole earth monitoring project and the idea that in doing so the earth itself became an ecological "artefact" – an idea we shall return to in the following section when we have considered the process by which Google Earth came into being.

How Did Google Earth Come Into Being?

Claims as to the origins of Google Earth as both an idea and as a program are frequently made and contradicted. Like many media forms, the idea of a fully simulated digital earth preceded by quite sometime the reality of its existence. We can see precursors to Google Earth in both the educational video *Powers of Ten*, made by Ray and Charles Eames in 1977 and in the "earth" programme featured in the 1992 novel, *Snow Crash*, by Neil Stephenson. Indeed, it has been claimed by technicians (Mark Aubin and John Hanke) originally working on the prototype to Google Earth that both of these references were influences,[4] but besides disagreements as to which was more significant, the fact is somewhat moot. More relevant is the wider developmental and political circumstances surrounding Google Earth's rise to prominence and what this tells us about the mediation of our environment through this program.

Google Earth was not just a panoptical project in the loose metaphorical sense of the term; it was literally panoptical in its origins within the defence and surveillance establishment. Originally called Keyhole (hence the computer coding acronym "KML" which stands for "keyhole mark-up language" still used by developers creating new "skins" for Google Earth), Google Earth originated as a spin-off project from a games company in 2001 and funded by In-Q-Tel (the tech project funding arm of the CIA) aiming to envision the earth as a dynamic model. The United States defence establishment had also deployed a spy satellite system in 1976 named Keyhole. In 2004 Google bought Keyhole and renamed it Google

[4]See paticularly: http://mattiehead.wordpress.com/2008/11/24/google-earth-inspiration-from-space-to-your-face/

Earth. Importantly, and as Michael Goodchild argues, Google Earth proved to be immediately popular because it adhered to the "child of ten" rule; that it was easy enough for a child of ten to use. In the GIS (Geographic Information Systems) community this is said to have been a disappointment for those who hoped for a more scientifically accurate, specialist piece of software. As Goodchild argues however, "the dominant paradigm of Google Earth remains visualisation – the manipulation of a virtual body whose appearance matches that of a real Earth as closely as possible" (2008: 36). Here, then, visualisation and ease of use are the driving forces rather than scientific accuracy or professional specificity. Central to this paradigm, is the fact that Google Earth operates as a platform applicable to the widest range of potential users and therefore potential media. Small wonder that the program has been so quickly adopted by news media around the world to provide immediate spatial content for breaking stories.

It is interesting that, in common with the Eames educational film, Google Earth should premise its development philosophy upon the concept of the "child of ten". All the more so when we consider that in 1998, foreseeing the moment at which the transition to a digitally remediated form of whole earth representation was likely drawing near, Al Gore started to aggressively rally support for the project to create a digital earth. In a much-commented on speech (Foresman 2008; Shupeng and Gendren 2008; Guo et al. 2010; De Longueville et al. 2010) Al Gore outlined his vision for a future digital earth. While he described it in terms similar to those of the Beijing declaration,[5] he also outlined a vision for its future that is particularly telling in retrospect and worth quoting at length:

> Imagine, for example, a young child going to a Digital Earth exhibit at a local museum. After donning a head-mounted display, she sees Earth as it appears from space. Using a data glove, she zooms in, using higher and higher levels of resolution, to see continents, then regions, countries, cities, and finally individual houses, trees, and other natural and man-made objects. Having found an area of the planet she is interested in exploring, she takes the equivalent of a "magic carpet ride" through a 3-D visualization of the terrain. Of course, terrain is only one of the many kinds of data with which she can interact... she is able to request information on land cover, distribution of plant and animal species, real-time weather, roads, political boundaries, and population. She can also visualize the environmental information that she and other students all over the world have collected as part of the GLOBE project. (Gore 1998)

Clearly then, Gore had more than a simple classroom tool in mind here. Given his subsequent foray into the climate change debate in *An Inconvenient Truth*, it seems apparent that Gore envisioned a tool that could not only educate children but also help shape and enhance their sense of connection to the earth's environment. Like many past discourses surrounding educational technology, the digital globe here was being envisaged as a partial answer to current social, political, cultural and ecological ills through the "education" of a future generation. Perhaps most

[5] As a means of aggregating multiple sources and forms of geo-located information and data, or as he summarised it, a "multi-resolution, three-dimensional representation of the planet, into which we can embed vast quantities of geo-referenced data".

notable about Gore's vision of a digital earth is that it was particularly global in outlook and civic in nature. Careful not to assert that such a project should be the exclusive domain of a single public enterprise or a single private company, Gore argued "obviously, no one organization in government, industry or academia could undertake such a project. Like the World Wide Web, it would require the grassroots efforts of hundreds of thousands of individuals, companies, university researchers, and government organizations" (Gore 1998). In the event, one organization (and a private one at that) did undertake such a project. As Lisa Parks has noted of Google Earth in another context:

> Google Earth transforms the sovereign territories of all of the world's nation-states into visual, digital, navigable and privatized domains (largely) owned by one US corporation, Google . . . Google claims copyright ownership of all of these frames unless they contain images originally classified as public domain. (NASA satellite images, for instance, remain public domain when they are part of Google Earth's databases.) In Google Earth the satellite image may be obscured or undated, but the Google brand is never lost. (Parks 2009: 541–542)

With this in mind it is particularly revealing to consider what happened to Al Gore's vision and ask why it was overtaken by Google Earth. Not least, such an analysis tells us a great deal about the current course and politics of environmental mediation.

In 2000 the incoming Bush administration is said[6] to have seen the digital earth moniker as a "political liability" and dropped Gore's vision in favour of a private, corporate delivery of a digital globe. Thus, with the Bush administration, the digital globe was dropped as a phrase and a concept. Ironically, however, its use continued in the wider scientific and academic community and was given formal state support in China where the International Society for Digital Earth was founded in 2006. On first site the Bush administration's rejection of the digital earth on the grounds of its association with the previous Democratic administrations Vice President looks like a geo-political error; especially given the fact that it provided the means by which China could step in to support the project. However, such an analysis fails to account for the fact that the digital earth that subsequently emerged in the US was far more in line with the neo-conservative preference for market led, corporately owned cultural, technological and economic innovation. As a corporate and privately owned media form, Google Earth may not have represented a publicly driven civic project (as NASA's digital earth – the World Wind program – would have) but it did reinforce the free market capitalist ideology of the neo-conservatives. The significance of this distinction could not be more profound for the way in which the environment and the simulation of the environment is mediated in the digital realm. This is not because it results in a different or edited account of the changes taking place and subsequently visualised, but because it is symptomatic of broader features that characterise the current approach to the environment in the world's most industrialised country.

[6]For specific detail on this point see the website of the 5th International Symposium of the Digital Earth available at: http://www.isde5.org/history.htm

Google Earth and Environmental Representation

Citing David Harvey's description of neoliberalism in her analysis of the "Crisis in Darfur" layer available on Google Earth, Lisa Parks points out that the intellectual property previously classified as public domain undergoes an ambiguous transition in status when accessed through the privatised interface of Google Earth. As she states, "public and private intellectual properties are intermixed within Google Earth so that their ownership status is unclear and becomes relatively indistinguishable" ((Parks 2009): 541–542). All of this is, she suggests, symptomatic of the neoliberal project to convert public property rights into privatised property rights. What Parks argues of the "Crisis in Darfur" layer in Google Earth is equally true of the data-collection and visualisation of the earth's environment in the program. Google Earth symptomatizes a wider privatised, neo-liberal and uncoordinated approach to the climatic changes taking place upon the earth. To paraphrase Parks, with Google Earth the digital corporation as opposed to the state, an international agency, or an NGO becomes the primary mechanism for distributing and visualising information about the environment.

More important than the public or private origin of the environmental data visualised in Google Earth's interface, then, is the fact that *it is presented as information now visualised by a private, corporate entity*. That is to say that Google's capacity as a private corporate entity to visualise the earth's past and future environments sends the message that such concerns are the natural domain of a market based cultural media form. When we consider that Microsoft owner, Bill Gates, is now one of the largest current contributors to geo-engineering research[7] (and a private individual at that) one might wonder if there is a correlation between the planet's climate, its mediation/remediation as a techno-economic IT industry project and the functional structure of neo-liberal capital. Indeed, at the opposite end of the ecological scale, Eugene Thacker has argued that a very similar correlation exists in the form of the global genome: a complex interrelation between biotech, infotech and corporate capital (Thacker 2005).

In a political environment in which governmental administrations do not maintain support for public calls to construct a digital globe, it should come as no surprise that they should also fall behind on the public coordination of, and debate over, geo-engineering to an extent that leads wealthy private individuals to conclude that they must step in to fill the gap. Of course it is more complex than this; the debate around geo-engineering research is fraught, to say the very least, with concerns that public and governmental agencies must take into account but which private individuals can more easily sidestep by operating outside the regulatory framework associated with public money. Nevertheless Google Earth has emerged as a cultural object subject to the development and ownership of a private corporation. If nothing else,

[7]See especially: http://www.guardian.co.uk/environment/2012/feb/06/bill-gates-climate-scientists-geoengineering

this is symptomatic of a context in which the planet's environment is symbolically mediated within the primacy of the neo-liberal market.

It would be problematically simplistic, however, to suggest that Google Earth as an interface, and as a machinic platform (Gurevitch 2013, 2014) functions solely according to the logic of a large global corporation. Indeed, the reception of Google Earth videos and Google Earth 'play tours' is an area deserving further research. The machinic nature of the Google Earth interface makes it a platform open to many uses and many readings by users across the world. The Google Corporation itself has set up a 'Google Earth outreach' arm with the stated aim of increasing access to, and diversifying use of, its programme. As Google's promotional/tutorial video (that has already passed two million views) on YouTube makes clear, its outreach programme is aimed at increasing non-profit making organisational access to the communicative capacities of its interface.[8] Beyond Google's own auspices, there have been numerous occasions in which the interface's potential to function to the detriment of neo-liberal structures of governance has gathered mainstream media interest. Military complaints that national security and military secrecy has been compromised (Scutro 2007) to broader governmental concerns that the programme could function to the advantage of terrorists planning attacks (Open Source Center 2008) suggest Google Earth is hardly monolithic in its capacity to structure the use and the readings of its interface.

To return to the individuated consumption of Google Earth however, there is another level of privatisation inherent in the interface that reconnects us to the distinction between the original whole earth photography genre and the specificity of the way in which contemporary digital earth software works. This distinction revolves around the centrality of the virtual camera in the digital earth interface. As Mike Jones (2007) has explained in his work on the virtual camera, its function marks a very different development of the concept of image capture from the camera forms that went before it and should not be confused as being a simulated equivalent of its physical predecessor. Without detailing the many productive distinctions that Jones lays out, it is worth noting first and foremost that the virtual camera is to some degree a by-product of games culture (another media form that powerfully connects visualisation and the child user). As such, objects do not exist in a physical or virtual reality before the virtual camera captures them (as with for instance the earth in the Blue Marble photograph of 1972), but instead are called into being as and when they are required. The socio-cultural and philosophical implications of this are considerable when dealing with the digital earth for a number of reasons.

Where the photograph of the earth was a cultural artefact subject to the logic of mechanical reproduction (Benjamin 2007) in which all consumers of the

[8]Though even here we are still returned to the neo-liberal context in which Google Earth has flourished: a tab in the outreach section marked 'grants' leads the user to a page that encourages organisations and individuals ('developers') to apply for grants for 'enterprise level products'. The question of Google's sovereignty over the allocation of this funding is entirely absent as are related questions: what kinds of 'organisations' qualify as worthy, which individuals will be deemed quality 'developer' material, or even what constitutes an 'enterprise level product'?

representation beheld the same image, the computer program works according to a very different logic. The virtual camera sets up a relationship between the viewer/user and an image form that is structured around the delivery of specifically requested, individuated, visualised information, which means that the "object" that comes into being for the user is precisely that; an object, manufactured in real time, solely for the purposes of the viewer's consumption and disposable at the point of satisfaction. What this means for users of the digital globe is that they are constructed by its interface as scopic gods, presiding over an earth object that can be visualised, viewed, reconfigured and disposed of at a whim. While the question of disposal should not be presented in too deterministic a light (a user is hardly likely to assume that disposal of a digital earth at the end of a session equates to the disposability of the earth itself), the question of human computer interaction is ripe for further analysis. To paraphrase Jones, with Google Earth the user of the digital globe has transitioned from a spectatorial position characterised by a shift from "eye to 'I'" (Jones 2007: 228). Furthermore, the user of the digital globe becomes accustomed, not only to a privatised interface (on multiple levels) but also to a representation of the earth as an engineered object fundamentally reconfigurable according to the logic of the computer program. Sea levels can rise, but they can also drop. Ice caps can melt but just as quickly grow. Everything is potentially subject to the logic of the undo button.

Whether geo-engineering has been researched and implemented or not, the philosophical process whereby the planet's environment is both simulated and gamed by users (whether they be adults or classroom children) is now being played out millions of times a day. What this means for the social and cultural acceptance of geo-engineering only time will tell (for more on this see Gurevitch 2013, 2014). It seems a supreme irony that whilst many in the environmental movement have argued for years that the combination of consumer capitalism and western individualism poses the greatest threat to the earth's long-term environmental future, such a future should now be represented via an interface so integrated with that philosophical reality. With this we are brought, by way of conclusion, to the future of environmental representation in the digital globe.

Future Eventualities of the Mediated Environment and the Digital Globe

When considering the future possibilities for the digital globe and their possible effects upon environmental representation, it seems unlikely that Google will retain its primacy indefinitely. There are many public offerings already in existence and there are many new digital globes coming online, supported by states (such as China and India) that have space programmes in their own right. How this plays out in the future could determine a great deal about the future of whole earth environmental representation. Similarly, whether states support new digital globes or not, it is inevitable that a programming task that was once technologically, economically and computationally expensive will become easier as time (and Moore's law of exponential return) goes on. With this in mind we can, to continue a theme of the chapter, say that digital globes are currently in their infancy. Like all futurism, attempting to predict exactly where they will go from here would be both unwise and of questionable scholarly usefulness anyway. As we have seen throughout this chapter, however, it is not so much the digital globe itself that is of most significance so much as the way in which the earth's environment is visualised and represented. Rather, it is the competing cultural, political, economic agents that constitute the context in which the digital globe comes to function that in turn led to the visualisation of the environment. It is hard to know where the neo-liberal project of the past two decades will end up but if the past half decade have been anything to go by, it seems unlikely that it will continue to function according to "business as usual". Already GIS research scientists are envisaging futures for the digital globe in which ubiquitous computing allows for people to function as a crowd-sourced human centred "nervous system" observing and recording environmental changes to the earth in real-time (De Longueville et al. 2010). Such future visions open up the possibility that a future digital earth could be, as Al Gore optimistically envisioned in the mid 1990s, "a 'collaboratory' – a laboratory without walls for research scientists seeking to understand the complex interaction between humanity and our environment." (Gore 1998). Equally, however, such developments could simply mark the widening and deepening of the anthropocene; the contemporary era so named to describe the human induced changes being carved into the planet's geology. Whatever transpires for the future of the digital earth and the mediation of the planet's environment, it is clear that the interactions between data-collection, representation and reconstitution of the earth's ecology has changed indefinitely from the whole earth photographs that ushered in a generation of mediated ecological thinking. As McLuhan would have argued, our planet is increasingly an industrialised, informational artefact.

References

Benjamin, W. (2007). *Illuminations*. New York: Schocken Books.

Brotton, J. (1999). Terrestrial globalism: Mapping the globe in early modern Europe. In D. Cosgrove (Ed.), *Mapping* (pp. 71–89). London: Reaktion Books.

Cosgrove, D. (1994). Contested Visions: One-World, Whole-Earth, and the Apollo Space Photographs. *Annals of the Association of American Geographers*, 84(2), 270–294.

Castels, M. (2010). *The power of identity*. Chester: Blackwell.

Dekker, E. (2007). Globes in Renaissance Europe. In D. Woodward (Ed.), *The history of cartography* (Vol. 3, Part 1, pp. 135–159). London: University of Chicago Press.

De Longueville, B., Annoni, A., Schade, S., Ostlaender, N., & Whitmore, C. (2010). Digital Earth's nervous system for crisis events: Real-time Sensor web enablement of volunteered geographic information. *International Journal of Digital Earth, 3*(3), 242–259.

Foresman, T. W. (2008). Evolution and implementation of the Digital Earth vision technology and society. *International Journal of Digital Earth, 1*(1), 4–16.

Franklin, S., Lury, C., Stacey, J. (2000). Spheres of life. In *Global nature, global culture*. London: Sage.

Gore, Al. (1998). *The digital Earth: Understanding our planet in the 21st century*. http://www.isde5.org/al_gore_speech.htm. Accessed 1 July 2013.

Guo, H. D., Liu, Z., & Zhu, L. W. (2010). Digital Earth: Decadal experiences and some thoughts. *International Journal of Digital Earth, 3*(1), 31–42.

Gurevitch, L. (2013). The digital globe as climatic coming attractions: From theatrical release to theatre of war. *The Canadian Journal of Communication, 38*(3), 333–356.

Gurevitch, L. (2014). Google warming: Google Earth as eco-machinima. *Convergence Journal, 20*(1), 85–107. doi:10.1177/1354856513516266.

Hale, J. (2007). Warfare and cartography, ca 1450 to ca. 1640. In W. David (Ed.), *The history of cartography* (Vol. 3, Part 1., pp. 719–737). London: University of Chicago Press.

Haraway, D. (1997). *Modest witness: Feminism and technoscience*. London: Routledge.

Jones, M. (2007). Vanishing point: Spatial composition and the virtual camera. *Animation: An Interdisciplinary Journal*, 2(3), 225–243. London: Sage.

Kagan, S., & Schmidt, B. (2007). Maps and the early modern state: Official cartography. In W. David (Ed.), *The history of cartography* (Vol. 3, Part 1). London: University of Chicago Press.

Lovelock, J. (1995). *The ages of Gaia: A biography of our living Earth*. London: Norton Books.

Lovelock, J. (2006). *The revenge of Gaia: Why the Earth is fighting back and how we can still save humanity*. London: Penguin.

McLuhan, M. (1962). *The Guttenberg galaxy: The making of typographic man*. Toronto: University of Toronto Press.

Morse, V. (2007). The role of maps in later medieval society: Twelfth to fourteen century. In W. David (Ed.), *The history of cartography* (Vol. 3, Part 1). London: University of Chicago Press.

Parks, L. (2009). Digging into Google Earth: An analysis of "Crisis in Darfur". *Geoforum, 40*, 535–545.

Poole, R. (2008). *Earthrise: How man first saw the Earth*. London: Yale University Press.

Redfern, M. (2003). *The Earth*. Oxford: Oxford University Press.

Roberts, M. (1998). Baraka: World cinema and the global culture industry. *Cinema Journal, 37*(3), 62–82.

Scutro, A. (2007). Satellite imagery raises security questions. *Navy Times*. Available at: http://www.navytimes.com/news/2007/08/navy_insecurity_070819w/. Accessed 1 Aug 2012.

Shupeng, C., & van Gendren, J. (2008). Digital Earth in support of global change research. *International Journal of Digital Earth, 1*(1), 43–65.

Thacker, E. (2005). The Global Genome: Biotechnology. London: MIT Press.

The Open Source Center. (2008). *The Google controversy—Two years later.* [Report]. URL: http://www.fas.org/irp/dni/osc/google.pdf [1st June 2013].

Watts, P. (2007). The European religious worldview and its influence on mapping. In W. David (Ed.), *The history of cartography* (Vol. 3, Part 1). London: University of Chicago Press.

White, F. (1998). *The overview effect: Space exploration and human evolution*. Virginia: The American Institute of Aeronautics and Astronautics.

Woodward, D. (2007). Cartography and the renaissance: Continuity and change. In W. David (Ed.), *The history of cartography* (Vol. 3, Part 1). London: University of Chicago Press.

GIS as Media Through the Geoweb

12

Daniel Sui and Bo Zhao

Introduction

This chapter first traces the evolution of GIS from its earlier focus on automated cartography (organizing geospatial information) to its current status as media, with an emphasis on the geoweb (organizing information geospatially). Using Adams' (2009) tetradic framework on geographies of media, this chapter then presents a preliminary analysis on geographies of media and mediated geographies as manifested through the ever-growing geoweb in the age of big data and open science. As the geoweb increasingly becomes both the container of and the contained by spaces and places, fruitful research can be carried out at the intersection of GIScience and communication geography by examining spaces and places as both contents and contexts for the geoweb.

Using the nascent evidence that emerged in the late 1990s, Sui and Goochild (2001) speculated that geographic information systems (GIS) were rapidly becoming part of the mass media. Based upon the proposition of GIS as media, we were able to link GIS with theories in media studies such as Marshall McLuhan's laws of media, which consider modern media as modifiable perceptive extensions of human thought (Sui and Goodchild 2003). Remarkable conceptual and technological advances in GIS have been made during the past 10 years (Sui 2014). The goal of this chapter is to provide an update on the "GIS as media" argument made over 10 years ago, and to discuss its implications on the geographies of media and mediated geographies using recent developments in communication and media geography.

As geographic information systems moved from earlier models of running on stand-alone desktop computers or workstations to the World Wide Web, their

D. Sui (✉) • B. Zhao
Department of Geography, The Oho State University, Columbus, OH, USA
e-mail: sui.10@osu.edu; jakobzhao@gmail.com

S.P. Mains et al. (eds.), *Mediated Geographies and Geographies of Media*,
DOI 10.1007/978-94-017-9969-0_12

primary function as a tool for sharing and communicating knowledge about the Earth's surface became more obvious. Currently, we have thousands of websites offering a variety of mapping or geospatial services. Indeed, the launching of on-line mapping tools such as Google Maps/Earth, Microsoft's Virtual Earth/Bing Maps, Apple Maps, NASA's World Wind, and crowdsourced OpenStreetMaps validated our speculation (Ball 2005; Sui 2005). The fact that GIS have also increasingly been recognized as media by software-tool developers and vendors is indicated by the names they choose for their products: GeoMedia, Spatial Media, Map TV, or MapTube. Supported by a corps of volunteers (Cowen 2007), NavTeq's Map Reporter program offers further evidence that GIS have become media both metaphorically and literally.

The accelerated incorporation of location into digital information and communication technologies has fostered major shifts in GIS and geographical practices. Different neologisms have been invented to describe these phenomena, including "volunteered geographic information" (VGI) (Elwood et al. 2011; Goodchild 2007), "neogeography" (Turner 2006; Warf and Sui 2010; Wilson and Graham 2013), "spatial media" (Elwood and Leszczynski 2013; Wilson and Stephens 2015), the "geospatial web" (Scharl and Tochtermann 2007), and "the geoweb" (Elwood and Leszczynski 2011; Leszczynski 2012). In this chapter, we use the term geoweb in a very broad and inclusive sense – to indicate the merging of geographical (location-based) information with the abstract information that currently dominates the Internet. Defined this way, the term geoweb includes the following: GIS, spatial media, locative media, geomedia, and geospatial web. Folded into the explosive growth of the geoweb, space and place information contributed by users through various APIs (application programming interfaces) and graphic user interfaces for general users, GIS are no longer tools used exclusively by trained experts. Instead, GIS – now increasingly part of the geoweb – have become media for the general public for constructive dialogues and non-verbal interactions about important social issues through websites like mapstory.org, storymap.knightlab.com, and storymaps.arcgis.com.

Furthermore, the past 5 years have also witnessed the accelerated convergence of GIS with social media – this is something Sui and Goodchild (2001) did not anticipate when they initially tried to reconceptualize GIS as media around 2000. This new role of GIS as social media can be understood from two perspectives (Sui and Goodchild 2011). First, various users and contributors of on-line mapping sites have formed their own virtual community for exchanging information. Google Maps, Bing Maps, Apple Maps, and Yahoo! Maps have attracted user communities from tens of thousands to millions. Within 2 years of moving ArcGIS on-line, ESRI's ArcGIS.com website has attracted a community of over 300,000 world-wide (Dangermond 2011). Second, interactions of on-line GIS users or neogeographers (Turner 2006) or neocartographers (Liu and Palen 2010) are not confined to cyberspace. A growing number of these actions have resulted in meetings in person and activities in real places. For example, participants of OpenStreetMap (OSM) in both North America and Europe have been organizing mapping parties over weekends to work together and map the road networks for their communities. OSM even gives specific instructions on how to organize these mapping parties (http://wiki.openstreetmap.org/wiki/Mapping_Weekend_Howto).

GIS as media constitutes a fundamental paradigm shift in GIS, from the old model of an intelligent assistant serving the needs of a single user seated at a desk, to a new role as part of the geoweb in which GIS act as media for communicating and sharing knowledge about the planet's surface with and among the masses. During that process, the geoweb not only brings people together in cyberspace but also attracts people to meet in person for the common good of their community. The paradigm change also implies a simultaneous shift of technical focus, from local performance to network bandwidth, and increases interest in issues of semantic interoperability in place of earlier concerns with syntactic interoperability. In other words, sharing requires a common understanding of meaning, as well as a set of common standards of format.

In retrospect, the concept of GIS as media only captures half of the story. During the past 5 years media in general, and social media in particular, have also become increasingly equipped with mapping and location-based features. In other words, media are increasingly becoming part of the geoweb. This new trend of media as geoweb can be understood from two perspectives. First, the mainstream media (TV, newspapers/magazines, radios etc.) are increasingly relying on geoweb and geospatial technologies to report news and to tell their stories to the general public. Nowadays, Google Earth or Bing Maps are almost an integral part of the TV broadcasting of everything from weather and traffic conditions to major stories. News organizations of every size use GIS (Herzog 2003), and a growing number of news media have incorporated geospatial servers and mapping functionality as part of their websites: the Chicago Tribune and Time magazine use ESRI's MapStudio (recently renamed as MapShop), while the Seattle Times and Guardian use Google Maps. Using the geospatial mapping server hosted by the Guardian, for example, the public can create custom maps from data disclosed by Wikileaks (http://www.guardian.co.uk/media/wikileaks).

Media as geoweb can also be understood from a second growing perspective, that is social media are increasingly location-based. Social media, led by Facebook, Twitter, LinkedIn, etc., have been described as one of the defining characteristics of Web 2.0 technologies. Moving beyond the static pages of earlier Web sites, Web 2.0 technologies are more interactive by enabling users to contribute content in real-time. The phenomenon of social media is not only transforming the scene of computing but also stimulating social change of various kinds (Qualman 2012). The development of location-based social media during the past 2 years has moved social media from cyberspace to real place, or at least knit the two together in novel ways. Most location-based social media allow users to know and see on a map where their friends are physically located at a particular time. Geo-tagged tweets can now be used to create Twitter heatmaps depicting global Twitter activity during the social media week (for details, http://socialmediaweek.org/blog/2011/02/17/twitter-heatmaps-depicting-global-twitter-activity-during-social-media-week). Social media's convergence with location-based services (primarily based upon social media users'

GPS-enabled cell phones and also personal status tracking devices such as Nike fuelband, Jawbone Up, Fitbit Flex, Smart Activity Tracker etc.) represents the latest development of the so-called locative social media or location-based social networking (Thielmann 2010) for sharing personal location information, which can be grouped into three major categories: A. Social check-in sites (e.g., Foursquare, Glympse, Jiepang, and NextDoor); B. Social review sites (e.g., Yelp, Tellmewhere, Groupon, and Blippy); C. Social scheduling/events sites (e.g., Plancast, Meetup, GonnaBe, Eventful, Upcoming, Geoloqi, WePopp, and Willo). We concur with Lapenta (2011, 2012) that the convergence of location-based services with social media is forming a new social navigation system that will transform society profoundly along multiple fronts.

More broadly speaking, the recent convergence of traditional mass media with geoweb can be seen as an extension of the long history of journalism cartography (Monmonier 1999). Journalism is a professional activity that is inherently embedded in space and time (Gasher 2015). Where has always been one of the fundamental questions guiding journalists, along with who, what, when, why, and how. So we are not surprised to see what may amount to a spatial turn in journalism and the traditional media. Journalist Krissy Clark (2011) recently observed that "The best journalism is like a map. It shows where you are in relation to others; it provides a sense of topography, a glimpse into a new world, or a better understanding of a familiar one. Ideally, journalism helps citizens and communities discover where they are, so they can better decide where they are going (YouTube video, no page number)." As demonstrated by the news organizations and location-based media so far, geoweb and on-line mapping have played a very important role in location-based storytelling (http://storymaps.esri.com/home).

This chapter represents the first attempt to systematically examine the geoweb in general and GIS in particular as media from the perspectives of media/communication geography. The chapter is structured as follows. The first section provides a short update on the meaning of "GIS as media" and recent convergence of GIS with geoweb and media. The second section presents an overview of the tetradic framework developed by Adams (2009, 2011), followed by a preliminary analysis of geographies of the media and mediated geographies in the context of the geoweb using Adams' framework. The last section contains further discussions and conclusions.

Understanding the Geoweb: Adams' Tetradic Framework

One of the major break-throughs in media and communication geography is the publication of Adams' (2009) *Geographies of Media and Communication*, in which Adams developed a quadrant framework (Fig. 12.1) for studying geographies of media and mediated geography. According to Adams (2009), space and place provide both the content and context for various media through which we can possibly know our world (and ourselves). Following Adams' (2009) lead, we argue that geographic studies on emerging geoweb should incorporate perspectives of

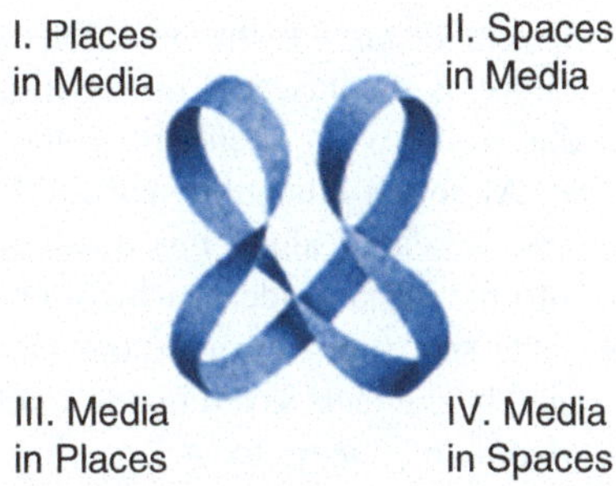

Fig. 12.1 Tetradic framework for geographies of media [Source: modified after Adams (2009)]

space and place (the horizontal dimension of Fig. 12.1) and coding/representation and spatial organization (the vertical dimension of Fig. 12.1).

According to this framework, questions related to geographies of media can be addressed through quadrant I (media in places) and quadrant II (media in spaces) whereas issues related to mediated geographies can be studied effectively through quadrant III (places in media) and quadrant IV (spaces in media). As Sack (1980) and Casey (1993) have so cogently argued, conceptualizations of space and place have a rich history and do not lend themselves for easy summary. According to Tuan (1977), space has often been associated by geographers with freedom, movement, flows, distance, potential, and abstraction whereas place often implies confinement, stability, proximity, meaning, and the concrete. To understand the geography of the world, we need both the space and place perspectives as they intersect with media and communication.

Adams (2009) argues that both media and communication play crucial role in process of both space- and place-making and at the same time both space and place influence the content and outcome of communication (Fig. 12.1). Space, place, and media are indeed mutually constituted, as Adams (2009) observed that "without space and place as the (a) priori frameworks for experience communication is meaningless, yet without communication we would not be able to conceive of space and place (p. 5)." The double Mobius strip at the center of the figure is used to signify the mutual constitution among space, place, and media. When media is examined from both space and place perspectives, four core themes can be raised: (I). places in media; (II). spaces in media; (III). media in places, and (IV). media in spaces.

Places in media: this quadrant focuses on how place or place images are represented in media. Place is a center of meaning and attention. Place formation is often a result of social interactions over time and their sedimented layers of meaning, some of which derive from mediated experiences (e.g. hearing about a park on the news) while others derive from direct, embodied encounters (e.g. having a picnic in the park). Further, people's sense of place typically invokes an emotional and affective dimension, which is often conveyed via a hybrid of words, pictures, audios and videos. Place in media has been a dominant theme in media and communication geography studies.

Spaces in media: this is a functional topological space – also known in the literature as mediascape, information space, or information landscape. "Space in media" focuses its attention on the virtual geographies of what is connected to what. Although intangible and fluid in most cases, "space in media" is real in its effects. Research along this quadrant aims to tease out the topology of the flow of information and ideas, which tends to be stable despite relocation of nodes in physical space. For example one can go on vacation but the social network space defined by contacts saved in one's cell phone stays the same.

Media in places: focus on what kind of communication occurs at particular place and what kind of actions or reactions that particular kinds of communication and media events can trigger at a particular locale. Media cannot be understood without making note of how they are materially encountered as objects, devices, substances and interfaces that people sense and manipulate in particular places. The material presence of a book, telephone, radio or computer is part of place's rich mixture of subjectivity, power, emotion, and affect. Media in place often involves practices, performance, boundary creation, and non-representational (or more than representations) aspects of communication.

Media in spaces: this quadrant is concerned with the geographical layout of communication infrastructure and the space of flows created by signals moving through infrastructure – sometimes referred to as footprints of the communication infrastructure. Precisely where the nodes and links are located at any given time, and the coordinates through which information and communication flow, are key issues to be addressed in themes related to "media in space."

Media are contained by the spaces in which they develop, and yet they also produce spaces of human development. Spaces and places contain media yet are also contained by media. Adams' framework artfully imbeds the complementary views, which are the key to addressing the major questions in media geography research. Adams (2011) tried further to articulate a fifth approach that is capable of transcending the dichotomies of space/place and content/context because it is often difficult to maintain distinctions between space and place and content and context. According to Adams (2011), instead of the neatly delineated four quadrants, we should also think about what lies at the center of these four interaction quadrants. By focusing on the intersecting center, we can move away from assumptions of scale that always implicit in the space/place dichotomy and the presumptions of causality that underlie the content/context dichotomy. Instead of hierarchy, the center of the four quadrants also impels us to think in terms of heterarchy whereby something might be contained but also at the same time be a container – a situation that applies to communications as well as to both spaces and places.

Geographies of Media and Mediated Geographies: The Case of the Geoweb

Using the tetradic framework developed by Adams, this section conducts a preliminary analysis of the geoweb.

Places in Media

Until about 2000, the development of GIS has been predominantly driven by "space" perspectives. Features on the earth surfaces are represented as points, lines, and polygons, inheriting a cartographic, perpendicular view. With GIS gradually transformed into media, "place" perspectives have gained ascendancy during the past 20 years (Goodchild 2011; Sui 2011). Both virtual places and real places are increasingly presented in the geoweb. Figure 12.2a is a map of on-line communities, from which one can detect the fundamental changes of various virtual places in recent years. For real places nowadays, if you are searching Columbus, Ohio in Google Maps, you will not only see major road networks in the city, but also a vast array of geo-tagged on-line information, such as photos, videos, Google Street View, Wikipedia entries, and many other forms of social media and user-generated content. In sharp contrast to the dry spatial information by GIS of the previous era, the geoweb/geomedia present users with much richer, multimedia platial information (Fig. 12.2b). Indeed, more than anything else, Google's new motto – "Google Maps = Google in Maps"- is rapidly transforming the entire web into a searchable database and presenting the search results in map form. Or in other words, Google has shifted from organizing geospatial information (the original focus of Google Maps) to organizing information geospatially (the focus of Google IN maps). As a result, we have an abundant data source to study and understand various places and place images in the new media.

Similar to the traditional media, how place(s) are being presented or described in the geoweb/geomedia can be controversial due to a variety of historical, cultural, and geo-political reasons, especially when territorial disputes are involved. Sometimes even how a place name is labeled on some of the on-line maps can be hotly contested. For example, the disputed islands in East China sea near Taiwan, depending on which version of the Google Maps one uses, can be labeled Daoyudao (China), Dianyutai (Taiwan), or Senkaku (Japan) (Fig. 12.2c). In late November of 2013, China also issued a map showing what was considered by China as the air recognition zone, which has not been recognized by the international community. In particular, Japan and the U.S. have made a united front strenuously opposing the idea of a new air recognition zone proposed by China. The Japanese government has, in fact, requested that local authorities and state-run universities stop posting Google maps on their websites because of some issues with the naming of several disputed islands. Japan is taking issue with the non-Japanese names that Google has used to refer to the islands that are still being contested by Japan, China, South Korea, and Russia. Included in the names being contested are the Senkaku Islands mentioned above, the Takeshima Islands in the Sea of Japan, which is currently controlled by South Korea and known to them as Dokdo Islands, and four islands off Hokkaido called the Northern Territories by Japan but referred to by Russia, who still hold the islands, as the Southern Kurils. Also involved in the dispute are the efforts of South Korea to rename the Sea of Japan as the East Sea. Japan also issued notices in early 2013 that electronic maps using the term East Sea are "incompatible" with the country's stance over these islands.

Fig. 12.2 Places in media (Source: screen capture from the web by the authors) (2a Source: http://www.huffingtonpost.com/2010/10/08/map-of-online-communities-2010_n_755545.html)

Trying to be neutral in territorial disputes, Google has said they only act on the "fair information" they receive from the states about their stance on several territorial issues, but they refused to comment on specific cases. The Japanese government has informed its state agencies to use maps that were compiled by the Geospatial Information Authority of Japan, but most have not done so because using Google Maps is still more convenient than any other existing mapping programs. As part of its corporate policy, Google has aimed to avoid getting involved in international territorial disputes, and whenever possible, has been trying to produce customized maps and labeling according to each individual country's claims. For example, the maps of the political boundaries of the Himalayan region provided by Google Maps in the US are quite different from those provided by Google Maps in China and India. Depending on where the geographic query is issued (primarily based upon their IP addresses), users can expect different results about how the Himalayan region or the Diayudao/Senkaku islands as a place are being represented in geoweb (Goodchild et al. 2012).

Spaces in Media

It has been known to geographers and scholars in other disciplines for quite some time that media/communication forms its own functional topological space. By zooming in who is communicating with whom, the "spaces in media" quadrant can help us better understand the formation of community without propinquity. Regardless the specific types of media – whether oral, print, or electronic, a particular medium's footprint can always be understood by the topology of communication (Adams 2010). For example, based on data and communication patterns as depicted in the Old Testament, researchers at Crossway.org were able to produce a map using IBM's state-of-the-art visualization program ManyEyes and data mining techniques (Fig. 12.3a). By probing into the topology of communication patterns of the characters in the new Testament with this interactive social-network mapping tool we gain a new perspective on the relative positions of Jesus' disciples and their network proximity to other Biblical figures (http://www.crossway.org/blog/2007/01/mapping-nt-social-networks).

We have also been following the discussions and debates on PM 2.5 (particulate matter in the air that is 2.5 μm or smaller in size) on Sina WeiBo – Chinese version of microblogging service similar to Twitter. Using a purpose-built crawler, we collected about 250,000 statuses, re-posts and comments from about 127,000 WeiBo users from October, 2012 through June 2013 with a 6 day gap in January, 2013. These quarter million WeiBo users are from a large crowd all over China, and they formed their own community – a formidable force pressuring the government to make meaningful policy changes regarding PM2.5 standards and monitoring. Again, to understand their communication typology, we took Adams' "space in media" perspective and visualized the WeiBo data using the OpenOrd tools (Fig. 12.3b). The curves refer to the communications between two users. These communications can be reposts and comments. The dots refer to WeiBo users who

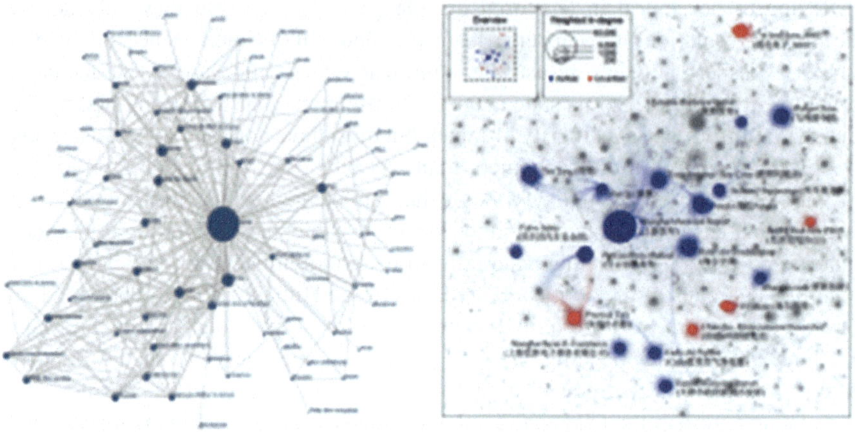

a. social networks in new testament
b. (Source: Crossway)

b. the virtual communities in Sina Weibo debating PM2.5 (Source: authors)

Fig. 12.3 Spaces in media

have published posts (tweets) which were collected through the WeiBo crawler procedure. Basically, they are collected through the Sina WeiBo search-topic API, which requires that the topic must contain a hashtag term (a term with hashtags at the front and end, such as #XXXX#), and also, one of the keywords (pm2.5, air quality, air pollution, haze and etc.) must be within the hashtag term (such as #pm2.5 in Beijing#). Moreover, the size of the dot is based on the value of the degree (the degree stands for the frequency of communication between the given dot and other ones). It is also interesting to notice that in this WeiBo media space, the important role government agencies and some corporations played as opinion leaders. We found that while micro-blogs are capable of empowering citizens to advance an environmental cause, social media have been successfully used by the Chinese government as a tool for social monitoring/control and by big corporations as a platform for profiting from air pollution (Kay et al. 2014).

Due to the lack of a central perspective and the flux-like state of information and interactions, Sina WeiBo user interaction resembles Deleuze and Guattari's (1987) rhizome: a structure with no end and no beginning where all points interrelate and where there is no privileged perspective. Instead of focusing on the topography on the ground or relative to the earth's surface, "spaces in media" perspective is concerned with complex topologies knitted together by a medium and the opportunities it provides to get from here to there by following inter-textual/hypertextual links or connections. We believe that the growing interest in studying "spaces in media" coincides with the recent turn from topography to topology in geographical thoughts, and the post-mathematical topological thoughts could lend us new approaches to studying various new dimensions of "spaces in media" (Secor 2013; Martin and Secor 2013;).

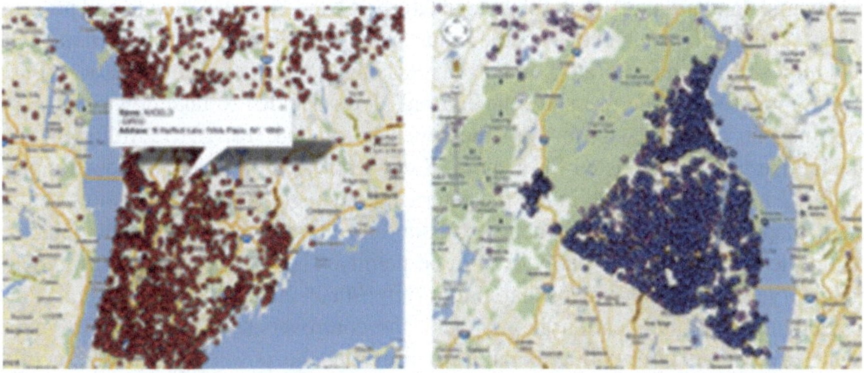

Fig. 12.4 Media in places in: the gun owner next door (Source: http://www.lohud.com)

Media in Places

The media in places quadrant in Adams' framework is concerned with the effects of media at particular places and what kinds of actions or reactions a particular kind of communication or media event can trigger at a particular locale. Amid renewed national calls for tougher gun control in the wake of the horrific school shooting in Newtown, Connecticut, some Lower Hudson Valley residents in New York, perhaps reflecting a sentiment of concerned citizens elsewhere, would like lawmakers to expand the amount of information the public can find out about gun owners. To meet this public demand, a local newspaper – The Journal News – published a report on Dec. 23, 2012 on gun ownership in the Lower Hudson Valley and an accompanying interactive online map displaying information on gun permit holders in Westchester, Rockland, and Putnam counties (Fig. 12.4).

Immediately after the publication of the maps, described by journalists as the gun owner next door, we witnessed a sensational public outcry about the appropriateness for publishing these maps. The controversy provides us a potent case study on "media in place." More specifically, "media" in this is the on-line interactive gun-owner map and the "place" is the Lower Hudson Valley. The actions and reactions taken after the on-line interactive map is unveiled really demonstrates the power of media in place.

While some residents in those two counties covered by the map expressed gratitude, opposition to the publication of these maps and the database seem to have dominated the debate. Gun permit holders argue that their privacy was violated; law enforcement agencies were concerned that these maps with individual gun owners' addresses could encourage burglary; gun-lovers fear that these maps serve no useful purpose except stigmatizing the mostly law-abiding gun owners in the country. The newspaper staffers were harassed by threatening emails and a suspicious package containing white powder was sent to the newspaper. As a result of these threats, the newspaper had to hire armed guards to protect their staff (see the Gun Debate section at www.lohud.com for details about the controversy).

By mid-January of 2013, the state legislature of New York weighed in and passed a new state law to give gun owners the right to remove their names from the database and impose a 120-day moratorium before releasing the gun permit information. Due to the mounting public pressure, the Journal News decided to take down the map and the gun permit database from their website on Jan. 18, 2013. Although the interactive map only existed for a short 27 days, it was viewed and searched more than 1.2 million times. Although the interactive maps have been taken down, the static maps showing the gun owners next door (and all related comments and debates about these maps) will probably exist in perpetuity. These gun maps may go down in history as one of those maps that matter (http://mapsthatmatter.blogspot.com).

For media/communication geographers, the controversies surrounding the gun map serve as an informative example to illustrate the powerful idea of "media in places." In this case, the gun map produced in the local newspaper could trigger diverse actions and discussions – primarily in and about a particular place, as well as smaller places (residences) within that place. There are other examples that potentially illustrate similar points on "media in places." The former social networking app "Girls Around Me" hosted by Apple App Store triggered a humiliating public condemnation due to concerns about stalking and other criminal activities and assaults on women whereas a similar Chinese product known as MoMo has been in wide use in China and criminal activities (e.g. for use in prostitution) have already been reported (http://news.gmw.cn/newspaper/2013-11/21/content_2466744.htm). Whereas websites such as harrasmaps.org do raise awareness for sexual harassment in Cairo, they also attract certain women who aim to gain attention from men to congregate in those areas that have a higher incidence of harassment. Even more ironic is that the report that the massive number of surveillance cameras installed in Chinese cities became dysfunctional due to the heavy smog, which led to the terrorist suicidal attack on Tiananmen Square on Oct. 28, 2013 (http://www.dailymail.co.uk/news/article-2488075/Chinese-smog-terror-risk-Pollution-bad-security-cameras-sensitive-sites-longer-film-haze.html).

All the examples show that media can have powerful impacts on what happens in particular places, although what happens (or does not happen) is contingent upon multiple social, cultural, and even environmental factors more traditionally associated with place.

Media in Spaces

The transformation of GIS from a stand-alone mapping/spatial analysis tool to incorporation in web-based geomedia for story-telling and communication is made possible by a combination of technological advances, including, but not limited to, high-speed broadband Internet infrastructure, cloud storage and computing, high resolution remote sensing, smart phones, and 4G networks. So, it is also very important to understand the quadrant "media in space" in Adams' framework – a geographical layout of communication infrastructure and the signals created by moving through that infrastructure.

As some parts of the world are flooded by big data and people are increasingly connected in a shrinking world, we must also be keenly aware that this world remains a deeply divided one – both physically and digitally. Nowhere is this digital divide more visible that the geographical distribution of the global IT infrastructure, especially viewed from the "media in spaces" perspective. In terms of the distribution of nodes (root server), submarine cables (links), and global information flows (Fig. 12.5), we can tell a consistent story at the global level – dominance of North America and Europe with growing influence of Asia (especially East Asia led by China and Japan) – more or less is the expected story of the global economy.

The divide in physical infrastructure also manifests itself in accessibility of the Internet at the global level. While a large majority of people in North America and Europe have access to the Internet (with Internet penetration rates at 78.3 % and 58.3 % respectively by the end of 2011), two-thirds of humanity do not have access to the rapidly expanding digital world; the world average Internet penetration rate is 30.2 % with Asia (23.8 %) and Africa (11.4 %) trailing at the bottom. The geographical distribution of new digital data stored in 2010 reflects both the digital divide and uneven development levels across the globe, with the developed world or Global North (North America and Europe) having 10–70 times more data than the developing world or Global South (Africa, Latin America, and Asia) (Manyika et al. 2011).

In the context of geographic information (and to some extent other types of data as well), the biggest irony remains that Murphy's law is still at work – information is usually the least available where it is most needed. We have witnessed this paradox unfolding painfully in front of our eyes in the Darfur crisis in northern Sudan in 2006, the aftermath of the Haiti earthquake (2010), and the BP explosion in the Gulf of Mexico in 2011. Undoubtedly, how to deal with big data in a shrinking and divided world will be a major challenge for GIS and Geography in the years ahead. The strengths, weaknesses, opportunities, and threats (SWOT) of VGI for improving the spatial data infrastructure (SDI) are quite different in the two global contexts of North and South (Genovese and Roche 2010).

Furthermore, as Gilbert and Masucci (2011) show so clearly in their recent work on uneven information and communication geographies, we must move away from the traditional, linear conceptualization of a digital divide, concerned primarily with physical access to computers and the Internet. Instead, we must consider the multiple divides within cyberspace (or digital apartheid) by taking into account the hybrid, scattered, ordered, and individualized nature of cyberspaces (Graham 2011). Indeed, multiple hidden social and political factors are at play for determining what is or is not available on-line (Engler and Hall 2007). Internet censorship (Warf 2011; MacKinnon 2012), power laws (or the so-called 80/20 rule) (Shirkey 2006), homophile tendencies in human interactions (De Laat 2010; Merrifield 2011), and fears of colonial and imperial dominance (Bryan 2010) are also important factors to consider for the complex patterns of digital divide and uneven practices of geoweb at multiple scales on the global scene.

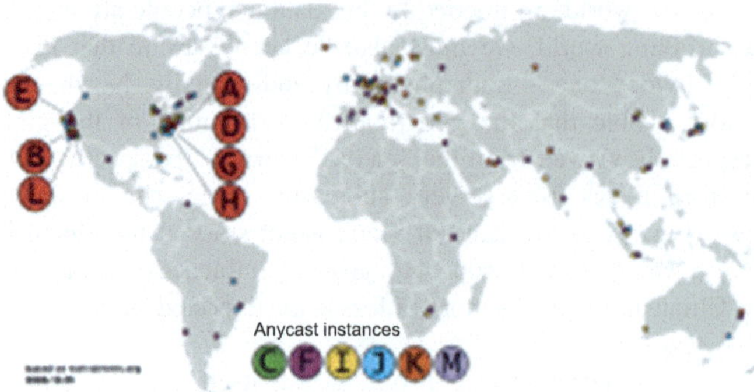

a. Global distribution of root servers (source: Matthäus Wander http://en.wikipedia.org/wiki/File:Root-current.svg)

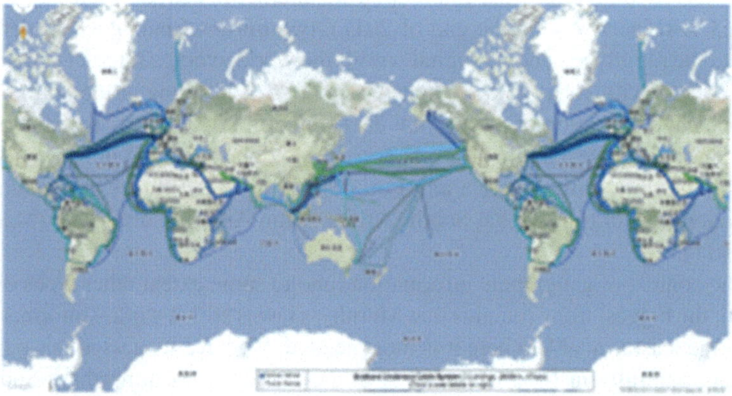

b. existing submarine internet cable routes (source: Gregory's map)

c. global internet traffic (source: Telegeography)

Fig. 12.5 Media in spaces

Summary and Conclusion

Geography as an intellectual enterprise has undergone many fundamental changes during the first decade of the twenty-first century. Among these changes, growing interests among geographers in media and communication studies – the communicational turn (Adams 2009) – have been gaining momentum as evidenced by the formation of the communication specialty group of the AAG (www.communication-geography.org) and the publication of new geography journals and textbooks devoted exclusively to media/communication geography (www.aetherjournal.org). Geography's communication turn has been paralleled by a "spatial turn" in media studies during the past 10 years (Morley 2006; Döring and Thielmann 2009), focusing on the complex interaction among people, space, and place as mediated by various media (Jansson 2007, 2009). Ground-breaking work has been reported by scholars in multiple disciplines under the general rubric of the spatial turn in media studies, ranging from the highly technical work of harvesting social-network data to the search for emerging geographical patterns of new social interactions enabled and revealed by social media. The tetradic framework developed by Adams artfully links these dual trends (geography's communication turn and media studies' spatial turn). This chapter has shown that fruitful research can be conducted using this framework for the geoweb.

The convergence of GIS with media in geoweb will continue to transform GIS in fundamental ways that are intellectually exciting, technologically sophisticated, and socially relevant. In this chapter, we have conducted a preliminary analysis on theme of this edited volume – geographies of media and mediated geographies – using geoweb as the primary example. We firmly believe that one productive way for GIScience to proceed is to ride on the discipline of geography's communication turn and the spatial turn in media studies. This chapter has demonstrated that fruitful research may be conducted if GIScience research is more intellectually engaged with media and communication geography. The geoweb has increasingly become both container and the contents, with regard to both spaces and places. GIScience can benefit greatly by using communication geography to examine the content and context of the geoweb.

Acknowledgement We would like to express our gratitude to Paul Adams, Samuel Kay, and two anonymous reviewers for their constructive comments, which have significantly improved this chapter. Thanks are also due to Jens Blegvad of the Ohio State Center for Urban and Regional Analysis for making our data collection possible. Research for this article is a result of a collaborative research project, "A GIScience Approach for Assessing the Quality, Potential Applications, and Impact of Volunteered Geographic Information," funded by the US National Science Foundation (2009–1014), award # 1048100. The usual disclaimers apply.

References

Adams, P. C. (2009). *Geographies of media and communication*. Chichester: Wiley-Blackwell.

Adams, P. C. (2010). Networks of early writing. *Historical Geography, 38*, 70–89.

Adams, P. C. (2011). A taxonomy for communication geography. *Progress in Human Geography, 35*(1), 37–57.

Ball, M. (2005). Google verifies that GIS is media. *GeoWorld,* December Issue, 5.

Bryan, J. (2010). Force multipliers: Geography, militarism, and the Bowman Expeditions. *Political Geography, 29*(8), 414–416.

Casey, E. S. (1993). *Getting back into place: Toward a renewed understanding of the place-world*. Blooming: Indiana University Press.

Clark, K. (2011). Stories everywhere. http://www.youtube.com/watch?v=KPEcbesTeUs. Accessed 1 Dec 2013.

Cowen, D. (2007). *Why not a geo-wiki corps*. Paper presented at the 2007 NCGIA Workshop on VGI. Available on-line at: http://ncgia.ucsb.edu/projects/vgi/docs/position/Cowen_paper.pdf. Accessed 25 Apr 2014.

Dangermond, J. (2011). Living maps – making collective geographic information a reality. http://www.youtube.com/watch?v=vIbGwY3aG0g. Accessed 1 Dec 2013.

De Laat, P. B. (2010). How can contributors to open-source communities be trusted? On the assumption, inference, and substitution of trust. *Ethics and Information Technology, 12*(4), 327–341.

Deleuze, G., & Guattari, P. F. (1987). *A thousand plateaus: Capitalism and schizophrenia* (Vol. 2). Minneapolis: University of Minnesota Press.

Döring, J., & Thielmann, T. (2009). *Mediengeographie*. transcript Verlag. Bielefeld: Germany

Elwood, S., & Leszczynski, A. (2011). Privacy, reconsidered: New representations, data practices, and the geoweb. *Geoforum, 42*(1), 6–15.

Elwood, S., & Leszczynski, A. (2013). New spatial media, new knowledge politics. *Transactions of the Institute of British Geographers, 38*(4), 544–559.

Elwood, S., Goodchild, M. F., & Sui, D. Z. (2011). Researching volunteered geographic information: Spatial data, geographic research, and new social practice. *Annals of the Association of American Geographers, 102*(3), 571–590.

Engler, N. J., & Hall, G. B. (2007). The internet, spatial data globalization, and data use: The case of Tibet. *The Information Society, 23*(5), 345–359.

Gasher, M. (2015). Geographies of the news. In S. P. Mains, J. Cupples, & C. Lukinbeal (Eds.), *Mediated geographies and geographies of media* (pp. 127–140). Dordrecht: Springer.

Genovese, E., & Roche, S. (2010). Potential of VGI as a resource for SDIs in the North/South context. *Geomatica, 64*(4), 439–450.

Gilbert, M. R., & Masucci, M. (2011). *Information and communication technology geographies*. Kelowna: Praxis (e) Press.

Goodchild, M. F. (2007). Citizens as sensors: The world of volunteered geography. *GeoJournal, 69*(4), 211–221.

Goodchild, M. F. (2011). Formalizing place in geographic information systems. In L. M. Burton, S. P. Kemp, M.-C. Leung, S. A. Matthews, & D. T. Takeuchi (Eds.), *Communities, neighborhoods, and health: Expanding the boundaries of place* (pp. 21–35). New York: Springer.

Goodchild, M. F., Guo, H., Annoni, A., Bian, L., de Bie, K., Campbell, F., & Jackson, D. (2012). Next-generation digital earth. *Proceedings of the National Academy of Sciences, 109*(28), 11088–11094.

Graham, M. (2011). Time machines and virtual portals The spatialities of the digital divide. *Progress in Development Studies, 11*(3), 211–227.

Herzog, D. (2003). *Mapping the news: Case studies in GIS and journalism*. Redlands: ESRI Press.

Jansson, A. (2007). Texture: A key concept for communication geography. *European Journal of Cultural Studies, 10*(2), 185–202.

Jansson, A. (2009). Beyond "other spaces": Media studies and the cosmopolitan vision. *The Communication Review, 12*(4), 305–312.
Kay, S., Zhao, B., & Sui, D. Z. (2014). Can social media clear the air? A case study of the air pollution problem in Beijing. *The Professional Geographer* (accepted, in press). doi:10.1080/00330124.2014.970838
Lapenta, F. (2011). Geomedia: On location-based media, the changing status of collective image production and the emergence of social navigation systems. *Visual Studies, 26*(1), 14–24.
Lapenta, F. (2012). Geomedia based methods: Exploring the theoretical and methodological tenets of the localization and visualisation of mediated social relations with direct visualisations techniques. In S. Pink (Ed.), *Advances in visual methodology* (pp. 131–150). London: Sage.
Leszczynski, A. (2012). Situating the geoweb in political economy. *Progress in Human Geography, 36*(1), 72–89.
Liu, S. B., & Palen, L. (2010). The new cartographers: Crisis map mashups and the emergence of neogeographic practice. *Cartography and Geographic Information Science, 37*(1), 69–90.
MacKinnon, R. (2012). *Consent of the networked: The worldwide struggle for Internet freedom.* New York: Basic Books.
Manyika, J., Chui, M., Brown, B., Bughin, J., Dobbs, R., Roxburgh, C., & Byers, A. H. (2011). Big data: The next frontier for innovation, competition, and productivity. http://www.mckinsey.com/Insights/MGI/Research/Technology_and_Innovation/Big_data_The_next_frontier_for_innovation. Accessed 25 Apr 2014.
Martin, L., & Secor, A. J. (2013). Towards a post-mathematical topology. *Progress in Human Geography*. doi:10.1177/0309132513508209.
Merrifield, A. (2011). Crowd politics, or, 'Here comes everybuddy'. *New Left Review, 71*, 103–114.
Monmonier, M. (1999). *Maps with the news: The development of American journalistic cartography*. Chicago: University of Chicago Press.
Morley, D. (2006). *Media, modernity and technology: The geography of the new*. New York City: Routledge.
Qualman, E. (2012). *Socialnomics: How social media transforms the way we live and do business.* Hoboken: Wiley.
Sack, R. D. (1980). *Conceptions of space in social thought: A geographic perspective*. Minneapolis: University of Minnesota Press.
Scharl, A., & Tochtermann, K. (2007). *The geospatial web: How geobrowsers, social software and the Web 2. 0 are shaping the network society*. Berlin: Springer.
Secor, A. J. (2013). Topological city. *Urban Geography, 34*(1), 430–444.
Shirkey, C. (2006). Power laws, weblogs, and inequality. In J. Dean, J. W. Anderson, & G. Lovink (Eds.), *Reformatting politics: Information technology and global civil society* (pp. 35–42). New York: Routledge.
Sui, D. Z. (2005). G for Google? What can we learn from Google's success so far. *GeoWorld, 15*, 18–22 (March Issue).
Sui, D. Z. (2011). GIS: From space to place. *GeoWorld,* March Issue, 13–15
Sui, D. Z. (2014). Opportunities and impediments for open GIS. *Transactions in GIS, 18*(1), 1–24.
Sui, D. Z., & Goodchild, M. F. (2001). GIS as media? *International Journal of Geographical Information Science, 15*(5), 387–390.
Sui, D. Z., & Goodchild, M. F. (2003). A tetradic analysis of GIS and society using McLuhan's laws of media. *The Canadian Geographer/Le Géographe canadien, 47*(1), 5–17.
Sui, D. Z., & Goodchild, M. F. (2011). The convergence of GIS and social media: Challenges for GIScience. *International Journal of Geographical Information Science, 25*(11), 1737–1748.
Thielmann, T. (2010). Locative media and mediated localities: An introduction to media geography. *Aether: The Journal of Media Geography, 5a*, 1–17.
Tuan, Y. F. (1977). *Space and place: The perspective of experience*. Minneapolis: University of Minnesota Press.
Turner, A. (2006). *Introduction to neogeography*. Sebastopol: O'Reilly Media.
Warf, B. (2011). Geographies of global internet censorship. *GeoJournal, 76*(1), 1–23.

Warf, B., & Sui, D. Z. (2010). From GIS to neogeography: Ontological implications and theories of truth. *Annals of GIS, 16*(4), 197–209.

Wilson, M. W., & Graham, M. (2013). Situating neogeography. *Environment and Planning A, 45*(1), 3–9.

Wilson, M. W., & Stephens M. (2015). GIS as media. In S. Mains, J. Cupples & C. Lukinbeal (Eds.), *Mediated geographies/Geographies of media*. Berlin: Springer.

13 GIS as Media?

Matthew W. Wilson and Monica Stephens

...media are increasingly becoming like GIS (Sui and Goodchild 2011, p. 1739).

Introduction

In early 2012, Google announced Project Glass, a device worn as glasses that augments the user's vision through communication with the user's mobile device. The glasses display images for the user's eyes and listens for audible commands. In a promotional video released by Google to announce the device (2012), the user navigates urban spaces augmented by the device, including city streets, the public transportation network, and even the internal, private spaces of a bookstore. While moving around the city, the user takes phone calls and responds to instant messages, taking photos of what he sees and enters into a video chat. The device functions through minimal gestures and human interaction, responding through voice commands and the framing of the user's gaze. Project Glass anticipates a future where the seemingly immaterial online spaces created through the internet are made material through embodied manifestations of the mobile user. In this present

Here, we continue to think through this question posed by Dan Sui and Mike Goodchild in 2001 and again in 2011.

M.W. Wilson (✉)
Department of Geography, University of Kentucky, Lexington, KY, USA
e-mail: matthew.w.wilson@uky.edu

M. Stephens
Department of Geography, University at Buffalo (SUNY Buffalo), Buffalo, NY, USA
e-mail: mgstephe@buffalo.edu

S.P. Mains et al. (eds.), *Mediated Geographies and Geographies of Media*,
DOI 10.1007/978-94-017-9969-0_13

future, media are both social and spatial, and understanding the phenomenality of life requires hybrid approaches and new conceptual footings.

While the promotional video displays new forms of device interactivity, the types of technology engaged (SMS, navigation, video phone calls, etc.) are becoming more common forms of mobile device functionality. Here, social and spatial mediation of life is represented as everyday, part of the mundane experience of urban, continuously connected living. In the context of these socio-technological developments, we consider the following question: in the wake of emerging social/spatial analytical tools for the study of this media saturation, how might thinking GIS as media help to better understand the social implications for digital information technologies? Indeed, geographers and sociologists are recognizing the increased importance of online social media in material, everyday relationships. As a result new methods of analysis and representation are emerging to both better understand and visualize patterns of social and spatial interactivity. In this chapter, we examine the reconceptualization of spatial technologies such as GIS *as* media.

To better understand the relationship between GIS and contemporary forms of social media, this chapter proceeds in three sections. First, we trace forward a long-standing debate around the ontological status of GIS as a tool or as a science by suggesting that a third position – GIS as media – might be productive (following Sui and Goodchild 2001, 2011). Second, we discuss the possibilities of bridging GIS and SNA in a study of mediation, although we recognize that a full treatment of these intersections are beyond the scope of this chapter (see Radil et al. 2012). Third, we briefly propose the political implications of such a move, before finally offering some concluding thoughts, to open a discussion of the relationships between GIS and media.

GIS: Science, Tool, Media?

Scholars engaged with GIS are more recently self-identifying in more diverse ways that just a decade ago. Now, with the development of the digital humanities and the solidification of critical GIS as a research and teaching subfield, critical human geographers and scholars within the humanities more generally are joining critical social scientists in the application of GIS toward understanding complex social-spatial relations. For instance, scholars like Franco Moretti and Lev Manovich are mapping literature and creating interactive digital media to better understand cultural artifacts traditionally understood through techniques like close readings (Moretti 2005; Manovich 2002). And geographers are increasingly hailed by the humanities to bring a technical sensibility to the traditional artifacts of the humanities (Bodenhamer et al. 2010; Dear et al. 2011).

But it has been a long, winding road toward transforming multi-disciplinary collaborations and working to create spaces within the academy which foster rigorous, if alternative, imaginations for geographic information technologies.

Indeed, the rebranding of GIS as a science (compare O'Sullivan 2006; Wright et al. 1997; Pickles 1997) further locks-in the notion of geospatial technologies as a tool of the scientific method, carrying with it all sorts of ghosts that have caused the critical humanities and social sciences to stand at a distance. In discussion of the posturing of GIS as a science, Wright et al. argue:

> first, the driving technology must be of sufficient significance; second, the issues raised by its development and use must be sufficiently challenging; third, interest in and support for research on those issues must be inadequate in the existing disciplines; and fourth, there must be sufficient commonality among the issues to create a substantial synergy. (1997, p. 357)

Here, the question of 'GIS as a science' is answerable only with regard to the positioning of 'GIS as a tool', instead. In reading Pickles's response to this article, we note the implications of this rub:

> the heuristic categorization of tool use, tool making, and science mis-specifies the issue. Modern science is now so thoroughly shot through with technical apparatus and so closely tied to various ways of controlling nature and society that it is thoroughly technological. Such a relation cannot be avoided or overcome by wishing or defining it away, an effort in futility we would have to characterize as idealism. (Pickles 1997, p. 364)

Pickles attempts to establish the grounds necessary to characterize GIS (either tool or science) as a technoscience, thoroughly saturated by technologies. As such, the move to rebrand GIS begs interrogation of the motives of this branding: what work does it enable? What work does it necessarily delegitimize?

Indeed, the struggle over GIS within the discipline is largely centered around the degree to which proponents recognize the epistemological perspectives enabled by the use of the technology, as well as the particular ontological fixes made necessary. The roots of this struggle might be found in the GIS wars of the early 1990s (Schuurman 2000), or even earlier in the late 1960s and early 1970s post-quantitative critiques in the embrace of a rising marxist geography (Wilson and Elwood forthcoming). And, Pickles (1997) suggests that GIScientists might look toward these debates in the wake of radical geography as a particular continuity.

For instance, Bill Bunge's work in *Fitzgerald* sought to resist the calculating gaze of spatial science, to actually deeply invest in the subjects studied by treating scholarship as action. Bunge sought to understand the everyday lives of those in Detroit, to put a human face on the realities of class and race struggle. Quantitative spatial analysis was a distraction, a disembodied over-simplification of complicated, saturated phenomena. As radical geography created a space in which to launch research programs that interrogated the intersections between political, economic, and social structure in the 1970s and 1980s, geospatial tools became more technically sophisticated, providing visual power to the modeling of physical and human geographies. These developments further entrenched notions that there were disparate and irreducible camps in geography departments that informal chats around the water cooler could not patch over (Taylor 1990; Openshaw 1991, 1992; Taylor and Overton 1991; Smith 1992; Lake 1993; Pickles 1993; Sheppard 1993).

Instead, GIScience and critical geography gain significant (if largely separate) ground in the 1990s, but opportunities for cross-pollination (or anything beyond collegiality) was largely hit and miss (however, see Pickles 1995; Sheppard 1995, 2005). GIS & Society gave way to participatory GIS and critical GIS, and later qualitative GIS (Wilson 2009). These permutations found new radical utilizations of GIS (for instance, Pavlovskaya 2002; Elwood 2006; Knigge and Cope 2006), but largely left questions as to the conditions and implications of the emergence and ubiquity of GIS unanswered by GIScientists (although, see Curry 1998; Pickles 2004). Furthermore, the question of how to engage and intervene in GIScience remains (Schuurman and Pratt 2002; Leszczynski 2009a, b; Crampton 2009b).

Regardless of its status as science or tool, these technologies are taking on new relevance in the context of 'big data', giving rise to the spatial humanities (Bodenhamer et al. 2010), a resurgence in maps as radical artistic expression (Bhagat and Mogel 2008; Harmon 2004), and the utilization and study of the geoweb (Turner 2006; Crutcher and Zook 2009; Stephens forthcoming; Zook and Graham 2007; Crampton et al. 2013, also see FloatingSheep.org). However, the *significance* of the questions posed by the GIS & Society movement did not simply evaporate. With these new permutations of geographic information technologies and a proliferation of their uses should come a deepened commitment to understanding their implications and conditions of development (Wilson and Graham 2013). And perhaps a recognition of GIS as but one technology that mediates social-spatial life is an alternative approach that has been underexplored. Indeed, GIS can be thought as a form of media.

We must also recognize that GIS is enrolled as a tool to study media (e.g. the FloatingSheep.org project). That GIS is both an *object of* and *tool for* study is not new. But the study of social media using GIS places GIS within an established set of practices and technologies that have been utilized to study media (and the networks established by media systems) for decades. Therefore, to understand the implications for thinking GIS as a mediator of social relationships, we turn to the development and contemporary relevance of one such established technique: social network analysis (SNA). SNA represents a significant area of scholarship that underscores the importance of social relations as a key factor in understanding the spatiality of social media.

Mediated Relationships

Online social networking websites like Facebook have undoubtedly altered social relations for some, documented in various ways by scholars in a number of disciplines including the information sciences, cultural and media studies, anthropology and sociology, as well as geography. Indeed, the ubiquity of online social networking has generated ripples across the web, where many websites are turning to Facebook Connect as a way to personalize experiences of the internet. As a result, unprecedented data are being generated through these networks. And sociologists are enrolling this data, with increased processing power and new analytics, to better understand the complex ways in which social life is organized.

Therefore, if we are to think GIS as media, how might we study the mediation that occurs through these new forms of social organization? Media evolution is an important hinge in this discussion: media have evolved from a one-way form of communication, from the information source to the audience (e.g. newspaper/television) into a two-way (where the audience communicates directly back to the source through technologies like email/phone). Most recently, internet scholars argue that as media have further evolved into a group dynamic where everybody can be both the source and consumer (prosumer) of media, we need a different way to understand the dissemination of information (Hogan and Quan-Haase 2010; Fuchs 2009; Ritzer and Jurgenson 2010).

These evolving dynamics amid the rise of internet communication technologies (ICTs) have prompted a need to analyze and provide order to large data sets of social information giving rise to big data and general challenges in the study of the internet as a mediator. These analyses included new SNA arguments such as "scale-free networks" (Barabási 2009) that imply, despite geographic implications, a sense of order to what seems like random social relationships. These new arguments speak back to as well as extend a well-trod argument in geography that implied that "everything is related to everything else, but near things are more related than distant things" (Tobler 1970, 236). Geography used this proposition to inform quantitative techniques that analyze heterogeneity and statistical relationships, forming one foundation of GIScience today (Goodchild and Haining 2004; Miller 2004).

SNA continues to evolve in multiple fields as anthropologists use SNA to understand kinship structures and sociologists use it to understand social structures (Borgatti et al. 2009). As a result, we have many ways of looking at the connections between entities in both the spatial and social realm. The tools of SNA allow us to understand the structure and constraints present in human interactions as well as the agents that change human interactions (Wellman and Berkowitz 1988).

Meanwhile, scholars within the digital humanities are also noting the ways in which social media is impacting everyday life, albeit using methods less understood as a "science of networks" (Watts 2004). This research examines the ways in which social media is enabling new composition practices (Bono et al. 2012) and constituting notions of place through narration (Ralston 2012) as well as examining identity online, such as transgender performance (Foster 2005) and the ways in which bodily experience is figured through the internet more broadly (White 2006).

Furthermore, the effects of the rise of social media are also felt within the academy, as a reconfiguration of mainstream media outlets impacts the work of scholars both in terms of research and pedagogy (Wilson and Starkweather forthcoming; Campbell 2010). The production of academic research is being pushed to new audiences using social media, as journals and other academic publishers use online social networks to increase readers/clicks/downloads amid increasing competition among a handful of publishing house conglomerates.

Just as social media plays a constitutive role in contemporary society (Boyd and Ellison 2007), the rise of spatial media in the form of location-based services and the geoweb has brought about new questions as to the ways in which digital information technologies condition everyday experiences (Wilson 2012). And while earlier versions of these kinds of questions were central to the GIS & Society movement of the 1990s, they are brought into greater focus given the ubiquity of digital spatial information technologies, as these geospatial technologies mediate.

Here, we understand spatial media as technologies that serve to constitute new relations not only to other individual users, but to place as well. Facebook Places, Foursquare, Google Latitude all serve to connect users to each other in place, where place is the driving analytic that enables social relations. Search engines like Bing, Yahoo, and Google use location as a central criteria in the production of search results. This mediatization of space thus draws upon spatial technologies like GIS, without positioning the GIS as a central interface. Therefore, while the GISciences and new spatial media are indeed related and co-implicated, they are not entirely overlapping developments.

While GIScientists are well-versed in the machinations of geospatial data, for instance in terms of ontologies and interoperability (Schuurman 2005, 2006), expertise and decision-making (Nyerges et al. 2006), and spatial analysis, more generally (Goodchild and Haining 2004), they have only recently begun to consider the role of the geospatial web. This has been a slow evolution for GISc, as the central research questions that were considered cutting edge in the early part of the last decade, questions around how to create and study the affordances of web-based GIS (cf. Peng and Tsou 2003), were re-configured by the emergence of mapping APIs and what was generally described by Crampton (2009a) as maps 2.0.

In other words, while entire mapping industries were being reinvented seemingly overnight, the GISciences were stepping carefully, so as to secure their expertise (see Crampton 2010). Meanwhile, the geoweb steadily proliferated alongside much greater developments in 'big data', on the backs of expanding social networking sites such as Facebook and Twitter. Geographers interested in the entrenchment of cyberspace, such as Kitchin (1998), Graham (1998), Dodge (with Kitchin 2001), and Warf (2001), were beginning to be joined by geographers interested in the particular spatial relationships constituted by social and spatial media (Goodchild 2007; Zook and Graham 2007).

As a result, a new specialization of geography scholars was occurring, in the development of methods of qualitative inquiry alongside geoweb data analysis and representation. These geographers, attuned to the social, political, and economic conditions and implications of the internet, created avenues of research that challenged GIScience to better grapple with the diversity of applications and developments around geospatial software, technology and data structures.

The emergence of 'volunteered geographic information' therefore marks an emergent area of inquiry within the GISciences, responding to less academic developments within a field of hobbyists and entrepreneurs called 'neogeography'

(Wilson and Graham 2013). VGI describes an area of GISc that recognizes the opportunities of social and spatial media in producing massive amounts of information that can and should be leveraged to understand human as well as physical geographies. As Sui and Goodchild (2011) document, these developments underscore a blurring of boundaries between GIS and media.

• • • •

As an aside, we recognize that some readers will be interested in the possibilities for drawing together SNA and GIS. When using these technologies to map a social network the object of study is the same but the 'map' depicts a different relationship. In other words, the use of GIS prioritizes Cartesian spatiality as a representational frame for the study of relationality, while SNA prioritizes the network. Social networks frequently depict non-spatial relationships, such as friendship through dyadic mobile phone connections (Eagle et al. 2009); kinship ties that determine social order (Emirbayer and Goodwin 1994); email networks that increase social capital (Wellman 2001); knowledge clusters that succeed or fail based on structural characteristics (Graf 2011); and the relationships and power dynamics embedded in organizations (Rowley 1997). With statistical techniques to identify and quantify each type of social relationship, it is unsurprising that SNA is more common than GIS as a method to conceptualize online social media.

SNA has also been used to understand social media and to graph the relationships among users (Boyd and Heer 2006). The social structures of popular sites, like Facebook have produced a plethora of research across multiple disciplines (Wilson et al. 2012): for instance, the racial clustering among Facebook users is telling about the impact of social media on xenophobia (Lewis et al. 2008). Unfortunately the integration of SNA and GIS to understand the complex social and spatial relationships of social media is rare.

Nonetheless, the possibilities for integration remain a fruitful area of scholarship. For instance, Steve Radil et al. (2012) enrolled SNA with GIS to analyze gang relations in Los Angeles, discovering a dynamism in the networks that structure the expression of spatial relationships (such as violence) between gangs. This study highlights that understanding social phenomena as it occurs in specific locations is aided by an understanding of the social relationships that are co-implicated in specific geographic locations. SNA is not an exclusive domain of analysis along these lines, as qualitative methods and analysis have provided similar contextualizations, but SNA brings a particular computational efficiency to the study of large and dynamic relationships.

Maps as Media: A New Map Politics?

Syria is Iran's only ally in the Arab world. It's their route to the sea. (Mitt Romney to Barack Obama in the final presidential debate, 22 Oct. 2012).

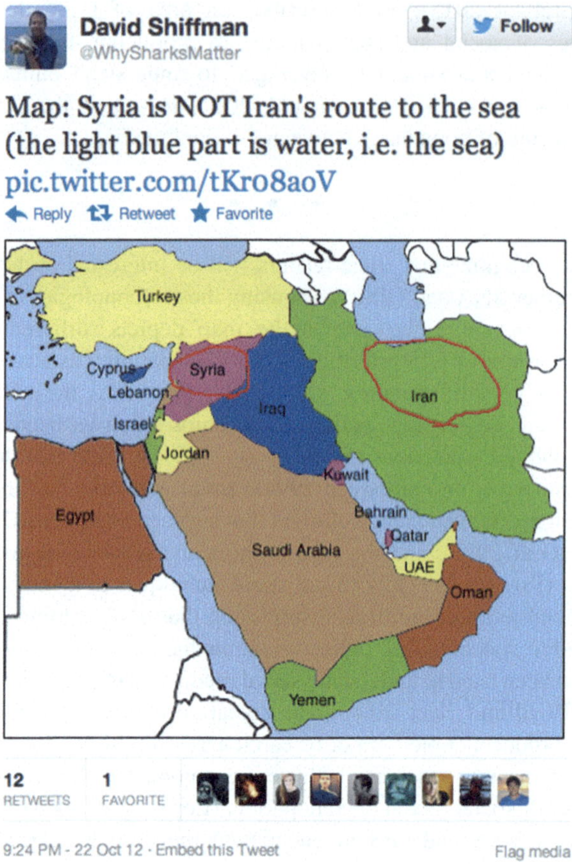

Social and spatial media are changing the relationship between the map and the map reader, towards a model of map interactivity instead of map communication. These shifts occur as maps are increasingly considered expressions of media. In the above image, a Twitter user named David Shiffman (@WhySharksMatter) reacts during the final presidential debate between Mitt Romney and Barack Obama. Romney argues that Syria is Iran's route to the sea, a comment that quickly trended on Twitter as users slapped together maps to provide an alternative geography to that offered by Romney.

This incident highlights a new relationship surrounding mapping – where maps are understood as manipulated media for political voice, drawing upon the authority and neutrality of the map (as an object that is undeniable) in order to participate in and intervene in the mediation of everyday life. These map interactions occur in the context of a dominant conceptualization of the map as a neutral communicative device. The map communication model (MCM), developed by Arthur Robinson in the mid-twentieth century (as an extension of Claude Shannon's mathematical theory of communicative invariance), rendered cartography as a scientific practice –

dominating the bulk of cartographic scholarship and map design thought (Crampton 2010). The MCM establishes the centrality of the map reader in cartographic design, a recognition of the importance to track how map readers *receive* the messages in the map. This meant a further standardization of map-making practices, through concepts like visual hierarchy and visual variables, a quite literal scientific method of experimentation and observation, to develop the most efficient, invariant, and 'bias-free' method of map-based communication.

New spatial media are reconfiguring this relationship, as is illustrated by the Syria-Iran maps that proliferated during and following the final presidential debate. The reader is no longer (perhaps they were never) a simple percipient of maps. The reader interacts within the map, querying and adjusting the scale and scope of the cartographic representation. In some cases, the map is reconfigured and layered with new meanings (Roth 2012). This emerging map interactivity has been largely described as maps 2.0, with map practices and map products increasingly driven by users – many are self-identified 'neogeographers'. These neogeographers and neocartographers rework the MCM model.

This reconfiguration from map communication toward map interactivity underscores the mediation that occurs through GIS, and should provide some pause for the map integrations figured by projects like Google's Project Glass. As maps increasingly become the vehicle for a plurality of expression and multiple documentations of 'truth', their embeddedness in everyday life can mean radical interventions in social interaction, in being in place.

And as the implications for these changes in the model of map interactivity are still being examined, we briefly sketch a few implications that might guide further research into the collisions between GIS and social media. First, the map is undergoing a political transformation in that the power of the map is increasingly complicated and no longer (was it ever?) a secured artifact of legitimacy and authority. In other words, the politics of the map that guided participatory GIS in the late 1990s is now being replaced by a politics of the map guided by crowdsourcing, where the map is entirely mutable toward multiple interests, including the underrepresented and dispossessed. Second, the map is increasingly being put into motion. By this we mean that the assumed temporal fixity of the map is giving way to new media interactions with cartography that represent the dynamisms of space and place. Third, the map is but one representation of a whole assembly of spatial and social relations within a complicated media environment. Here, social network analysis is placed alongside GIS as two toolsets for quantifying and visualizing a multiplicity of relationships.

Conclusions

GIS and social media are increasingly co-implicated in their development, marketing, and proliferation. The logics that enable the corporatization of online social media overlap with the logics of GIS as a new spatial media. One can simply observe the rise of location-based services on mobile devices to begin to recognize these overlaps (Kelley 2011; Wilson 2012). However, what is to be gained by conceptualizing GIS as media? There are certainly business logics that drive the thinking of GIS as media, but what are the intellectual impacts of this re-conceptualization?

In this chapter, we have outlined developments that begin to address these questions, particularly by examining, the emergence of GIS and SNA as tools to better understand, organize, and visualize social-spatial phenomena. As mapping tools like GIS are understood as spatial media, the opportunity to recognize the constraints of GIS is greater. While GIS are excellent for examining the absolute 'spatialness' of phenomena, the more relational understanding of the spatiality of everyday life requires different toolsets and, indeed, different epistemologies. SNA is one such toolset that examines relationships, recognizing very simply (and not unproblematically) that social proximity is distinct from spatial proximity. Networks and spaces abound in the combination of these toolsets.

The social and spatial relationships developing through interactions with technology rely neither on a social or a spatial framework alone. Society is increasingly organized and influenced by geolocalized information and networks have become more local and mobile: witness for example the rise of location-based services as a vehicle to refine the granularity of targeted, strategic marketing. As we return to the advertisement for Google Project Glass (2012), we can begin to witness the convergence of spatial and social media, and in its wake, the refashioning of GIS *as* media.

References

Barabási, A. L. (2009). Scale-free networks: A decade and beyond. *Science, 325*(5939), 412–413.

Bhagat, A., & Mogel, L. (Eds.). (2008). *An Atlas of radical cartography*. Los Angeles: Journal of Aesthetics and Protest Press.

Bodenhamer, D. J., Corrigan, J., & Harris, T. M. (Eds.). (2010). *The spatial humanities: GIS and the future of humanities scholarship*. Bloomington: Indiana University Press.

Bono, J. J., Hisayasu, C., Sayers, J., & Wilson, M. W. (2012). Standards in the making: Composing with metadata in mind. In D. Journet, C. E. Ball, & R. Trauman (Eds.), *The new work of composing*. Logan: Computers and Composition Digital Press/Utah State University Press.

Borgatti, S. P., et al. (2009). Network analysis in the social sciences. *Science, 323*(5916), 892–895. doi:10.1126/science.1165821.

Boyd, D., & Ellison, N. B. (2007). Social network sites: Definition, history, and scholarship. *Journal of Computer-Mediated Communication, 13*(1), 210–230.

Boyd, D., & Heer, J. (2006). *Profiles as conversation: Networked identity performance on friendster*. Paper presented at the Proceedings of the Hawai'i International Conference on System Sciences (HICSS-39), Kauai.

Campbell, D. (2010). The new ecology of information: How the social media revolution challenges the university. *Environment and Planning D: Society and Space, 28*, 193–201.

Crampton, J. W. (2009a). Cartography: Maps 2.0. *Progress in Human Geography, 33*(1), 91–100.

Crampton, J. W. (2009b). Being ontological: Response to "postructuralism and GIS: Is there a 'disconnect'?". *Environment and Planning D: Society and Space, 27*, 603–608.

Crampton, J. W. (2010). *Mapping: A critical introduction to cartography and GIS*. Malden: Wiley-Blackwell.

Crampton, J. W., Graham, M., Poorthuis, A., Shelton, T., Stephens, M., Wilson, M. W., & Zook, M. A. (2013). Beyond the geotag: situating 'big data' and leveraging the potential of the geoweb. *Cartography and Geographic Information Science, 40*(2), 130-139. doi: http://dx.doi.org/10.1080/15230406.2013.777137

Crutcher, M., & Zook, M. A. (2009). Placemarks and waterlines: Racialized cyberscapes in post-Katrina Google Earth. *Geoforum, 40*, 523–534.

Curry, M. R. (1998). *Digital places: Living with geographic information technologies*. London: Routledge.

Dear, M., Ketchum, J., Luria, S., & Richardson, D. (Eds.). (2011). *GeoHumanities: Art, history, text at the edge of place*. London: Routledge.

Dodge, M., & Kitchin, R. (2001). *Mapping cyberspace*. London: Routledge.

Eagle, N., Pentland, A., & Lazer, D. (2009). Inferring friendship network structure by using mobile phone data. *Proceedings of the National Academy of Sciences of the United States of America, 106*(36), 15274–15278. doi:10.1073/pnas.0900282106.

Elwood, S. A. (2006). Beyond cooptation or resistance: Urban spatial politics, community organizations, and GIS-based spatial narratives. *Annals of the Association of American Geographers, 96*(2), 323–341.

Emirbayer, M., & Goodwin, J. (1994). Network analysis, culture and the problem with agency. *American Journal of Sociology, 99*(6), 1411–1454.

Foster, T. (2005). *The souls of cyberfolk: Posthumanism as vernacular theory*. Minneapolis: University of Minnesota Press.

Fuchs, C. (2009). Information and communication technologies and society: A contribution to the critique of the political economy of the internet. *European Journal of Communication, 24*(1), 69–87. http://ejc.sagepub.com/content/24/1/69.short

Goodchild, M. F. (2007). Citizens as sensors: The world of volunteered geography. *GeoJournal, 69*, 211–221.

Goodchild, M. F., & Haining, R. P. (2004). GIS and spatial data analysis: Converging perspectives. *Papers in Regional Science, 83*(1), 363–385.

Google. (2012). Project glass: one day... [YouTube]. Available at: http://youtu.be/9c6W4CCU9M4

Graf, H. (2011). Gatekeepers in regional networks of innovators. *Cambridge Journal of Economics, 35*(1), 173–188.

Graham, S. (1998). The end of geography or the explosion of place? Conceptualizing space, place and information technology. *Progress in Human Geography, 22*(2), 165–185.

Harmon, K. A. (2004). *You are here: Personal geographies and other maps of the imagination* (1st ed.). New York: Princeton Architectural Press.

Hogan, B., & Quan-Haase, A. (2010). Persistence and change in social media. *Bulletin of Science, Technology & Society, 30*(5), 309–315.

Kelley, M. J. (2011). The emergent urban imaginaries of geosocial media. *GeoJournal, 78*, 181–203.

Kitchin, R. M. (1998). Towards geographies of cyberspace. *Progress in Human Geography, 22*(3), 385–406.

Knigge, L., & Cope, M. (2006). Grounded visualization: Integrating the analysis of qualitative and quantitative data through grounded theory and visualization. *Environment and Planning A, 38*, 2021–2037.

Lake, R. W. (1993). Planning and applied geography: Positivism, ethics, and geographic information systems. *Progress in Human Geography, 17*(3), 404–413.

Leszczynski, A. (2009a). Poststructuralism and GIS: Is there a 'disconnect'? *Environment and Planning D: Society and Space, 27*, 581–602.

Leszczynski, A. (2009b). Rematerializing GIScience. *Environment and Planning D: Society and Space, 27*, 609–615.

Lewis, K., Kaufman, J., Gonzalez, M., Wimmer, A., & Christakis, N. (2008). Tastes, ties, and time: A new social network dataset using Facebook.com. *Social Networks, 30*(4), 330–342.

Manovich, L. (2002). *The language of new media* (1 MIT Press pbkth ed.). Cambridge, MA: MIT Press.

Miller, H. J. (2004). Tobler's first law and spatial analysis. *Annals of the Association of American Geographers, 94*, 284–289.

Moretti, F. (2005). *Graphs, maps, trees: Abstract models for a literary history*. London: Verso.

Nyerges, T. L., Ramsey, K., & Wilson, M. W. (2006). Design considerations for an Internet portal to support public participation in transportation improvement decision making. In S. Dragicevic & S. Balram (Eds.), *Collaborative geographic information systems* (pp. 208–236). Hershey: Idea Group.

O'Sullivan, D. (2006). Geographical information science: Critical GIS. *Progress in Human Geography, 30*(6), 783–791.

Openshaw, S. (1991). A view on the GIS crisis in geography, or, using GIS to put Humpty-Dumpty back together again. *Environment and Planning A, 23*(5), 621–628.

Openshaw, S. (1992). Further thoughts on geography and GIS: A reply. *Environment and Planning A, 24*(4), 463–466.

Pavlovskaya, M. (2002). Mapping urban change and changing GIS: Other views of economic restructuring. *Gender, Place and Culture, 9*(3), 281–289.

Peng, Z.-R., & Tsou, M.-H. (2003). *Internet GIS: Distributed geographic information services for the internet and wireless networks*. Hoboken: Wiley.

Pickles, J. (1993). Discourse on method and the history of discipline: Reflections on Dobson's 1983 automated geography. *The Professional Geographer, 45*(4), 451–455.

Pickles, J. (Ed.). (1995). *Ground truth: The social implications of geographic information systems*. New York: Guilford.

Pickles, J. (1997). Tool or science? GIS, technoscience, and the theoretical turn. *Annals of the Association of American Geographers, 87*(2), 363–372.

Pickles, J. (2004). *A history of spaces: Cartographic reason, mapping, and the geo-coded world*. New York: Routledge.

Radil, S. M., Flint, C., & Tita, G. (2012). Spatializing social networks: Using social network analysis to investigate geographies of gang rivalry, territoriality, and violence in Los Angeles. *Annals of the Association of American Geographers, 100*(2), 307–326.

Ralston, D. F. (2012). Where ya at? Composing identity through hyperlocal narratives. In D. Journet, C. Ball, & R. Trauman (Eds.), *The new work of composing*. Logan: Computers and Composition Digital Press/Utah State University Press.

Ritzer, G., & Jurgenson, N. (2010). Production, consumption, prosumption: The nature of capitalism in the age of the digital 'prosumer'. *Journal of Consumer Culture, 10*(1), 13–26.

Roth, R. E. (2012). Cartographic interaction primitives: Framework and synthesis. *The Cartographic Journal, 49*(4), 376–395.

Rowley, T. J. (1997). Moving beyond dyadic ties: A network theory of stakeholder influences. *Academy of Management Review, 22*(4), 887–910.

Schuurman, N. (2000). Trouble in the heartland: GIS and its critics in the 1990s. *Progress in Human Geography, 24*(4), 569–590.

Schuurman, N. (2005). Social perspectives on semantic interoperability: Constraints on geographical knowledge from a data perspective. *Cartographica, 40*(4), 47–61.

Schuurman, N. (2006). Formalization matters: Critical GIS and ontology research. *Annals of the Association of American Geographers, 96*(4), 726–739.

Schuurman, N., & Pratt, G. (2002). Care of the subject: Feminism and critiques of GIS. *Gender, Place and Culture, 9*(3), 291–299.

Sheppard, E. (1993). Automated geography: What kind of geography for what kind of society? *The Professional Geographer, 45*(4), 457–460.

Sheppard, E. (1995). GIS and society: Towards a research agenda. *Cartography and Geographic Information Systems, 22*(1), 5–16.

Sheppard, E. (2005). Knowledge production through critical GIS: Genealogy and prospects. *Cartographica, 40*(4), 5–21.

Smith, N. (1992). History and philosophy of geography: Real wars, theory wars. *Progress in Human Geography, 16*, 257–271.

Stephens, M. (2013). Gender and the GeoWeb: divisions in the production of user-generated cartographic information. *GeoJournal, 78*(6), 981–996.

Sui, D. Z., & Goodchild, M. F. (2001). GIS as media? *International Journal of Geographic Information Science, 15*(5), 387–390.

Sui, D. Z., & Goodchild, M. F. (2011). The convergence of GIS and social media: Challenges for GIScience. *International Journal of Geographical Information Science, 25*(11), 1737–1748.

Taylor, P. J. (1990). GKS. *Political Geography Quarterly, 9*, 211–212.

Taylor, P. J., & Overton, M. (1991). Further thoughts on geography and GIS. *Environment and Planning A, 23*(8), 1087–1090.

Tobler, W. R. (1970). Computer movie simulating urban growth in Detroit region. *Economic Geography, 46*, 234–240. //A1970Y186900003.

Turner, A. J. (2006). *Introduction to neogeography*. Sebastopol: O'Reilly.

Warf, B. (2001). Segueways into cyberspace: Multiple geographies of the digital divide. *Environment and Planning B: Planning and Design, 28*(1), 3–19.

Watts, D. J. (2004). The "new" science of networks. *Annual Review of Sociology, 30*, 243–270.

Wellman, B. (2001). Computer networks as social networks. *Science, 293*, 2031–2034. //000171028700058.

Wellman, B. & Berkowitz, S. (1988). Introduction: Studying social structures. In *Social structures: A network approach* (pp. 1–14). Cambridge: Cambridge University Press.

White, M. (2006). *The body and the screen: Theories of internet spectatorship*. Cambridge, MA: MIT Press.

Wilson, M. W. (2009). Towards a genealogy of qualitative GIS. In M. Cope & S. A. Elwood (Eds.), *Qualitative GIS: A mixed methods approach* (pp. 156–170). London: Sage.

Wilson, M. W. (2012). Location-based services, conspicuous mobility, and the location-aware future. *Geoforum, 43*(6), 1266–1275. doi: http://dx.doi.org/10.1016/j.geoforum.2012.03.014

Wilson, M. W., & Elwood, S. A. (forthcoming). Capturing: Mapmaking. In R. Lee, N. Castree, R. Kitchin, V. Lawson, A. Paasi, S. Radcliffe & C. Withers (Eds.), *Sage handbook of human geography*. London: Sage.

Wilson, M. W., & Graham, M. (2013). Neogeography and volunteered geographic information: A conversation with Michael Goodchild and Andrew Turner. *Environment and Planning A, 45*(1), 10–18.

Wilson, M. W., & Starkweather, S. (forthcoming). Web presence of academic geographers: A generational divide? *The Professional Geographer*.

Wilson, R. E., Gosling, S. D., & Graham, L. T. (2012). A review of Facebook research in the social sciences. *Perspectives on Psychological Science, 7*(3), 203–220. doi:10.1177/1745691612442904.

Wright, D., Goodchild, M. F., & Proctor, J. D. (1997). GIS: Tool or science? Demystifying the persistent ambiguity of GIS as "tool" versus "science". *Annals of the Association of American Geographers, 87*(2), 346–362.

Zook, M. A., & Graham, M. (2007). The creative reconstruction of the Internet: Google and the privatization of cyberspace and DigiPlace. *Geoforum, 38*, 1322–1343.

Crowd-Sourced Augmented Realities: Social Media and the Power of Digital Representation

14

Matthew Zook, Mark Graham, and Andrew Boulton

Crowd-Sourced Augmented Realities

Everyday lives are increasingly experienced with reference to, and produced by digital information. This information ever more includes crowd-sourced and social media content that mediates interactions with and between places and individuals. Ranging from Yelp reviews of restaurants, to check-ins to points of interest, to uploading image and video records of events, these digital practices and performances have emerged as key moments in the process of place-making.

This "social" element of digital annotations and interactions is linked intrinsically with other discourses and practices comprising and reflecting the contemporary urban (and increasingly suburban and rural) experience. The specific forms that these mediations take – the processes and politics in and through which content and code work socially and spatially – are complex and multifaceted. While digital technologies are implicated in the (re)production of exceptional spaces of routinized surveillance such as airports and international borders (Dodge and Kitchin 2004), social media and consumer technologies mediate more fundamentally the mundane practices of urban life, shaping and reflecting the collective experience as particular versions of events and locations are vigorously promoted or sidelined (Graham 2013; Zook and Graham 2007).

M. Zook (✉)
Department of Geography, University of Kentucky, Lexington, KY, USA
e-mail: zook@uky.edu

M. Graham
Oxford Internet Institute, University of Oxford, Oxford, UK
e-mail: mark.graham@oii.ox.ac.uk

A. Boulton, Ph.D.
Humana, Inc., Louisville, KY, USA
e-mail: andrew.boulton@gmail.com

S.P. Mains et al. (eds.), *Mediated Geographies and Geographies of Media*,
DOI 10.1007/978-94-017-9969-0_14

In this chapter, we sketch the contours of a media geography concerned with what we term crowd-sourced augmented realities. In particular, we set out a methodological framework within which to consider the articulations between power-laden social and coded processes working in and through augmented realities. We use the term *augmented reality* in reference to the indeterminate, unstable, context dependent and multiple realities brought into being through the subjective coming-togethers in time and space of material and virtual experience. The social component emphasizes the significance of social media – Twitter, Facebook, Foursquare and other more-or-less mass-market platforms – in shaping engagement with and understandings of place. Aside from the important backdrop of software code and coded rationalities shaping cities as "coded spaces" (cf. Dodge and Kitchin 2005, 198), there is a clear case that can be made around the disciplining effect of "geomedia" (Lapent 2011) – multimedia representations or annotations with a locational dimension – in engendering particular ways of conceiving of and interacting with place. The specific bodily orientations implicit in the usage of handheld (smartphone) devices and the performative theatricality of social media (self-) representations reshape in significant ways the experiential and discursive boundaries of engagement with urban places (Boulton and Zook 2013). More generally, the role of code as an algorithmic ontology circumscribing and limiting representational practice remains underexamined even as the unprecedented openness of social media platforms/practices confers opportunities for the democratization of geospatial representation. Augmented realities then include personalized and social components – content and practices – in the configuration of individualised, time/space-specific and power-laden material/virtual nexuses mediated through technology, information, and code (Graham et al. 2013).

This approach to social media and augmented realities more generally calls attention to the power-laden social-technological processes and relations by which geographically referenced content may acquire a persuasive rhetoric of authority, and thus the duplicity of code (Graham et al. 2013) in naturalising the uneven production and consumption practices inherent within representations of urban and increasingly, rural places (Allen 2003). To better illustrate the process by which these augmented realities are jointly produced and contested we provide a case study of the crowd-sourced annotations created via Google Map Maker and Google Places reviews at the site of Osama Bin Laden's compound in Abbottabad, Pakistan on May 2, 2011 shortly after the raid in which he was killed.

Content, Social Media and the Duplicity of Code

Rather than envisioning the ontologically distinct constructs of cyberspace and material space that were popular in the 1990s, we need to recognise that 'virtual' and material spaces have always been inextricably linked (Nagenbourg et al. 2010; Firmino and Duarte 2010; Graham 2010). We urge a broadened and socio-spatially nuanced concept of augmented realities which emphasizes the potent ways in which virtual information – in particular, geographically referenced content – intersects and helps shape the relationships that undergird our lived geographies.

Unpacking the Duplicity of Code

Dodge and Kitchin (2005) draw a distinction between 'code/space' and 'coded space': that is, between code/spaces in which 'code *dominates* the production of space' (original emphasis, p. 198) and 'coded space' in which code is part of, but incidental to, the production of space. Such a distinction is useful in emphasizing how code acts in more-or-less visible ways, sometimes contingent upon other coded processes and infrastructures, to produce or 'transduce' space via continuously 'reiterated digital practices that create space anew' (Wilson 2011). As Thrift and French (2002) point out, few, if any, aspects of contemporary urban and economic infrastructure are *not* ultimately reliant up on, or in part produced by, code.

Code delimits behaviour in a number of contexts – from requiring adherence to accepted procedures at electronic checkouts to pre-screening 'risky' airline passengers – and engendering behaviours so mundane as to go unquestioned, until the code 'fails' via an error or unexpected rupture. But its action is more subtle than simple compulsion as code does not deterministically produce space, but works through broader "technologies of power" (Foucault 1988, 18) deployed in the management of bodies and populations. In the case of airport security, for example, code becomes part of an ensemble of tactics through which "power is actually deployed" (Rose-Redwood 2006, 474) both in the production of the law-abiding, global traveller, and in the biopolitical totalization of a statistically known travelling public. Moreover, under the guise of convenience and functionality many of the finest-grained, most sustained and intrusive relationships with coded surveillance in contemporary social life are entered into ostensibly willingly by locative media consumers. Specifically, online personalization and segmentation – where search results, advertisements and "suggestions" derive from previously recorded online behaviour – is vastly enhanced by the addition of high-resolution locational information such as that derived from (and required by) GPS tracking inherent to social mobile applications. While the extent to which locative media consumers and social media participants critically consider the privacy implications (and questions of authorship and ownership more generally) vis-à-vis their use of superficially innocuous social platforms, a trend towards pernicious surveillance is both complemented and ameliorated by an increasing participation in practices of "sousveillance" (Dodge and Kitchin 2007).

It is precisely the invisibility of code and the ambivalence of its authorship that make its deployment unnoticed and its operation so hegemonic (Budd and Adey 2009). Although code must continuously be enacted in order to work in the world, a generalized lack of awareness and engagement with code per se arguably alienates citizens a potentially 'key source of creative power' (Dodge et al. 2009, 1284). Instead users' spontaneity is sated within the bounds of preordained customization options and creative potential. Nonetheless, the pervasiveness of social media entails a dramatic broadening of access to and usage of the "tools of the powerful" vis-à-vis (potentially) powerful and socially affective representations of place. Simultaneously, search engines' powerful emphasis on personalization portends a situation in which each individual is presented (however problematically) with a best guess amalgam of their proclivities, prejudices and preferences – "filter bubbles" (Pariser 2011) – derived from their recorded locational and social interactions rather than a universal narrative or set of representations.

The algorithmic and social spaces of user-generated spatial data provide an important arena in which to examine the character of the uneven and power-laden practices through which digital representations of place are authored, ordered, and momentarily stabilized in the production of augmented realities.

Power Through Social Content and Code

All spatial representations are both the products and producers of specific configurations of power relations (Harley 1989; Allen 2003; Pickles 2004), and thus a key question is whether the ways power in augmented reality is constructed and exercised is novel. Our account of power in augmented realities draws on the Harleian tradition of cartographic critique in asserting the contingent and incomplete nature of the map as socially constructed and embedded discourse. While Harley cannot have anticipated emerging geoweb/geomedia and social media phenomena we nevertheless draw attention to the prescience of his anti-foundationalist, Foucault-infused accounts of cartographic power (Crampton 2001). His insistence, for example, on breaking down the categorical distinction between propaganda maps and other (truthful) maps represents a necessary first step towards recognizing that all representations of place – including digital, crowd-sourced maps – are products of and productive of, social relationships and associated power relations. Building upon Harley's foray, cartographic scholars have further developed the critique of the notion of a unitary author of a map imposing a vision on a reader without resistance and emphasized the situated and contingent process by which representations are constructed by tracing the "genealogy of power discourses" associated with maps (Crampton 2001, 243).

In the context of "social" representations of place – the net outcome of multiple users' contributions and the coded processes by which these are combined, promoted, ordered, etc. – it makes less sense to think explicitly in terms of individual authorial intent than in terms of the work done by maps and mapping practices in supporting and creating particular kinds of subjects (Wood et al. 2010). In such an approach, while power is productive – of mapping subjects from amateur cartographers (Wilson 2011) through Foursquare users "checking in" at locations throughout the day, to the casual user performing a Google search from a cell phone – it is not inherently oppressive or negative (Elwood et al. 2012). Indeed, in the case of user-generated geographical data we note the broadening of mapping publics with access to modes of cartographic representation *and* a marked socio-spatial unevenness obscured by this veneer of democratization. As such it is important to carefully think through the overlapping ways power relations are differentially manifest over time and space through digital augmentation.

In the remainder of this chapter, we suggest four kinds of power manifest differentially over space and time for different individuals in the coming-together of material and virtual spatialities: first via primarily social actors in the cases of *distributed power* and *communication power*, and second via the action of software in the cases of *code power* and *timeless power*. While these categories are separated by dynamic and porous boundaries they provide a useful heuristic for understanding some of the ways in which power is enacted within augmented realities.

Distributed Power

Distributed power refers to the complex and socially/spatially distributed authorship of the geospatial content that forms the basis of augmented realities and consumer locative media products. Additionally, it recognizes the similarly diffuse power of users to view, promote, rank, comment, redistribute (etc.) geographically referenced information. While distributed power points to the relative and potential democratization of spatial data production and consumption under the auspices of crowd-sourced spatial data, it also recognizes the potential that a veneer of democratization – the notion that because everyone can contribute, all voices will be heard – leads to a potential depoliticization of code's work and the power relations inherent in geospatial authorship.

Significant portions of the content comprising augmented realities is derived from social and crowd-sourced/collaborative endeavours. Places can, for instance, be represented as: a Wikipedia article; myriad mentions on Twitter; multiple photos uploaded to Flickr and Picasa; and YouTube videos created by locals or tourists, just to name a few. Some argue that this explosion in user-generated content fundamentally challenges traditional gatekeepers of knowledge (Shirky 2010) and potentially gives voice to any of the two billion people online. While this libertarian vision of technological equality is attractive, reality falls far short of it (Graham 2010; Graham and Zook 2011). For example, user-generated content is characterised not only by occasional unreliability and poor quality (e.g. (Carr 2007)), but also the potential for relatively untraceable manipulation of representations of place such as the CIA editing Wikipedia articles about Iran (Fildes 2007).

Of more concern, however, is that the reliance upon a volunteer labour force means that a majority of the world's population are excluded from the technologies and connectivity needed to create content about place: a group dubbed by Manuel Castells (1998) as the "fourth-world". In short, some people are simply better positioned and hence more likely to engage in content production due to their available resources (i.e. access to information technologies and greater amounts of free time). Less obvious, but equally important, is that even among those with time and access, only a small minority tend to engage in the production of geospatial content (Crutcher and Zook 2009). Research has shown that the internal technical and social structures of content platforms such as Wikipedia privilege some contributors over others, e.g., the cultural convention of a relatively aggressive style in arguments has privileged the work of men over women (O'Neil 2009). This results in a relatively small group of people authoring representations in augmented reality (Glott et al. 2010) and a correspondingly high power to influence representations of places.

While access to this group is relatively open compared to historical practice, it does not mean that its representations are neutral or even draw from stakeholders relevant to the topic at hand. Moreover, despite its promise, the distributed power of user-generated content also makes it extremely challenging to understand what the embedded biases in content are, and to work to address them (Boulton 2010). And at its most extreme, the expectation within distributed power that because everyone can contribute all voices will be heard could lead to a depoliticisation of geospatial content. The expectation that all viewpoints contribute to the production and reproduction of spatial representation, is belied by disproportional power wielded by those with the time, inclination, education, resources, and network positionalities necessary to make their ideas visible.

Communication Power

Communication power refers to the ways in which augmented representations are brought into being: the ways in which particular representations gain prominence, while others may be unheard by, or incommensurable with, prevailing modes of representation. Communication power refers then to the differential capabilities of particular groups, individuals or interests to assert and to stabilize particular representations of place, and the practices by which those capabilities are realized.

The ability to filter or prioritize content creates considerable power to claim or stabilize particular visions of place. This communication power is enacted in two principle ways. First is the power that comes with the ability to effectively use channels of communication to promote a specific goal of representation. In Castells' (2008, 47) formulation of communication power this is the role played by 'programmers' who continually program and reprogram the goals of any particular network, in this case the networks of authorship behind geocoded content. A particularly important limitation of programming in communication power is the assumption of a common language. Geographic information in Estonian, for instance, is meaningless for most Thais and thus places can have separate (and conflicting) representations in either language. Moreover, smaller linguistic groups would have less communication power relative to *lingua francas* such as English which have the ability to create much more visible representations.

The second aspect of communication power is the ability to not only create and interpret content, but also to recirculate, repackage, and even contest it: Castells' notion of the 'switchers'. Without an ongoing nurturing of attention – via (re)linking, (re)blogging, (re)tweeting – any bit of geospatial information can be deprioritised in the dynamic remaking of augmented reality. It takes not only well-networked connections to give visibility to particular bits of information, but also the ability to package information in a way appealing to the those that are most likely to circulate it. Furthermore, on jointly authored platforms (such as Wikipedia) there is a need to not just create information, but also to continuously justify its existence to others that might seek to dismiss it as irrelevant. For instance, Kenyan Wikipedia editors created an English-language article about a popular Kenyan superhero Makmende. Since Western gatekeepers had never heard of Makmende, the article was repeatedly deleted until the international press used the case as an example of the encyclopedia's bias (Ford 2011).

Communication power thus allows certain people and groups to filter and promote the representation and circulation of a stabilized meaning of place. It provides a way for some information to be amplified and other information to fade from the spotlights of attention. Rather than simply being blank spots on a map–*terra incognito*–the representations of place turn on the constellation of programming and switching power of those tied to its network with the power to enact and re-enacted content.

Code Power

Code power is used to refer to the autonomy of software code to regulate actions, or mediate content – whether by proscribing or requiring particular actions (see Graham 2005), or ordering representations in particular ways. In other words, code power refers the role of code in transducing space (Kitchin and Dodge 2011): that is, bringing (representations of) space into being via reiterated processes and in conjunction with other actors.

There is a rich vein of scholarship exploring the power of code to regulate conduct ranging from fixed rules governing actions (e.g. (Lessig 1999)) to performative acts that rely on both human and technological rituals (Chun 2008). In reference to space, Graham (2005) argues that we now move through software-sorted geographies: landscapes that are often shaped, formed and mediated by invisible lines of code. Dodge and Kitchin (2005) illustrate that the technicity of code (its power to influence action) is able to influence our spatial experiences and actions through processes of transduction (constant remaking and re-enactions).

Code power works in what are frequently opaque ways to order representations of place. For example, the data presented to a particular user via the Google Maps interface is a product not only of the particular search terms used, but also complex algorithmic judgments based on the user's past tracked online activity, their geographical location, and any number of inferred characteristics based on segmentation of user populations, or even derived from the recorded behaviors and preferences of individuals in users' social networks (Miller 2011). This is to say nothing of the characteristics of the actual content served which is, in turn, valued and weighted according to myriad more-or-less documented factors, increasingly including locational relevance (proximity) and currency (time).

Although code power is inseparable from the performances and politics of its distributed authors/consumers who stabilize and claim particular representations within the ephemeral and continuous reiterations of augmented realities, we also draw attention to the fact that code power is, often, exercised in a very centralized and hidden manner.

Centralization of code power entails a hard limit on the creativity or power of a system's switchers – ordinary users and their deliberate/unwitting promotion of content within digital representations of place – imposed arbitrarily by a sovereign, such as Google. A clear example of the latter is the deliberate editorial cleansing of so-called Google bombs created by distributed users. In high profile cases, such as the mass-utilization by liberal activists of the phrase "miserable failure" in reference to U.S. President George W Bush, countered by similar actions by conservatives directed towards President Jimmy Carter and documentary maker Michael Moore, Google chose to claim that the defusing of the bomb was "completely algorithmic" (Sullivan 2007). Similarly, no editorial intervention was claimed in the relegation of anti-Semitic "Jew Watch" from its top ranking for the term "Jew" (Bar-Ilan 2006). The fact that lower profile public figures are still targeted in such ways – see, for example, the neologism coined around U.S. Senator Rick Santorum's last name – belies the claim of algorithmic neutrality, even as it provides a source of optimism that some algorithmically determined democracy/openness may exist, unless or until the algorithmic neutrality is perceived to threaten the search engine's reputation.

Through the increasing employment of personalised and opaque code, geocoded content, and even our movements through material space, are ever more fragmented into individualized representations that ultimately enable the construction of self-reinforcing information cocoons. As such, code power makes a largely opaque contribution to the representation of places resulting in augmented realities that are increasingly contingent on every person's individual positionalities in time, space and society.

Timeless Power

Timeless power refers to the ways in which digital representations of place reconfigure temporal relationships, particularly sequence and duration, between people and events. Although cartography has always entailed the synthesis of multiple temporalities – whether in the form of (hidden) data collection or surveying processes taking many months or years, or in the form of labels, references or imagery relating to historical events/significances – digital representations of place arguably entail a deepening of this effect. Although digiplaces are products of distributed individuals, coded rules, and various more-or-less visible practices of social- and software-sorting (Graham 2005), they are also positioned in an ambiguous temporal relationship to their consuming audiences. Not only are familiar representations of place such as Google Earth or Street View imagery, as well as the Flickr photographs, user-generated points of interest and Wikipedia articles served up as current, of-the-moment representations of place, they are put to work as such by individuals who act upon and in relation to these representations. Manuel Castells

(1996) describes this flattening of time as timeless time, wherein society is installed within an infinite ephemerality of continuous re-cycling and copresence.

Thus the production of augmented reality is an exercise of the power of what Castells (1996, 467) refers to as "timeless time" where time is dissolved "... by disordering the sequence of events and making them simultaneous, thus installing society in an eternal ephemerality"; ephemeralities that are even more pronounced when combined with those discussed in the previous sections. This collapse of time is particularly remarkable given that much of this geospatial content has a time stamp associated with its creation: when the photo was taken, when it was uploaded to an online platform. However, this temporal data is routinely ignored in the enactment of augmented realities in the effort to construct seamless representations of place. Implicated within timeless power are the constraints imposed by code power, e.g., the software design that make it extremely difficult to access time metadata within augmented reality interfaces.

Distributed, Communication, Code, and Timeless Power

It is important to reiterate that the practices of power within augmented reality are often situated on the borders of these four contexts. Google's PageRank, for example, takes the communication and distributed power that results from interlinkages, reposting and promotion, and filters it through the coded power of its search algorithm to derive a particular representation of a place via a ranked Google Maps search. Likewise the distributed power possessed by active Wikipedia editors is merged with timeless power when a geocoded Wikipedia entry is placed seamlessly on a real time digital map.

Moreover, because the separation between powers is blurred, the way in which one can respond to the representations within augmented reality is similarly contingent. For example, Google's coded power is opaque, but the company has a knowable presence. It can be criticized, lobbied, sometimes even debated with. In contrast, representations within Wikipedia, which in theory is far more open, possesses a diversity and diffuseness allowing a very different sort of power to be enacted. It is hard to disagree with an amorphous cloud if the cloud simply discounts one's arguments or if the dispute is channelled into separate and non-interacting arenas of deliberation.

The Augmented Realities of "Osama Bin Laden's Hideout Compound"

In order to explore these issues in more detail, this chapter reviews the augmentation of Abbottabad, Pakistan. Abbottabad is a city of close to a million and half people located 100 km north of the capital of Islamabad and achieved international attention as the location in which Osama bin Laden had been hiding and where he was killed by a U.S. military raid in the early hours of May 2nd, 2011. Overnight Abbottabad became a site through which the practices of power exercised in crowd-sourced augmented realities (particularly as manifested via Google Map Maker and Google Places reviews) can be easily discerned.

While any number of software services can contribute to crowd-sourced data, Google Map Maker is among the most potentially transformative as it exorts users to "Enrich Google Maps with your local knowledge." The stated goal of Google Map Maker is to leverage user generated content to fill voids in maps (which were famously labelled "unknown" on maps in earlier centuries) that may ultimately be deployed in Google Maps as "places of interest". Map Maker is particularly targeted to country contexts where data is scarce, constrained by security concerns or both, making it extremely relevant to developing country contexts (see Boulton 2010) for a detailed review and critique). Google Places reviews (rebranded as Google+ Local in 2012) is a service that allows Internet users to comment on "places of interest" within Google Maps. While ostensibly for reviews of business, users made extensive use of the reviews to comment on various sites within Abbottabad.

A quick review via Google Maps of the area of Abbottabad in which bin Laden lived (see Fig. 14.1) reveals the standard formatting and coloration familiar to users of Google's mapping services. The only thing that stands out is the slightly incongruous label "Osama bin Laden's Hideout Compound" in the lower middle of

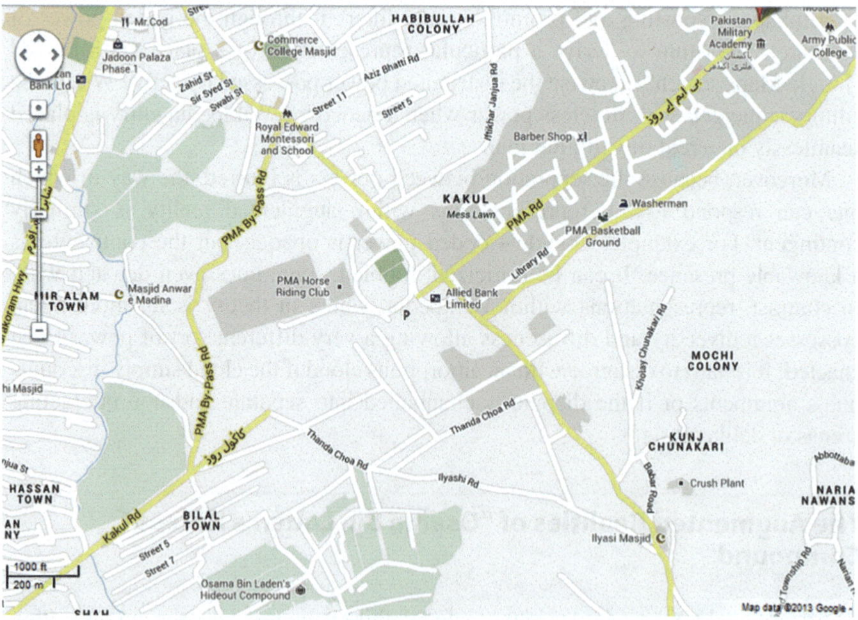

Fig. 14.1 Google Maps screen shoot of Abbottabad, Pakistan, 2013 (Source: author screenshot from June 12, 2013). Note, Osama bin Laden's hideout compound on the *lower lefthand* side and the Pakistan Military Academy at Kabul at the *upper right*. The Margis Arif Hospital is not visible in this screen shot but is located in the *upper left* of the map

the map that marks the Google Maps defined place of interest where the founder of Al Qaeda died and the focal point of crowd-sourced annotation. Also of note is the Pakistan Military Academy in the upper left which also emerged as a target in the geospatial content created after May 2nd, 2011.

Distributed Power

While distributed power reflects the possibility of the democratization of spatial data production, even a brief review of the crowd-sourced augmented content layering the bin Laden compound site provides a clear counter example. Almost immediately after news reports of the raid on the compound, multiple user defined places of interest were findable via a Google Maps search. In Fig. 14.2, each placemark represents a separate crowd-sourced augmentation to this location. While most of these were placed within the compound walls are were relatively matter of fact, e.g., Placemark D, Osama Bin Laden Compound, some took an incriminatory or derogatory tone such as the “Bin Laden pot farm” highlighted in Fig. 14.2. While this is unsurprising given bin Laden’s role in the 9/11 terrorist attacks, it highlights

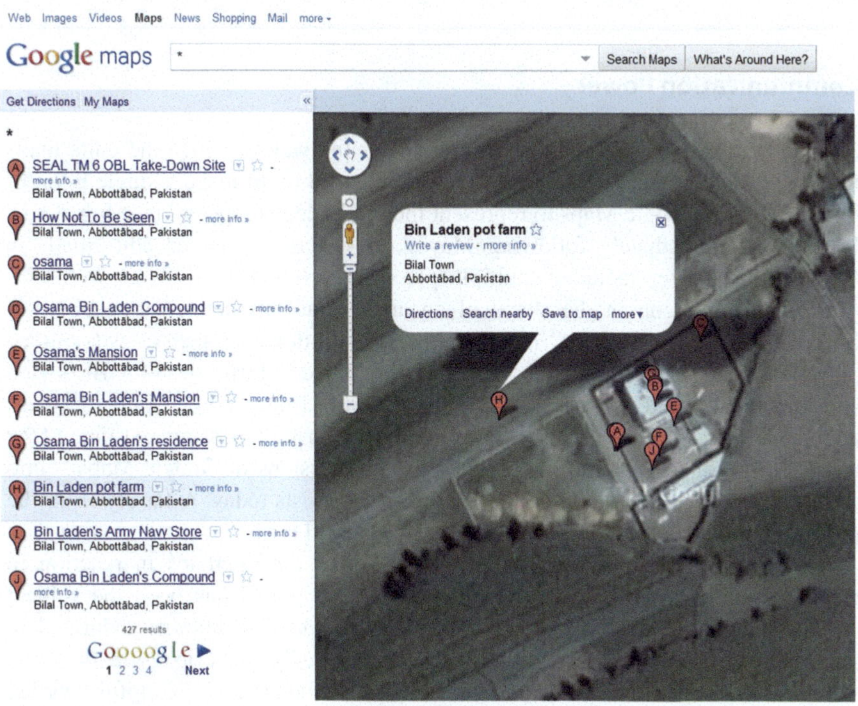

Fig. 14.2 Results of a Google Maps search of the bin Laden compound, 2011 (Source: author screenshot from May 2, 2011 at 4:30 pm US Eastern Time)

how distributed power allows a non-local population to construct representation of point of interest that is literally thousands of miles away, the antithesis of local knowledge. Moreover, the brush used by non-locals to make these representations are extremely broad and capture points of interest that share nothing with this compound besides spatial proximity. For example, nearby businesses have received reviews referencing bin Laden, e.g., "some 6′5″ guy… kept walking around and trying to recruit people saying something about 72 virgins".

The nearby site for the "Pakistan Military Academy" provides an interesting contrast in the enactment of distributed power. While multiple reviewers of this point of interest (especially within the first few weeks of the raid on the bin Laden compound) provide strong critiques of the Pakistani military, e.g., "Usama's compound was 100 yards from your front gate. You're the best? At what?!", this evolved into a more contested discourse, with comments evenly split between denigrating and supporting the Pakistani military. For example, in late 2012 a user commented, "if Pakistan's military was so good they would've rolled tanks into the compound killed the guy. surprised they even have a military that wasnt [sic] destroyed already" and was met with a response (albeit an asynchronous one that came 3 months later) that said in part, "the pak [sic] army which is the best army in the world had checked all abbottabad everytime before the passing out parade. This was a trick by the US just to decrease the popularity of Pakistan and its people."

Communication Power

An excellent example of communication power the way in which the edits made in Google Map Maker on May 2, 2011 were cleaned up to make a single point of interest with in Google Maps to represent the bin Laden compound (see Fig. 14.1). In the days immediately following May 2, 2011, crowd-sourced edits made in Google Map Maker included multiple placemarks (see Fig. 14.2) as well as a user created polygon in the shape of an arrow that was defined as a place of interest named "He is here!" (see Fig. 14.3). Additionally other crowd-sourced places of interest included point locations for "Jihad Tshirt Sales", "UBL's Bed and Breakfast", and "SEALS heliport" (see Fig. 14.4). While some of these augmentations appeared in Google Map searches during this time (see Fig. 14.2), they were ultimately stabilized – largely by the decisions of Google Maps – into the representation of a single point of interest that exists today. A far cry from the multiple and wide-ranging representations in May 2011.

In concert with the activity within Map Maker (and perhaps in reaction to the reduction to a single point of interest) the compound has been the subject 1,482 reviews that generally echo the sentiments and dark humour exhibited in the creation of points of interest. Reviewers repeatedly make jokes in the style of fake hotel reviews to complain about noisy helicopters and unhelpful bearded men as well as much more derogatory comments. While salutatory reviews also exist, "He (is) a great man" they are far out-numbered by this process of crowd-

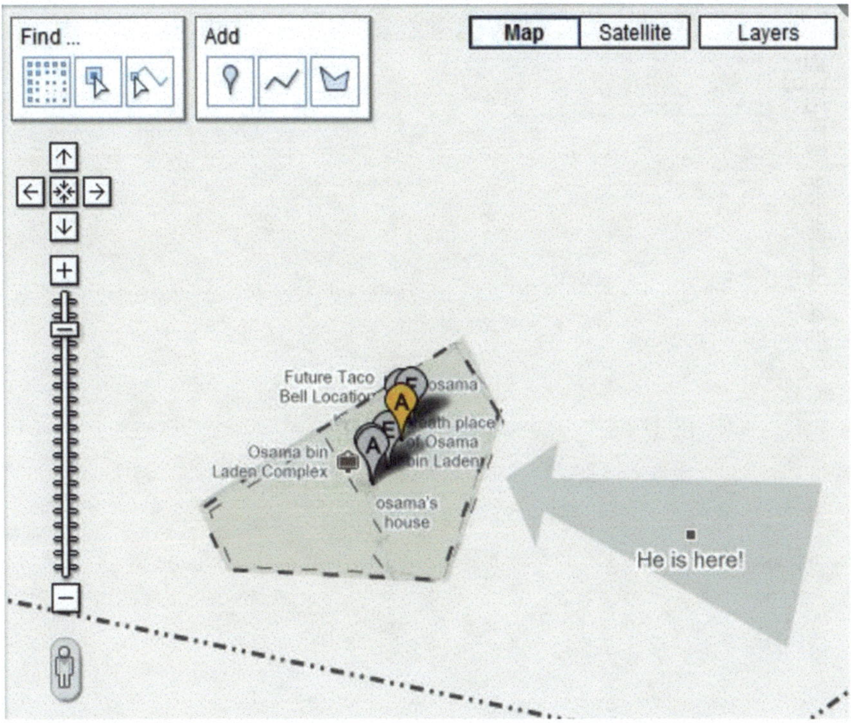

Fig. 14.3 Screenshot of the bin Laden compound within Google Map Maker, May 2, 2011 (Source: author screenshot from May 2, 2011 at 4:30 pm US Eastern Time). Note: the large *shaded arrow* pointing to the compound with the label "He is here!" that has been added to Google Maps as distinct place of interest

sourced communication power which seeks to reinforce a specific, dark-humoured representation of the compound as a very bad hotel.

Perhaps an even better instance of communication power is the evolving representation of the nearby Margis Arif Hospital at which reviewers reference bin Laden's supposed need for dialysis as well as his death. For example, on May 2, 2011 a reviewer quipped, "Great for bullet wounds and shrapnel injuries! If you happen to use dialysis, you might be privileged to use the same equipment as the great man". Another review posted on the same day and by a user with the name Osama bin Laden stated, "Good doctors, friendly nurses. Ample supply of dialysis equipment". While Google provides a mechanism for the removal of reviews by allowing a business owner to flag comments as inappropriate, these reviews have remained for more than 2 years. While the reason for this is unknown – clearly these are inappropriate reviews, the hospital managers have not had availed themselves of this despite the promise of distributed power.

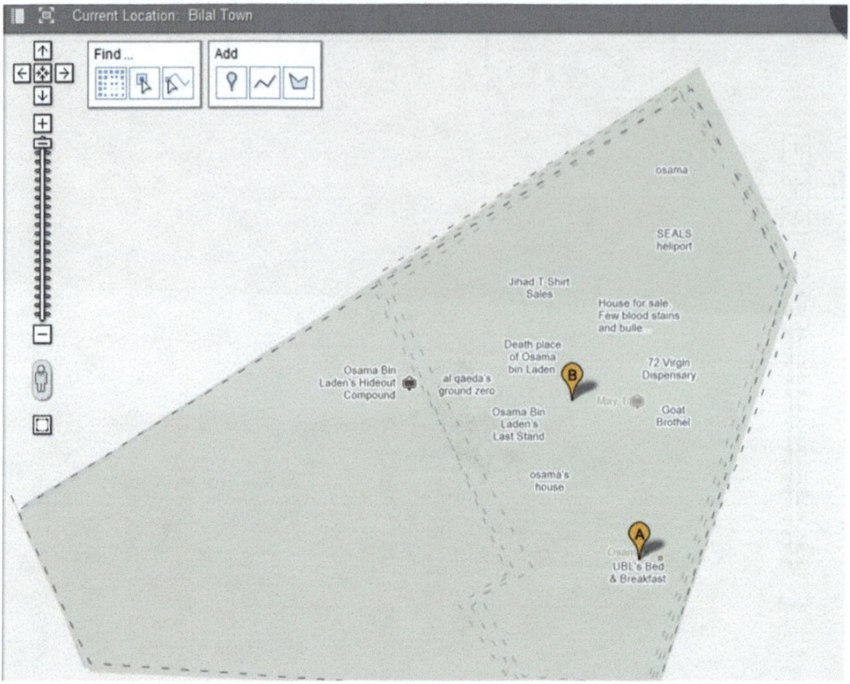

Fig. 14.4 Screenshot of the bin Laden compound within Google Map Maker, May 2, 2011 (Source: author screenshot from May 2, 2011 at 4:30 pm US Eastern Time). Note: Some of the labels read "Jihad Tshirt Sales", "al qaeda's ground zero", "Goat Brothel", "UBL's Bed and Breakfast", and "SEALS heliport"

Code Power

The ability of coded algorithms to shape crowd-sourced augmented realities is manifest throughout this case study. Even the basic functions of interacting with augmented realities, e.g., submitting an edit to Google Map Maker or a review, are structured by user interface design, classification categories and rating scales that compel users to standardize their input. A more overt example of code power is the process by which multiple listings in a map are ordered (e.g., a search for mosque in Abbottabad) or reviews about a specific point of interest within Google Maps (e.g., the bin Laden compound) are prioritized (Zook and Graham 2007). For example, when one examines the 1482 reviews associated with the compound there are four ordering options including highest score, lowest score, latest and most helpful.

This latter category of Most Helpful is a good example of code power as the process by which "Most Helpful" is calculated is extremely opaque. The only information provide by Google Maps is "When you sort the reviews by" Most helpful, "this takes into consideration many aspects of each review's text and author, as well as feedback from users who vote on whether the review was helpful

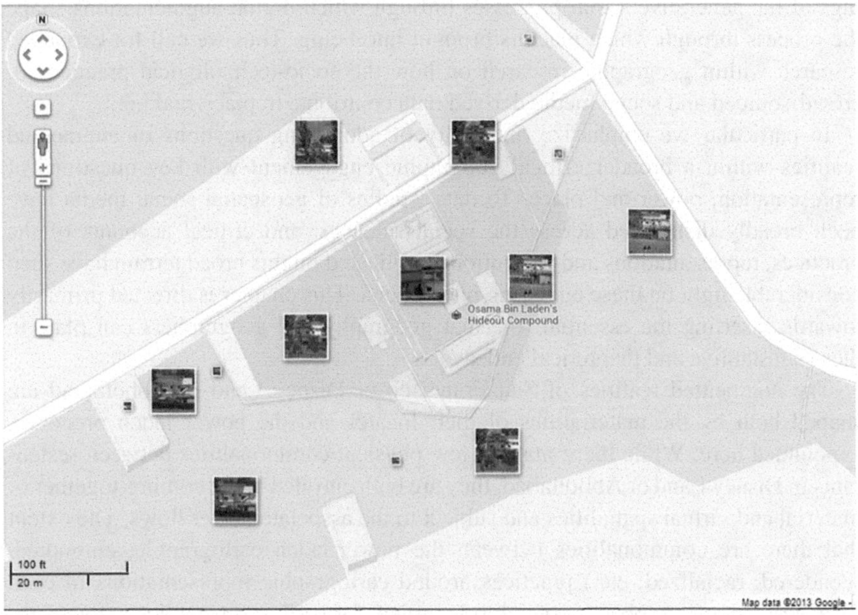

Fig. 14.5 Screenshot of the bin Laden compound within Google Map with the Photos option, 2013 (Source: author screenshot from June 12, 2013)

or not" (Schneider 2012). This brief statement indicates at least three points of evaluation – text, author and feedback – that produce this metric which provides and important ordering of these augmented representations. Thus it remains that the "Most Helpful" ranking process remains largely a black box despite its powerful role in creating representations.

Timeless Power

Timeless power is also observable within this case study as users continue to post reviews of the compound – albeit at a much reduced pace – that contribute to a representation of this location that flattens time into an infinite May 2, 2011. This process also finds support in the dozens of photos of this location taken from an array of angles in the immediate aftermath of the raid that provide a static representation of this location (See Fig. 14.5).

Critical Geographies of Augmented Realities

The lines of separation between these four different manifestations of power are undoubtedly blurred, but by constructing this typology and illustrating it via this case study we seek to provide a useful beginning for developing further understandings of the generative social processes through which digital augmentations shape the process through which place is brought into being. Thus we call for expanded research within geographic research on how the socio-technological practices of crowd-sourced and social media derived data contribute to place-making.

In particular we emphasize the utility of addressing questions of augmented realities within a broader critical geographic engagement with key questions of representation, power and place. To date, studies of geospatial social media have been broadly distributed across the social sciences, and critical accounts of the practices, representations and institutions implicated on this broad terrain have shed considerable light on these emerging phenomena. This chapter is directed primarily towards asserting the essential role that geography and geographers can play, in these substantive and theoretical endeavors.

The augmented realities of San Francisco or Disney Land or Abbottabad are shaped both by the materialities of their locales and the power laden processes expounded here. While there may be few physical commonalities between restaurants in Disney Land or Abbottabad, they are both enrolled in the coming-together of material and virtual spatialities and subject to the associate power flows. The extent that there are commonalities between the power-laden cartographic, embodied, (gendered, racialized, etc.) practices around cartographic representations in each case, critical geographers' crucial role going forward is to explicate the ways in which crowd-sourced augmented realities are constructed, experienced and contested in various contexts.

References

Allen, J. (2003). *Lost geographies of power*. Malden/Oxford: Blackwell.

Bar-Ilan, J. (2006). Web links and search engine ranking: The case of Google and the query 'Jew'. *Journal of the American Society for Information Science and Technology, 57*(12), 1581–1589.

Boulton, A. (2010). Guest editorial—just maps: Google's democratic map-making community? *Cartographica, 45*(1), 1–4.

Boulton, A., & Zook, M. (2013). Landscape, locative media and the duplicity of code. In N. Johnson, R. Schein, & J. Winders (Eds.), *The Wiley-Blackwell companion to cultural geography* (pp. 437–451). Oxford: Wiley-Blackwell.

Budd, L., & Adey, P. (2009). The software-simulated airworld: Anticipatory code and affective aeromobilities. *Environment and Planning A, 41*, 1366–1385.

Carr, N. (2007). The ignorance of crowds.http://www.strategy-business.com/article/07204. Accessed 25 May 2013.

Castells, M. (1996). *The rise of the network society*. Oxford: Basil Blackwell.

Castells, M. (1998). *End of millennium*. Oxford: Basil Blackwell.

Castells, M. (2008). *Communication power*. Oxford: Oxford University Press.

Chun, W. (2008). On 'sourcery', or code as fetish. *Configurations, 16*, 299–324.

Crampton, J. (2001). Maps as social constructions: Power, communication and visualization. *Progress in Human Geography, 25*(2), 235–252.

Crutcher, M., & Zook, M. (2009). Placemarks and waterlines: Racialized cyberscapes in post-Katrina Google Earth. *Geoforum, 40*, 523–534.

Dodge, M., & Kitchin, R. (2004). Flying through code/space: The real virtuality of air travel. *Environment and Planning A, 36*, 195–211.

Dodge, M., & Kitchin, R. (2005). Code and the transduction of space. *Annals of the Association of American Geographers, 95*, 162–180.

Dodge, M., & Kitchin, R. (2007). Outlines of a world coming into existence: Pervasive computing and the ethics of forgetting. *Environment and Planning B, 34*(3), 431–445.

Dodge, M., Kitchin, R., & Zook, M. (2009). How does software make space? Exploring some geographical dimensions of pervasive computing and software studies. *Environment and Planning A, 41*, 1283–1293.

Elwood, S., Goodchild, M., & Sui, D. (2012). Researching volunteered geographic information: Spatial data, geographic research, and new social practice. *Annals of the AAG, 102*, 1–20.

Fildes, J. (2007). Wikipedia 'shows CIA page edits. BBC News. http://news.bbc.co.uk/1/hi/6947532.stm. Accessed 15 May 2013.

Firmino, R., & Duarte, F. (2010). Manifestations and implications of an augmented urban life. *International Review of Information Ethics, 12*, 28–35.

Ford, H. (2011). The missing Wikipedians. In G. Lovink & N. Tkacz (Eds.), *Critical point of view: A Wikipedia reader*. Amsterdam: Institute of Network Cultures.

Foucault, M. (1988). Technologies of the self. In L. Martin, H. Gutman, & P. Hutton (Eds.), *A seminar with Michel Foucault* (pp. 16–49). Amherst: The University of Massachusetts Press.

Glott, R, Schmidt, P. & Ghosh, R. (2010). Wikipedia survey – overview of results. http://www.wikipediasurvey.org/docs/Wikipedia_Overview_15March2010-FINAL.pdf. Accessed 17 Mar 2011.

Graham, M. (2010). Neogeography and the palimpsests of place: Web 2.0 and the construction of a virtual Earth. *Tijdschrift voor Economische en Sociale Geografie, 101*, 422–436.

Graham, M. 2013. "The Virtual Dimension". In Global City Challenges: debating a concept, improving the practice, eds. M. Acuto and W. Steele. Palgrave. pp. 117–139.

Graham, M., & Zook, M. (2011). Visualizing global cyberscapes: Mapping user generated placemarks. *Journal of Urban Technology, 18*(1), 115–132.

Graham, M., Zook, M., & Boulton, A. (2013). Augmented reality in urban places: Contested content and the duplicity of code. *Transactions of the Institute of British Geographers, 38*(3), 464–479.

Graham, S. (2005). Software-sorted geographies. *Progress in Human Geography, 29*, 562–580.

Harley, J. B. (1989). Deconstructing the map. *Cartographica, 26*, 1–20.

Kitchin, R., & Dodge, M. (2011). *Code/space: Software and everyday life*. Cambridge, MA: The MIT Press.

Lapent, F. (2011). Geomedia: On location-based media, the changing status of collective image production and the emergence of social navigation systems. *Visual Studies, 26*(1), 14–24.

Lessig, L. (1999). *Code and other laws of cyberspace*. New York: Basic Books.

Miller, C. (2011). Google search results get more social. *New York Times*. 17 February. http://bits.blogs.nytimes.com/2011/02/17/google-search-results-get-more-social/. Accessed 17 Mar 2011.

Nagenbourg, M., Albrechtslund, A., Klamt, M., & Wood, D. (2010). On ICT and the city. *International Review of Information Ethics, 12*, 2–5.

O'Neil, M. (2009). *Cyber chiefs: Autonomy and authority in online tribes*. London: Pluto Press.

Pariser, E. (2011). *The filter bubble*. New York: Viking.

Pickles, J. (2004). *A history of spaces*. London: Routledge.

Rose-Redwood, R. (2006). Governmentality, geography and the geo-coded world. *Progress in Human Geography, 30*, 469–485.

Schneider, V. (2012). FAQ: Google+ Local [READ THIS.]. Posted on May 31. http://productforums.google.com/forum/#!msg/business/oRvAQC_IakA/vsfNhuMzURYJ. Accessed on 10 June 2013.

Shirky, C. (2010). *Cognitive surplus: Creativity and generosity in a connected age*. London: Allen Lane.

Sullivan, D. (2007). Google Kills Bush's miserable failure search & other Google bombs. Search Engine Land.http://searchengineland.com/google-kills-bushs-miserable-failure-search-other-google-bombs-10363. Accessed 5 June 2013.

Thrift, N., & French, S. (2002). The automatic production of space. *Transactions of the Institute of British Geographers, 27*, 309–335.

Wilson, M. (2011). 'Training the eye': Formation of the geocoding subject. *Social and Cultural Geography, 12*(4), 357–376.

Wood, D., Fels, J., & Krygier, J. (2010). *Rethinking the power of maps*. New York: Guilford.

Zook, M., & Graham, M. (2007). The creative reconstruction of the Internet: Google and the privatization of cyberspace and DigiPlace. *Geoforum, 38*, 1322–1343.

Section IV

Placing New Media Platforms

Science Blogging Below-the-Line: A Progressive Sense of Place?

15

Jonathan Mendel and Hauke Riesch

Introduction

Drawing on our research on the 'badscience' blog network (Riesch and Mendel 2014), this paper will discuss the construction of this community and how this case study can contribute to our understanding of the geographies of new media and of social research methodologies more broadly. We will look at the spaces that this blogging community interacts with – the physical national location of the network, its interactions with the existing mainstream media spaces in the UK and the virtual above- and below-the-line spaces that allow bloggers and commentators to construct their community (as well as delineating outsiders).[1] We argue that both positive and negative interactions between bloggers and commentators can enhance community networks, bootstrap individual bloggers to higher prominence and thus

We are grateful to the bloggers who answered our questions, commented on and discussed our work and otherwise helped us. This paper would not have been possible without their assistance. We are also grateful to Stuart Allan for his helpful comments on this chapter and to Lorraine Allibone, Alice Bell, Petra Boynton and Simon Locke for their useful suggestions re this broader project. Thanks are due to participants in discussions of this project at the 2010 Science and the Public conference, to students who discussed these ideas as part of Dundee University's 2014 Key Ideas in Geography module, and to the editors and reviewers of this volume.

[1]"Below-the-line" refers to the space in a blog or increasingly online newspaper article, where people can comment and discuss the issue in the main blog post or article ("above-the-line"); these comment spaces are usually situated below the main post or article.

J. Mendel (✉)
School of the Environment, University of Dundee, Dundee, Scotland, UK
e-mail: J.M.Mendel@dundee.ac.uk

H. Riesch
Department of Social Sciences, Media & Communications, Brunel University, London, UK
e-mail: Hauke.Riesch@brunel.ac.uk

S.P. Mains et al. (eds.), *Mediated Geographies and Geographies of Media*,
DOI 10.1007/978-94-017-9969-0_15

enhance their readership and general visibility within and eventually beyond the network. Despite recent worries about uncivil comments and 'trolling' (Brossard and Scheufele 2013), we will argue that the below-the-line interaction on blogs can be a sign of vitality both for the community and the individual site – and can be important for building a progressive sense of place.

Science Blogging

Science blogging is increasingly prominent in academic discussion, and considerable hopes are attached to it. Articles on the virtues of science blogging have appeared in several 'high impact' scientific journals (e.g. Bonetta 2007; Schmidt 2008; Nature 2009); in addition *Nature* featured a regular column (now discontinued) 'from the blogosphere'. The Royal Geographical Society (with IBG) guide to Communicating Geographical Research Beyond the Academy (Gardner et al. 2010) discusses blogging as one option, giving David Campbell's online work on photography, politics and conflict as an example. More recently, Kitchin et al. (2013) discuss the virtue of their own academic blogging in geography as a public engagement tool. There are many reasons why scientists and researchers blog: blogging has been viewed as an inexpensive and accessible outlet for academics to communicate their research directly to the public, bypassing the stringent demands of publishers in terms of turnaround time and editorial control. As such, blogs are increasingly being written by academics as part of their daily activities either on a personal or on an institutional level. In his response to Kitchin et al. Graham (2013b: 79) however outlines some reservations about whether academic blogging really can create new public geographies, due to for example uneven access to the internet and the fact that a very few established voices will always be able to be "much louder, present and visible than others" (see also (Hindman 2008), below). While there is scope for much interesting work on arts and social science blogging, this chapter will focus on science blogging, drawing on our aforementioned research on the 'badscience' blogging community (see (Riesch and Mendel 2014)).

Alongside the increasing prominence of science blogging there are also real concerns and risks linked up with science blogging. Bell (2012: 259) discusses "the characterization of science blogging as a space of cliques and clubs". There are worries that the spaces of science blogging might be exclusionary in problematic ways and that some voices might be silenced. In his survey of political blogs in the US, Hindman (2008) has found that although anybody with the necessary time and resources can set up a blog, blogs' readerships and therefore influence follow a power-law such that the relatively few very influential blogs receive almost all of the attention. The social make-up of these influential bloggers, Hindman found, is remarkably elitist: they were even more likely to be white, middle-class, male and Ivy-league educated than the average mainstream newspaper columnist. This means that initial high hopes over the democratic tendencies of blogging offering a voice for all sections of society are quite unrealised, prompting Hindman to title his study "the myth of digital democracy". While a large number of people are able to set up a blog, getting people to read and interact with a blog requires significant resources – for example, it can be invaluable if the blogger knows the right people and has the kind of job which allows them to spend considerable time on their online activities.

Similar issues apply to science blogs. The science blogs which get by far the most traffic and – not unrelated to this – have the most influence tend to be written by people, though not necessarily scientists, who either had their employers' approval (such as Ed Yong[2]) or who have enough of a developed media profile to draw substantial traffic (such as Ben Goldacre[3] or Martin Robbins[4]). Breaking into this very prominent group of science bloggers is itself a difficult task. Indeed, many of the most prominent science bloggers have a mainstream media presence already (such as Goldacre again) and additionally science bloggers such as Robbins, "Grrlscientist"[5] and Jon Butterworth[6] now have regular blogs at *the Guardian* (a national UK broadsheet). This suggests that the line between mainstream media and blogs is very blurred at least at the successful end of the spectrum – in content, readership and the profile of the bloggers themselves.[7]

Within the very competitive environments surrounding blogging, constructing networks is crucial if a new blogger wants to have their voice heard. Science blogging networks can function as powerful social advancement tools for aspiring bloggers, bringing their blogs more visibility – because when your voice is being listened to (and crucially, linked to) by more prominent writers then the chances of getting heard rise. For this reason, spaces below-the-line are often important in the development of blogging communities: commenters (who very often maintain their own blogs themselves) interact with the blogger, establish a relationship and in some cases become acquaintances and friends, not just with the blogger but with other commenters, and therefore the interactions between bloggers and their commenters merit special attention, both in terms of the physical and virtual spaces[8] they are situated in and in the way these contribute to a wider identity for the community of bloggers.

---

[2]http://blogs.discovermagazine.com/notrocketscience/

[3]http://www.badscience.net/

[4]http://www.guardian.co.uk/science/the-lay-scientist

[5]http://www.guardian.co.uk/science/grrlscientist

[6]http://www.guardian.co.uk/science/life-and-physics

[7]There is a long history of such blurring – one might note, for example, Allan's (2006: 13–4) discussion of Time Magazine's moves towards "electronic dialogue" in the 1990s.

[8]A note on our terminology; we use 'physical' spaces to refer to spaces in the bloggers' offline world (for example, which national borders they fall within) and 'virtual' spaces to refer to the spaces in the online world (e.g. blogs, blog-networks, online newspapers and their below-the-line spaces on which users can comment), which the bloggers use to write their contributions and interact with other bloggers or commentators. Clearly, these are not discrete categories: instead, we would follow Zook and Graham (2007a) in acknowledging the existence of "complex and hybrid spaces made up of multifarious entangled elements of the virtual and physical environments". However, this is the best – or the least bad – way we have found to distinguish such spaces.

'Badscience' Blogs

The 'badscience' blog network developed in around 2007 as a community of bloggers from the below-the-line commentators (and later the attached community forum) of the UK-based blog 'badscience' by science blogger and Guardian columnist Ben Goldacre. People who commented on Goldacre's blog often either already had their own blog or were encouraged – by Goldacre and other members of the developing community – to start their own. Our research is ethnographic in the sense that both of us have been active in this community since the inception, and we have interacted with the community through contributing our own comments, starting our own blogs as well as developing relationships and used these connections to enhance our own blogging and academic networks. As well as this ethnographic approach, we draw on an e-mail-based qualitative study with bloggers from the community (see Riesch and Mendel 2014) for further details of our approach).

While below-the-line interactions are important for this community, such interactions are not always – and perhaps not even mostly – positive interactions between a blogger and their readers. This is likely to be especially true for a science-blogging community that foregrounds active conflict with an assortment of 'quacks', science-illiterate public figures and other purveyors of 'bad' science. Negative comments and reader reactions create certain problems for and risks to bloggers. Risks are not necessarily to bloggers' physical safety, though some comments do threaten violence or death.

There is also risk to the safety of the blog or to the blogger's way of life more broadly, i.e. through threats of libel action or complaints to the bloggers' professional body (which has been a very concrete issue for several 'badscience' bloggers: see our analysis of the risks of science blogging in Riesch and Mendel (2014: 66–67)). However, while we acknowledge these problems and risks – and, clearly, would not argue that threats of violence or death are positive ways to interact online – the chapter will look at how the community analysed here can serve as an example of the generative potential of spaces below-the-line.

Gender and Online Abuse

In addition to more general risks, there are clearly gendered aspects to much online abuse. The activist, journalist and blogger Laurie Penny (2011) sums this up vividly when she describes how:

> You come to expect it, as a woman writer, particularly if you're political. You come to expect the vitriol, the insults, the death threats. After a while, the emails and tweets and comments containing graphic fantasies of how and where and with what kitchen implements certain pseudonymous people would like to rape you cease to be shocking... An opinion, it seems, is the short skirt of the internet. Having one and flaunting it is somehow asking an amorphous mass of almost-entirely male keyboard-bashers to tell you how they'd like to rape, kill and urinate on you.

Helen Lewis (2011) describes a range of abuse received by women bloggers, arguing that:

> While I won't deny that almost all bloggers attract some extremely inflammatory comments – and LGBT or non-white ones have their own special fan clubs, too – there is something distinct, identifiable and near-universal about the misogynist hate directed at women online ... What does it feel like to be subjected to regular rape threats or death threats? To have people send you emails quoting your address, or outlining their sexual fantasies about you? That's the reality of what many female bloggers experience.

More recently, the journalist Caroline Criado-Perez has faced extremely hostile, threatening and abusive misogynistic comments on Twitter following her involvement in the successful campaign to have more historically important women represented on the back of British banknotes (Criado-Perez 2013).[9] There have been significant discussions of questions around gender and science blogging as a reaction to the way that men outnumber women in many spaces, and how such spaces and networks might shut out female voices (see Munger 2010). In a particularly unfortunate example, one such discussion occurred in association with the behaviour of a prominent science blogger (Raeburn 2013). Clearly, science blogging does not escape the problems of abusive behaviour online, nor the gendered nature of much of said abuse. However, while this may be a problem within the wider science blogging community, the theme of explicitly gendered abuse within the 'badscience' community did not arise from our research.

Clearly, though, we would not want to disregard the issues caused by such abuse. It is therefore worth reflecting on how our own positionality may have influenced what we did and did not find. Morrowa et al. (2015) note that in their online research work they "were all already positioned – by virtue of privileges across age, education, language, and geographic location, among others – to feel comfortable and literate in the online environments to which we had access". Likewise, we were positioned by a range of factors from education to gender to location to feel relatively comfortable in the online environments discussed in this paper. Morrowa et al. (2015) argue that "the researcher's relationship to online data cannot be seen as one-way (i.e. the researcher is 'taking' data from online spaces that function like archives of discourse) ... the researcher's position at points of connection between virtual and material worlds must also be taken into consideration"; they call for a "virtual-material positionality". With this in mind, we should acknowledge that our own positions in the communities discussed here and in broader academic and other communities may have limited what we found – for example, as men we may have been less likely to encounter certain types of gendered abuse (ranging from our physical appearance being criticised to threats of sexual violence) and research participants may have felt less comfortable sharing concerns about such abuse with us.

We are aware that there are significant ethical issues here – it is likely that privilege associated with our genders means that we have not fully experienced some of the gendered abuse and threats that can spring from below-the-line, and

[9] Criado-Perez (2014) has given examples of some of the abuse she faced in a recent blog.

clearly we do not want to write a defence of threats of rape, murder etc. However, as argued below we also see real progressive potential in online comment spaces. Penny (2014) acknowledges that "yes, the internet is dangerous for women. But that doesn't mean that the answer is to sit quietly with your legs and laptop shut like they want you to do." We would hope that very real problems with online abuse can be viewed as issues to be resolved in lively discussion spaces – for example, with abusive commenters being actively challenged, being excluded from discussion spaces, or being reported to law enforcement agencies for certain types of threats – rather than such abuse being viewed as a reason for closing down these spaces entirely.

Geographies of Science Blogging

Science blogging has led to the development of some interesting geographies. Graham (2013a: 179) criticises how the metaphor of cyberspace "constrains, enables and structures very distinct ways of imagining the interactions between people, information, code and machines through digital networks." For Graham (2013a: 179) "'Cyberspace' [is] both an ethereal alternate dimension which is simultaneously infinite and everywhere (because everyone with an Internet connection can enter), and as fixed in a distinct location, albeit a non-physical one (because despite being infinitely accessible all willing participants are thought to arrive into the same marketspace, civic forum, and socialspace). 'Cyberspace' then becomes Marshal McLuhan's... 'global village'." Graham (2013a: 177) argues that "geographers should take the lead in employing alternate, nuanced and spatially grounded ways of envisioning the myriad ways in which the internet mediates... experiences." It is in that spirit that we hope these geographical analyses of science blogging can contribute to the academic literature.

Globalisation has been described as "the accumulating consequences of 'the annihilation of distance', i.e., the improvement in techniques of, and the rapidly reducing costs of, transportation and communication" (Dore 2001: 6). For McLuhan (1994: 3), "we have extended our central nervous system itself in a global embrace, abolishing both space and time as far as our planet is concerned". McLuhan (1994: 93) goes on to argue that "[o]ur specialist and fragmented civilization... is suddenly experiencing an instantaneous reassembling of all its mechanized bits into an organic whole. This is the new world of the global village". Cairncross (1997) hailed the 'death of distance' more than 15 years ago, viewing "the Communications Revolution" as life-changing. Dodge and Kitchin (2001: 5) note arguments that that "network technologies of cyberspace are forging connections and virtual groups that potentially subvert the primacy of national boundaries. These borders are relatively meaningless to logical connections and data flows that operate on a global scale. The question in these cases is therefore: 'How much sense do existing political borders of the material world make when mapping cyberspace?'" These arguments and questions may leave one uncertain as to whether online networks such as the one analysed here offer a move beyond (conventional) geography.

However, distance, nationality and material political borders do continue to have an influence in online spaces. Graham et al. (2012) have mapped the uneven geographies of Internet access in different countries and the same group (2013) have begun to look at the countries where georeferenced tweets were sent. In terms of more specific case studies, one might note for example that Hine (2000: 115) finds that links between online spaces (for example forums, Usenet newsgroups) and offline geographies (for example, of national identity) were sometimes – though not universally – made in online discussions of the Louise Woodward case,[10] showing that there is no clear binary opposition between online and offline spaces. Paulussen and D'heer (2013) find that online media can allow 'hyperlocal' news coverage. For Thurman et al. (2012) there remains a demand for local news websites, although there are questions about whether organisations working on a 'big media' scale can deliver this. Research in the 'badscience' blogging community supports this idea that – if distance is dead – it remains a rather lively corpse. The bloggers from the 'badscience' community who responded to our survey were overwhelmingly UK-based and largely based in one part of the UK (see Riesch and Mendel (2014: 59). The concerns they discuss are often UK-based – for example, the regulation of 'alternative' medicine in the UK. While some in this community would describe themselves as sceptics/skeptics – and some practices are similar to what is described as skepticism in the US – there are also significant differences which seem to fall along geographical lines. For example the 'bad science' community, in contrast with many US skeptics, tends to show little interest in debunking unconventional beliefs linked to ghosts and 'psi', while creationism – a huge topic on US science blogs – remains rather marginal in Britain. Geographically specific activities and norms can persist, even in communities which very much rely on the virtual spaces of the Internet. We would thus follow Graham's (2013a: 181) argument that while the "internet is characterised by complex spatialities which are challenging to understand and study . . . that does not give us an excuse to . . . ignore the internet's very real, very material, and very grounded geographies."

Territory, Borders and Space

This aforementioned influence of distance, nationality and material political borders means that what is seen in this community is not a straightforward deterritorialisation – not 'just' a move to virtual spaces. Instead of networks replacing territory or acting in a deterritorialised way, they interact with and overlay territories in complex ways (Mendel 2010: 740–1; Painter 2009). For example, many members of this largely UK-based blogging community share a common interest in UK law and regulation. This is not entirely surprising, as being based inside UK national

[10] Louise Woodward was a British au pair working in Massachusetts, who was controversially convicted in 1997 of the murder (later reduced to involuntary manslaughter) of the baby in her care. The case attracted a lot of media interest in Britain at the time.

territory means that they and those close to them may be especially affected by this: vulnerable to the libel laws in place in different parts of the UK, and relying on the protection of UK law. As Smith (2005: 51) argues, then, "[p]ower is never deterritorialized; it is always specific to particular places. Reterritorialization counters deterritorialization at every turn". It is thus the case that "movements of deterritorialization and processes of reterritorialization [are] always connected, caught up in one another" (Deleuze and Guattari 1998: 10).

There are also virtual and material boundaries that define the community and which provide it with a sense of identity which strengthens community cohesion as well as delineating outsiders. In the case of the 'badscience' blogs, one of these boundaries is implicit in the name itself, with 'bad science' situated as external (though it should be noted that the name was chosen by Goldacre himself long before the community established itself and therefore reflects more of an implicit rather than consciously chosen boundary/identity). The 'badscience' blogs distinguish themselves from a wider community of science blogs (with whom they nevertheless interact) by their activist focus of fighting against perceived 'bad' science (see Riesch and Mendel 2014: 63–6). The emphasis on direct activism and campaigning against 'bad' science therefore turned this blogging network into a particularly active locus for boundary work, which Gieryn (1999) analysed as the collective effort of scientific and other communities to differentiate themselves from outside groups and thus establish a collective epistemic authority.

Our research in the 'badscience' community showed that, for some community members, distance is also an issue insofar as it is possible to meet with other members. While a number of members were either unable to meet others face-to-face or preferred not to, face-to-face meetings were one aspect of this community. Proximity (to allow face-to-face meetings) is still an important part of community construction.

Our research thus indicates that, rather than the virtual spaces of this blogging community floating freely away from non-virtual territories and spaces, material distance continues to play a role – we have certainly not seen its annihilation. Instead, it would be better to think about an interplay or interaction between online and offline spaces. Virtual spaces can overlay the offline: enabled and constrained by the material but also opening up new possibilities.

Place

In this context, place remains important. Blog posts reacting to particular political events, or to particular national news media and celebrity figures, provide a common focus on particular physical places. This is important for the commenters' own ideas and blogs to be heard. The connection between reacting to nationally important issues and the commenter having their own voice amplified and thus building up their own place within the blogging network is important because of the networked construction of authorial credibility that the bloggers rely on to distinguish themselves from the overwhelming majority of blogs that are rarely read.

We argue in Riesch and Mendel (2014) that a networked construction of credibility can offset potential credibility loss when often-anonymous blog-authors cannot (or do not want to) appeal to their professional expertise to establish their authority to discuss a subject. The mutual bootstrapping of blogs within the network that we described above works in this particular national context, and as noted above several bloggers have indeed found a voice that then transmitted onto national newspapers through either the occasional comment piece,[11] a regular blog or column or even a transitioning into a professional or semi-professional science writer.[12] These developments have stayed substantially national in character through the media outlets that they broke into (*the Guardian* featuring very heavily here because it was the outlet of Goldacre's original column, through a very loosely left-of-centre political outlook shared by the majority of 'badscience' bloggers, and through *the Guardian's* active championing of blog-style commentary, especially in their Science coverage).

This transitioning from blogger to more conventional media commentator is intricately linked with the place provided by the community network and its facilitation of particular types of action. This is very much a socially and relationally constructed idea of place, with certain blogs being seen as increasingly credible places for debate due to their place in a broader network of links and authority. Successful writing for conventional news platforms can in turn brings more success and visitors to the original blog, starting a virtuous cycle. In this sense the virtual places of below-the-line blog comments and community forums may allow the hoped-for democratic advantages of new media, such as giving a space to marginalised voices – if they were really present to start with – to be partially abandoned for the well-known and still far more influential spaces of traditional media. Therefore we might reconceptualise new media spaces and places as emerging not just in competition with the harder-to-access spaces of conventional media, but also as a catalyst to engagement with them.

One might then question to what extent the virtual places of science blogging can offer what Massey describes as a progressive sense of place. For Massey (1993: 67) one can view a place through "not some long internalized history but the fact it is constructed out of a particular constellation of relations". The type of virtualised place discussed here, while not without history, is clearly relational. However, the nationally- and regionally-focussed and often-anonymous nature of this community means that it might lack some other characteristics of a global place.

Massey (1993: 66–7) draws on the example of Kilburn High Road, and argues that it is "impossible even to begin thinking about Kilburn High Road without bringing in half the world and a considerable amount of British imperialist history. Imagining it in this way provokes in you . . . a really global sense of place".

[11]For example, David Colquhoun has written various Guardian pieces – including http://www.guardian.co.uk/science/2011/sep/05/publish-perish-peer-review-science and http://www.guardian.co.uk/science/2013/jun/02/medical-cure-health-quackery-david-colquhoun

[12]For example, Martin Robbins or Frank Swain.

One question coming out of our research on the 'badscience' community – raised by the community itself – was that of diversity and outreach. This leads back to Bell's (2012: 259) discussion of "the characterization of science blogging as a space of cliques and clubs". The very name of the 'badscience' community defines it in opposition to bad or pseudo science, emphasising an outside. The focus on British concerns might also represent a relative narrowing of the sense of place. Because so many of the community members blog pseudonymously, it is often difficult to get a sense of where people are from and broader cultures – so this might appear to be a place which is both exclusionary and geographically limited.

However, this community is also – as part of its campaigning nature – outwards-looking: aiming to engage with 'quacks', but also to engage with some of the harms done by 'bad science'. It is possible that particular constellations here may feed into progressive actions which move this beyond being a place for any closed clique. However, this is not a straightforward, tightly-managed move to academic 'impact' or participatory democracy. Instead, this is an outwards-looking approach that is shaped by the community's somewhat conflictual place-making. Such conflict is especially apparent when one looks at 'trolling'.

"Trolling Makes You More Stupid": Spaces, Places and Methodologies Below-the-Line

The important role of comments and below-the-line spaces in our case study links to concerns springing from Brossard and Scheufele's (2013) *Science* commentary on uncivil online comments. This work has been used to question whether 'web 2.0' spaces and places are 'good' ones for science communications, and also led to discussions of whether allowing comments on science communication blogs and other sites is conducive to what is perceived as good understanding and discussion. However, what generally qualifies as desirable understanding within science communication is still a contested topic in the field (for an overview see Bauer et al. 2007).

The 'badscience' blogging community emerged from discussions in the comments below Ben Goldacre's *Bad Science* blog and, subsequently, from discussions in a forum associated with that blog. This is particularly interesting given the current concerns about comments and online communication, and the type of research in which these concerns find their strongest justification. Anderson et al. (2013) argue that while "[o]nline communication and discussion of new topics such as emerging technologies has [sic] the potential to enrich public deliberation . . . online incivility may impede this democratic goal." This research study – and concerns springing from it – has been widely discussed in diverse fora from *the Guardian* (Bell 2013) to *Science* (Brossard and Scheufele 2013). Perhaps the most striking response was (as flagged up by Bell's Guardian piece) the headline on the *Liberal Conspiracy* blog proclaiming that "New study shows trolling makes you more stupid" (Hundal 2013). However, we found that 'trolling' can have positive effects relating to community cohesion and collective identities, as the community gathers around shared norms and values and shared, sometimes stereotyped, conceptions of the outsider (Hogg and Abrams 1988) and constructs particular boundaries (Gieryn 1999). This, in turn, strengthens the individual blogger in their confidence and ability to broadcast their message more broadly through the stronger intergroup identity. This includes ‘trolling’ in both directions: as well as being the ‘victims’ of ‘trolling’ on their own blogs and forums, ‘badscience’ bloggers – through their direct activist focus on ‘quacks’ and other ‘bad’ scientists – were also often seen as trolls themselves. This is worth analysing further both as a counter-example to some common concerns about online comment spaces and as a contribution to methodological debates.

The Great British Sport of Moron Baiting

One of the ‘badscience’ community sub-forums promises “action, and all the fun of the fair: quackery, scare stories, miracle cures, iffy adverts, passing banter and the great british [sic] sport of moron baiting . . . ”[13] While community members would not view what they do as trolling – and the content-rich and analytical nature of even quite conflictual engagements is at odds with general views of trolling – some of the targets of their activism and some of the readers did experience their actions as uncivil (concerns which in extreme cases materialised as threats of libel or other significant action).

This in turn creates a community spirit that rallies around the risks posed by “moron baiting”, and also allows the individual blogger to distribute risks they might have been reluctant to take on had they been blogging on their own (which we discuss in more detail in Riesch and Mendel 2014). An offended party may be less likely to threaten libel action against 20 bloggers than against one, while the community provides help and advice should individual bloggers nevertheless find themselves in trouble.

There is also a novel type of place-making and territoriality in play here: online spaces become seen as something to defend, for example from libel threats. There are progressive aspects to making and defending places where ideas can be criticised in ways which might otherwise be blocked by, for example, the UK’s then libel laws.

“Chicken-Flavoured Nipple Biscuit”: The Afterlife of Uncivil Comments

Alongside a community ethos which can feed into quite conflictual online engagements, robust interactions also took place the other way round: with, for example, uncivil comments on ‘badscience’ bloggers’ sites. One interesting example (briefly discussed in Riesch and Mendel 2014: 71) is a comment on one of the ‘badscience’ blogs – *jdc325* was asked “ARE YOU A CHICKEN-FLAVOURED NIPPLE BISCUIT”.[14] This comment was clearly not a useful contribution to the debate in

[13] http://badscience.net/forum/

[14] http://jdc325.wordpress.com/2008/09/12/dodgy-supplements-for-serious-diseases/

any conventional, substantive sense. The post in question was not about nipples, biscuits or chicken, and there is no good reason to think the author might be any of these things. In context, it was also clearly uncivil – for example, the contributor chose to give "yo [sic] momma [sic] sucks eggs out [sic] leemer [sic] bung holes" as their name when posting the comment. It would thus seem to be a clear example of 'trolling'. However, this comment had a life beyond its initial posting and therefore the effects of the incivility were not so negative as one might expect.

When the comment was initially posted, it was quickly absorbed into a larger discussion thread – if anything, helping to generate wider discussion. The community was able to take on this comment as a running joke: helping to enhance community cohesion. When we shared Riesch and Mendel (2014) with the community, two members expressed their pleasure that we had included this example in the article – suggesting the role that such trolling/being trolled plays in community cohesion. Being able to joke about this comment also helped feed into broader discussions about social research and related themes.

Engaging with this trolling also helped to build a particular type of place for discussion: one where, in common with other aspects of the community ethos (e.g. 'moron baiting'), swearing and incivility is something to engage with – in often humorous ways – rather than necessarily to censor. This is in contrast to some more 'official' spaces of science communication, which despite the currently dominant two-way engagement rhetoric are often still very much dominated, owned and hence censored along a top-down model.

It is thus the case that a clearly uncivil comment which – insofar as it is useful to write about 'trolling' – offers a clear example of this practice did not have a straightforwardly negative impact. Clearly, this incivility may have affected readers' interpretations of the blog post in question (and there are not-unreasonable concerns which might prompt many to censor this type of comment on a science blog post or to avoid reading this type of comment). However, such comments and incivility also have life beyond where they are posted – they can feed into particular types of community action.[15]

"[T]he Tulips Were Actually Quite Lovely": The Good Old Days of Trolling

The above example illustrates how 'trolling' comments can sometimes be recirculated and sprout new activities in unexpected ways. A further instance of this is how, when discussing the above example of 'trolling' on Twitter, we were surprised that this discussion elicited a type of nostalgia.[16] Discussing 'trolling' in the past, another blogger (who frequently writes about science, medicine and technology)

[15] In the spirit of self-reflexivity, one should note that this paper itself can be seen as drawing on various below-the-line discussions and is also offering a different space to recirculate this 'trolling'.

[16] See https://twitter.com/JoBrodie/status/400401592366546946

referred to her "favourite rude comment".[17] Again, the comment she is referring to seems a clear example of trolling insofar as one can talk about such a practice: rather than engaging with the substance of the blogger's writing, the comment calls her "fat" and an "ugly fucker" and tells her that her "blog is utter shit and nobody cares about your pictures of glittered tulips".

However, this comment once again does not have a purely negative effect. At the time, it also attracted a more supportive comment for the blogger. However, more striking is that about a year later it seemed to be a source of nostalgia: cited as a favourite example of 'trolling', and leading into discussions of how "the tulips were actually quite lovely" and the use of an older Twitpic widget. While clearly such comments are potentially upsetting and may affect how readers interpret blog posts they can also – as shown in this example – feed into the construction of stories about community history. By becoming part of a shared history and allowing nostalgic reminiscence, nipple biscuits and glittered tulips feed into the development of community identities that can persist over time.

Trolling Methodologies: Trials and Observation

The two case studies in this chapter thus offer a helpful input to methodological debates: serving as additional evidence of how qualitative, ethnographic and observational work can provide different information to quantitative work which relies on randomised trials as its key data source, and of why it is important to use such qualitative work in order to develop a fuller understanding of our social reality. Even if – when tested in trial situations – uncivil comments do polarise views in problematic ways (see Anderson et al. 2013) social media does not only lead to once-only interactions. Instead, social media offers user engagement in which trust (or antipathy) is built up over time and in which users become content-generators as well as consumers. Ethnographic approaches have considerable potential for analysing such places and interactions: as indicated for example by Hine's (2000) aforementioned work on online ethnography, Parr's (2011: Chapter 6) work on virtual communities, self-help and mental health and Brown, McGregor and Laurier's (2013) development of ethnomethodology to analyse mobile device use 'in the wild'.

As shown in our research, many of those engaging in (more or less civil) debate below-the-line will also be active in different ways in these and other spaces: in the context looked at here, everything from starting their own science blogs to activism such as filing ASA (Advertising Standards Authority, UK) complaints about problematic science claims in advertising. Rather than 'trolling' simply making one more stupid, it is plausible that arguing and 'shouting' online can be part of activism and of some aspects of (atypical) public engagement. Community –

[17] http://brodiesnotes.blogspot.co.uk/2011/12/watch-out-for-some-pr-about-nuratrim.html?show Comment=1337961933934#c3515788642144422809

and the medium- and long-term patterns of interactions which help constitute community – are thus important factors that often get insufficient attention when ‘trolling’ and incivility is being discussed. The ‘badscience’ community strengthens the individual blogger when they engage in an aggressive, activist, manner or when they face incivility, and in turn the community is strengthened in some ways.

While there is clearly a place for research that focuses on single interactions with discussions of science – for example, trialling differing conditions for viewing online science articles – this paper therefore also adds to the evidence that there is a need for complementary research that provides a longer-term view of the social context. Interactions with online discussions of science need to be theorised as very much as interactions – as ongoing processes that take place in a broader social context, and in which readers can also be or become content producers – rather than as one-time situations in which readers respond to online material.

Trolling Places

Our case study of the ‘badscience’ blogs illustrates that these blogging networks develop not only through above-the-line posts and the virtual spaces afforded by below-the-line commentary and other interactive forums, but also linked in with territorialised national spaces in ways which can lead commenters to take part in largely national rather than international interactions. Even with such online communities, then, distance is clearly not dead – it is better to think about complex interactions between the virtual and material. Zook and Graham’s (2007a: 468) concept of “DigiPlace” is useful insofar as it

> provides a focus on the ways in which the physical, tangible world combines with virtually accessible information and creates not a fixed setting for interaction, but a lived, fluid, and subjective space, shaped by space, time, and information. In other words, DigiPlace represents the simultaneous interaction with software (information) and ‘hard-where’ (place) by a individual. It is a way to conceptualize the scales of everyday life, and simultaneously to imagine the differences and interdependencies of place.

Everything from geographical shaping of search engine results (see Zook and Graham 2007b) through to common interests shaped by physical places can perpetuate a tendency for the users to stay roughly within their geographical area in the material they read online. For Zook and Graham (2007b: 1323) “while Internet users can in theory circumnavigate any and all discourses they encounter, they are highly likely to utilize hubs, e.g., search engines that have enclosed the Internet via their ranking and indexing methodology.” The ‘badscience’ blog and forum community can itself be seen as a part of this construction of place: directing the other members of the community largely to debates relevant to their geographical location.

This chapter has illustrated how ‘trolling’ and being ‘trolled’ helped build and fortify the community: by providing rallying points, a shared history and common enemies. This community stayed largely UK-based because these community-building processes were mostly focussed around things happening in the UK rather than elsewhere.

Conclusions

There are diverse aspects to the spaces and places of science blogs and their below-the-line and forum discussions, as illustrated by the case study used here. This can build up communities and enhance the voices of individual bloggers by attaching these voices to other blogs, broadening readership and social networks. Simultaneously, though, these spaces can be used as platforms for those who vehemently disagree and would like that space or activity to be shut down. Such blogging can also provide places in which 'trolls' can play: for example, seeking to provoke a response from communities or simply to offend. To understand such interactions, it is important to move beyond trials of single-instance below-the-line interactions in order to consider ongoing and potentially-productive developments.

A key question here, and one that we do not have a firm answer to, is what this means in terms of the places of science blogging: whether there is the development of a progressive sense of place or of narrow, exclusionary fighting. Uncivil comments can seem to make online spaces more exclusionary, and in some cases may quite clearly have this effect: for example, a web page (or academic article) referring to nipple biscuits may be inaccessible to some people behind web filters in settings such as schools or people may close down blogs or Twitter accounts after torrents of abuse. However, we would also argue that there is more positive and progressive potential in spaces that allow some degree of 'trolling' and incivility. Thinking back to Massey's metaphor of the bustle of Kilburn High Road, we would defend the noise, arguing and joking – and, indeed, the 'trolling' – which finds a place in the below-the-line spaces of science blogging. An empty, quiet, ordered road might be less likely to offend than the bustle of a busy city high street, but it also lacks much of what makes life in these communities so appealing. Likewise, it may be in the noise and shouting – and, sometimes, vulgar abuse – of above-the-line blogging and below-the-line comments that the joy of a community is found, rather than in attempts to construct calmer and more sterile spaces designed for various conceptualisations of online participatory democracy. While Penny (2014) acknowledges the dangers of gendered online abuse, she also emphasises the opportunities presented by online spaces where she "realised that I wasn't alone. There were other people out there who felt quite like I did. There were other weird kids, other queer kids, other angry feminists, other nerds and anarchists and people who wanted to rearrange the world to suit our notion of what was just and right." We would emphasise the value of diverse, noisy and messy online spaces where people can meet and act and organise.

The academic literature on science communication has moved from a Public Understanding of Science model towards a Public Engagement with Science one that foregrounds scientists' need to engage and enter into a meaningful dialogue with the public and where the very concept of a boundary between the public and the experts is seen as problematic and blurred (Bauer et al. 2007). By closing off comments – by trying to limit what branches off from science discussion online and blocking critical, uncivil or 'trolling' statements – we see a move back towards a more hierarchical approach to science communication, where communicating 'balanced' information is prioritised over more active engagement. Such an approach offers fewer spaces for shouting, off-topic rambling and offence – while potentially 'cleaner' looking, it lacks the excitement and bustle of a thriving high street or community, and offers fewer opportunities to build a more progressive and open sense of place.

We would, then, argue that there may be more progressive potential in the noise of less-regulated online spaces than in more closed and censored places without the same space for comment. This bustle may keep places active, draw people in, challenge ideas and open new opportunities. As with a bustling high street, the occasional raised voice or expletive is as much a part of the life of a place as more restrained and 'serious' conversations. It is out of this below-the-line bustle that a more progressive sense of place and of community might (or might not) develop.

References

Allan, S. (2006). *Online news*. Maidenhead: Open University Press.

Anderson, A. A., Brossard, D., Scheufele, D. A., Xenos, M. A., & Ladwig, P. (2013). The "Nasty Effect:" Online incivility and risk perceptions of emerging technologies. *Journal of Computer-Mediated Communication*. doi:10.1111/jcc4.12009.

Bauer, M. W., Allum, N., & Miller, S. (2007). What can we learn from 25 years of PUS survey research? Liberating and expanding the agenda. *Public Understanding of Science, 16*(1), 79–95.

Bell, A. (2012). "ScienceBlogs is a high school clique, Nature Network is a private club": Imagining the communities of online science. *The Canadian Journal of Media Studies* (Special Issue Fall 2012), 240–226.

Bell, A. (2013, January 16). Do online comments hurt – or aid – our understanding of science? *The Guardian*.http://www.guardian.co.uk/commentisfree/2013/jan/16/online-comments-hurt-aid-understanding-science?CMP=twt_fd. Accessed 22 June 2013.

Bonetta, L. (2007). Scientists enter the blogosphere. *Cell, 129*(3), 443–445.

Brossard, D., & Scheufele, D. A. (2013). Science, new media and the public. *Science, 339*(5115), 40–41.

Brown, B., McGregor, M., Laurier, E. (2013). iPhone in vivo: Video analysis of mobile device use. *CHI' 13*. http://www.ericlaurier.co.uk/resources/Writings/Brown-2013-iPhone-in-vivo.pdf. Accessed 9 Jan 2014.

Cairncross, F. (1997). *Death of distance: How the communications revolution will change our lives*. Cambridge, MA: Harvard Business School Press.

Criado-Perez, C. (2013, October 22). Managing negativity in the face of internet abuse: Take the high road. *The Guardian*.http://www.theguardian.com/women-in-leadership/2013/oct/22/managing-negavitivy-take-high-road. Accessed 3 Dec 2013.

Criado-Perez, C. (2014). Twitter: Optimised for abuse. *Week Woman*. http://weekwoman.wordpress.com/2014/05/11/twitter-optimised-for-abuse. Accessed 16 May 2014.

Deleuze, G., & Guattari, F. (1998) *A thousand plateaus: Capitalism & schizophrenia* (Trans. B. Massumi). London: Athlone.

Dodge, M., & Kitchin, R. (2001). *Atlas of cyberspace*. Edinburgh: Pearson Education.

Dore, R. (2001). *Making sense of globalisation*. London: Centre for Economic Performance.

Gardner, R., Dodds, K., Souch, C., & McConnel, F. (2010). *Communicating geographical research beyond the academy: A guide for researchers*. London: Royal Geographical Society (with IGB). http://www.rgs.org/NR/rdonlyres/00208CCD-BA4A-4DAB-828D-B2890EAB8AEB/0/RGSCommunicatingGuide2013.pdf. Accessed 21 June 2013.

Gieryn, T. F. (1999). *Cultural boundaries of science: Credibility on the line*. Chicago: University of Chicago Press.

Graham, M. (2013a). Geography/internet: Ethereal alternate dimensions of cyberspace or grounded augmented realities? *The Geographical Journal, 179*(2), 177–182.
Graham, M. (2013b). Social media and the academy: New publics or public geographies? *Dialogues in Human Geography, 3*(1), 77–80.
Graham, M., Hale, S., & Stephens, M. (2012). Featured graphic: Digital divide: The geography of Internet access. *Environment and Planning A, 44*, 1009–1010.
Graham, M., Stephens, M., & Hale, S. (2013). Featured graphic: Mapping the geoweb: A geography of Twitter. *Environment and Planning A, 44*, 100–102.
Hindman, M. (2008). *The myth of digital democracy*. Princeton: Princeton University Press.
Hine, C. (2000). *Virtual ethnography*. London: Sage.
Hogg, M. A., & Abrams, D. (1988). *Social identifications*. London: Routledge.
Hundal, S. (2013). New study shows trolling makes you more stupid. *Liberal Conspiracy*.http://liberalconspiracy.org/2013/01/10/actual-evidence-shows-trolling-makes-you-stupid/.
Kitchin, R., Linehan, D., O'Callaghan, C., & Lawton, P. (2013). Public geographies through social media. *Dialogues in Human Geography, 3*(1), 56–72.
Lewis, H. (2011). "You should have your tongue ripped out": The reality of sexist abuse online. *New Statesman*.http://www.newstatesman.com/blogs/helen-lewis-hasteley/2011/11/comments-rape-abuse-women.
Massey, D. (1993). Power-geometry and a progressive sense of place. In J. Bird, B. Curtis, T. Putnam, & L. Tickner (Eds.), *Mapping the futures: Local cultures, global change* (pp. 59–69). London: Routledge.
McLuhan, M. (1994). *Understanding media: The extensions of man*. Cambridge, MA: MIT Press.
Mendel, J. (2010). Afghanistan, networks and connectivity. *Geopolitics, 15*(4), 726–751.
Morrowa, O., Hawkins, R., & Kern, L. (2015). Feminist research in online spaces. *Gender, Place & Culture: A Journal of Feminist Geography, 22*(4), 526–543.
Munger, D. (2010). Blogging out of balance. *Seed*. Accessible at http://seedmagazine.com/content/article/blogging_out_of_balance/
Nature. (2009). Filling the void (editorial). *Nature, 458*, 260. 19 March.
Painter, J. (2009). Territoire et Réseau: Une Fausse Dichotomie? In M. Vanier (Ed.), *Territoires, Territorialité, Territorialisation: Controverses et Perspectives*. Rennes: Presses Universitaires de Rennes.
Parr, H. (2011). *Mental health and social space: Towards inclusionary geographies*. Oxford: Wiley-Blackwell.
Paulussen, S., & D'heer, E. (2013). Using citizens for community journalism: Findings from a hyperlocal media project. *Journalism Practice*. doi:10.1080/17512786.2012.756667.
Penny, L. (2011, November 4). A woman's opinion is the mini-skirt of the internet. *Independent*. Accessible at http://www.independent.co.uk/voices/commentators/laurie-penny-a-womans-opinion-is-the-miniskirt-of-the-internet-6256946.html.
Penny, L. (2014). Is the internet intrinsically sexist? *The Debrief*. Accessible at http://www.thedebrief.co.uk/2014/05/is-the-internet-intrinsically-sexist.
Raeburn, P. (2013). Scientific American blog editor admits to sexual harassment. *Knight Science Journalism Tracker*. Accessible at http://ksj.mit.edu/tracker/2013/10/scientific-american-blog-editor-admits-s
Riesch, H., & Mendel, J. (2014). Science blogging: Networks, boundaries and limitations. *Science as Culture, 23*(1), 51–72.
Schmidt, G. (2008). To blog or not to blog? *Nature Geoscience, 1*(4), 208.
Smith, N. (2005). *The endgame of globalization*. New York: Routledge.
Thurman, N., Pascal, J.-C., & Bradshaw, P. (2012). Can big media do big society? A critical case study of commercial, convergent hyperlocal news. *International Journal of Media & Cultural Politics, 8*(2-3), 269–285.
Zook, M. A., & Graham, M. (2007a). Mapping DigiPlace: Geocoded Internet data and the representation of place. *Environment and Planning B: Planning and Design, 34*, 466–482.
Zook, M. A., & Graham, M. (2007b). The creative reconstruction of the Internet: Google and the privatization of cyberspace and DigiPlace. *Geoforum, 38*(6), 1322–1343.

From Webcams to Facebook: Gay/Queer Men and the Performance of Situatedness-in-Displacement

16

Ken Hillis and Michael Petit

Online a Lot of the Time

MIT's August 2012 issue of *Technology Review* presents a snapshot of an important aspect of the contemporary zeitgeist: the sense of dis-ease engendered by the rapid rise of information technologies and the social changes that have accompanied their deployment, coupled to relentless hype that we must embrace these same technologies as progress incarnate even as we wait impatiently for their next iteration and further development. The magazine's cover, a remix of the publicity poster for the film *The Social Network*, David Fincher's 2010 examination of Facebook's early years, features a close-up head shot of founder Mark Zuckerberg staring directly into the reader's eyes. WHAT FACEBOOK KNOWS is emblazoned across his face, and immediately beneath: "It has collected more personal data than any other organization in human history. What will it do with that information?" A boxed section on the cover's upper right corner lists more of the issue's content. The first title promises to answer "Why you will wear Google Goggles." These titles pivot between the present and the future—what Facebook knows now and why augmented reality as offered by Google, coming soon, will be irresistible. Their

K. Hillis (✉)
Department of Communication, University of North Carolina at Chapel Hill, Chapel Hill, NC, USA
e-mail: khillis@email.unc.edu

M. Petit
Department of Arts, Culture and Media, University of Toronto, Scarborough, Toronto, ON, Canada
e-mail: mepetit@utsc.utoronto.ca

S.P. Mains et al. (eds.), *Mediated Geographies and Geographies of Media*,
DOI 10.1007/978-94-017-9969-0_16

juxtaposition points to and is part of the dis-ease—can Facebook be trusted with this much power?—and the hype—you must have it!—on display throughout much of contemporary culture.[1]

Technology Review's titles also recognize, indeed are premised upon the assumption, that key social relationships take place online in social media's virtual settings, and they offer the expectation that living in such highly-mediated settings, conceived of as a series of spaces both separate from yet parallel to and part of our broader reality, will increasingly become the norm. Current statistics confirm a propensity to be online a lot of the time, and on social networks for much of it. During March 2012, Americans spent, on average, 38.8 h online (ComScore 2012:5), and as of October 2011, social networking was the most popular category worldwide of all web activities, accounting for 19 % of all time spent online (ComScore 2011:4). Indeed, Facebook has been able to amass its vast targeted-advertising treasure trove of user information precisely because so many choose, often enthusiastically, to mediate their lives through it.

Each iteration of IT innovation naturalizes the contemporary ideal, however implicit, of being online a lot of the time. Facebook's Timeline, for example, allows users to construct a complete digital record of their entire lives, both online and off. Facebook automatically adds to each user's Timeline all status updates, postings, photos, videos viewed, sites visited and liked, and so forth; users are encouraged to fill out the "Way Back" section before they joined. "It's fun and easy," Zuckerberg reportedly offered at Facebook's f8 conference in September 2011. "It's a way to show off who you are, what you do and where you've been. It's . . . a visual history of everything you've ever done, all the way to when you were born" (Parr 2011). Seemingly, one has the responsibility if not moral duty to oneself to spend a lot of time online and labor to record, in as much detail as possible, one's actions in toto. Enter Google's goggles (aka "Project Glass") as another innovation that also naturalizes the "need" to be online. "Don't Leave Home Without It" takes on new meaning in a world of Google goggles, which we will supposedly wear because "It simply will be better to have a [wearable computer] that is hooked onto your body" as "functional jewelry" that presents information drawn from the Web "relevant to what you're doing at any given point" (Manjoo 2012:73–74). Whether "better" or otherwise, this overlay of information, augmented onto the human sensory field of sight, further sutures online and offline space, and thanks to Timeline we will have the ability to document their merger.

The nexus of dis-ease/hype engendered by information technologies is in dynamic relationship with other aspects of the contemporary zeitgeist: a desire for the next thing, the killer app, the belief that the progress myth can, even

[1]This dis-ease/hype nexus is also apparent in the other titles listed on the cover: "Will Germany's clean-energy gamble sink Europe?" and "Creating human organs on a microchip." The rhetorical question implies that renewable/green technologies could very well sink Europe, but human organs on a microchip are on the way, just around the corner, without question or social or cultural import save that they'll be "better."

should, be achieved through technology alone, a feeling encouraged by corporate and academic spin alike that we stand on the threshold of a breakthrough moment when "this changes everything." Yet, as Carolyn Marvin points out in *When Old Technologies Were New* (1990), no electronic communication technology "changes everything"; rather, patterns of use echo and at times reify existing social structures: "the early history of electric media is less the evolution of technical efficiencies in communication than a series of arenas for negotiating issues crucial to the conduct of social life; among them, who is inside and outside, who may speak, who may not, and who has authority and may be believed" (4). Innovations such as Timeline and Google's goggles constitute new arenas for this ongoing set of always contentious negotiations, and while such innovations continue to shift the conceived relationship between the online and the offline, their use is as much a continuation of older forms and uses of media as it is a break with them. It therefore remains important to examine and historicize what it means to live in and through media, why individuals might wish to do so, and what the negotiation of power dynamics suggests about the conduct of social life "therein." If, courtesy of Facebook and Google, social interactions are to increasingly occur within a mediated geography that merges offline and on, 3-D embodied and 2-D pixelated space, then it is instructive and timely to look at their precursors. In the remainder of this chapter, we examine the web camera—an old technology that once was new—and its use by some early adopters, gay/queer men, to broadcast themselves on the internet live. These men, we argue, provide a useful case study through which to interrogate the dynamics of lifecasting—of living online a lot of the time—and in doing so to trace one component leading to the current technological conjuncture.[2]

Webcams-R-Us

The ability to be online a lot if not *all* of the time is intrinsic to webcam technology. It is internet lore that webcams started with the Cambridge University computer scientists Quentin Stafford-Fraser and Paul Jardetzky, who wrote software in 1991 that allowed them to watch, via a local area network, the collective coffee pot located elsewhere in the building where they worked. The coffeecam was connected to the internet in 1993 as the first live, 24-h webcam site, and when taken down in 2001 had amassed more than 2.4 million visitors.[3] In 1994, the U.S. computer company Connectix, later purchased by Logitec, introduced the first commercial webcam. The QuickCam produced a grey-scale image of 320 × 240 pixels and allowed users to broadcast somewhat murky video of themselves at 15–60 frames per second.

[2]Portions of the following summarize arguments made in Hillis 2009.

[3]"Farewell, Seminal Coffeecam," *Wired*, March 7, 2001, http://www.wired.com/culture/lifestyle/news/2001/03/42254. Accessed August 12, 2012. See also Stafford-Fraser's 1995 essay, "The Trojan Room Coffee Pot." http://www.cl.cam.ac.uk/coffee/qsf/coffee.html. Accessed August 12, 2012.

Shortly after its introduction, Danni Ashe, in late 1994/early 1995, taught herself HTML and launched her 24/7 porn-centric site, danni.com. The first non porn-centric lifecaster, Jennifer Kaye Ringley, went online with the JenniCam in 1996 (closed 2003) and began broadcasting daily activities live from her dorm room at Dickinson College in Carlisle, Pennsylvania.[4]

Launched January 1997 (closed 2001), SeanPatrickLive holds the likely distinction of being the first gay/queer lifecasting site. Its owner, Sean Patrick Williams, garnered considerable media attention at the time. Interviewers were perplexed as to why so many viewers (Williams' site received between 35,000 and 80,000 unique views per day) found the largely banal images of Williams eating, sleeping, and looking at his computer so compelling. In one interview he offered, "It's sort of to let people know that everybody's not as different as we like to think" (Long 1998), and in another he stated that the internet "is of particular benefit to gay activism because it empowers closeted gays to 'publicly stand up for themselves in a private way'" (Aravosis 2000). Williams and other early gay/queer adopters of lifecasting used the Web to announce, in essence, "We're here, we're queer—*in media*."

Webcams allowed these men to construct a networked performance space that facilitated their ability to transmit to viewers a degree of public visibility frequently denied them in public settings organized according to the dictates of heteronormative forms of interpersonal expression. Geographically isolated from one another and subject to discrimination and violence, they had the ambivalent though potentially liberating experience of circumventing forms of social marginalization by performing idealized gay/queer identity constructions for one another in online settings. These identity constructions—like those of most online users who choose to construct one—reflect the experiential psychic realities of the individuals who perform them. In the case of gay/queer men, these realities are inflected by their experience of having been frequently positioned by heteronormative discourse as "never here"—never in the central social, psychic, and material spaces that heteronormativity claims for itself—and always on the margins at best. In the late 1990s, then, these men's desire for voice and presence was actualized as a moving online image or persona that said, in ritualistic fashion, "Look at me, in truth I exist." Staking an ontological political claim to exist in the here-and-now through a mediated form of virtual space in which one is conceptually detached from one's body abounds with spatial ironies, yet doing so is also a tactical response to different forms of heteronormative silencing of queer lives. The askance performances of online gay/queer identities—situated "here" yet at the same time displaced "there," out of sight—constitute a pragmatic Goffmanesque "front stage, back stage" response to the social shunning and potential for physical violence that these men experience and anticipate experiencing.

Sean Patrick Williams recognized that "there are a lot of people who are closet on the Internet, a lot of people who just normally wouldn't do anything . . . you know, when you go to a list of hate crimes and you see how many people get killed . . . when

[4]For further discussion of Ashe's and Ringley's sites, see Senft 2008 and White 2003.

it constantly is happening over and over you kind of go 'oh wait, they kill gay people on the streets these days'" (Snyder 2002:192). While Williams chose to be out, many gay/queer men at this time preserved their closet on the Web. Billy's Perv Website—A Cool Gay Site for Servicing Your Dick (1997–1999) invited visitors to "click here" to "vote which tanline should I go for? speedo or thong . . . click here to have virtual sex with my image. And please, send me email, but don't ask for sex because I don't sleep around. I'm only out on the net, shy in real life." Being "out" only on the Web is a spatial strategy that depends on understanding the Web as a relational space available for such purposes as displacing identity and presence that may be conceived as too difficult and too dangerous—"they kill gay people on the streets"—to negotiate in the real world. Taking one's place in a virtual public realm allows one to seemingly mitigate or negotiate on one's own terms how one "enters into dialog" with the potential and actual forms of psychic and physical violence that attend to what heteronormativity continues to insist is "too much" gay/queer public audibility and visibility. The phenomenon of *telepresence*—the experience and simulation of the sense of presence achieved through the use of a communications technology—allowed these men, and those who have followed in their footsteps, the utopian experience of being virtually "out" in the relatively anonymous discursive space of a webpage, chat room, or webcam. This self-created mediated space seems to be without the psychic and physical risks they must incorporate into their daily routines as part of being in the material world.

Through telepresence's indicative magic, an operator's closet "here" in the real world may seem more bearable through designing and performing an idealized gay/queer persona "there" online. The closet, always a relative spatial metaphor, is for many men in the gay/queer diaspora the outcome of a still resilient internalized homophobia that is rationalized as a survival strategy when faced with the harsh reality of bearing the brunt of rituals of social excommunication (Carey 1998), geographic and social isolation, and ongoing violence and hate crimes. In creating a telepresent persona, therefore, we suggest that the webcam settings these men produced allowed them to ritualistically update early anthropological understandings of the fetish as "an inanimate object worshipped . . . on account of its supposed inherent magical powers, or as being animated by a spirit" (OED) and produce what we term the *telefetish*. Like the archaic fetish, the telefetish, in taking the form of an operator's webcam image, garners unto itself an ironic power: it is skillfully contrived by its operator, made as a form of art, animated by spirit and desire, and at times worshiped in its own character (see Hillis 2009). This persona/telefetish performs for both its author and for others from the virtual space of a website where the author also can feel personal empowerment by being out. In an interview Williams stated, "You know . . . I used to think [being gay] wasn't this defining part of you, it wasn't a big part of you, but it is, I mean, I think that as long as other people will judge you on it, I think it is a defining part because you have to deal with it" (Snyder 2002:192). For Williams and many other gay/queer men, virtual spaces such as those provided by webcams serve as a "dream surrogate for better values" (Apter 1993:3). Today we see the updating of this magical and wishful thinking conjoined to utopian desires about information technologies on display in *Technology Review*'s explanation of why we will all wear Google goggles—"It simply will be better" (Manjoo 2012: 73).

Coping mechanisms such as Billy's pornutopic strategy of being "out only on the web," however, are not entirely unproblematic for those younger digital queers who see themselves as post-gay or even post-queer. Though a site such as Billy's Perv Website can form a bridge from the closet to a fuller self-acceptance, and while it should also be understood as a vehicle for a young gay/queer man to test, explore, and experiment with different ways of being (Knopp 2004:123), it also may confirm a decision made in the claustrophobic logic of fear that perhaps one need not—or should not—be all that out in real life. Given heteronormativity's pressing dictates, for such men the mediated space of the Web may sometimes seem enough. The need to be fully out "here" may feel lessened when a fragment or emblematic trace of one's self—a telefetish—can connect through digital networks to other digitally queered hybrid identities "there" that are, to varying degrees, also fractured from flesh.

Being "out" only on the Web complements and extends the logic of the "virtual equality" noted by Urvashi Vaid in 1995, 2 years before SeanPatrickLive launched. The logic of this virtual equality is rendered hegemonic through Web practices that inadvertently or otherwise position "equality" for gay/queer people in such a way as to render as invisible as practicable those concrete differences inherent in the multiple realities of gay/queer lives. While the glass ceiling of embodied acceptance is higher than it was a generation ago, it remains the case that contemporary queers often are treated as virtual equals only if they largely eschew performing queer difference. Many digital queers only out on digital networks therefore have moved to craft virtual forms of the ghetto as a partial turning away from or tactical accommodation to those who would deny these men or repress the truth of the array of subjectivities achieved through decades of organization both *against* homophobia and its practices as well as struggle *for* articulating a different *and* complementary collective sociopolitical and cultural identity. As Donald Snyder observes, "A virtual world without closets does not necessarily help recreate the real world into a place where closets are no longer needed" (Snyder 2002:179).

Belief or desire that the Web might constitute some form of actual space occurs in tandem with the reality that it remains taboo in many if not most parts of the wired and wireless world to, for example, kiss or hold hands with one's same-sex partner at the mall.[5] And while it is acceptable for a handful of secondary sitcom characters to be out on mainstream network U.S. TV, viewers see little enactment of actual gay/queer experience or sexuality on these channels. The ontological acceptance of the idea of gay/queer cedes to, in Eve Sedgewick's memorable phrase,

[5]As of 2012, there remain 80 countries worldwide where homosexual acts remain illegal (notably throughout the Middle East, South Asia and in most of Africa, but also in much of the Caribbean and Oceania), including five countries that carry the death penalty. On June 13, 2012, Bonisiwe Mtshali, 29, was beaten unconscious by security guards at the Carlton Centre Mall in Johannesburg, South Africa after guards saw her kissing her girlfriend goodbye. IOL News.

the epistemology of the closet. This is so because any human experience of feeling "in" place requires both the actual space within which one "takes place," along with an experience of this space that endures over some period of time. If gay/queer men frequently fail to attain a positive experience of duration in heteronormativized material space because—apart from any demand for affirmative recognition—they are disavowed physically (positioned as always over there, in some ill-defined liminal "queer" space and therefore never fully here), then it follows that these men rarely experience full acknowledgment as being "in" this here place together with and at the same time as non–gay/queer individuals. These individuals would have had to move beyond the experiences of disavowal or indifference in order for a face-to-face full acknowledgement to happen.

As a form of virtual politics that responds to emplaced disavowal, "I'm only out on the Web, closet in real life" is similar to the ritualized use of a recited story or myth to keep a fear at bay—"let's produce a virtual ideal that both responds to and turns its back on the problem." This virtual ideal and desire can, however, become a locus of tantalization for some and, ironically perhaps, mimetically mirror the disavowal the ideal was enacted to overcome. Within shifting, even flexible, but often still hostile contexts, making sense of the world through the Web becomes more than just another gay/queer option, for these men have lived by the betwixt-yet-between compartmentalizing and askance spatial logic inhering in telepresence long before its technical application to digital contexts such as Facebook and Google goggles. Their psychic reality parallels the spatial reality established by webcams of being neither here nor there yet somehow both at once, which is one factor explaining these men's predisposition to the early adoption of webcam technology as a way to perform a complex and affective online trace of their lived realities.

Living with Situatedness-in-Displacement

A separate, interrelated factor contributing to the early rise of gay/queer webcam sites is their continuation of the sense of "front stage, back stage" afforded by gay/queer spaces that long precede the Web. Such spaces allowed for the segregation of aspects of gay/queer lives into sanitized public fronts and more private forms of interaction. The hustler bar, the leather bar, the western bar, the drag bar, and so forth aestheticize and embody understandings of identity expression linked to performing commodity forms. Other spaces such as men's Radical Faerie gatherings are avowedly noncommercial. Typically set in isolated rural environments, ritualized activities take center stage as a means of promoting an alternative range of gay/queer/faerie identities separate from that of gay/queer male consumer culture. Such spaces are "backstage" and largely free from the prying eyes of heteronormative policing practices. Henri Lefebvre's (1992) argument that spaces of representation locate identity practices and are always produced and designed is useful here in considering that webcams, like gay bars and faerie gatherings, are produced representations of space that reflect the lived and imagined spaces of the men who design them. While the ersatz spaces of webcam sites are formally different from concrete spaces, participants in the material spaces of commercial bars and faerie gatherings and in the virtual settings of webcams share in ritualized exchanges that trade in visually based and highly artistic conventions. These exchanges require mutual agreement on the part of those present and telepresent for something akin to living in art (and through media in the case of virtual settings) in ways that build social capital through the performative aestheticization of their own bodies. These are the same bodies that still are often judged as possessing little to nothing of value by heteronormative standards.

Crafting identity in the spaces of representation that Lefebvre identifies depends for any success, then, on the effectiveness and competency of the performances as well as the artistry inherent in the settings, clothing and props chosen by the participants. Jason Weiderman, operator of As You Gaze Upon Me (1999–2000) and one of the few gay/queer men to have published a self-reflexive personal account of his experiences as a webcam operator, writes that "A great 'performance' occurs when the aura of a hidden camera is produced despite careful attention to each shot. Often, whole evenings were 'posed' so the end result appeared to be a natural flow of images—the subject 'caught' unaware by the camera, performing intimate acts such as sleeping, reading, picking one's nose" (2003:16). A fine line must be walked, and fear of walking it unsuccessfully online is often expressed through the recurring concern of early webcam operators that visitors not find their webcam performances "boring." "Ritual," notes Victor Turner, "is a declaration of form against indeterminacy" (1995:77), and to be found boring by one's fans means an uncertain and therefore indeterminate future as a webcam operator. Adhering to the logic of the disposable commodity, the frequent updating of sites and the performance of intimacy on camera for fans are ritualistic activities that maintain and extend the meaning and shelf life of the webcam operator-as-telefetish. Indeterminacy is kept at bay. The early insistence that webcam operators refresh their images, promoted at the time on sites such as Jasbits.com, which during the late 1990s featured "Top5 CAM Current Guy and GirlCam Rankings," reverberates today in the expectation that one should, in the moral sense of the verb, post frequent updates for "friends" on Facebook and other social media sites.

Larry Knopp (2004), examining queer "quests for identity," draws attention to the fact that gay/queer men have long engaged in journeys from one place to another as part of their quest to find social and material spaces and places more supportive of who they are and who they may wish to become than the places they have left behind or felt compelled to flee. They therefore possess "a corresponding ambivalent relationship to both placement and identity" (2004:124) and as a result can find pleasure in being "simultaneously in and out of place" (ibid). Knopp's assessment of gay/queer men's relationships to place and identity is applicable to the Web. We suggest the notion of "situatedness-in-displacement" as a way to understand the spatio-psychic reality of the gay/queer webcam experience that celebrates neither being here nor there but both at once—to be, in other words, in the space of flow and mobility. Movement and mobility are widely prized in the contemporary technological conjuncture and clearly gay/queer men are not alone in privileging them. They do, however, differ in that while mobility is increasingly a normalized expectation of capitalized social relations, these men, because they are discursively positioned as elsewhere, have been discursively rendered as mobile, denied fixity, and forced into flow, without advance consultation—in other words, potentially against their will. Yet Knopp also details the ways by which such men experience pleasure in movement and displacement as a result of feeling both in and out of place at the same time. Ironically indicative of a kind of situationally-enforced cosmopolitan perspective, this situated-in-displacement is a queered outcome consonant with both an understanding of the Web as a site of movement and flow, and with the construction of Web-based personae or telefetishes to perform within it.

Fetishizing living in and through media—of becoming one with the networked digital stage upon which one performs one's virtual identities—engages a spatial strategy that has allowed gay/queer men to use the Web as a means to achieve forms of connection frequently eluding them in actual space. Jason Weidemann was able to achieve a kind of situatedness and fixity in movement through his webcam performances.

> ...I had deep reservations about my own body image . . . At the start of my webcam, I wasn't consciously interested in exhibitionism. Not until I realized how much I enjoyed being on camera did I begin to see myself as an exhibitionist who depended on the comments of others to form opinions about himself. I began each night wondering who was watching and who was passing me by. But as the e-mails rolled in—sometimes four or five a minute—telling me how beautiful I was, how aroused my images were making them, I felt emboldened. Feeling beautiful turned me on, and so I masturbated, garnering even more praise. Masturbating . . . never became a clinical act. Pleasing others amplified the pleasure of masturbating on-line. (2003:17)

The sense of mastery achieved by Weidemann arguably extended to the broader local conditions of his life, and in gaining comfort through a process of continual surveillance by strangers, he also evolved a coping strategy for dealing with the sense that many people now experience of having little to no control over events taking place in the so-called global arena, which nonetheless affects each of us at the local level.

"Broadcast Yourself!"™

Any tactical response to an exigent situation can have ambivalent outcomes, and situated-in-displacement, pleasurable though it may be for some, is a necessarily ambivalent survival strategy. Equally, however, living in art and through media cannot resolve the exigencies of the real world, and these early adopting gay/queer men often found themselves positioned as ersatz saviors by their many fans. They achieved visibility through telepresence, and their living life so openly online made a difference in the lives of gay/queer male viewers. Will, a self-identified middle-aged Bostonian, wrote on his blog in 2006 that "What I admired about [Sean Patrick Williams'] cam site was how powerful and courageous a statement it was for a gay man to take the ultimate 'out' position of living a gay life on line for the whole world to see."[6] Yet the demands of niche celebrity status for Williams and others ultimately became overwhelming. Williams refused to answer the hundreds of emails he received daily and cautioned that his site was "a window, not a voice" into his life (Snyder 2002:187). He told fans not to consider him their personal friend. "You ultimately should go out and talk to other people. . . . It's not about developing this personal relationship with me. You tune in and you watch, but get out the door and talk to somebody. My camera is on 24 h a day, but if you think that's my real life, it's not. You're not getting everything, you're getting a picture every 30 s" (Long 1998). The owner of another gay/queer website, Queerz 'n Jox 'n Sox (1998–2000),[7] in explaining why he was shutting the site, wrote to followers,

> I am overwhelmed by the expectation that I am qualified or capable of being [an] online healer . . . no more email . . . do not let the web shield you from your destiny. boys need destiny . . . do not count on a magic combination of computer keystrokes to deliver your soul-mate to your door. while that strategy might deliver a pizza, the odds are against the internet satisfying hunger of the heart. turn off your computer, go outside . . . do not believe that beauty is what photographers see . . . though we can share experiences on the net we are also separated by it. like television, the web is an exacting replica of reality, but the farthest thing from it.

In formulating a tactical response to the problem of the present, the alienation these early adopters experienced in the material world was ameliorated on the Web, but their strategy in time introduced its own hybrid forms of alienation.

This earlier generation of gay/queer men turned to a then-new technology—the webcam—to stake an ontological claim to exist in the here-and-now, and the brief history we offer here suggests that the social and geographic marginalization and isolation that favors the emergence of situatedness-in-displacement as a queered form of self-experience migrated to and transmographfied within online settings. In the decade and more since these men closed their sites, what was once novel is now commonplace for those with the means of access. YouTube's "Broadcast Yourself!™" is the slogan of our time. Situatedness-in-displacement, a psychic mechanism developed as a tactical response to marginalization, has become an everyday reality for millions of people online, the majority of whom are not gay/queer. While webcams remain popular with many gay/queer men, and sites such as Justin.tv provide easily accessible platforms for lifecasting oneself, significant components of online gay/queer men's identity performances have moved to mobile apps and websites such as Gaydar, Gridr, Recon, Scruff, and Tumblr. These platforms allow users to (re)create themselves as online telefetishes performing various aspects of gay/queer identity across the online diaspora *and* to locate and meet other gay/queer men for embodied encounters in the actual here and now. These newer technological forms, like Facebook's Timeline, suture more closely expectations of online and offline space, and the situated-in-displacement that

[6]DesignerBlog, June 6, 2006. http://designerblog.blogspot.ca/search?q=sean+patrick+williams. Accessed August 12, 2012.

[7]Site name changed at owner's request.

Google goggles promises will only up the ante in the merger of the here and there into one encompassing network. Under the circumstances, it seems worthwhile to consider the caveats of the operators who closed their webcam sites precisely because of the new forms of networked alienation they discovered. Living in and through media, as their example shows, risks misrecognizing the telefetish as equivalent to the embodied being it both represents and seemingly supersedes. As the broader society continues to renegotiate the relationship between online human presence and actual bodies, a relationship crucial to the conduct of online social life, we do well to heed these operators' experiences.

Postscript

Senior *GQ* magazine editor Mickey Rapkin writes in the April 2012 issue, "what this technology really offers is faux intimacy: It's the perception of intimacy with the added hangover that comes with waking up in a queen-size bed next to nothing but your laptop. Seeing each other every night—but not being able to touch each other—is its own form of punishment." Rapkin's lament refers to being apart from his boyfriend for some months after "his dude" accepted a job on the American West Coast. Despite Rapkin's articulating a gay/queer positionality throughout the article, the accompanying image evokes nostalgia for an idealized form of heteronormative pillow talk that serves to visually negate gay/queer subjectivity. To the illustration's right, a pajama-clad man lounges in bed while holding the speaker of a candlestick phone to his mouth. To his left, separated by the requisite jagged line of desire and distance that telephony introduces into pop culture representations, a perfectly coiffed woman, head on her pillow, eyes closed, listens dreamily. The article and its accompanying illustration serve to indicate that while gay and queer people have achieved considerable momentum toward a fuller recognition of their subject selves, we continue to be positioned by heteronormative discourse as somehow never quite "here," not quite part of the picture. And so it goes.

References

Apter, E. (1993). Introduction. In E. Apter & W. Pietz (Eds.), *Fetishism as cultural discourse* (pp. 1–12). Ithaca: Cornell University Press.

Aravosis, J. (2000). StopDRLaura. *NewRepublic* (Oct 23). www.wiredstrategies.com/tnr_sdl.htm. Accessed 12 June 2015.

Carey, J. W. (1998). Political ritual on television: Episodes in the history of shame, degradation and excommunication. In T. Liebes & J. Curran (Eds.), *Media, ritual and identity* (pp. 42–70). New York: Routledge.

ComScore.com. 2011. It's a social world: Top 10 need-to-knows about social networking and where it's headed (December 21). http://www.comscore.com/Press_Events/Presentations_Whitepapers/2011/it_is_a_social_world_top_10_need-to-knows_about_social_networking. Accessed 1 Aug 2012.

ComScore.com. 2012. State of the internet: 1st quarter 2012 (June 14). http://www.comscore.com/Press_Events/Presentations_Whitepapers/2012/State_of_US_Internet_in_Q1_2012 Accessed 1 Aug 2012.

Hillis, K. (2009). *Online a lot of the time: Ritual, fetish, sign*. Durham: Duke University Press.

Knopp, L. (2004). Ontologies of place, placeless, and movement: Queer quests for identity and their impacts on contemporary geographic thought. *Gender, Place and Culture, 11.1*(March), 121–134.

Lefebvre, H. (1992). *The production of space* (D. Nicholson-Smith, Trans.). Oxford: Basil Blackwell.

Long, T. (1998). Livecam web sites bring us the excitement, boredom of others' lives. *Detroit News* (August 3). http://detnews.com Accessed 2 Feb 2008.

Manjoo, F. (2012). You will want Google goggles. *Technology Review, 115.4*(August), 72–74.

Marvin, C. (1990). *When old technologies were new: Thinking about electric communication in the late nineteenth century*. New York: Oxford University Press.

Parr, B. (2011). Facebook timeline redefines user profiles. *Mashable Social Media* (September 22). http://mashable.com/2011/09/22/new-facebook-profiles/#27009Timeline Accessed 1 Aug 2012.

Rapkin, M. Honey, this face time is torture." *GQ* (April 2012). http://www.gq.com/news-politics/mens-lives/201204/how-to-have-long-distance-relationship-men-video-chatting Accessed 1 Aug 2012.

Senft, T. (2008). *Camgirls: Webcams, live journals and the personal and political in the age of global brands*. New York: Peter Lang.

Snyder, D. I. (2002). 'I don't go by Sean Patrick': On-line/off-line/out identity and SeanPatrick-Live.com. *International Journal of Sexuality and Gender Studies, 7*(2/3), 177–195.

Turner, V. (1995 [1969]). *The ritual process: Structure and anti-structure*. New York: Aldine de Gruyter.

Vaid, U. (1995). *Virtual equality: The mainstreaming of gay and lesbian liberation*. New York: Anchor Books.

Weidemann, J. (2003). Confessions of a webcam exhibitionist. *The Gay & Lesbian Review Worldwide, 10*(4), 13–17.

White, M. (2003). Too close to see: Men, women, and webcams. *New Media and Society, 5*(1), 7–28.

Placing Satellite Radio in the Contemporary Radio Landscape

17

Sara Beth Keough

Introduction

It is a Monday morning, 7:30 am EST, and as I get in the car to go to work (and as is my habit), I turn on the radio, hoping that during my 15 min commute National Public Radio's (NPR) "Morning Edition" will give me something to use as a teaching moment in class that day. While this act has been part of my morning routine for many years, it is also a cultural practice involving delineations and compressions of time, space, and place. "Morning Edition" is produced in Washington D.C. and syndicated by my local NPR station. By the time I leave for work, I am not hearing "Morning Edition" broadcast live. I do, however, hear live broadcasts of regional news and announcements made by local my public radio station; thus, the broadcast is simultaneously live and pre-programmed, local and global (through syndication and streaming over the web). My car stereo receives the terrestrial signal from the local NPR station, a signal with defined spatial boundaries and limits. The frame of my car defines another spatial aspect of my morning radio routine: the listening space. If my car windows are open, the listening space expands beyond the confines of the car's frame. The broadcast becomes part of my social space as I reflect on a particular story with a colleague later that day. Finally, as a listener to this program, I am connected to different places (spaces with meaning): some distant, some close and familiar. I am also connected to place through the regional accents of reporters and announcers, especially those from my local station. Finally, I am connected to place through the musical interludes during "Morning Edition," which are chosen to reflect the story just reported. Even if I catch a "Morning Edition" broadcast in another context (e.g. hearing it coming from

S.B. Keough, Ph.D. (✉)
Department of Geography, Saginaw Valley State University, University Center, MI, USA
e-mail: sbkeough@svsu.edu

S.P. Mains et al. (eds.), *Mediated Geographies and Geographies of Media*,
DOI 10.1007/978-94-017-9969-0_17

a colleague's office) and hear those eight chords that mark the transition between broadcast segments, I immediately picture myself in a specific, familiar place: as a listener in my car.

Radio's terrestrial signals define listening space and allow the radio to become what media geographer Paul Adams (2009) refers to as "media in space." Despite moving as I listen in the car, my link to the program remains even as my location changes, as long as I stay within the boundaries defined by the station's terrestrial signal. This is an example of Adams' "space in media." The representational images that news reports and music create exemplify how radio can create "places in media." Finally, my listening to the radio in the confined space of the car is an example of Adams' "media in place" because the boundaries of the broadcast are defined by the closed vehicle I am driving. This closed listening space provided by my car is also an example of the kinds of private listening spaces that mobile media technologies like the Walkman and now the cell phone allow us to enjoy (Bull 2004).

Spatialities change, however, as the type of signal changes, such as the case of satellite (and internet) radio. One must ask, then, is satellite radio still radio? This chapter answers that question by exploring the changing spatialities of satellite radio in the contemporary mediascape through an examination of how geographic concepts of landscape, space, place, commodities, and communities are negotiated in this medium. First, satellite radio is placed in an historical context as I trace the evolution of political power and agency in radio broadcasting. This history is followed by a brief description of SiriusXM satellite radio, the company I use for examples of these geographical concepts. Then, I "place" satellite radio within geographical and media theories by considering the cultural landscape of radio, the satellite radio's place in the global mediascape, the deterritorialization of media spaces in satellite broadcasting, the importance of place in satellite radio broadcasts, the commodification of radio, and new communities of satellite radio listeners.

Historical Evolution of Power in Radio Broadcasting

In 1895, radio officially became the first wireless technology after Guglielmo Marconi purchased Thomas Edison's patent for electrostatic coupling and became the first person to transmit information wirelessly. Radio was initially adopted as point-to-point technology (called ham radio) and by the 1920s, radios had become popular and affordable, and had achieved commercial status in the United States. Radio was also profitable for owners, as the advertisers paid money to radio stations (or their parent companies) for advertisements aired on the radio, a concept called pay-for-broadcast. In this way, stations began to cover the costs of broadcasting. Broadcast networks (groups of stations) emerged in 1922, and thus programs could be aired to several potential markets at once. By the mid-1940s, three major networks (NBC, ABC, and CBS) owned 90 % of the radio stations in the United States. With the emergence of networks, broadcasting began a transition from bottom-up control to one of top-down control, as parent companies began to broadcast similar (or the same) content to all the stations they owned (Keith 2010).

Radio's popularity had consequences, however, because too many stations were using the same frequency resulting in widespread interference and frustration for listeners and broadcasters alike. Although the federal government had tried (unsuccessfully) to regulate radio communications earlier, by the 1920s the need for regulation of the radio medium was apparent. The government response began with the Radio Act of 1927, which created the Federal Radio Commission (later the Federal Communication Commission or FCC) for the purpose of licensing stations, issuing frequencies, and determining power and operating schedules (Keith 2010). Federal regulation meant, however, that the top-down control assumed by networks became partially relinquished to federal agencies that now controlled radio frequencies and how they were used.

Prior to World War II, radio was largely an entertainment venue with broadcasts of space-specific events like concerts and sports competitions, and other entertainment options like music, radio dramas, and talk shows. Sporting events, such as the 1922 World Series, were the first programs simultaneously broadcast to different regions (thus expanding the spatial boundaries of radio's reach) because of the emergence of radio networks. During the Great Depression, shows like "Amos 'n' Andy" brought humor to the despair of the time. Franklin D. Roosevelt's "Fireside Chats" became a critical component of his efforts to reach and inform the American people. The fact that radio was free made it immensely popular during the Depression (Keith 2010). Shows like "Amos 'n' Andy" and Roosevelt's speeches indicated that individuals still had some power to choose what was broadcast, even though the networks were responsible for airing the programs.

During World War II, radio stations in the United States became key disseminators of information for Americans wanting daily updates on the war effort. It was during this time that broadcast news became an important element in radio broadcasting. Although the FCC placed an embargo on new radio stations during the War, the radio grew in popularity, as did industry profits (Keith 2010).

In the post-World War II era, when competition with television threatened radio's future, radio entered a new phase of broadcasting: format radio, where disc jockies (DJs) assumed some control over what was broadcast and the focus on niche audiences began (a condition that eventually became more narrow with the advent of satellite radio). In programming for niche audiences in this era, DJs were not only focused on genre, they also assumed their listening audience was tied to a particular place defined by the extent of the terrestrial signal. Initially, free-form programs were the norm, where DJs chose the musical numbers they wanted to play and the order they wanted to play them. This free-form nature of radio programming became an art form for some DJs, who earned recognition for the way their radio music shows were constructed. The result was a mix of popular and original music, some of which the audience had never heard before. Previously marginalized musicians, like Bob Dylan for example, got airtime and saw their popularity grow. Dick Biondi, a Top 40s DJ known for his screaming antics on the air in major radio markets, was the first to play Elvis Presley. Although Top 40 hits were broadcast at the beginning of the post-war era, they had yet to dominate the broadcast format. It was also during this era in the United States that the Corporation for Public Broadcasting was established and National Public Radio emerged in 1970 (Alper 2006).

Profits were of particular concern to radio stations because the growing popularity of television broadcasting in the 1940s and 1950s initially resulted in revenue loss for radio stations. While radio does not attract the advertising dollars today that it did in the first half of the twentieth Century, the industry profits are higher now than they were 60–70 years ago. Most of the advertising dollars in commercial radio come from local businesses (rather than networks or national advertisers). Thus, the juxtaposition between local and global emerged in radio, as stations played music determined to be "popular" on a national and global scale, while relying on local advertising dollars for profit (Keith 2010).

By the 1980s, however, networks and station owners realized the profit potential that could be gained through the commodification of radio. Stations hired consultants to establish prime listening (and thus marketing) times of the day and week and poll the listening audience. Although Billboard Charts had been used since the 1940s, it was during this era that the playlists of popular songs became standard use by radio stations. The sound produced by commercial radio stations became homogenized across the country as DJs from New York to San Francisco selected music from the same playlists, playlists created by corporations to determine the most popular songs. No longer could marginalized musicians get airtime on commercial radio. The consequences to profits were too risky (Alper 2006).

The 1990s could be called the era of de-regulation for radio. The FCC relaxed many of its rules, and soon there was no limit on the number stations a company could own nation-wide. The corporate station buying and selling binge that occurred in the beginning of this decade resulted in large debts when the economy slowed. To account for profit reductions, broadcasters began to form local marketing agreements (LMAs) whereby stations lease time and/or facilities from a station in another market (or another area). The loss of regional music identity in radio broadcasting that began in the 1980s continued through LMAs because a single company leasing airtime from stations across the country would broadcast the same content nationwide (Keith 2010). Regions of the United States famous for the evolution of particular music genres (like country music in the South and grunge music in the Pacific Northwest) no longer held the monopoly on particular music genres.

A classic example of this evolution from free-form programming to playlisting can be seen if once considers the history of radio station KSAN in San Francisco. This station, which began as classical music station KSFR, became a free-form rock station in 1968 after hiring striking employees from another Bay Area radio station who had the strike and their jobs. The station became a trend-setter in the 1970s, playing music by the Rolling Stones, the Grateful Dead, Bob Dylan, and more, as other stations across the United States followed suit. During this decade, KSAN reached the top position in the country in the 18–34 year old demographic, as its counter-culture broadcasts about the Vietnam War, the Nixon Administration, and marijuana use appealed to the country's angry youth. By 1980, however, due to declining ratings, the station changed its free-from format to playlisted country music, and its popularity returned. Station format changed again on July 3, 1997 to classic rock playing, ironically, much of the music that has made the station popular in the 1970s (Bay Area Radio Museum 2013; Crosper 2009).

The emergence of digital audio broadcasting (DAB) in the 1990s, which had become the industry standard by the 2000s, further changed the nature of radio broadcasting, as it made analog signals out-moded. Music recordings moved to MP3s (from tapes and CDs), and broadcasting began to occur through the internet and via satellite. These new broadcasting formats threatened what was left of the local nature of radio in the United States, yet in other countries that lacked the infrastructure for coverage and the financial means to expand broadcasting, the new formats were welcomed (Keith 2010). With the move to digital, a new mode of marketing and listening present in other industries became important in radio: a focus on niches.

These changes in radio broadcasting, and the role that radio played in the decline of the music industry in the United States, are lamented in the *Frontline* documentary, "The Way the Music Died" (Kirk 2004). Michael Guido, music industry attorney interviewed for the documentary, calls what has happened to music and radio "the perfect storm." Record labels became consolidated, the industry preference moved from album creation to single tracks for download, and the radio stations that played this music became consolidated, especially after 1996. "There's essentially, I don't know, two or three radio stations for all intents and purposes right now," Guido stated, "which limits the ability of music, different music, to get out there. Program lists are being devised on an almost national basis for certain kinds of formats. The independent thinker as used to exist in the radio station is no longer allowed to exist." (Guido Interview 2013).

In general, the evolution of radio technology and broadcasting has moved from broad to narrow in focus. Terrestrial radio is generally considered broad in focus, even though some stations have specific music formats in their mandates. Internet radio, especially sites that select songs for the listener, are narrower. Satellite radio, however, may bridge the gap between the two. It could be considered "narrow-casting" because the channels are genre-specific, but at the same time, this technology can be considered broad, as channels can include any music recorded in that genre, not just the top-40 hits. The music-by-decade channels are a good example. The 1980s Music channel broadcasts popular hits from that decade, enough to satisfy listeners who like music from that era, but because the listener is already a fan of the music, programmers can include recordings that were not at the top of the charts in the 1980s. The commercial-free, 24/7 nature of broadcasting on satellite radio motivates the inclusion of such music because time must be filled by something.

SiriusXM Satellite Radio

This chapter uses one specific satellite radio company to illustrate key ideas. SiriusXM, which resulted from the merger of two independent satellite radio providers (Sirius and XM) in 2008, is the only satellite radio provider in the United States and Canada. While satellite companies exist in other parts of the world (Eutelsat in Europe, and Worldspace in Europe, Africa, and Asia), these providers offer both television and radio services via satellite. For the purposes of this chapter, SiriusXM is used because the company offers exclusively radio listening packages.

In the United States, a subscriber to SiriusXM satellite radio receives more than 130 commercial-free channels of music, sports, talk shows, comedy, and other categories for $14.49 USD per month. Channels are often genre-specific, and broadcasting occurs 24 h a day, 7 days a week. Reception is available nation-wide (as well as in Canada), unless the satellite signal is blocked by a physical feature such as a mountain, and the signal extends 200 miles off-shore. More expensive packages are available that include more channels and an annual subscription option. Subscribers must have a satellite radio receiver or access to the internet.

As of March 2013, SiriusXM had 24 million subscribers in the United States and 2 million in Canada, making it the largest radio provider by revenue north of the Mexican border (SiriusXM Corporate Website 2013). Carolyn Lin remarked that in 2006, the diffusion of subscriptions had become stagnant, which partly explains the merger of Sirius and XM companies in 2008 (Lin 2006). Subscription-based companies need to grow subscriber numbers (or increase fees) in order to increase profits. Thus, to continue profit expansion, SiriusXM began partnering with all major automobile manufacturers in the United States so that receivers are installed in cars and satellite radio signals are provided in most new vehicles and many pre-owned vehicles for a limited time, after which the owner has the option to continue the subscription. Offering the same channel packages through the internet also helped expand the subscription base as users can access the channels through most devices (including a number of smart-phone models) that can connect to the internet. Thus, while the title of this chapter indicates that the focus is on satellite radio, the relationship to, and influence of, internet radio must also be considered.

Placing Satellite Radio

In her 1996 memoir *Radio On*, journalist Sarah Vowell claims that "radio is a landscape, a place inhabited by heroes and villains" (Vowell 1996, p. 1). While satellite radio had not yet emerged as a popular medium at the time Vowell wrote this memoir, the opening statement to her book is still quite relevant. Satellite radio makes an imprint in the media world that reflects some of the values of American corporate and listener culture. The landscape of satellite radio is a territory with boundaries. It has been deterritorialized and reterritorialized by both local and global forces. It is a place (i.e. a space with meaning) shaped by political and economic forces (the heroes and villains to which Vowell refers). Corporations, content producers, and listeners alike place value on the radio landscape, making satellite radio a cultural commodity. The nature of satellite radio's new "imagined communities" (Anderson 1991) shares some traits with terrestrial radio, but adds new dimensions as well. The rest of this chapter explores these elements of the satellite radio landscape.

Satellite Radio as Landscape

Landscapes have long consumed the attention of geographers because of the meaning that can be derived from their analysis. People create cultural landscapes by leaving marks on the earth that give clues to their intentions, tastes, and desires. The landscapes defined and created by SiriusXM satellite radio reflect both corporate and listener culture in the United States. For subscribers, listening to satellite radio is part of everyday life, and should be considered in the context of daily activities (Orgad 2007). Furthermore, what listeners hear has been filtered by SiriusXM corporate practices and decisions. Therefore, the satellite radio landscape has been purposefully constructed by the corporation complete with holes (eliminated content) and content portrayed as that demanded by subscribers. Content broadcast by satellite radio is interpreted by listeners and new spaces of communication in everyday life are created (Silverstone 2004).

One example of satellite radio as media landscape was found on St. Patrick's Day 2013. On this weekend, SiriusXM offered "Irish" music (as defined by the station) on channel 142, available to all listeners. The new, temporary, option was announced on the homepage of the company's website and encouraged listeners to "Celebrate your Irish heritage (real or imaginary)." In this way, SiriusXM defined "Irish-ness" for its listeners and chose what music would represent the cultural heritage of this region. This definition was filtered through the music chosen for broadcast on St. Patrick's Day weekend. Although there are other channels on satellite radio that focus on music of a particular ethnicity as defined by the company (such as Korean, Mexican, and Latin), the temporary availability of "Irish" music on channel 142 implied that connection to one's "Irish-ness" is also a temporary condition. The decision to alter satellite radio's musical/cultural landscape by offering "Irish" music exclusively on one channel for one weekend also implies that the company anticipates listener desire for such music beyond those with familial connections to Ireland. Thus, SiriusXM defines "Irish-ness" by musical preference. This musical preference, or taste, is an important element of the cultural landscape of satellite radio because one can understand both the listening audience and corporate culture better by examining how taste is reflected in radio programming.

Part of the landscape of satellite radio is its soundscape: the sounds created by the act of receiving a satellite transmission and the expansion of these sounds outward from the point of reception. Unlike terrestrial radio, where boundaries are defined by the reach of the terrestrial signal, elements of the physical landscape define the soundscape boundaries of satellite radio transmission. If a listener's receiver is in the car, driving through a mountainous region can temporarily move the listener out of signal range. In a similar vein, if the listener has subscribed to SiriusXM's internet-only option, then the boundaries of the listening landscape are defined by the availability of internet connections. A listener may move across these borders many times throughout the day. For example, people traveling by subway in New York City may exist within the soundscape as they walk from their apartment to the subway station listening to a satellite radio broadcast through their smartphone, but then step outside the soundscape of satellite radio as they descend to the subway platform and travel underground through the city. Once they emerge again at street level, the listeners enter the soundscape again.

The method of listening creates its own boundaries as well. If a person listens in his/her car, the car frame serves as a barrier to the expansion of the broadcast to secondary listeners (those outside the vehicle), similar to the example provided in the opening paragraph of this chapter, unless, of course, the vehicle's windows are down. A listener using ear buds restricts the expansion of sound to the individual and creates the smallest possible listening space.

Satellite Radio in the Global Media Landscape

The global media landscape of the 2010s is marked by a move toward transnational media. Satellite radio is an example (Hilmes 2004). In North America, satellite radio subscriptions are sold in both the United States and Canada, but no interruption in service is experienced as one crosses the international border between the two countries. Theoretically, the technology involved in satellite radio broadcasting could allow SiriusXM to expand subscriptions (and thus service) to other countries as well.

One of the key characteristics of the global media landscape is deregulation. A combination of capitalist and entrepreneurial initiatives driven by profit and competition have insisted that growth is only possible if governments remove regulations that restrict corporate initiatives (Morley and Robins 1995). The Telecommunications Act of 1996 in the United States is a quintessential example of deregulation by the federal government. This Act removed many government restrictions on ownership concentration, and the result was a massive reduction of individual ownership of radio stations. As national networks bought more and more individual radio stations, control over the content became centralized, even though distribution of content was decentralized (Castells 2009).

In the case of satellite radio, the Telecommunications Act made it possible, 12 years later, for the satellite radio companies in the United States to merge in 2008. Subscription-based radio depends on a company's ability to grow the number of subscribers, and when Sirius and XM saw subscriptions level-off by 2006, a merger was the only option for either company to survive. The merger, in some regards, gave satellite radio's "direct-to-consumer" services an edge over broadcast radio, as listening options for subscribers expanded. The growth of satellite radio, however, increased the importance of local content in terrestrial broadcasts. It is perhaps for this reason that satellite radio tries to "appear" local by using place identification in station identification announcements, a condition that I elaborate on later in this chapter.

The Deterritorialization of Radio Spaces

The arrival of the digital age, and satellite radio in the early 2000s, came with a power shift, to some extent, back to the individual. With the new focus on niche marketing, listeners can now listen to exactly what they want to, provided a specialized channel exists for their desired type of music. Instead of the radio introducing listeners to new, innovative, and marginalized music, satellite and internet radio provide listeners with the ability to listen to more of what they already know they like. Listeners no longer have to wade through music on a band's CD that they didn't want in order to play the song they desire. Instead, they can download a single song electronically and add it to a collection of music carefully chosen to reflect their tastes. They can load MP3s of their favorite tunes onto MP3 players, which they can plug into their car stereo systems, or download the files to their computers. Music websites like Spotify.com create playlists for listeners based on what other songs the listener likes (and what other listeners like).

The renewed power of listeners in the digital age has created "me-focused" listening, a type of deterritorialized listening experience. Listeners download songs and rate them on internet sites. Recommendations for new songs are made based on what other listeners liked. The experience is not completely deterritorialized, though, because there is an indirect, causal relationship between listeners (a song recommendation is not made until the current listener and other listeners have both agreed on and rated a common song). One variation on this concept is Pandora's music genome project. Here, company employees classify music by sound, instrument, or theme. Recommendations are made to listeners based on the classification of a listener's music choice, not on what other listeners to that piece also liked.

Pandora's approach to music selection creates greater opportunities for listeners to hear new music. One could argue that satellite radio shares this feature as well. Although the music broadcast on satellite radio is less spontaneous (the next song is not based on the listener's opinion of the current song), and the music is still genre-specific, the listener has the potential to hear songs from that genre that they have not heard before. In other words, some listeners may choose Pandora or Satellite radio because they like (and might miss) the anticipatory nature that traditional terrestrial radio offered: the thrill of not knowing what is coming next and the serendipity of listening to something new to them.

The soundscape of satellite radio has also become deterritorialized as the SiriusXM broadcast is broken up among more than 100 channels, and signals can be received by listeners through a variety of devices (receivers, internet, phones). This schizophonetic (Murray Shafer 1969) soundscape of satellite radio is created when the original musical piece, sporting event, or talk show is transmitted via satellite signal and then received by a listener via any number of methods. The geographic and acoustic distance between original sound and reception of that sound (by the listener) has increased with satellite technology, yet at the same time, the listener has more immediate access to music, sporting events, or talk shows through personal transmission devices like smartphones.

The Importance of Place in Satellite Radio Broadcasting

Despite the deterritorialization of satellite radio, place remains important. Consideration for how place is represented in radio broadcasting can illuminate those "media in place" connections (Adams 2009). Longan (2013) studied how places are represented in and by the World Wide Web. In my own satellite radio listening experience, DJs of channels that play music with a regional hearth in the United States indicated in their commentary between music segments that the channel was broadcasting from the hearth. For example, the Broadway Music channel DJs talked about being in Manhattan. The Bluegrass channel DJs said they were broadcasting from Nashville. While I was not able to determine if indeed the DJs were actually broadcasting from these places (satellite technology makes it possible to broadcast from anywhere), this rhetoric indicates two important ideas about place. If indeed some channels base their broadcasts out of a musical hearth, then SiriusXM must recognize the advantage of locating broadcasts from a place where a specific musical genre was born and currently enjoys popularity. If, however, the DJs say they are broadcasting from Nashville, for example, but they are really sitting in a cubical in Manhattan, this demonstrates that SiriusXM acknowledges that place is important at least to the listener. Perhaps the company feels that presenting the perception of being located somewhere specific adds credibility to the broadcasts in the eyes (and ears) of the listeners.

Regardless of which scenario is accurate, the rhetoric of place demonstrates that satellite radio provides an important bridge between the global and the local. SiriusXM is a transnational company, with subscribers in Canada, and the border between the United States and Canada is irrelevant in signal acquisition. The satellite technology allows content to be broadcast from any location, yet references to specific places situate certain channels in locales, similar to terrestrial stations. While terrestrial commercial stations use local advertisers to indicate the station's location, commercial-free satellite radio must use a different tactic. Nevertheless, place references in DJ narratives suggest that this element of terrestrial radio broadcasting is still important with satellite radio technology.

The assumption that terrestrial radio remains local while satellite radio is national or global in focus was challenged in 2004. FCC licensing agreements mandated that the content broadcast by satellite radio companies must be broadcast nationally. This agreement did not prevent XM and Sirius from broadcasting city-specific content, like traffic and weather updates; it simply mandated that this information be broadcast nationwide and not just to subscribers in one city. Politicians and terrestrial radio broadcasters alike grew concerned that the practice of broadcasting city-specific content by XM and Sirius would threaten the life of "free" terrestrial radio stations. XM and Sirius responded to these concerns by confirming that city-specific content was indeed part of national broadcasts, and thus no license violation existed. Furthermore, the two companies claimed that they used repeaters solely to circumvent physical barriers to the satellite signal, not to broadcast content to one geographic market that was not available to another market. At the same time, media access proponent groups argued that under the First Amendment, anyone who wanted to broadcast local content should be able to do so (Zarek 2004).

Despite the controversy, the merged company SiriusXM prevailed in their efforts to broadcast local content. In 2013, SiriusXM had traffic and news channels for most large urban centers in the United States (SiriusXM Corporate Website 2013). The traffic information is more useful to drivers with an online subscription or navigation system installed in their vehicle, than those drivers with just an audio receiver, due to the mapping feature involved. In addition to regularly updated traffic reports, this navigation feature also includes information on local gas prices and weather conditions. In accordance with licensing requirements, this information is available to all subscribers who pay the extra fee and have the hardware to support the feature.

This example demonstrates the position between the local and global that satellite radio occupies. The controversy itself speaks to the importance of local content in broadcasting, especially the satellite radio company's claim that listeners want local content (Zarek 2004). The importance of local content is clear because terrestrial commercial radio stations felt their local content gave them a competitive advantage over satellite radio and Internet music sites like Pandora. *USA Today* reports that the new units in automobiles that provide riders with access to satellite radio, HD listening options, and other "apps" are further forcing terrestrial commercial stations to provide quality local content to remain competitive (Woodyard 2013). This trend supports geographer Ronald Abler's 1973 prediction that instead of homogenized communication spaces, "media innovations may promote cultural diversity and spatial differentiation" (Abler 1973, p. 186).

The traffic and weather feature now available on SiriusXM, however, diminishes terrestrial radio's role as provider of local information. Furthermore, subscribers to SiriusXM Traffic can get information about an urban area through which they are traveling long before they reach the boundaries of that city's terrestrial radio stations carrying similar information. In other words, SiriuxXM Traffic transforms and compresses space and time (Harvey 1990; Castells 2000) because information about locations can be accessed quickly and at any time by someone who subscribes to the feature with the location-sensitive technology of satellite radio that permits access almost anywhere.

Satellite Radio as a Commodity

Materials become commodities when they hold value to individuals or groups. Bill Campbell, co-owner of Blue River Communications, a broadband technology provider in Indiana, laments what he refers to as the commodification of radio: the fact that radio stations and companies can be bought and sold for profit by brokers working on commission or purchased through auctions. In either case, those involved in the sale are "people who sometimes have little appreciation or understanding of what radio is really all about," and the mission of radio stations and companies to inform and entertain the public has been lost (quoted in Keith 2010, p. 21). In other words, Bill Campbell criticizes those who decide and benefit from the value of radio stations because they often care only about profit and have little attachment to the station and its purpose.

The cost of obtaining radio stations has increased as well. In the 1980s, the average price of an AM radio station was $450,000. Today, the price tag of AM stations is in the tens of millions of dollars, and FM stations are even more expensive (Keith 2010, pp. 21–22). This rise in cost happened despite the introduction of Internet and satellite radio, although these new radio broadcast methods (often commercial-free) did force terrestrial stations to generate revenue in new ways besides just the sale of advertisements.

The fact that SiriusXM is a subscription-based option that broadcasts or communicates cultural content (especially music) demonstrates one way that radio has become a commodity. Access to content on satellite radio is considered something of value that must bought. Sirius/XM is an example of "an organized system of machines which produce symbolic goods that are transmitted to its consuming publics" (Brunner 1992, p. 21). In this way, satellite radio is a cultural commodity because it "turns society [the consuming public] into a market" to be exploited for profit (Martín-Barbero and Janer 2000, p. 28), at least more directly than does commercial radio, as with satellite radio there is no need for advertising—subliminal or blatant.

Martín-Barbero and Janer (2000) encourage scholars to move away from traditional definitions of culture industries and commodities suggested by Horkheimer and Adorno (1944) that standardize cultural components and mass-produce them to a large group of passive receptors (i.e. the public). Instead, the definition of culture industry should consider:

> places of condensation or interaction of multiple cultural webs, as cross- roads of different areas of social production, made up of complex devices that are not merely technological, mercantile, or political and in which affiliations weigh less than alliances, heavy machines of fabrication less than sinuous trajectories of circulation, and [that take into account] both appropriation stratagems and the logic of property. (Martín-Barbero and Janer 2000, p. 28)

In other words, when considering satellite radio as a culture industry or commodity one must include how satellite radio becomes a medium for interaction between listeners and content, how that content is diffused or circulated, ideas of ownership as conceived by both company owner and listeners alike, and the value associated with these interactions and ideas.

Thus, SiriusXM satellite radio is a cultural commodity because it makes cultural content, in this case music, sports, news, and talk, available to those willing to pay for it through subscriptions and the upfront cost of purchasing a receiver. There is value placed on commercial-free broadcasting in American culture, and such a benefit is something listeners are willing to pay for. SiriusXM connects listeners to content regardless of their location, as long as they are in listening range of their receiver or have an Internet connection (for Internet subscribers). Listeners can interact with this content as they hear broadcasts and "share" things they like via social media like Facebook and Twitter. While SiriusXM categorizes content by genre, listeners have the opportunity to hear new content on channels to which they regularly listen, since the content is regularly updated, or by hearing content on channels to which they do not normally listen. The company itself has owners, but listeners themselves are involved in levels of ownership as they obtain receivers or Internet connections. Listeners can also create customized channel guides, a form of "ownership" that encourages listeners to think of SiriusXM as "theirs."

New Imagined Communities of Satellite Radio

Several media scholars (McLuhan 1964; Cavell 1999; Longan 2013, p. 129) have studied how technologies of space produce and reproduce society. We have already established how satellite radio creates and emphasizes the importance of space and place. This technology produces society because new "imagined communities" (Anderson 1991) of listeners are formed. These imagined communities in turn reproduce a fan base for particular musical genres, talk shows, or sporting events.

The "imagined communities" concept aptly applies to terrestrial radio because as a form of mass media, listeners to terrestrial radio can assume a larger pool of interested listeners for a station exists, otherwise the station would not be able to draw the kinds of advertisers (or supporters in the case of community radio) needed for the station's survival (Keough 2009, 2010). The concept applies to satellite radio as well, but in some ways, these imagined communities are different than the ones that exist for terrestrial radio. While there are certainly some overlaps in these communities (listeners who both subscribe to SiriusXM and also listen to terrestrial stations), satellite radio is sub-culture based and niche-focused, narrower in scope than terrestrial radio, because individual channels have a defined format that is more specific than most terrestrial radio channels. Subscribers to satellite radio listen to specific channels that match their tastes, thus becoming part of the sub-culture that the channel represents, and will never access other channels. Terrestrial radio listeners can assume that the imagined community of listeners has some connection to the place of the radio station, whether they live in the city or are just passing through on their way somewhere else. Satellite radio listeners cannot assume that their imagined community is connected to place, but rather to genre. A listener to the Grateful Dead channel on SiriusXM, for example, assumes that enough fans of the band exist in North America for SiriusXM to justify an entire channel that plays only music by that band. If a Washington Redskins game is broadcast on one of the sports channels, listeners can assume that fans of either the Redskins or the opposing team exist, but that these listeners could be anywhere on the continent (including at the game itself). Thus, the subculture groups, or "minicultures" (Abler 1973), of satellite radio may not share territory, but they share an affinity with particular musical groups, sounds, or eras. Furthermore, Abler (1973) would argue that the variety of niche-focused channels on satellite radio creates more diversity among these communities due to the number of listening options.

It is also possible that the act of listening as both an individual and as a collective practice is a central component of a community identity/relationship. The SiriusXM presence on Facebook provides an example. Accurately measuring the size of a radio station's community is impossible due to the nature of the technologies involved, but Facebook has made it possible to estimate. At the time of this writing, over 577,000 people had "liked" the SiriusXM Corporation Facebook page, and in the moment I checked the page, 3,286 people were talking about the satellite radio company. While the page is designed so that only the administrators (most likely employees of the company itself) can make large posts, people who "like" the page can make comments. Comments made the day prior to my accessing the page mostly consisted of complaints about customer service and channel changes, or technical questions. Posts made by the company, however, were mostly advertisements about events occurring on various channels, but each post was accompanied by an image, thereby adding a visual element to radio programming that isn't available when a listener accesses the service through the receiver in their car. Individual channels have their own Facebook pages as well, with their own communities. Most of the posts made by the company employees on these pages get comments from listeners. On music channel pages, there are a large number of song requests posted by listeners. On sports pages, listeners comment on sports news and results posted by company employees. It is difficult, however, to call the Facebook communities "imagined" because specific names (and sometimes locations) are directly connected to each individual who has "liked" the page. One member of the community can list other members by name using the Facebook page. The "imagined" part of the community should come in the assumption that other listeners exist without having direct access to them. In this way, Facebook has taken imagined communities and made them real.

Media technologies shape everyday life by changing or adding to the material world (Silverstone 2004). Listeners to satellite radio can be considered a community or society because they share elements of their material worlds: satellite radio receivers, computers with Internet connections, and smart phones that can access streaming content from the SiriusXM website. These elements of material culture connect listeners to content, although unlike terrestrial radio's talk shows and request programs, this material culture does not facilitate connections to other listeners. Listeners to satellite radio do not participate in these broadcasts in the way that listeners do in traditional radio. Thus, while listeners may subscribe to satellite radio for similar reasons (Lin 2006), the everyday uses of satellite radio are different from the participatory nature of terrestrial radio.

Conclusion

In the afterward to the Michael C. Keith's *The Radio Station*, Jay Williams, Jr., a radio broadcast consultant from Charleston, South Carolina, notes that the term "radio" had changed in meaning since its inception. Originally, the term "radio" referred to "AM and FM stations licensed to a limited number of cities in a given city or market" (in Keith 2010, p. 329). Today, however, technological changes mean that the term now encompasses a variety of transmission methods. Some methods, such as Internet programs like Pandora and satellite radio, do not use the radio wave technology from which the term was originally derived. This change necessitates radio scholars to consider whether or not satellite radio is, indeed, "radio."

On the technical side, satellite radio shares some, but not all, of the components involved in terrestrial radio broadcasting. Patsiokas (2001) identifies four main segments of satellite radio systems: ground, space, repeater, and technology. Satellite and terrestrial radio both share the ground segment (where content is acquired or produced) and the repeater segment (which uses devices that repeat the satellite signal around physical barriers, like tall buildings, in areas of high population density). Unlike terrestrial radio, however, the space segment is necessary in satellite radio, whereby a signal is transmitted from the ground to two geostationary satellites. The technology segment for satellite radio includes some components that are not present in terrestrial radio, like certain hardware and software used to operate the system. The mobile antenna, however, is one element of technology that satellite radio shares with terrestrial radio, especially since satellite radio receivers are now installed in most new automobiles produced in the United States. This agreement between SiriusXM and car manufacturers indicates that these companies recognize the consumer demand that drives the cultural practice of radio listening, as well as the potential profits gained by recruiting listeners to satellite radio who might not have gotten a subscription on their own.

Hilmes (2004) and Lin (2006) indicate that in purpose, satellite radio is still "radio." Lin concludes that traditional radio has simply been "refurbished . . . to fit in a new delivery system." Hilmes (2004, iii) uses the term "re-staging" to refer to the transformation from terrestrial radio to satellite and digital formats. Furthermore, Lin concludes that listeners choose satellite radio for some of the same reasons that they listen to terrestrial radio: listening provides a diversion, and the act of listening had become a habit. In other words, the term "radio" can refer to the purpose for using the technology, in addition to (or instead of) the technological features of the delivery system.

Despite the fact that satellite radio has been publically available for more than a decade, there is (at the time of this writing) surprisingly little empirical research on this form of media, and much of this research was published prior to 2008, before the Sirius and XM companies merged. Alper (2006) was optimistic about the future of satellite radio because he felt that subscribers still held some power: if the corporation tightened music programing to focus only on popular songs and added commercials to increase revenue, subscribers would cancel, thus forcing the company to revert back to original programming practices. The merger of XM and Sirius in 2008, however, indicates that the plateau of new subscriptions was compromising each company's profits. There remains, then, many unanswered questions about the geo-cultural elements, impact, and significance of satellite radio. How is music chosen for broadcast on each channel? How much control do channel DJs have in music selection? How often is new music introduced and what sources provide this new music? How has audience reception and adoption habits changed since Lin's (2006) study? What is the balance between online subscribers and those listening via a receiver? How have the spatial dynamics of satellite radio changed in the last 10 years compared to terrestrial radio? How does satellite radio negotiate content at the global and local levels for a seemingly diverse national audience? Answers to these questions and others will help media geographers further explore satellite radio's place in the contemporary media landscape.

Acknowledgements I would like to thank Dr. Robert Drew for the immensely productive brainstorming sessions that helped shape this chapter, and for his helpful comments on earlier drafts. I would also like to thank Dr. Scott Youngstedt and Dr. Thomas Bell for their helpful suggestions on drafts of this work.

References

Abler, R. (1973). Monoculture or miniculture?: The impact of communications media on culture in space. In D. Lanegan & R. Palm (Eds.), *An invitation to geography* (pp. 186–195). New York: McGraw-Hill.

Adams, P. (2009). *Geographies of media and communication*. Malden: Wiley-Blackwell.

Alper, G. (2006). XM reinvents radio's future. *Popular Music and Society, 29*(5), 505–518.

Anderson, B. (1991). *Imagined communities: Reflections on the origin and spread of nationalism*. Rev. and extended 2nd edition. London: Verso.

Bay Area Radio Museum. (2013). KSFR radio 94.9, San Francisco: A broadcast day begins. http://bayarearadio.org/audio/ksfr/index.shtml. Last Accessed 3 Jul 2013.

Brunner, J. (1992). *América Latina: Cultura and modernidad*. Mexico City: Grijalbo.

Bull, M. (2004). 'To each their own bubble': Mobile spaces of sound in the city. In N. Couldry & A. McCarthy (Eds.), *MediaSpace: Place, scale and culture in a media age* (pp. 275–293). New York: Routledge.

Castells, M. (2000). *The rise of the network society* (2nd ed.). Malden: Blackwell.

Castells, M. (2009). *Communication power*. Oxford: Oxford University Press.

Cavell, R. (1999). McLuhan and spatial communication. *Western Journal of Communication, 63*, 348–363.

Crosper, A. (2009). Part two: San Francisco, Oakland, San José Radio (1950s–2000). *San Fransicso-Bay area radio history*. http://www.playlistresearch.com/sfradiohistory2.htm. Last Accessed 3 Jul 2013.

Guido Interview: http://www.pbs.org/wgbh/pages/frontline/shows/music/interviews/guido.html. Last Accessed 10 Jul 2013

Harvey, D. (1990). *The condition of postmodernity: An enquiry into the origins of cultural change*. Cambridge, MA: Blackwell.

Hilmes, M. (2004). Forward: Transnational radio in the global age. *Journal of Radio Studies, 11*(1), iii–vi.

Horkheimer, M., & Adorno, T. (1944). *Dialectic of enlightenment*. Stanford: Stanford University Press.

Keith, M. C. (2010). *The radio station: Broadcast, satellite and internet* (8th ed.). Burlington: Focal Press.

Keough, S. (2009). Internet radio and cultural connections: A case study of the St. John's, Newfoundland radio market. In T. Bell & O. Johannson (Eds.), *Sound, society, and the geography of popular music* (pp. 185–204). Hampshire: Ashgate.

Keough, S. (2010). The importance of place in community radio broadcasting: A case study of WDVX, Knoxville, Tennessee. *Journal of Cultural Geography, 27*(1), 77–99.

Kirk, M. (Writer, Producer, Director). (2004). *The way the music died*. Frontline. Available at: http://www.pbs.org/wgbh/pages/frontline/shows/music/

Lin, C. (2006). Predicting satellite radio adoption via listening motives, activity, and format preference. *Journal of Broadcasting and Electronic Media, 50*(1), 140–159.

Longan, M. (2013). Landscape as media, media as landscape: Representing Northwest Indian as landscape on the World Wide Web. *Aether: Journal of Media Geography, 11*, 127–147.
Martín-Barbero, J., & Janer, Z. (2000). Transformation in the map: Identities and culture industries. *Latin American Perspectives, 27*(4), 27–48.
McLuhan, M. (1964). *Understanding media: The extensions of man*. New York: Signet Books.
Morley, D., & Robins, K. (1995). *Global media, electronic landscapes, and cultural boundaries*. London: Routledge.
Murray Shafer, R. (1969). *The new soundscape: A handbook for the modern music teacher*. Toronto: BMI Canada.
Orgad, S. (2007). The internet as a moral space: The legacy of Roger Silverstone. *New Media & Society, 9*, 33–41.
Patsiokas, S. (2001). XM satellite radio technology fundamentals. *SAE Technical Paper 2001-01-1328*, doi: 10.4271/2001-01-1328.
Silverstone, R. (2004). Regulation, media literacy and media civics. *Media, Culture & Society, 26*, 440–449.
SiriusXM Corporate Website. http://www.siriusxm.com/. Last Accessed 10 Mar 2013.
Vowell, S. (1996). *Radio on: A listener's diary*. New York: Martin's Press.
Woodyard, C. (2013, 25 Mar). Newfound rivals jam car radios. *USA Today*, p. 1B
Zarek, C. (Summer 2004). Catching static: Satellite radio companies XM and Sirius come under fire for broadcasting local traffic and weather reports. *The News Media and the Law*, 38–39.

Section V

Image/Imagining Media Events and Spaces

18 Visibility, Media Events and Convergence Culture: Struggles for the Meaning of 9/11

Kevin Glynn

> *It matters profoundly, we are convinced, that the horrors of September 11 were designed above all to be visible . . . .*
>
> *September's terror . . . was premised on the belief (learned from the culture it wishes to annihilate) that a picture is worth a thousand words . . . .*
>
> *At the level of the image (here is premise number one) the state is vulnerable; and that level is now fully part of, necessary to, the state's apparatus of self-reproduction. Terror can take over the image-machinery for a moment—and a moment, in the timeless echo chamber of the spectacle, may now eternally be all there is—and use it to amplify, reiterate, accumulate the sheer visible happening of defeat.*
>
> (Boal et al. 2005: 25–8)

Introduction

Media events have become important sites of political activity, affective engagement and cultural struggle. They involve (often spectacular) visibility, the articulation and circulation of meaning, and the formation of powerful discourses and counterdiscourses. The imaging technologies at the heart of media events shape our encounters with place and our geopolitical imaginaries. This chapter explores the realm of visual media events through the lens of the attacks of September 11, 2001. These attacks were profoundly mediated, and they generated complex reactions and ongoing political contestation across a diverse array of media realms. I aim to use this event to demonstrate the complicated and contingent politics of visual media at a time when media apparatuses have multiplied and saturated the world as never

K. Glynn (✉)
School of English and Media Studies, Massey University Wellington, Aotearoa, New Zealand
e-mail: k.t.glynn@massey.ac.nz

S.P. Mains et al. (eds.), *Mediated Geographies and Geographies of Media*,
DOI 10.1007/978-94-017-9969-0_18

before. In the age of global visual cultures and media convergence, images and discourses spill interactively across digitalized, networked platforms and a multitude of screens including televisions, computers, tablets and smart phones. The new media geographies to which they give shape become invested with affects and meanings as different groups and agents struggle to promote particular modes of sensemaking and political interests. These geographies define a terrain upon which dominant forces and interests work to establish and stabilize discursive control, but encounter image insurgencies and alternative knowledges that disrupt and contest such control through the disarticulation and rearticulation of its visual and narrative elements.[1]

I will first sketch the broad historical context of contemporary media events through a brief consideration of the centrality of vision and visuality within modernity and postmodernity, which has often been characterized as a state of image saturation and visual overload. I will then briefly consider the affective intensity and complexity of the 9/11 iconography, which mirrored familiar patterns from extant Hollywood templates, and note its relationship to the emergence of an increasingly convergent mediasphere marked by new technological mobilities and interactivities. Finally, I will turn to John Fiske's account of "democratic totalitarianism" and theory of media events as sites of discursive contestation between forces of domination and countervailing, democratizing energies. In the case of the 9/11 attacks, one of the most interesting formations of alternative discursivity and popular skepticism toward officialdom is that of the so-called 9/11 Truth movement, which alleges some degree of US government complicity with or involvement in the attacks. The 9/11 Truth movement has found resonance with some of the most historically marginalized elements of US society (African Americans, for example), and raises interesting questions about the production and circulation of alternative discursive and knowledge formations on the terrain of media convergence. I will conclude with a few closing remarks about the emergent geographies and topographies of the new convergent media environment.

Modernity and the Visual

The modern age gives a special place to visuality, to human vision, and to the process of visualizing, and this characteristic of modernity has played a prominent role in shaping the contemporary world as we experience it. Hence the defining philosophical movement of early modernity was significantly called the Enlightenment, a designation that highlights the importance of vision and visuality

[1] I intend "articulation" in the theoretical sense this term has acquired within cultural studies, where the simultaneity and conjoint productivity of its dual meanings (*contingent linkage* and *expression*) are given emphasis. When signifying elements or agents are contingently linked (articulated) to one another, this linkage generates particular meaning effects (articulations) and mobilizes particular actions and agents (to which these meaning effects are in turn articulated or contingently linked). See Grossberg (1996).

as central metaphors within the modern epistemologies of the West, for which the "mind's eye" becomes a central figure. Early modern philosophers such as Descartes and Locke placed visuality at the apex of our senses, where it stood as a badge of rationalism and clarity of thought (see Jay 1993: 85). Hence we refer to important thinkers as "visionaries" or "luminaries," we describe ways of understanding as "worldviews," we talk about the importance of "keeping things in perspective," of not "losing sight" of our objectives and priorities. When we're looking for solutions, the first thing we do is try to "shed some light on the problem." When we achieve mutual understanding with another we often proclaim "ah yes, I see," or "I get the picture." The predilection for the visual, which Jay (1993) and others call modernity's "ocularcentrism," impelled Martin Heidegger (1977: 115–154) to characterize the modern epoch as "The Age of the World Picture." In contrast, medieval thinkers such as Thomas Acquinas maintained a hermeneutic of suspicion in relation to the visual (Mirzoeff 1999: 5). Heidegger writes that "a world picture . . . does not mean a picture of the world but the world conceived and grasped as a picture . . . . The world picture does not change from an earlier medieval one into a modern one, but rather the fact that the world becomes picture at all is what distinguishes the essence of the modern age" (p. 129–30, quoted in Mirzoeff, p. 5). Calling our attention to the distinctive modern nexus of visuality and subjecthood, Heidegger (1977: 131) observed that the ancient Greeks characteristically understood "man" as "the one who is looked upon by that which is" (though he went on to note that Plato's allegory of the cave foreshadowed the modern thinking that would not become dominant until more than a thousand years later).

Michel de Certeau (1984: 187) states that while premodern cultures required of their members a belief in what cannot be seen, contemporary ones insist upon accession to all that can. He argues that technologically driven processes of visualization and the forms of narration they entail thereby become core means for the creation of "facts" and so for the production and regulation of what we take to be "real" in contemporary times. For "what can you oppose to the facts?" he asks. "You can only give in." Hence, "the fabrication of simulacra . . . provides the means of producing believers," and "the establishment of the real" becomes "the most visible form of our contemporary dogmas" (de Certeau 1984: 186–7). Similarly, according to Nicholas Mirzoeff (1999: 6), the tendency to visualize the world and to conceptualize human understanding in terms of vision has in contemporary times become something of a compulsion. "Human experience is now more visual and visualized than ever before," he writes, so that "seeing is much more than believing. It is not just a part of everyday life, it *is* everyday life" (Mirzoeff: 1, emphasis added). As a consequence, contemporary societies invest astonishing levels of energy and resources into making things visible that otherwise wouldn't be. Such investments both symptomatize and intensify ocularcentrism, as an ever more dizzying array of new media technologies work overtime to deliver images of the insides of bodies, the furthest reaches of the galaxy, and the diverse spaces and surfaces of our planet, which are continually scanned, monitored and subjected to the power of vision by increasingly sophisticated surveillance and remote sensing technologies, whether for purposes of investigation, examination, security-enhancement, reconnaissance, or entertainment. In the process, the development of such image-generating apparatuses and practices reveals that the drive to extend our "world picture" knows few boundaries; this drive is transforming our existing media environments in fundamental ways as it swamps distinctions such as public/private, interior/exterior, and surface/depth in orgiastic "rituals of transparency" and ecstasies of communication (Baudrillard 1988).

For example, in 2003 the New York Times reported that the Pentagon is working on a radar-based surveillance device designed to identify individuals from 500 ft away on the basis of how they walk, using methods of "gait recognition analysis" and "3-D body tracking" under development at MIT. A US Congressional report subsequently confirmed that the goal of the surveillance program "is to identify humans as unique individuals . . . at a distance, at any time of the day or night, during all weather conditions, with noncooperative subjects, possibly disguised" (Dowd 2003). In short, this program dreams of making individual identity visible, even at a significant distance and under low-visibility conditions, and thus illustrates the use of visualization as a technique of security. While on one level such imagistic expansion of our contemporary "world picture" constitutes the development of new ways to exert and consolidate social control by intensifying the mechanisms of surveillance and reifying the identities it thereby establishes and catalogues (see, e.g., Andrejevic 2007), on another level the dialectics of this expansion lend to it a self-destabilizing dimension. Lisa Parks (2005: 183) notes that the multiplication and refinement of satellite imaging technologies, for instance, "amplifies the possibility of alienation or difference within the field of the televisual," as in "the haunting possibility that events on the earth could always be seen and encoded from anOther (nonhuman) point of view," as well as "by virtue of the satellite's own limitations in representing events on the earth," which "is, of course, experienced all the time in unseen, unknown, and untold ways." Along similar lines, Mirzoeff (1999: 8) notes that contemporary levels of image saturation, overload and hypervisuality in a world where pictures are increasingly made to circulate everywhere instantaneously, mean that we have surpassed the concept of the "world picture" in our development of a constantly swirling global image flow. We have thus entered "an era in which visual images and the visualizing of things that are not necessarily visual has accelerated so dramatically that the global circulation of images has become an end in itself, taking place at high speed across the Internet."

In order to stand-out amidst such seemingly self-justifying and self-perpetuating image flows (Mirzoeff 1999), any particular picture or set of pictures must work extremely hard to capture peoples' overloaded attentions. The images associated with the attacks of September 11, 2001 were supremely effective in this regard, having been engineered for maximum visual impact and constituting something like a two-billion gigawatt jolt to the global image circulatory system. At the World Trade Center, the timing and sequencing of the attacks was in itself enough to ensure their spectacular instantaneous global visibility. The impact of American Airlines Flight 11 with the Center's North Tower, more than 15 minutes prior to United Airlines Flight 175's collision with the South Tower, created an absolute guarantee that live video cameras would be in place, in abundance, and trained precisely on the second target in advance of its strike, and that the event would therefore be instantaneously visible—live and in real-time—from any position on the planet capable of electronic image reception. The attacks were in this sense produced by techniques of visualization as much as ones of terrorist violence, thus generating an image-event characterized by a strategy of visual terror, or of terrorizing visuality.

The Sublime Power of the Image

Such terrorizing visuality is reminiscent of the postmodern sublime and the technological sublime, which deploy excess, alterity and allegory to "represent the unrepresentable" and thus to express the awesome "complexity and horrors" of an age characterized by blurred categories and multiple boundary implosions (Kellner 2003: 141). The postmodern and the technological sublime build upon traditional Kantian notions of a "strong and equivocal" aesthetic through which, paradoxically, "pleasure derives from pain" (Lyotard 1984: 77) and powerful dread combines with longing in its exploitation of the "sensual immediacy" of the visual (Mirzoeff 1999: 15–6; Freedberg 1991). As Parks (2005: 176) writes, images of the attacks enlisted "the multiple discursive modalities of the televisual . . . in full force," as "commercial entertainment, public education, scientific observation, and military monitoring collided in . . . coverage that lasted not just for days but months," notably disrupting various familiar networks, circuitries and rhythms of everyday life.

If the sublime technological power conveyed through the live, instantaneous and global dispersion of images of the attacks combined pleasure with horror for some of their audiences, that pleasure led ultimately, and perhaps by way of the death drive and whatever requisite detours involving the serpentine and circuitous pathways of displacement and repression, to "the aesthetics of destruction" associated with visions of "wreaking havoc, making a mess" (Sontag 2004 [1965]: 102). Susan Sontag (ibid.: 101) notes that science fiction films (along with those of certain other genres) provide a kind of "sensuous elaboration" that is unavailable in written texts, whereby "one can participate in the fantasy of living, through one's own death and more, the death of cities, the destruction of humanity itself." According to Sontag (ibid.: 103), the spectacular and increasing "visual credibility" of contemporary films indexes the extent to which "modern historical reality has greatly enlarged the imagination of disaster, and the protagonists—perhaps by the very nature of what is visited upon them—no longer seem wholly innocent." Such dynamics of the destructive image complicate and potentially intensify the critical charge associated with events such as the hypervisual 9/11 strikes against central icons of US imperialism and global finance.

As Slavoj Žižek (2002b: 15–6) would write in a short piece that circulated widely through cyberspace in the days immediately following 9/11,

> to us, corrupted by Hollywood, the landscape and the shots we saw of the collapsing towers could not but be reminiscent of the most breathtaking scenes in big catastrophe productions.

> [. . .] Not only were the media bombarding us all the time with talk about the terrorist threat; this threat was also obviously libidinally invested—just remember the series of movies from *Escape From New York* to *Independence Day*. That is the rationale of the often-mentioned association of the attacks with the Hollywood disaster movies: the unthinkable which happened was the object of fantasy, so that, in a way, America got what it fantasized about, and that was the biggest surprise. [Žižek 2002b: 15–6].

Or, as Jean Baudrillard similarly put it,

> the fact that we have dreamt of this event, that everyone without exception has dreamt of it—because no one can avoid dreaming of the destruction of any power that has become hegemonic to this degree—is unacceptable to the Western moral conscience. Yet it is a fact, and one which can indeed be measured by the emotive violence of all that has been said and written in the effort to dispel it.
>
> At a pinch, we can say that they did it, but we wished for it. . . . Without this deep-seated complicity, the event would not have had the resonance it has, and in their symbolic strategy the terrorists doubtless know that they can count on this unavowable complicity. [Baudrillard 2003: 5–6.]

Notably, for Baudrillard (2003), the universality of complicit desire for this "'mother' of all events" (p. 4) does not require us to posit a death drive at its base. Rather, he asserts that the "countless disaster movies [that] bear witness to this fantasy" register instead an all-inclusive urge "to reject any system . . . as it approaches perfection or omnipotence," a universal "allergy to any definitive order" or power (pp. 6–7). This of course is not to negate the difference between the experience of such images in fiction/fantasy and that of their "real" counterparts generated by the 9/11 attacks, nor to deny the incredibly shocking and disturbing force of the latter. Geoff King argues that as the 9/11 images were repeatedly reedited and replayed in the days following the attacks, they were increasingly transmitted through familiar codes associated with Hollywood continuity editing and thus rendered less disturbing and perhaps even reassuring by their subjection to processes of ordering and control. In this way, television's treatment of the 9/11 images both provoked and assuaged intense shock and disruption (King 2005).

The San Francisco authorial collective that calls itself Retort (Boal et al. 2005: 26) suggests that the long-term geopolitical consequences of the worldwide circulation of images of a global superpower "afflicted" as on 9/11 are ultimately unpredictable. The animation of alternative possible eventualities depends in part upon the contingencies whereby such images are laden with meaning through their various subsequent discursive activations or, in other terms, subjected to the practices of semiotic struggle, which is why Žižek (2002a) exhorts us "precisely now, when we are dealing with the raw Real of a catastrophe," to "bear in mind the ideological and fantasmatic coordinates which determine its perception." But whatever narrative grooves may have been historically prepared in advance for the arrival of the attacks have also, through historical processes, been constituted in diverse and contested ways. For Žižek's "ideological and fantasmatic coordinates" are, as Boal et al. (2005: 26) put it, "bound up, in the longer term, with circuits of sociability—patterns of belief and desire, levels of confidence, degrees of identification with the good life of the commodity" and so forth, which are in their

Fig. 18.1 The Devil in the Towers

turn "aspects of the social imaginary still (always, interminably) being put together." They are, that is to say, always in process, subject to the conflicts and struggles of sociality, and thereby constituted in heterogeneous and contingent forms.

A key consequence of these struggles, heterogeneities and contingencies is that images of the attacks on the Twin Towers are widely available for a range of alternative discursive practices and enactments. For example, consider the famous "Satan in the smoke" (Phillips 2011) images that circulated widely on TV, in newspapers and through the Internet, such as Fig. 18.1, which was taken from CNN's live coverage of the attacks (a similar image, taken by photojournalist Mark Phillips, can be found online at http://www.guardianangel.in/ga/268-D-Obituary-Images-of-the-World-Trade-Center-fire-reveal-the-face-of-Satan.html).[2]

This "face of Satan" image is fundamentally ambivalent with regard to its capacity for insertion into alternative and competing discursive structures. Its sensual immediacy and richly polysemic potential to convey powerfully condensed meanings and ideologies is thus available for semiotic and affective mobilization by different social formations and struggles. On the one hand, for instance, this picture of the doomed World Trade Center (WTC) can be readily articulated to well-established Orientalist discourses that construct a racialized opposition between the wondrous achievements of Western civilization and the wicked barbarity of a demonic Other bent on bringing about its demise, all of which is registered in this and similar images of a spectacular yet ultimately fragile architectural grandeur under assault from the embodiment of absolute evil. Mark Phillips reports that he received more than 30,000 mostly emotionally laden messages after the widespread

[2]The CNN footage containing Fig. 18.1 is available at http://www.youtube.com/watch?v=k3nLbc_8wfM.

publication of his "Satan in the smoke" photograph. Some told Phillips that he "had been put on the earth to take this photo and that the photo showed who was really behind the attack" (Phillips 2011: n.p.). One wrote, for instance, that

> if I were you, I would feel privileged (unfortunately) that the Lord used you to shed "the light on" darkness. . . . He used your picture to stir up a lot of Americans, Christian and non-Christian, as to who was really behind this whole traumatic event. . . . We have become so complacent to the fact that there really is a devil loose on this earth and you better believe he was behind this whole thing! The faith of Islam is rebellion towards God that goes way back and what better way to divide people on this earth than to use "religion." [Quoted in Phillips 2011: n.p.]

On the other hand, Fig. 18.1 and similar images can likewise be mobilized within a very different set of discursive practices that associate the WTC with global financial hegemony, whose true face is brought to the fore only upon the breach of its superficial facades. Thus, the Catholic website "Guardian Angel" asks,

> don't these photos of Satan at the World Trade Center catastrophe tell us that the current seat of Satan's power is the World Trade Center? Don't these photos depict Satan being awakened from his hiding place in the World Trade Center? For it is the international bankers who operate from Fed, the CFR and the World Trade Center who create first, second and third world debt. Usury according to the Bible is Satan's method for enslaving the world under his priesthood, the accountants and bankers of the world (IMF, World Bank Group, WTO).[3]

Hence the visuality that is central to modernity is a site of contestation between groups engaged in discursive practices that promote competing understandings of the world and thus advance different social and political interests. This constitutes a *politics of representation* that comprises an important part of the terrain on which the meanings of place and control over space are struggled for. Therefore, the struggle to articulate images to one set of discursive practices or another is an extremely important one within the broader politics of a hypervisual and convergent global media culture (which constitutes, moreover, an increasingly contingent and open-ended cultural-political terrain).

Cultures of Media Convergence

My use of the concept of convergence here is intended to include the various levels of expanding interconnective practices that have become associated with the term in recent media studies (see especially Jenkins 2008; also Meikle and Young 2012). Hence convergence entails geographical, technological, social and political dimensions, as consolidating systems of top-down control intersect with the increasingly interconnected yet multivalent and participatory practices and emergent mobilities of grassroots media users, giving rise to new forms, levels and degrees

[3] Available at http://www.guardianangel.in/ga/268-D-Obituary-Images-of-the-World-Trade-Center-fire-reveal-the-face-of-Satan.html.

of transmediation. Convergence here also entails then the strange complicities and intersections noted above between, for example, Hollywood cinema audiences and Islamic fundamentalists (just as it entails the circumstance whereby in some cases these two identities are in fact the same convergent one). As Jenkins (2008: 3) notes, "convergence represents a cultural shift as consumers are encouraged to seek out new information and make connections among dispersed media content" and across different media platforms. Convergence brings cultural consumption and production together in ways that mobilize synergistic new regimes of "collective intelligence," which operate on the principle that, although no individual can possess complete knowledge, all have access to some; therefore, resource-pooling and task distribution across a broad base of people with unique and distinctive skills can generate alternative sources of media power (Jenkins 2008: 4). In an environment of intensifying media-cultural convergence, the field of potential discursive articulations into which images enter is a markedly expansive one. Thus, in relation to 9/11, as we'll see, new knowledge formations have exploited the affordances associated with media convergence to generate and share forms of collective (counter)intelligence regarding the nature and perpetrators of the attacks.

Jenkins (2008: 1–2) cites the case of Dino Ignacio, a US high school student who playfully created a photomontage that depicted Bert from *Sesame Street* cavorting with Osama bin Laden for his "Bert is Evil" website. As a consequence, an anti-American activist in Bangladesh wound up plastering images of Bert (and bin Laden) on placards and t-shirts that were distributed for use by marching protestors in numerous Middle Eastern countries. When global cable news networks transmitted the images of Bert and the protesters, *Sesame Street*'s producers were outraged at the apparent tastelessness of this IP infringement, which in turn led tech savvy observers to create new websites associating more *Sesame Street* characters with Al-Qaeda. Thus, "from his bedroom, Ignacio sparked an international controversy. His images crisscrossed the world, sometimes on the backs of commercial media, sometimes via grassroots media. And, in the end, he inspired his own cult following" (Jenkins 2008: 2). Here, media mobilities, images, discourses, platforms, connectivities, far-flung social collectives, political constituencies, articulations and complicities, combine and combust in ways that are anything but predictable and that challenge traditional ways of understanding power, control, and meaning production in media culture.

"Democratic Totalitarianism," Media Events and Image Insurgency

John Fiske (1998) has noted an expanding regime of "democratic totalitarianism" in the US that operates most intensively around racial difference and whose core attributes include rampant technologized surveillance, escalated policing and "appeals to moral totalism." Under such conditions, the power-bearing dimensions of visuality are exerted asymmetrically across racially differentiated populations. Fiske characterizes this social environment as "democratic totalitarianism" because its capacity to exert control depends upon the extent to which its key techniques of power can be operationalized "underneath the structures of democracy" (p. 69). For example, in November 2000, less than a year prior to the 9/11 attacks, the new millennium threw up its first major media megaspectacle: the intensively racialized Florida election debacle that swept Bush and Cheney into the White House by 537 votes, standing on the backs of tens of thousands of eligible African American would-be voters who were purged from the state's digital electoral rolls when a private computer firm hired by Jeb Bush falsely designated them to be convicted felons (Palast 2006: 240–6). Moreover, on election day, there were reports of the intimidating use of police roadblocks near polling places in Black neighborhoods of Miami, and seven whistle-blowers from the GOP-connected Sequoia Voting Systems, Inc., would reveal in 2007 that defective ballots designed to produce "hanging chads" were deliberately manufactured and shipped only to Palm Beach County, a large, heavily Democratic district with a significant nonwhite population (Glynn 2009: 229, 232). Thus would commence a decade that was soon to yield an even more disruptive media mega-event that would provide the basis and justification for manifold enunciations of moral totalism and media spectacle, and myriad extensions of imperializing surveillant, policing and war-making powers that reached across geographical space and multiple axes of social difference while extending the gulf between the haves and have-nots to levels unseen in the US since the Great Depression, and perhaps since the Gilded Age of the late nineteenth century Robber Barons.

In *Media Matters*, Fiske (1996) explores the characteristics of media events and their relationships to the complex currents of meaning that comprise contemporary media cultures. Updating Raymond Williams (1997), Fiske argues that a media culture can be likened to a river of discourses that includes dominant, residual and emergent streams that jostle, contest and unsettle one another. A calm surface may at times mask and belie the churning forces and complexities below, though unexpected turbulence may suddenly bring to the surface deep, powerful and well established currents that had previously been all but invisible. Spectacular media events become focal sites of discursive activity, maximal turbulence, and competing bids and counterbids for meaning and knowledge as they resonate powerfully with a culture's deepest fears, desires and anxieties; the most powerful media events may therefore lead to significant shifts in a culture's overall structure of feeling. And yet to do so, they must break through a surface that is more image-saturated than at any time in human history.

The culture of the contemporary US is one of "extreme multiplicity . . . of images, of knowledges, and of information technologies" (Fiske 1996: 239). Contemporary strategies for its hegemonization must therefore seek to exert control over technology, visibility, knowledge and information, and yet the multiculturalism implied by multiplicity means that such control can only ever be achieved in precarious, unstable and contested forms. In such an environment, countersurveillance might contest and disrupt the imperializing management of visibility, and counterknowledges might be assembled through the technologically mediated disarticulation and rearticulation of fragments of information or repressed facts, perspectives and images (Fiske 1996: 191–2). Practices of countersurveillance and the production of counterknowledge can become valuable weapons of the socially weak and of emergent social formations in the creation, expansion and defense of localized and increasingly networked social spaces against incursion by colonizing and controlling forces.

Media and cultural theorists have noted the decline of what Daniel Dayan (2009) has called "central TV" as a source of hegemonic consensus, widely shared spaces of discursivity, and a sense of coherent national cultural vision. In the shift from central television to post-broadcasting, post-national and convergent, networked digital media environments, multiplicity, contestation and discursive collage have expanded possibilities for the rearticulation of information and images into contestatory counterknowledges and alternative, countersurveillant practices of monstration and remonstration (Dayan 2009). Spectacularly disruptive media mega-events such as the 2000 presidential election and the 9/11 attacks may paradoxically re-center national attention for a time and, as Lynn Spigel (2004: 260) has observed, create opportunities for the mediated performance of "myths of reunification and nationalism"; but they also inevitably provoke counternarratives and establish new terrains of political contestation, negotiation and dialogue, particularly in the context of the "multitiered public sphere" (Dayan 2009) of the contemporary media environment of convergence, digitalization and post-broadcasting. While this necessarily gives rise to a degree of fragmentation often discussed in terms of media tribalism (Dean 2009) or referred to as "communities of dissensus" (Birchall 2006: 79), the contemporary media environment also creates new pathways for the assemblage of collective counterintelligence, new articulatory possibilities for alliance formation, and new techniques of countervisualization. The contemporary mediascape of "technostruggles" thus remains a politically vibrant terrain of contestation, where the motivation to participate in the formation and circulation of counterknowledges, alternative visibilities and articulatory alliances is driven by and through the social relations and positions occupied by those engaged in these practices (Fiske 1996). The means and ways of engaging in such technostruggles have expanded rapidly in the past decade.

If Al-Qaeda drew upon and rearticulated a deeply familiar repertoire of Hollywood narratives and images to orchestrate a hypervisible spectacle of vulnerability that lay at the very core of the world's most powerful empire—an empire advanced by image-power as well as by economic and military might—then this might be understood as a kind of *image-insurgency*. As Marc Redfield (2009: 3) writes, the "space inhabited by the World Trade Center was (and is) so heavily mediatized, so utterly penetrated by representational technologies of global reach, and so symbolically at the heart of the world's various political, financial, and semiotic webs of power that the destruction of the towers could not help being at once the ultimate media event and (therefore) a haunting image of the deracinating force of communicational technology at work." The Bush/Cheney regime's response to the WTC's collapse relied in turn upon the rearticulation of equally familiar and phantasmatic media images to imperializing discourses and narrative grooves capable of countering the event's disruptive force by reasserting Orientalist sense-making categories organized around moralistic binaries such as "civilization versus barbarism" and "good versus evil." The cultural work of this moral totalism was to at once performatively constitute, underwrite and justify the extension of new powers and expanded geopolitical practices of global and domestic policing and surveillance, and thus to initiate new modes of power-bearing visuality and control. The benefits and impacts of these measures were distributed in a wildly asymmetrical and disproportional manner across the social differences that regulate access to alliances that establish the corporate power-bloc whose relations and interests were most active at the center of the Bush/Cheney regime.

Popular Counterknowledge: The 9/11 Truth Movement

In the face of this imperializing hegemonization of the 9/11 media event, another alternative image insurgency has formed that also appeals to a certain familiar media phantasmagoria and to an established counter-reading of US history. The 9/11 Truth movement (9/11TM; see Bratich 2008) is a diverse global collection of people and organizations that reject the official accounts of the 9/11 attacks. Most adherents allege complicity of one sort or another in the planning and/or execution of the attacks by elements within the Bush Administration motivated primarily by the potential to reap a variety of political, economic and military opportunities and advantages in their wake. Some within 9/11TM articulate the spectacle of the collapsing towers to the counter-histories that have been generated in response to unresolved questions and suspicious circumstances around the assassinations of Presidents Lincoln and Kennedy, and of RFK and MLK, to evidence of the provocation of a Japanese attack on Pearl Harbor by the Roosevelt administration, and to suspected or actually documented "false flag" operations such as the attacks on the USS Maine and the Lusitania, the Gulf of Tonkin incident, and Operation Northwoods.[4] 9/11TM thus draws upon, appeals to, and expands an established stock of popular skepticism around the secretive machinations and treachery of the powerful. It does this by articulating well-established histories of power-bloc misdeeds and corruption, and of the exploitation and endangerment of the socially subordinated by the dominant, working through the established institutions and agencies of its domination, to the serious questions raised by willful gaps, shortcomings and deep flaws in the official investigation of 9/11, and to a substantial body of evidence, much though by no means all of it circumstantial, that has been amassed by a loosely affiliated group of 9/11TM researchers, including both lay *and* well-credentialed, expert investigators, a significant number of whom are university professors working within and beyond the US.

[4]Operation Northwoods was designed to involve terrorist attacks by the US government on major US cities and the false attribution of the attacks to Cuba as a pretext for the invasion of that country; the operation was approved by the Joint Chiefs of Staff in 1962 but rejected by JFK. See http://en.wikipedia.org/wiki/Operation_Northwoods.

The sprawling and expanding body of counterdiscourses and counterknowledges that are being generated and circulated through 9/11TM's articulatory practices and processes can be understood as both offensive and defensive weapons of sorts—weapons against the imperializing reach of hegemonic power, weapons of countersurveillance and countervisualization, weapons or tools for the expression of democratic impulses, of desires, anxieties and refusals, for the assertion of claims on behalf of particular normative visions of the world, and for the production of communal spaces, identities, affective energies, solidarities, and transformative popular imaginaries.

In this regard, 9/11TM has interesting affinities and points of overlap and intersection with the ambitions, networked informational flows and operational modalities of the multitude that articulated itself around and emerged expressively in the form of the Obama movement in 2008. As W.J.T. Mitchell (2009: 126) notes, Obama attained the presidency "on the crest of a wave of popular feeling that he helped to create, but that largely pre-dated his candidacy"; the "aura of a social movement being born" (Mitchell 2009: 126) that surrounded the spectacular rise of the Obama phenomenon was grounded in a deeply rooted and broad-based sense of popular longing for political transformation and for the rejection and expulsion of Bush-Cheney-ism. Obama was a product first and foremost of grassroots activism (Bobo 2009). Obama achieved deep resonance with many who felt alienated by the political system and cynical about the depths to which it and US society descended under the calamitously transformative Bush regime, which "in every conceivable way set the conditions for Obama's emergence" (Mitchell 2009: 128). Like the 9/11 Truth movement, the Obama movement made itself partly through mobilization of the affordances of convergent media (see, e.g., Everett 2009; Castells 2009; Cogburn and Espinoza-Vasquez 2011; Harris et al. 2010). Moreover, like widespread reactions to the 9/11 Truth movement (see Bratich 2008), conservative reaction to the social movement that emerged around candidate Obama figured the phenomenon as a collective failure of political rationality that threatens the body politic (Spicer 2010).

The production and circulation of popular counterdiscourses and counterknowledges around 9/11 has stepped up considerably since 2005, the year of the release of both "Bin Laden," by Mos Def, Immortal Technique and Eminem, and of the first of four versions of the feature-length film *Loose Change*, a kind of grassroots *Fahrenheit 9/11* for the convergence culture generation that was created on a laptop computer by 21-year-old Dylan Avery with a total budget of about $2,000.[5] The viral YouTube vid of "Bin Laden" features deep bass beats and a rapid-fire assemblage of miscellaneous images including heavy doses of TV journalists and Fox News commentators, Bush administration politicians, the WTC attacks,

[5] Mos Def, Immortal Technique and Eminem are hip-hop artists; "Bin Laden" is available at https://www.youtube.com/watch?v=RJ4iZE2yoLk. *Farenheit 9/11* is a film by Michael Moore that won the Palme d'Or at the 2004 Cannes Film Festival and went on to become the highest grossing documentary of all time.

corporate logos, third world military interventions and death squad victims; all of these are set to fiercely sardonic lyrics about "fake Christians" and "fake politicians . . . in mansions," the ongoing fight for survival in the ghetto, US war mongering and covert paramilitary actions, and the recurring lines: "Bin Laden didn't blow up the projects . . . Bush knocked down the towers." The feature-length *Loose Change* has circulated globally on the Web, where it has, as Avery told *ABC News Nightline* in 2010, "gone viral and back." It is, in the words of *Vanity Fair*, "the first Internet blockbuster" (quoted in Dean 2009: 156). The film has been translated into 26 languages, sold more than a million DVDs, aired on TV in the US, Portugal, Belgium, Ireland the Netherlands and elsewhere, and purportedly been watched more than 155 million times.[6] It is a central node in what Jack Bratich (2008: 135) calls the "conspiratology archipelago" that constitutes the discursive spatiality of 9/11TM, which has emerged as a rhizomatic set of sites, researchers, knowledges and practices linked through decentralized media distribution. Mark Fenster (2008: 278) suggests that among the core messages of the film "is that we are all Dylan Avery now—a laptop battalion prepared to go into battle, armed with information, insight, and an interpretive method that *Loose Change* has provided." By 2006, the virality of 9/11TM had made multiple incursions into media outlets such as CNN, HBO, Comedy Central and ESPN, as the movement's celebrity spokespeople (including Ed Asner, Rosie O'Donnell, Ed Begley, Jr., and Martin Sheen) and nationwide commercials imparted to it a different form and level of visibility (see Bratich 2008: 132–3). That same year, on the fifth anniversary of the collapse of three WTC towers, thousands marched on New York City from around the world wearing the movement's trademark black t-shirts emblazoned with the words "Investigate 9/11."

In November 2007, the third version of the film, entitled *Loose Change, Final Cut*, was released with an eye toward the 2008 US campaign season. In this third version, several speculative assertions about the 9/11 attacks from the film's first two editions were dropped and the film's arguments were honed in response to criticisms from within and beyond the 9/11 Truth movement. As well, the element of political critique directed against the Bush administration was sharpened. *Final Cut*'s closing lines of narration state that "the government designed by the people, for the people, has turned its back on us. Or have we turned our backs on it? They spy on us. They torture and imprison innocent civilians. Ask yourself: what's happening? Where are we headed? And would we be here today without 9/11?" The *Loose Change* films incorporate images from Google Earth and NASA satellites to enact a kind of countersurveillance that questions the *official* conspiracy theory of 9/11—the one about a conspiracy among 19 Islamic terrorists armed only with box-cutters who successfully demolished three WTC towers and a sizeable chunk of the Pentagon before any effort at a US military intervention could be mustered. *Loose Change: Final Cut* also samples and mashes up clips to illustrate the inadequacies of mainstream media coverage of the attacks and their 6-year aftermath, thus

[6]See http://www.loosechange911.com/about/faq/.

performing a kind of media criticism that might also be read as countersurveillance. The soundtrack over the final credits consists of a hip hop theme song whose chorus announces, "We say no more! We're gonna fight back! We want the truth out, we want our rights back! . . . Change is loose don't stop it now!"

Like the YouTube vid by Mos Def, Immortal Technique and Eminem, the *Loose Change* series draws on a remix aesthetic with a very long history that passes through the Caribbean, and includes the creation of Haitian vodou from a hybrid assemblage of West African, Indigenous Hispaniolan, and Northern as well as Southern European Christian religions, and to which some scholars attribute the social solidarities, communities and identities necessary for the eventual emergence and success of the Haitian Revolution, which gave the world its first Black people's republic in 1804. This long history of remix culture also passes through Jamaica and New York in the birth of hip hop, through the culture of vidders,[7] through that key site of Bush era social commentary, political critique and satire, the *Daily Show*, and through the Internet branches of the Obama movement throughout 2008. The multifarious instantiations of this variable remix culture must be understood as being rooted in the specific social relations of particular historical and geographical conjunctures and material conditions of existence. In their 2007 appearance together on *Real Time with Bill Maher*, Mos Def and Cornell West provide a glimpse into the articulation of material conditions of existence to the forms of 9/11-oriented popular skepticism that are expressed through remix practices in the "Bin Laden" vid:

Maher: You have to admit that there are people who do want to kill Americans.

Mos Def: Yeah, some of them are called the police. . .

Maher: But you don't want to get blown up by a bomb.

Mos Def: Listen, I'm Black in America. I live under *constant pressure*. I don't believe in that bogeyman shit. . . . I don't believe that was bin Laden [in a videotaped message] today. I don't believe it was ever him. . . . I'm from the projects—I *know* danger. . . .

Maher: You don't think bin Laden knocked down the World Trade Center?

Mos Def: Absolutely not! . . . Go to any barber shop. I am *so* not alone! . . . Highly educated people in all areas of science have spoken on the fishiness around the whole 9/11 theory. It's like the "magic bullet" and all that shit! . . .

Cornel West: I think that bin Laden . . . had *something* to do with the buildings, 'cause he said so, and I believe it. But the thing is that if at the

[7]"Vidding" involves the production of music videos through the appropriation of material from media sources such as TV shows and news reports. By recontextualizing music and media images in this way, vidders comment on the music, the imagery, or both. Vidding (the production of "vids") has become a common practice within contemporary fan cultures and has had an influence on practices of political remix videomaking: the appropriation and recontextualization of media source material to make political or critical arguments (see McIntosh 2010).

> same time, you have multiple sources of force, coercion and terror coming at you, which many Black Americans do—prison-industrial complex, racial criminal justice system, . . . disgraceful school systems—to what degree do you begin to think: my paranoia is actually justified, because if *they* can sustain this level of psychic and physical terror against *me*, and *they're* obsessed with the terror that's against *them*, then maybe they might not be believable and credible.

Geography of the New Media Environment

This leads me toward my final observations on the relationships between the different elements of the contemporary media environment—that is, of Dayan's "multitiered public sphere," with its remnants of "central TV" in the major networks, and its exuberantly expanding rhizomatic margins which, through their constant fragmentation and dispersal of attention, increasingly threaten to swamp the center out of existence. One key feature of this environment is the role of spectacular media events as rejuvenators of attention centers, where centralized dominant forces concentrate their efforts to re-stabilize hegemonic regimes of power, social formations and discourses, and to repress or vanquish marginal ones. The relatively socially weak forces and voices, in turn, radiate outwards toward the networked margins, which provide a kind of safe harbor of enclaves and opportunities for growth and development through processes associated with convergence culture (Jenkins 2008) such as collective intelligence and creative remix, which are increasingly important to the production of counterdiscourses and counterknowledges, to the practices of rearticulation and resignification, and to the generation of oppositional identities and the maintenance or protection of endangered solidarities. I want to suggest that there is a growing traffic between the rhizomatic, deterritorialized, networked margins consisting of podcasts, blogs, vid sites and social networks, on the one hand, and the mid-sized juncture points that provide what I would like to call (with pun intended) *medium visibility*, on the other. The latter might include, for instance, cable TV outlets such as HBO, Comedy Central, MSNBC and Fox News, as well as non-prime-time programming such as *Nightline* and *The View*. Though such sites of medium visibility may be consumed "tribally" by relatively homogenous niche audiences, they nevertheless expand visibility in relation to the smaller clusters on the deterritorialized margins and so provide opportunities for the formation of new alliances across social difference, and perhaps for eventual break-through emergence within the attention centers associated with "central TV."[8] The appearance of professional physicists, architects and engineers in media such as CNN and national network breakfast TV

[8]See, e.g., *Geraldo at Large* (13 Nov. 2012), available at http://www.youtube.com/watch?feature=player_embedded&v=pFPobKeSzKQ.

to discuss the publication in 2009 in a peer-reviewed scientific journal of an article offering strong evidence of the use of super- or nano-thermite, a high-tech explosive produced by only a handful of military contractors such as Livermore Labs, in the destruction of the three WTC towers represents one such moment of *medium visibility* capable of generating new articulations and political alliances across social differences.[9]

The first decade of the new millennium in US culture was in many ways book-ended and defined by two key, counterposed spectacular media events: the collapse of the WTC in 2001, and the twin moments of high spectacle associated with the hopeful transformation of the dark nightmares of the Bush/Cheney regime into the age of Obama: Obama's speech accepting his party's nomination for the presidency at Denver's Mile High Stadium during the 2008 Democratic National Convention, and his electoral victory speech in Chicago's Grant Park just over 2 months later.[10] The attacks of 9/11 constituted a media megaspectacle that generated struggles to put its images into discourse in a manner capable of shifting the US culture's central structure of feeling—whether this is understood to entail an effort by Al-Qaeda to shift it away from an interventionist, Orientalist, and American exceptionalist swagger toward a newfound sense of vulnerability and humility in the face of an angry world and God, or alternatively to entail the Bush/Cheney regime's efforts to shift this structure of feeling through shock and awe toward a sense of deep and widespread fear in the service of democratic totalitarianism and war without end. By contrast, the remix-driven counter-mobilizations of this media event by 9/11TM and the mashups and counterspectacles that constituted the high moment of the Obama movement (which re-formed subsequently in the Occupy movements) must be read as major popular efforts to reclaim and redirect the structure of feeling and to reconstitute the political conditions of possibility within and beyond the US for the new millennium.

Acknowledgement This research is supported by the Marsden Fund of the Royal Society of New Zealand, grant number MAU1108.

References

Andrejevic, M. (2007). *iSpy: Surveillance and power in the interactive era.* Lawrence: University Press of Kansas.

Baudrillard, J. (1988). *The ecstasy of communication* (S. Lotringer (Ed.); B. Schutze, & C. Schutze, Trans.). New York: Semiotext(e).

Baudrillard, J. (2003). *The spirit of terrorism* (C. Turner, Trans.). London: Verso.

[9]See, e.g., Richard Gage on Fox TV affiliate KMPH's breakfast show (May 2009), available at http://www.youtube.com/watch?v=oO2yT0uBQbM.

[10]These were arguably the two key moments of high media spectacle in the US in 2008. Obama's speech at Mile High Stadium on 28 August was delivered to an estimated crowd of 84,000 people in attendance, and his 4 November electoral victory speech was delivered to a Grant Park audience estimated at 240,000. Countless millions around the world watched both events on TV.

Birchall, C. (2006). *Knowledge goes pop: From conspiracy theory to gossip*. Oxford: Berg.
Boal, I., Clark, T. J., Matthews, J., & Watts, M. (2005). *Afflicted powers: Capital and spectacle in a new age of war*. London: Verso.
Bobo, J. (2009). Impact of grassroots activism. *Journal of Visual Culture, 8*(2), 158–160.
Bratich, J. Z. (2008). *Conspiracy panics: Political rationality and popular culture*. Albany: State University of New York Press.
Castells, M. (2009). *Communication power*. Oxford: Oxford University Press.
Cogburn, D. L., & Espinoza-Vasquez, F. K. (2011). From networked nominee to networked nation: Examining the impact of Web 2.0 and social media on political participation and civic engagement in the 2008 Obama campaign. *Journal of Political Marketing, 10*(1–2), 189–213.
Dayan, D. (2009). Sharing and showing: Television as monstration. *The ANNALS of the American Academy of Political and Social Science, 625*, 19–31.
de Certeau, M. (1984). *The practice of everyday life* (S. Rendall, Trans.). Berkeley: University of California Press.
Dean, J. (2009). *Democracy and other neoliberal fantasies: Communicative capitalism and left politics*. Durham: Duke University Press.
Dowd, M. (2003, May 21). Walk this way. *New York Times*. http://www.nytimes.com/2003/05/21/opinion/21DOWD.htm. Accessed 1 Dec 2013.
Everett, A. (2009). The Afrogeek-in-chief: Obama and our new media ecology. *Journal of Visual Culture, 8*(2), 193–196.
Fenster, M. (2008). *Conspiracy theories: Secrecy and power in American culture* (Rev. and Updated Ed.). Minneapolis: University of Minnesota Press.
Fiske, J. (1996). *Media matters: Race and gender in U.S. politics* (Revth ed.). Minneapolis: University of Minnesota Press.
Fiske, J. (1998). Surveilling the city: Whiteness, the black man and democratic totalitarianism. *Theory, Culture & Society, 15*(2), 66–88.
Freedberg, D. (1991). *The power of images: Studies in the history and theory of response*. Chicago: University of Chicago Press.
Glynn, K. (2009). The 2004 election did not take place: Bush, spectacle, and the media nonevent. *Television & New Media, 10*(2), 216–245.
Grossberg, L. (1996). On postmodernism and articulation: An interview with Stuart Hall. In D. Morley & K. Chen (Eds.), *Stuart hall: Critical dialogues in cultural studies* (pp. 131–150). London: Routledge.
Harris, H. E., Moffitt, K. R., & Squires, C. R. (Eds.). (2010). *The Obama effect: Multidisciplinary renderings of the 2008 campaign*. Albany: SUNY Press.
Heidegger, M. (1977). *The question concerning technology and other essays* (Trans. and with an Introduction by Lovitt, W.). New York: Harper Torchbooks.
Jay, M. (1993). *Downcast eyes: The denigration of vision in twentieth-century French thought*. Berkeley: University of California Press.
Jenkins, H. (2008). *Convergence culture: Where old and new media collide* (Rev. Ed.). New York: NYU Press.
Kellner, D. (2003). *Media spectacle*. London: Routledge.
King, G. (2005). 'Just like a movie'?: 9/11 and Hollywood spectacle. In G. King (Ed.), *The spectacle of the real: From Hollywood to 'reality' TV and beyond* (pp. 47–57). Bristol: Intellect Books.
Lyotard, J. (1984). *The postmodern condition: A report on knowledge* (G. Bennington & B. Massumi, Trans. and with a Foreword by Jameson F.). Minneapolis: University of Minnesota Press.
McIntosh, J. (2010, December 20). Interviewed by Henry Jenkins about DIY remix video. *Rebelliouspixels*. http://www.rebelliouspixels.com/2010/henry-jenkins-interview. Accessed 17 Dec 2013.
Meikle, G., & Young, S. (2012). *Media convergence: Networked digital media in everyday life*. Houndmills: Palgrave Macmillan.
Mirzoeff, N. (1999). *An introduction to visual culture*. London: Routledge.

Mitchell, W. J. T. (2009). Obama as icon. *Journal of Visual Culture, 8*(2), 125–129.
Palast, G. (2006). *Armed madhouse*. New York: Dutton.
Parks, L. (2005). *Cultures in orbit: Satellites and the televisual*. Durham: Duke University Press.
Phillips, M. (2011). *Satan in the smoke? A photojournalist's 9/11 story*. Brooklyn: South Brooklyn Internet. Ebook available from www.amazon.com
Redfield, M. (2009). *The rhetoric of terror: Reflections on 9/11 and the war on terror*. New York: Fordham University Press.
Sontag, S. (2004 [1965]). The imagination of disaster. In G. Rickman (Ed.), *The science fiction film reader* (pp. 98–113). New York: Limelight Editions.
Spicer, R. (2010). The Obama mass: Barack Obama, image, and fear of the crowd. In H. E. Harris, K. R. Moffitt, & C. R. Squires (Eds.), *The Obama effect: Multidisciplinary renderings of the 2008 campaign* (pp. 190–208). Albany: SUNY Press.
Spigel, L. (2004, June). Entertainment wars: Television culture after 9/11. *American Quarterly, 56*(2): 235–270.
Williams, R. (1977). *Marxism and literature*. Oxford: Oxford University Press.
Žižek, S. (2002a). Welcome to the desert of the real. *The symptom: Online journal for Lacan* 2. http://www.lacan.com/thesymptom.htm. Accessed 17 Dec 2013.
Žižek, S. (2002b). *Welcome to the desert of the real! Five essays on September 11 and related dates*. London: Verso.

The Case of the Missing Laureate: The Communication Geography of the 2010 Nobel Peace Prize

19

Paul C. Adams

The Nobel Prize ceremony for Liu Xiaobo was held on December 10, 2010, at the Oslo City Hall. The flower bedecked room was filled with some 1,000 guests, including King Harald, Queen Sonja, representatives from 46 different embassies, Chinese dissidents who had participated in the 1989 occupation of Tiananmen Square, the US Speaker of the House Nancy Pelosi, and even the Hollywood stars Denzel Washington and Anne Hathaway. Notably missing, however, was the laureate, Liu Xiaobo, who had just finished the first of year of an 11 year sentence for "inciting subversion of state power." The laureate's absence generated unusual dynamics of presence and absence that highlight various facets of communication geography. This chapter provides a concrete demonstration of the non-Euclidean terrain of communication using the 2010 Nobel Prize ceremony to explore the experiential continuum between space and place as it intersects the ontological paradox that communication is a container that is also contained (Adams 2009, 2011).

It is not particularly paradoxical that A would contain B and also be contained by B—consider for example a soap bubble which both contains and is contained by air. However, it is paradoxical if the opposite is *also* true: B also contains A and is contained by A. The only situation that fits this description is a "non-orientable two-dimensional manifold," which is to say a Klein Bottle (Fig. 19.1). We do not normally encounter Klein Bottles and images like Fig. 19.1 are only approximations of a mathematical abstraction. Everyday experience with the world teaches us that containment implies a one-directional relationship; we therefore assume that communications are either in spaces and places as arrangements of objects and flows, or else they contain spaces and places by functioning as environments of signs and symbols.

P.C. Adams (✉)
Department of Geography and the Environment, University of Texas at Austin, Austin, TX, USA
e-mail: paul.adams@mail.utexas.edu

S.P. Mains et al. (eds.), *Mediated Geographies and Geographies of Media*,
DOI 10.1007/978-94-017-9969-0_19

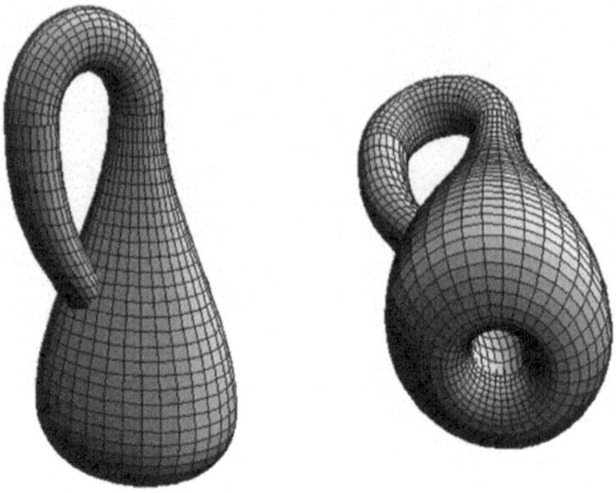

Fig. 19.1 Two views of a Klein bottle in three dimensions. The image is only an approximation since the neck of the bottle must intersect the body without either surface cutting or being cut by the other surface (Source: AugPi, http://commons.wikimedia.org/wiki/File:BouteilleKlein3d.PNG. GNU Free Documentation License/Creative Commons Attribution-Share Alike 3.0 Unported License)

This logical tendency to designate one entity as contents and the other as context leads to a rift within the media geography scholarship. Some researchers treat spaces and places as contents of communication, for example studying how the nation circulates symbolically within cartoons and magazines (Dittmer 2013; Sharp 2000), how landscapes circulate in film (Zonn and Aitken 1994), or how the earth itself flows through various media (Cosgrove 2001). Other researchers treat spaces and places as contexts of communication, for example studying how people struggle over inscriptions that will be left at sites of tragedy (Foote 2003), occupy and claim territories in order to send political messages (Wainwright 2007; D'Arcus 2003), arrange communication capital and infrastructure in space (Zook 2005; Graham and Marvin 2002), or circulate communication products through space (Christophers 2009). It is rare to find scholarship that engages with places and spaces as *both contents and contexts of communications* (although see MacDonald et al. 2010; Adams et al. 2001). Such a divide arises out of our unwillingness to let our ontologies wrap around and cross through themselves like a Klein Bottle. Nonetheless, we can explore the inside-out/outside-in terrain through direct encounters with phenomena, like a car driving along the surface of a Klein Bottle (http://www.youtube.com/watch?v=sRTKSzAOBr4&fmt=22).

Accordingly, this chapter will provide a demonstration of this kind of ontological duality, showing how we can study spaces and places as both contents and contexts of communication. We will consider, first, how communication depended on Oslo City Hall as people came together in a physical container to honor Liu Xiaobo's life, his activism, and his absence. Next we will touch on the ways in which China was represented in and by the discourses associated with this Prize. Following this, our attention turns to experiential movement through and beyond the website of the Norwegian newspaper *Verdens Gang*, using the website as a kind space in which to connect various ideas. Finally we consider the uneven distribution of information through space and how it set the stage for Liu's activism and the international relations enacted by his prize. The objective of the chapter is to use the act of giving the Prize, the motivation behind the Prize, and the circulation of the Prize through both place representations and communication technologies to reveal four distinct and complementary ontologies of communication geography. These have been characterized elsewhere as: communication in place, place in communication, space in communication and communication in space (Adams 2009). The objective of this taxonomy is to de-naturalize dichotomies between space and place and between communication content and context, and to reveal intriguing contradictions and complexities, crossings and links that are overlooked if we adopt only one or two of the four complementary perspectives (Adams 2011).

An Empty Chair in a Full Hall[1]

If one views the purpose of the Nobel Peace Prize solely as the handing out of a prize then the 2010 ceremony seems like a non-event. The prize could not pass into the hands of an awardee because he was in prison in China. However the offering of a prize to the absent laureate had the effect of offering Liu Xiaobo to the world as an exemplary prisoner of conscience. Bestowing a prize is an illocutionary act, a communication that transforms the status of the one who is spoken to; it has a *phatic* role, meaning that rather than simply describing or predicting something or transmitting information it causes something to be (Austin 1975; Malinowski 1936/1993). Place is essential to this kind of communication because what one "says" depends, in such cases, on where one "says" it.

The ceremony also intervened in the symbolism of incarceration as exclusion. It did this by countering Liu's legal exclusion from Chinese society with his inclusion in the ranks of Nobel laureates. It conferred a kind of presence in a global polity that was accentuated by, and even justified by, the recipient's bodily absence from the award ceremony. The capacious Oslo City Hall served as a container full of witnesses to a *very particular* absence. Rather than subtracting from a communication, like a word missing from a sentence, Liu's missing body was key to the communication as it was contextualized in this particular place. The place, empty in a particular way (of a particular body), broadcast a particular message regarding human rights, freedom of speech, justice, and peace.

The use of Oslo City Hall for this odd inversion inevitably linked Norway's national identity to a certain ideal of world peace. The walls of the venerable room are covered in murals depicting famous Norwegian Nobel laureates including the explorer Fridtjof Nansen and the author Bjørnstjerne Bjørnson. The hall is a synec-

[1] In what cannot but seem like a sick parody, the Republican National Convention of 2012 utilized an empty chair as a prop for Clint Eastwood's vitriolic attack on the previous year's Nobel Peace Prize laureate, Barack Obama.

doche of Norway, a micro-place standing for a macro-place, like the White House stands for the United States or the Kremlin for Russia.[2] Conversely, the micro-place was infused from above by the regional symbolism of Scandinavia as a peace-oriented "norm entrepreneur" (Ingebritsen 2002). Yet the Scandinavian norms are rooted more generally in the West, including Kant's theory of democratic peace and European enlightenment notions of individualism, equality, and civil rights. Further complicating this macro-micro dynamic, the Hall is redecorated each year in honor of the prize recipient's nationality. In 2010 the Hall was adorned with thousands of orchids, bamboo flowers, chrysanthemums and plum blossoms, associated in Chinese culture with the four seasons, and with an enormous portrait of Liu Xiaobo near the rostrum. With its various adornments, the place functioned not as a bounded container but as a means of activating very specific linkages to other places (Massey 1993). The murals situated Liu in a semi-mythical version of Norway, the Hall's traditional association with the Peace Prize evoked Western values, while the flowers situated the event within a realm of de-politicized Chinese culture.

Performances make place what it is (Thrift 2008) and, as such, the Prize ceremony can be broken down into distinct acts. First, an act embodying Norway itself: King Harald and Queen Sonja entered to a horn fanfare and were seated front and center. Next came the singing of Solveig's Song by Edvard Grieg (Norway's most famous composer) from the incidental music to Peer Gynt, the most famous play by Henrik Ibsen (who is Norway's most famous playwright). This performance positioned Norway as a bastion of Culture but additionally the song's lyrics evoked the separation of lovers, thereby folding Liu Xiaobo and his wife Liu Xia into a local mythology.

The two main acts of the day were the presentation speech by Thorbjørn Jagland, head of the Norwegian Nobel Committee, and the reading by Liv Ullman of Liu Xiaobo's final declaration "I have no enemies" from his Dec. 23, 2009 trial. Supporting acts included the musical selections performed before, between, and after the speeches, as well as audience applause, itself a performance of what Richardson in a slightly different context calls "presence in the face of absence [which] turns us from a factual given, from simply being here as objects occupying space, into a gift" (2001, p. 264). Nineteen countries heeded China's call to boycott the event, contributing their strategic silences that could not be heard because the audience seats were filled in any case. The award presentation speech spoke Liu Xiaobo into a new status, that of Nobel Laureate, while also referring to the relationship between economic prosperity and democracy, and the reasons that civil rights struggles should be seen as struggles for peace.

> We regret that the Laureate is not present here today. He is in isolation in a prison in northeast China. Nor can the Laureate's wife Liu Xia or his closest relatives be here with us. No medal or diploma will therefore be presented here today. This fact alone shows that the award was necessary and appropriate. (Jagland 2009)

[2]Synecdoche is a literary trope in which a part of something stands for the whole, for example referring to food as "daily bread" or workers as "hired hands."

Fig. 19.2 The empty chair that marked the symbolic position of Liu Xiaobo at the 2010 Nobel Peace Prize Ceremony (Photo courtesy of Ministry of Foreign Affairs, Oslo, Norway/Marta B. Haga)

Jagland's speech placed Liu in the company of other Nobel laureates who were previously prevented from coming to Oslo to receive the prize—Andrej Sakharov, Lech Walesa, and Aung San Suu Kyi, then attention was directed to a chair: "At this stage in the ceremony where we normally hand over the medal and the diploma I place them in the empty chair held by Liu Xiaobo." Norway's Foreign Minister, Jonas Gahr Støre, later commented: "The representation of the imprisoned Laureate by an empty chair is special and unique" while the Director of the Nobel Institute, Geir Lundestad, described the empty chair as "a tremendously powerful message about the human rights situation in China" (Rasch and Kvaale 2010). The empty chair (Fig. 19.2) became a kind of spatial and discursive nucleus. Liu's absence (from the chair, functioning as a synecdoche of the hall) evoked China's absence from the community of countries supportive of peace.

Jagland's presentation speech was followed by an intermission with music. The US-born, Chinese-American violinist Lynn Chang played three musical selections: the Chinese folk song *Jasmine Flowers* (Mo Li Hua), a piece by the Paris-educated Chinese composer, Ren Guang, and a piece by the English composer Edward Elgar. The choice of compositions, and the performer's embodied presence, both suggested blending of East and West, and in light of their sequence, a process of Westernization.

Next, actress/director Liv Ullmann read "I Have No Enemies." She entered the hall, bowed to the portrait of Liu Xiaobo, then gave voice to the words Liu Xiaobo had drafted for his 2009 trial but had not been allowed to deliver. Choosing someone to speak on Liu's behalf must have posed a dilemma for the organizers of the event. As UNICEF Goodwill Ambassador and honorary chair of the Women's Refugee Commission, Ullman had international credentials but the portion of the speech in which Liu tenderly expresses his love for his wife came across a bit strangely. Liu's absence was perhaps most jarring at this moment of the ceremony.

The final musical selection did not turn toward China in order to balance the Grieg selection at the beginning and the cultural hybridity of the interlude. Instead what followed were a Norwegian folk song and two more pieces by Grieg. These were sung by the children's choir from the Norwegian National Opera who were dressed in traditional Norwegian folk costumes.

Through all of these performances in place, Liu was situated virtually among the Nobel laureates. His absence was transformed into a kind of presence. Meanwhile, the Hall's appearance, its audience, and the performances that took place in it said as much about Norway as they did about China. Taken as a whole, then, the ceremony reveals how communications are integral to places rather than simply emanating from places; they depend on places for their meaning and they rework the meanings of places by embedding representations and performances in them.

China Viewed from Oslo

While places surround and contextualize communications, as demonstrated above, places are also embedded representationally in communications. We move on, then, to consider the range of Chinas which circulated in various media, before, during and after the Prize ceremony. Jagland's Speech foregrounded an image of China as a problematic mix of success and failure.

> Economic success has lifted several hundred million Chinese out of poverty. For the reduction in the number of poor people in the world, China must be given the main credit. We can to a certain degree say that China with its 1.3 billion people is carrying mankind's fate on its shoulders. If the country proves capable of developing a social market economy with full civil rights, this will have a huge favorable impact on the world. If not, there is a danger of social and economic crises arising in the country, with negative consequences for us all . . . [W]ithout freedom of expression, corruption, the abuse of power, and misrule will develop. (Jagland 2009)

We can understand this portion of the speech as intervening in a particular "discourse of democracy" (Bell and Staeheli 2001, p. 178). The place portrayal whereby China was a potential failure precisely because of its lack of democracy did not go unchallenged. Naturally Chinese authorities demurred, but in addition there were objections from the West. Arnulf Kolstad, a professor of social psychology at the Norwegian University of Science and Technology advanced an argument that articulated local peculiarity in terms of the cultural difference:

> In China, they do not even have a distinct word for 'me.' They think all the time in relation to other people. When we talk about freedom of speech what we mean is freedom of speech in our sense. China is realizing human rights in their own culture and at their own level of development. To claim that personal freedom of speech is the most important human right and a precondition for a peaceful world is a typically Western way of thinking. (Andersen et al. 2010)

On this account, China is a different kind of place than Norway, on a different path than the West in general. Kolstad's approach avoids the kind of Othering, often critiqued following Said (1978/1995), by which the East is portrayed as perpetually in need of discipline and correction by the West. However, by positing the lack of democracy as a natural condition in China Kolstad inadvertently echoed the Oriental despotism motif (Wittfogel 1957). Jagland's speech reflected an awareness of this tricky line between critical condescension and apologetics, challenging both through the unusual gesture of likening China to the United States.

> Many Americans were opposed to the award of the Nobel Peace Prize to Martin Luther King in 1964. Looking back, we can see that the USA grew stronger when the African-American people obtained their rights. Many will ask whether China's weakness—for all the strength the country is currently showing—is not manifested in the need to imprison a man for eleven years merely for expressing his opinions on how his country should be governed. (Jagland 2009)

From a Western perspective the harshness of the 2010 Prize's implicit critique of China is softened by linking it to America's past, although from a Chinese perspective this parallel might have added insult to injury. A hallmark of place representations is the fact that they often depend on geographical analogies, and analogies may have divergent implications depending on the contexts in which they are interpreted. Nonetheless, beneath disagreement one usually finds certain grounds of agreement; taken-for-granted elements bind together the place images employed in opposing sides of most debates.

Drawing at random from the online chat of "ordinary" Norwegians on the day immediately following the Nobel Committee's announcement we can find confirmation of this idea. In the Norwegian blog, *Ukorrigerte Meninger* (Uncorrected Opinions, http://ukorrigert.wordpress.com, October 9, 2010) there appeared a rather scathing assessment of China: "The Chinese are like fucking ants, tyrannical oppressors," (participant 7891) to which another participant going by the screen name of "ko" responded: "[C]onsidering the future of humanity over the long perspective, isn't the world's population served by having a high percentage who think like bees without concern for themselves?" Although they disagree with each other, both participants employ a symbolic framing whereby China is a kind of hive and its inhabitants are like insects. In disagreement they nonetheless agree on this image of place. Likewise, Jagland and Kolstad agree that China is a place where economic growth currently takes priority over democracy, even as they disagree about whether freedom of speech is necessary or appropriate for China.

Much more could be extracted with regard to representations of China by exploring the various grounds for agreement and disagreement among the texts generated in response to the 2010 Prize, but for our purposes it suffices to note that: (a) shared assumptions about a place combine with certain areas of disagreement, (b) these countervailing tendencies indicate the dynamic tension saturating discourses about place, and (c) place images both recognize and discount the Other. Therefore debates about places exist in a state of dynamic evolution but at the center of debates certain shared assumptions maintain a "common sense" about a place; they hold a place in place. Place-in-communication is therefore a type of representation that embraces both difference and sameness.

From Fish Trade to the Dalai Lama

As Nigel Thrift argues (2008), we must think not simply of representation but also of non-representation, that is, processes of encounter, hybrid assemblages, and nonverbal interactions among humans and nonhumans. I would argue that any such study is focused on communications. Seeing nonrepresentational geographies as communication processes could take us in many directions with regard to the 2010 Nobel Peace Prize. Here we explore how interlinked websites created what Thrift would call a "movement space" (Thrift 2008, p. 103).

Someone exploring the website of the Norwegian newspaper *Verdens Gang*[3] between October 8th and 13th, 2010 would have encountered some forty pages addressing Liu Xiaobo, the Nobel Prize, China, and related topics. These pages formed a space for exploration and encounter interlinking various types of information, discussion, debate, reflection, and visualization. Movement in this space was highly tactile and visual. To think of a space as existing in this way requires that we recognize not just the topology of links between various online places (Adams 1998), but the nature of movement which, by following certain paths in certain ways, continually reconstitutes people's experience of the world: "the critical importance of spatial distribution in flow architectures will produce an extended spatial vocabulary which will provide new opportunities for thinking the world, opportunities which will themselves be constitutive of that world" (Thrift 2008, p. 105). One moves through some spaces on foot (walking) and others using the hands and feet (pushing a gas pedal and turning a steering wheel) while one moves through yet others by hand (typing and using a mouse). In all cases, vision is crucial, but other senses come into play, and the human metabolism always responds to the play of senses. So encounters with online spaces are not disembodied but rather they are *differently embodied*.

Thus encountering online news about Liu Xiaobo and the 2010 Peace Prize was not simply "going online" but rather a matter of participating in the creation of a particular relational space involving many places (websites) joined by particular paths, which in turn presented certain opportunities and constraints to users. This movement space was in turn embedded within the larger space of the internet by a

[3] *Verdens Gang* is the Norwegian newspaper with the second largest circulation; its name means the Way of the World.

particular set of directed links to and from particular nodes within that space. Paths were accessed via highlighted words in the text and or via links in a box or sidebar.

One can map this space in various ways. One intriguing technique is "clique percolation," which identifies densely interlinked clusters of nodes within a larger population of nodes that is interlinked at lower levels (Palla et al. 2005). We can think of the densely interlinked webpages as constituting small spaces within larger spaces comprised of more sparsely interlinked webpages. The structure of online space is directly related to movement: moving within a cluster is easier than moving between clusters since within a cluster one has more options for reaching node B from node A. If node B can be reached from node A by only one link, then it helps constitute a large space of weakly interconnected nodes while those nodes that are well-connected create smaller, more densely connected spaces. On this account, a "space" is a cluster of densely interlinked nodes each of which is linked to every other node in the cluster by "many" links. What constitutes "many" can be set by a variable and we can raise or lower the value of this variable, altering the threshold at which nodes appear to be interconnected. This offers a reproducible way to discover the different modular structures within a set of interconnected nodes (Palla et al. 2005, p. 814).

Figure 19.3 applies this technique to the *Verdens Gang* website showing the five clusters that appear when we set the threshold variable as 7. In other words, we can see five groups of webpages that are all connected to each other by at least seven hyperlinks. These clusters all include the webpage labeled "VG-peaceprize-index" which was updated throughout the period from October 8 to October 13 and offered links to all of the relevant articles and information addressing the 2010 Peace Prize and the Peace Prize in general. Figure 19.3 also shows how the online movement-space was divided into five overlapping clusters or sub-spaces. Most of the articles appearing on October 8 were poorly connected; they could not include links to articles that had not yet appeared, and for some reason subsequent articles did not generally link to them. In contrast, at the upper center one can find three articles from October 10 (here labeled Oct10b, Oct10d, and Oct10f) that were included in four clusters, and towards the bottom center one can find three pages (Oct11a, Oct11f, and Oct11g) that are included in four (different) clusters. What this means is that a reader who accessed the news site on October 10 or 11 would have encountered many opportunities to meander through related articles, like a visitor to New York City who arrives at Grand Central Station while someone accessing the news site a few days earlier or later would have encountered fewer opportunities to move about within related stories. The diagram also reveals the proportion of each node's links that are in-coming. The least accessible pages are indicated by 0.0 in the parentheses (indicating that there was no way to get them except from the index page), while the most accessible page, the Peaceprize Index, has a ratio of 1.0, meaning that all of the other pages link to it. Figure 19.3 reveals a movement-space where opportunities and constraints in the form of links act upon a user of the website to structure his or her ability to access particular ideas, opinions, representations, and interpretations.

This movement-space also offered opportunities to leave the *Verdens Gang* site and explore other spaces. These links are indicated by icons around the edge of

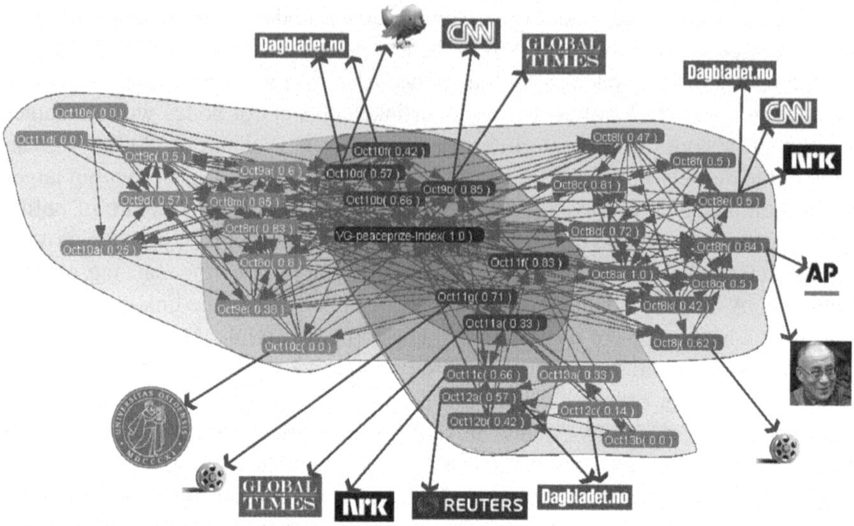

Fig. 19.3 A diagram of the space a Norwegian reader would have encountered if accessing news about the Peace Prize on the website of the newspaper *Verdens Gang* between October 8 and October 13, 2010. The sites are indicated by date and arbitrary letter label, followed by a ratio of inward-directed links. More central web pages on the diagram are those which are easier to reach (statistically more likely to be encountered). The clusters were identified using CFinder 2.0.5 (© 2005–2010 The Department of Biological Physics, Eötvös University, Budapest. Online file at: http://cfinder.org/wiki/?n=Main.HomePage. "For non-profit users the permission to use, copy, and modify this software and its documentation for any purpose is hereby granted without fee, provided that the above copyright notice appears in all copies and that both the copyright notice and this permission notice appear in supporting documentation")

Fig. 19.3 and include the website of another Norwegian newspaper (*Dagbladet*), the Norwegian broadcasting corporation (NRK), foreign and international news sources (CNN, AP, Reuters, and China's Global Times), and the websites of the University of Oslo, Twitter, the Norwegian Nobel Committee, and even the website of the Dalai Lama. (Videos are shown as external links, as well, although they were on the VG server, because of their different format and the fact that they were produced off-site).

What this shows is that webpages are not just texts but also points of connection within a space of encounter. Conceiving of this as a space requires that we understand the movement itself (browsing) that is required in order for the space to exist as such. We can conclude by tracing one possible movement through this space. Starting from the page we are calling Oct10e, at the upper left corner, one could progress to Oct10a, Oct9e, and then to the VG Peace Prize Index page. At this point, one could jump to Oct8h and finally to the website of the Dalai Lama, at the far right of Fig. 19.3. Translating headlines the movement is as follows:

- "Fisheries Minister is not afraid of being silenced by China," a story about the upcoming visit of Norway's Fisheries Minister to China.
- "Still no contact with Liu's wife," a short article indicating the disappearance of Liu Xia.
- "We are waiting by the phone," an article expanding on the issue of the Liu Xia's disappearance, from the point of view of an American lawyer.
- Index page for the Nobel Peace Prize: a column of article titles with short descriptions, divided into many separate pages.
- "Chinese authorities: They have given the prize to a criminal," an article describing the reaction of Chinese authorities to the selection of Liu Xiaobo for the Peace Prize.
- "Press Statement on Liu Xiaobo," a statement of support for Liu Xiaobo and for free speech in China on the official website of the Dalai Lama.

This path through online movement-space from fish exports to the Dalai Lama, represents only one route and is of course a miniscule part of the space created by all forms of communication about that prize, but it exemplifies space-in-communication. In general a set of communication opportunities provides means of moving to other communication opportunities, any of which can be engaged in by communicators, guided by their own objectives and interests. Communications thereby form small-scale spaces of habitation nested within larger-scale spaces of habitation and these spaces-in-communication in turn shape their "inhabitants" not unlike other environments. There remains a final intersection to explore.

Footprints and Trend Surfaces

Hubbard et al. (2002, p. 31) argue that geography's quantitative revolution conceived of space as: "a surface on which the relationships between (measurable) things were played out." The idea of space as a surface modulated by particular densities, in other words, a trend surface, has not been abandoned in later years; it remains useful for thinking through many geographical issues related to communication. We can follow this line of thought to consider quantitative and qualitative differences in access to information such as variations in freedom of the press and freedom of speech. "High points" on this surface are places with ample communication infrastructure and high levels of press freedom while "low points" are those with opposite characteristics. We could add maps showing the footprints of various types of technology, for example the positions of cell-phone towers (masts) or the layout of the Internet's fiber optic "backbone." In fact, the options for showing media as spatial patterns on the earth surface are so numerous that this section can offer only a hint, and little in the way of illustration, of this communication-in-space approach.

One glimpse of China as a communication space is provided by the fact that the number of Internet users in China is greater than the entire population of the United States (Talbot 2010). However, 79 % of Americans regularly go online versus 29 % of Chinese (Pew 2011), indicating a much lower social penetration of this technology in China. A similar but smaller gap exists with regard to literacy, which is rated as 92 % in China versus 99 % in the US (CIA 2011). We would envision China as lower than the West if we were to create a global density gradient of information access. Many measures could be used. For example, excluding Hong Kong, China had 592,000 Mbps (megabits per second) "lit" (used) bandwidth in 2005 while Norway had 429,000. Scaled to the difference in population, a typical Norwegian was 193 times better connected to the world than a typical Chinese.

Of course these gradients realign over time. A global survey of submarine telecommunication cables (Malecki and Wei 2009) has indicated that China and India are approaching the data transmission capacity of more economically developed places such as Japan, South Korea, Taiwan and Singapore. If the utilized bandwidth for Hong Kong is combined with the rest of China, the resulting total exceeds any other Asian destination except Japan. While this suggests a flattening of the information density gradient, Chinese internet traffic remains highly controlled. Chinese internet communications must pass through one of three international gateways that impose stringent government controls (Malecki and Wei 2009, p. 375).

An international NGO, Reporters without Borders, produces an annual Press Freedom Index. Although based on subjective evaluations by reporters in various countries, the index nonetheless offers some basis of comparison that goes beyond simplistic measures such as kilobytes per second. For example, seven countries (including Norway) tied for the honor of having the most press freedom in 2010 while China ranked near the bottom, 171 out of the 178 countries that were evaluated (Reporters without Borders 2010). For comparison, the UK and the US ranked 19 and 20, respectively. This organization's 2012 map of the world is shown in Fig. 19.4. Absolute access to data transmission bandwidth on a country to country

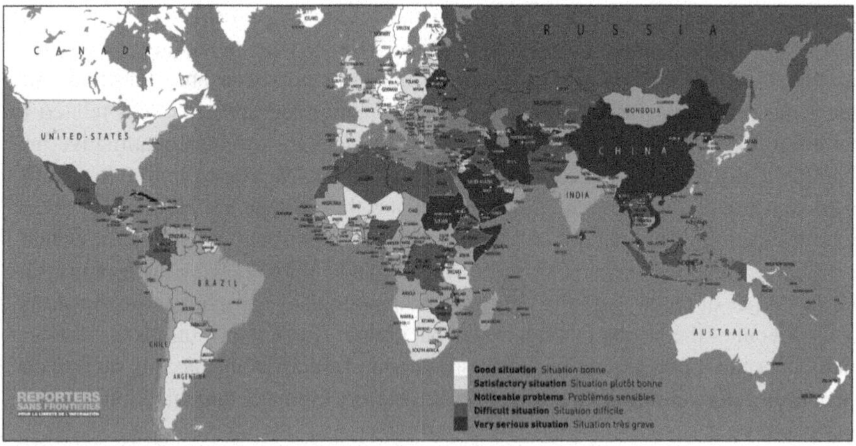

Fig. 19.4 Freedom of the press worldwide in 2012 according to reporters without borders (Map based on an online image at http://en.rsf.org/IMG/jpg/carte2012.jpg. Accessed 28 August 2012)

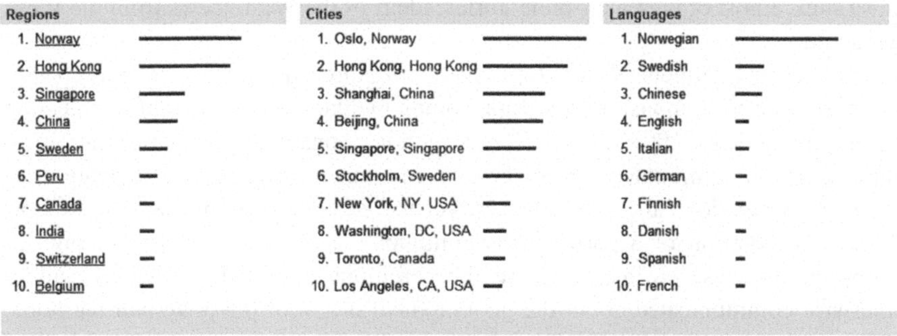

Fig. 19.5 The top regions, cities, and languages of internet users searching for the term "Liu Xiaobo" in 2010. Google Trends first identifies areas and languages from which the most searches are received using IP address information from server logs and records of the language of the Google site where the search originated. Bar lengths represent the number of searches for the chosen term relative to the total number of Google searches from the each location (Data from Google Trends, http://www.google.com/trends, accessed 28 August 2012)

basis is a poor indicator of people's actual access to information; per-capita figures in conjunction with an assessment of the degree of top-down information control or filtration provides a much richer basis of comparison.

Interestingly, data collected by Google Trends complicates this trend (Fig. 19.5). A higher percent of internet searches for Liu Xiaobo originated from Hong Kong, Shanghai and Beijing than from any other city in the world except Oslo. Aggregated at the national level, the greatest interest in Liu was manifested in Norway, but Chinese, both in Hong Kong and elsewhere on the mainland, demonstrated a keen interest in Liu and were willing to use the internet to purse that interest. It appears, however, that the Chinese language was less than a third as prominent in these searches as the Norwegian language was. Comparing these insights, it is evident that the trend surfaces of communication-in-space map several contrasting geographies, demanding further research.

Conclusion

It would have been easy to simply tell how it was that China was portrayed in the debates surrounding the 2010 Peace Prize, or alternatively to describe how Oslo City Hall served as a microcosm of Norway and the West. Drawing on the strands associated with infrastructure footprints and trend surfaces of information increases the complexity of the story and considering the topology of webpages as a mediated space of encounter complicates the picture even further. Yet if we overlook any of these geographical aspects of communication we run the risk of missing a clue that helps solve the mystery of precisely what "took place" in October 2010 between China, Norway, Liu Xiaobo, Thorbjørn Jagland, the Oslo City Hall, the ephemeral gathering in that Hall, a Chinese prison, a medal, countless websites, people surfing those sites, a host of laureates, their stories, ideas of the West, views from the West, and so on.

While a full treatment of these topics would require a book, with luck this chapter has been able to demonstrate the four complementary aspects of communication geography (Adams 2009, 2011). The inversions constituting communication-in-place, place-in-communication, space-in-communication, and communication-in-space seem paradoxical in the abstract, yet the examples offered here show how a full consideration of a particular communication event involves this range of issues. In this case we asked where the ceremony occurred, what was said or otherwise communicated about the geographical origin of the prize and the home of the recipient, what kinds of movements permitted messages to be linked to other messages, and finally what kinds of unevenness characterized the relationship between the Norwegians, the Chinese, and the rest of the world. A sense of the geographical role of the Nobel Peace Prize is incomplete without sensitivity to each of these four complementary aspects of communication geography. Naturally, many more questions could have been raised in each of the four sections and the questions themselves could have been pursued in much greater depth. Still, this brief illustration demonstrates in concrete terms the paradoxical relations that attend communication geography: space versus place and containment *of* communication, on the one hand, versus containment *by* communication, on the other.

References

Adams, P. (1998). Network topologies and virtual place. *Annals of the Association of American Geographers, 88*(1), 88–106.

Adams, P. (2009). *Geographies of media and communication: A critical introduction*. London: Wiley-Blackwell.

Adams, P. (2011). A taxonomy for communication geography. *Progress in Human Geography, 35*(1), 37–57.

Adams, P. C., Hoelscher, S., & Till, K. E. (Eds.). (2001). *Textures of place: Exploring humanist geographies*. Minneapolis: University of Minnesota Press.

Andersen, I., Gamst, M., & Sperstad, K. T. (2010). NTNU-professor: Totalt misforstått fredspris. *Verdens Gang* (October 8, 2010).

Austin, J. L. (1975). *How to do things with words* (2nd ed.). Cambridge: Harvard University Press.

Bell, J. E., & Staeheli, L. A. (2001). Discourses of diffusion and democratization. *Political Geography, 20*(2), 175–195.

Christophers, B. (2009). *Envisioning media power: On capital and geographies of television*. Lanham: Rowman & Littlefield/Lexington Books.

CIA 2011. CIA World Factbook. (2011). Online file at: https://www.cia.gov/library/publications/the-world-factbook. Accessed 3 Feb 2011.

Cosgrove, D. (2001). *Apollo's eye: A cartographic genealogy of the earth in the western imagination*. Baltimore: The Johns Hopkins University Press.

D'Arcus, B. (2003). *Boundaries of dissent: Protest and state power in the media age*. New York: Routledge/Taylor & Francis.

Dittmer, J. (2013). *Captain America and the nationalist superhero: Metaphors, narratives, and geopolitics*. Philadelphia: Temple University Press.

Foote, K. E. (2003). *Shadowed ground: America's landscapes of violence and tragedy*. Austin: University of Texas Press.

Graham, S., & Marvin, S. (2002). *Telecommunications and the city: Electronic spaces, urban places*. London/New York: Routledge.

Hubbard, P., Kitchin, R., Bartley, B., & Fuller, D. (2002). *Thinking geographically: Space, theory and contemporary human geography*. London/New York: Continuum.

Ingebritsen, C. (2002). Norm entrepreneurs: Scandinavia's role in world politics. *Cooperation and Conflict: Journal of the Nordic International Studies Association, 37*(1), 11–23.

Jagland, T. (2009). Nobel Prize award ceremony speech. (10 Dec, Oslo).

MacDonald, F., Hughes, R., & Dodds, K. (Eds.). (2010). *Observant states: Geopolitics and visual culture*. London: IB Tauris.

Malecki, E. J., & Wei, H. (2009). A wired world: The evolving geography of submarine cables and the shift to Asia. *Annals of the Association of American Geographers, 99*(2), 360–382.

Malinowski, B. (1936/1993). The problem of meaning in primitive languages. In J. Maybin (Ed.), *Language and Literacy in Social Practice* (pp. 1–10). Clevedon: The Open University.

Massey, D. (1993). Power-geometry and a progressive sense of place. In J. Bird, B. Curtis, T. Putnam, G. Robertson, & L. Tickner (Eds.), *Mapping the futures: Local cultures, global change* (pp. 59–69). London: Routledge.

Palla, G., Derényi, I., Farkas, I., & Vicsek, T. (2005). Uncovering the overlapping community structure of complex networks in nature and society. *Nature, 435*(7043), 814–818. doi:10.1038/nature03607.

Pew Foundation. (2011). Updated: Change in internet access by age group, 2000–2010. Pew Internet and American Life Project. Online file at: http://www.pewinternet.org/Infographics/2010/Internet-acess-by-age-group-over-time-Update.aspx. Accessed 3 Feb 2011.

Rasch, J. S., & Kvaale, V. K. (2010). Den tomme stolen er et enormt sterkt budskap. *Dagbladet* 10 Dec 2010. Online file at: http://www.dagbladet.no/2010/12/10/nyheter/nobels_fredspris/innenriks/liu_xiaobo/kina/14652114. Accessed 3 Feb 2011.

Reporters without Borders. (2010). Press freedom index 2010. Online file at: http://en.rsf.org/press-freedom-index-2010,1034.html. Accessed 5 Jul 2012.

Said, E. (1978/1995). *Orientalism*. London: Penguin.

Sharp, J. P. (2000). *Condensing the cold war: Reader's Digest and American identity*. Minneapolis: University of Minnesota Press.

Talbot, D. (2010). *China's internet paradox* (Technology Review). Cambridge, MA: MIT Press.

Thrift, N. (2008). *Non-representational theory: Space | politics | affect*. London/New York: Routledge.

Wainwright, J. (2007). Spaces of resistance in Seattle and Cancun. In H. Leitner, J. Peck, & E. S. Sheppard (Eds.), *Contesting neoliberalism: Urban frontiers* (pp. 179–203). New York: The Guilford Press.

Wittfogel, K. (1957). *Oriental despotism: A comparative study of total power*. New Haven: Yale University Press.

Zonn, L. E., & Aitken, S. C. (1994). Of pelicans and men: Symbolic landscapes, gender, and Australia's *Storm Boy*. In S. C. Aitken & L. E. Zonn (Eds.), *Place, power, situation, and spectacle: A geography of film* (pp. 137–159). Lanham: Rowman & Littlefield.

Zook, M. (2005). *The geography of the internet industry: Venture capital, dot-coms, and local knowledge*. Malden: Blackwell.

From Bolt to Brand: Olympic Celebrations, Tourist Destinations and Media Landscapes 20

Susan P. Mains

On Your Marks: To di World

> Olympic 100 metres champion and world record holder Usain Bolt has made dancing a hobby… Nuh Linga in Standard English means not wasting time, which Bolt displayed in his world and Olympic record run of 9.69 seconds.
>
> Bolt, who has celebrated his wins with dances, also gave a glimpse of 'To the World', a dance, composed by RDX. (Jamaica Gleaner 2008)

The date is August 8, 2008, which seems an auspicious moment from which to begin a global media tale: in the Chinese calendar, the number 8 is traditionally thought of as encouraging good fortune, being influenced by its similarities to the characters for achieving wealth. This date also marks the commencement of the Summer Olympic Games held in Beijing, which have now become iconic in Jamaica, having resulted in 11 medals being awarded to the national team. This success more than doubled the previous number of medals garnered by the Caribbean island at earlier Olympiads and substantially increased the profile of Jamaica as a serious contender in international athletic competitions. One of several stars emerging from this successful campaign was 100 and 200 m runner, Usain Bolt—who, by winning both events, confirmed his title as the world's fastest man. Bolt not only propelled himself into the halls of sports success through his athletic talents, but also cemented his reputation as a consummate entertainer. Playing with camera close-ups and following successful races with the "Nuh Linga" dances and now infamous "To di World" pose (or "Lightning Bolt"/"Bolting" pose)—an archery stance where he aims an imaginary bow at his next inevitable target (see Fig. 20.1)—Bolt cultivated a distinctive media persona. He demonstrated an awareness of, and ability to "speak"

S.P. Mains (✉)
Geography, School of the Environment, University of Dundee, Perth Rd, DD1 4HN, Dundee, Scotland, UK
e-mail: s.p.mains@dundee.ac.uk

S.P. Mains et al. (eds.), *Mediated Geographies and Geographies of Media*,
DOI 10.1007/978-94-017-9969-0_20

Fig. 20.1 The trademark "To di World" pose ("Usain Bolt, Anniversary Games, London 2013" Photograph taken J. Brichto at the London Anniversary Games. Licensed under Creative Commons Attribution-ShareAlike 3.0 via Wikipedia -http://en.wikipedia.org/wiki/File:Usain_Bolt,_Anniversary_Games,_London_2013.jpg#mediaviewer/File:Usain_Bolt,_Anniversary_Games,_London_2013.jpg)

with a wide international audience, while also meeting an ever hungry international media, continually on the hunt for "new stars" and appealing personalities to "flesh out" the more nuts and bolts stories of physical talent.

The tale of Usain Bolt's sporting success, is not, however, as straightforward as it initially appears. It is not simply a story of a rural Jamaican boy who has talent, does well at school, wins athletic competitions and goes on to become a global star—which, in itself, is quite exceptional. Bolt's story—or rather, stories—is a series of competing narratives, including: the affirmation of the human body as a unique "machine" able to move at remarkable speeds; the promotion of a relatable, humourous and talented mobile subject (i.e., Bolt); the creation of an (unofficial) national hero for the Jamaican populous; the provision of a hopeful story of success (despite all the odds and "humble beginnings"); the development of a regional and national tourism mantra; and, the inexorability of a Jamaican *and* transnational "global brand." In this chapter, I focus on Usain Bolt, to illustrate that central to complementary and competing stories of renowned success, there are interconnected media representations and practices that are sustained by the fuel of celebrity, and interwoven with multi-layered notions of place and consumption.

Geographers have been somewhat hesitant to engage with "celebrity" geographies—perhaps reflecting an apprehension about exploring what may be viewed as "less academic" areas of media studies (rather than perhaps thematic/genre/economic studies). Political and cultural geographers have, to some extent, undertaken a more direct engagement with famous political figures (e.g., via critical historiographies of national leaders), explorations of celebrity landscapes (e.g., Alderman's exploration of Martin Luther King, Jr.'s urban geographies (2008), fictional heroes in popular culture (e.g., Dittmer's discussion of Captain America (2012)), and have only more recently begun examining connections between entertainment celebrities and geopolitical causes (see, for example, Benwell et al. 2012). The topic of sports celebrities, and their related places, is, however, an area that offers significant potential in terms of understanding the ways in which these "exceptional" geographies become mapped onto and challenged through a variety of contexts—not only in relation to sporting events. Studies of the fictional film *Bend it like Beckham*, for example (Chacko 2010; Algeo 2007), begin a move towards connecting the "otherworldly" media spaces of David Beckham's football fame into the context of schools and teaching cultural geography, and this could be taken further by exploring how sports celebrity (as well as other forms of infamy) cuts across and negotiate convergent media to produce new gendered and racialised social and material spaces.

Although work in the area of celebrity and popular culture is somewhat limited in human geography, it has been a substantial area of research within cultural and media studies for some time. The intersections between celebrity, race and gendered imagery have formed an important feature of bell hooks' ongoing studies of popular US media culture for several decades (e.g., (1992)). Joshua Gamson's (1994) work exploring fame and celebrity during the twentieth century also helps to situate fame and its implications within specific historical contexts. The importance of the "familiar" figure and celebrity is examined in Turner's (2014) taxonomy of celebrity, the social roles of celebrities and how these intersect with media convergence. While Brockington (2014) specifically explores the role and impacts of celebrity advocacy in the context of international development. This work grapples with the ubiquitous, and yet continually slippery concepts of celebrity, fame and their related tensions. At the same time, there still appears to be significant potential for making connections between this research and their understudied geographies.

Celebrity mediascapes are "star maps" to specific spaces and practices of communication, control and consumption. As part of interrogating the mediated spaces of Usain Bolt, this chapter, therefore, also charts out the ways in which terrains of fame permeate day-to-day landscapes. These mediated landscapes not only serve as forms of entertainment, but also frame, filter and fight with popular ideas about identity, nationhood and financial autonomy/inequality. This chapter illustrates the ways in which certain skills, identities and landscapes are represented and valued (or excluded), and highlights that Usain Bolt's ubiquity in Jamaica (and beyond) is not only a story about an individual (which is what initial appearances suggest), but is part of a series of ongoing narratives negotiating unequal mobilities. These are stories that are intertwined with/through converging media and represent dynamic and complex cultural landscapes.

This journey into the promotion of Usain Bolt as a global brand, and Jamaica as a tourist destination, examines interwoven voyages—or media stories—each suggesting the desires, opportunities and fragilities embodied by celebrity, tourism, and consumption. I draw on two case studies in order to explore specific snapshots of Bolt's mediated geographies: (1) the Jamaican viewing and reception of Bolt's appearance at the 2008 Summer Olympic Games; and, (2) the 2012 Jamaica Tourism Board's highly publicized advertising campaign—*Pose*, *Speed* and *Stand Still*—featuring Usain Bolt. These tales hint at the distinctive and competing narratives that are a crucial component of celebrity identities and their associated geographies while offering a window into the dynamic relationships between place, representation and identity.

Get Set: Bird's Nest to Half Way Tree

> Traffic came to a halt in Half Way Tree (HWT) Square yesterday morning as hundreds of motorists and pedestrians paused from their usual duties to watch their own Usain Bolt smashing records and grabbing gold in the men's 100 m finals at the 29th Olympiad in Beijing, China.
>
> A television screen was erected at the bus stop at Mandela Park in HWT, across from the Jamaica National building and the transportation centre.
>
> Vendors, commuters, pedestrians and the patrons waiting outside the building all stopped what they were doing to view the 'big race'.
>
> Several men took seats in the middle of the road, but the police were too busy to notice because their eyes were stuck to the television screen. (The Gleaner 2014, 4A)

International athletics stars, or easily recognizable stars more generally, are often broadcast dislocated from their homelands or framed within a specific activity or event. Each of these figures, of course, hail from particular places—indeed, there may be more than one "homeland"—and, in those locations the experience of witnessing a local success story on a global scale can throw-up a range of emotions and re-imaginings of space and community. Rather than beginning Usain Bolt's media story from the striking setting of an Olympic Stadium or multi-media advertising campaign, I begin unearthing these interwoven mediated spaces "on the ground" at Half Way Tree, a central intersection and hub of entrepreneurial (and informal) activities in Kingston, Jamaica.

Half-Way-Tree encompasses some of the main arterial transportation routes from West to East Kingston and between Downtown and Uptown (the latter two areas often seen as signifying, lower income and higher income communities respectively, although the streetscape delineations are more complex). The area consists of many retail outlets, a relatively new transportation hub (opened in 2008), and high through traffic. One of the more popular shops found in the area's Tropical Plaza shopping plaza, is that of the Puma store—the first major sponsor and supporter of Usain Bolt's athletic career, and key destination for those seeking Bolt (and Jamaica) branded Puma sports products. In addition to the more mundane activities of commuting and shopping, Half Way Tree also became an integral site for the collective witnessing of Jamaican athletes participating in the 2008 and 2012 Olympic Games.

Prior to the highly anticipated finals of the 100 and 200 m races in 2008, representatives from the private sector erected a large multi-media screen in Half Way Tree adjacent to Mandela Park and the nearby Transportation Centre (in addition to screens in Mandeville and Montego Bay). These screens enabled passers-by to view the Olympic events as part of live broadcasts, which were being shown on national terrestrial television channels (Jamaica Information Service 2008). The Ministry of Information, Culture, Youth and Sport (a department of the Jamaican government), in partnership with Visual Vibes, the Sports Development Foundation, and the Jamaica Amateur Athletics Association, also erected similar screens in St. William Grant Park (Downtown Kingston) (Jamaica Information Service 2008), but it was the Half Way Tree site that became emblematic of the aspiration and celebration of Jamaican athletic success, which continued in later Olympic events. As news commentator Kwesi Mugisa (2012) noted in relation to Jamaica's athletes—Usain Bolt, Yohan Blake and Warren Weir's gold, silver and bronze medals (respectively) in the 200 m race at the London 2012 Olympic Games, Half Way Tree had (literally) cemented its reputation as a nodal point for experiencing such multi-media events:

> The vibrations were felt across the island, but perhaps nowhere more so than Half Way Tree, the hub of Jamaica's capital Kingston, where thousands of people gathered to witness the event just as they had for the 100 m final four years ago in Beijing.
>
> As it has throughout the Games, the giant screen erected in the middle of the bustling city centre served as a window on the Olympic world for the eager spectators crammed into the square, clad in various shades of black, green and gold, and waving flags of all sizes.
>
> Considered in many ways the heart of the city, the venue pulsated on Thursday. The sound of blaring plastic horns and clanking saucepan lids, now customary instruments of celebration, gradually rose as the sea of onlookers willed the trio on before triumphantly displaying three-finger salutes.

The Half Way Tree screenings provided the opportunity to collectively experience highly emotive events in a period in which the use of mobile technologies had become increasingly individualized (although household television ownership in Jamaica is slightly below the Caribbean average, mobile phone use and personal computers are relatively higher (although still below that of other Caribbean states) (World Bank 2014). This interactive viewing also pointed to an important spatial process—the significant role of convergent media (including the ongoing importance of television) and the use of technology in mediating, and placing, popular national narratives. Although it had been suggested that increasing mobile phone ownership and the diversity of internet usage may undermine a sense of shared media experience, connections made between media formats during the Olympics illustrated the opposite. For example, phone calls to family and friends exchanging stories of former school friends now competing as athletes (school athletics receives substantial alumni and local media support, including the annual 'Champs' competition (Dyachkova 2014)); terrestrial news footage of Jamaican sports figures, which was discussed on social media sites such as Facebook, and later, Twitter; US and UK television news shown via cable in Jamaica; front page stories of athletic success via national print newspapers (e.g., *The Gleaner* and *Jamaican Observer*) which were responded to and revised in relation to online comments, and various related social media links on Youtube, Twitter and Facebook. Bolt himself has been an enthusiastic user of various media formats, and has demonstrated a nuanced understanding of his media persona, reflected in: the canny cultivation of his To di World pose (reproduced by large numbers of Kingston, Jamaican and international fans); his discussion of Jamaican Dancehall music in media interviews; and his role as a celebrity endorser for Puma, Digicel, Virgin Media, Gatorade, Hublot, Visa, and most recently, Soul Electronics (Badenhousen 2012). Each of these contexts have intersected to reinforce a sports "brand" that is grounded in the image of Bolt's celebratory victory salute—a thoroughly embodied depiction—while at the same time, successfully promoting the notion of a transcendent body that knows no (or very few) limits (physical or virtual).

The large screen format described above is intertwined with the spectacle and performances of sports events—e.g., the physical and social performance of star athletes, audience chants and street dances, and radio and television commentaries by sports pundits—to (literally) bring together bodies and disparate viewers via an idealized notion of community. Boyle (2014, 5) notes the contradictions of increasing content choice and ubiquitous sports events:

> In an age when technological change, in part unleashed through a lighter regulatory framework, is restructuring how people watch and think about television, the ability of sports at major events to pull together fragmented audiences remains compelling, even if they disagree over what they appear to be watching, and are engaging in second or even third screen activity and conversations.

The ongoing importance of viewing an event as part of a "group," and in real time, as evidenced at Half Way Tree, suggests the ongoing significance of television broadcasting both for the (re)production of celebrity figures and for facilitating a shared experience via the context of sports. As Boyle (2014, 7) reflects in relation to the 2014 football World Cup:

> The 2014 World Cup in Brazil offers an opportune moment to reflect on the continuities as well as the changes that characterize television's relationship with the global game of football. The dazzle of screens can at times lead us to become too technologically determinist in linking technology and social change. It is worth emphasizing that it is the interaction of technology and its social usage and context that drives change, often in very unpredictable ways. The television screen is not going away, and will remain at the centre of the networked media sporting environment for some time yet.

It is also interesting to note the convergence of different sports events through celebrity endorsement: Usain Bolt featured prominently in Visa's recent Brazilian World Cup sponsorship advertisements; he is shown travelling from the athletic training ground at the University of the West Indies-Mona (Kingston, Jamaica) to the Maracanã football stadium, via Copacabana, all apparently courtesy of visa's "contactless" payments (the advertisement, shown widely on television broadcasts during the event, is linked at the end of the chapter (Tracktor 2014)). But, while

Bolt may have "transformed" from a sprinter into a world class footballer—and the beaches and favelas he dances in while effortlessly chasing a football metamorphose into slightly more picturesque versions of their everyday selves—the surrounding inequalities of Jamaica and Brazilian urban life are left fundamentally unchanged by, as Boyle puts it, the "dazzle."

It would be a mistake, however, to suggest that media images of such events (and their celebrity promoters), reflect a merely superficial depiction of place. As Boyle (2014) also hints at, social change and the increased awareness of exclusionary practices can be part of unanticipated or strategically coordinated multi-media campaigns. Such activities were noted in the run up to and during the Brazil World Cup through, for example: social network discussions critiquing FIFA (Watts 2014), inequality and host nations' overstretched resources; media coverage of street protests in a variety of formats; politically strategic tweets via various organizations and national political leaders; and, international access to a range of (previously, largely domestic and nationally oriented) television and radio sports coverage of football related activities. At the same time, the public shared viewing experience is not only a recent phenomenon: for example, McCarthy (2001) notes the impacts of televisions being introduced to US taverns in the mid-1940s, and the various strategies and advice disseminated in order to maximize this "new" social and business experience.

While analyzing the use and potentially transformative impact of large screens, Papastergiadis et al. (2013, 2) explore a range of "alternative models" of usage including: public space broadcasting, civic partnership, and art. This discussion suggests a particularly relevant question: how did the Half Way Tree screenings/celebrations represent, construct and challenge conventional (i.e., prevailing and commonplace) mediated geographies? On the one hand, the screenings were removed from the physical constraints of living rooms and bars, and offered a more fluid physical space: this occurred literally, in the midst of traffic and pedestrians who challenged traditional vehicular/pedestrian regulations and policing, and produced public sites of collective triumphant expression. Street viewers were embodying celebration as well as engaging in active spectatorship. And on the other hand, the collective and performance spaces of Dancehall music—created in streets, gardens and nightclubs throughout the island—appeared to be recreated and re-channeled through these public demonstrations of sporting success. Dancehall is a Jamaican musical genre emerging in the 1950s/1960s, growing through the following decades, and merging local sound systems, dance, the increasing visibility of DJs (known as "selectors") and digital media production. It is also a musical form that is viewed as resisting the status quo and repressive divisions between private and public, particularly in urban Jamaica, where dancehall events have been hosted in Kingston streets (for a more detailed discussion of dancehall spaces see Stanley Niaah 2010). This connection between what initially appear to be quite distinct public spaces (e.g., a busy city street and a nightclub or a sports arena and a Dancehall "bashment" (street party)), can also be seen in the popularity of Usain Bolt's Olympic Nuh Linga dance described at the opening of this chapter, where Dancehall (often viewed as disruptive and morally challenged by various Jamaican public figures) becomes the local (and global) signifier. As part of the Olympic Games, the Nuh Linga dance (released in 2008 by the Dancehall performer, Elephant Man), resonated with Jamaicans via international broadcasts: it communicated a recognizable performance that had successfully "gone to the world," while marking Bolt as a unique figure who was legitimizing Dancehall through the globally viewed sports arena, claiming a space for what was widely regarded as a working class medium. Although fleeting (literally!), Bolt and other Jamaican athletes, were "representing;" hailing Jamaica through dance and in the process merging informal and formal mobilities. To watch Bolt performing his brief post-race dance celebration in Asia while being part of an audience in the Caribbean—a distinct transnational/transcontinental embracement of Dancehall—provided a border crossing moment that escaped the limitations of pre-packaged corporate dictated celebrity endorsements. It may have become co-opted into commercial and tourist campaigns soon after, but Nuh Linga connected Beijing's tracks to Kingston's streets via spontaneous and evocative performances.

In tandem with these street-based activities, Jamaican national terrestrial news channels, TVJ and CVM, aired footage of Half Way Tree, thus self-referencing the celebratory sporting and street events, and further consolidating a wider sense of national success. As noted in Mugisa's (2012) comment above, the notion of Kingston, and in particular, Half Way Tree, symbolising the "heart of the city" and the nation, became reinforced through print, television and digital news coverage. And that heart was viewed as being tangibly demonstrated through the emotive performances of Half Way Tree screening participants and their imagined virtual audiences (see, for example, Serwer 2012).

At the same time, the highly visible and widely broadcast Half Way Tree screenings reproduced the promotion of Kingston's urban spaces as central to the story of Jamaican nationhood and new national heroes, such as Bolt (and in contrast to his and other successful athletes' rural upbringings). A story of the city as the catalyst for success and recognition has also been mirrored in contemporary popular culture, for example, the iconic Jamaican film, *The Harder They Come* (Henzell 1973) and Dancehall music, the latter of which is a genre ubiquitous in the island and in the nightclubs where feted sports stars may unwind. Dancehall is largely sold to Jamaican audiences while "old school" reggae and idealized rural and coastal landscapes are promoted to overseas tourists. However, for many islanders, that urban space of inclusion and opportunity seems far removed from everyday life: with limited physical and social mobility, increasing austerity measures under IMF designed loan agreements, and longstanding concerns about corruption and violence, the screenings provided a temporary (if welcome) escape. Half Way Tree was "halted" temporarily (and ironically, given critiques of the overstretched public transportation centre based nearby), but the fundamental inequalities it briefly masked—or in some cases highlighted for those who could not access a television to view the events—may have been the key catalyst for such vociferous joy.

Ready: Place, Pace and Representing Jamaican Tourism

> In tourist promotions, places are represented in a kind of patois. Destinations are referred to, epigrammatically, as "the sunshine coast," "the city of discovery," "a world of difference, "cultural capital" .... (Hughes 1998, 19)

Tourism is ubiquitous in the Caribbean, both in terms of its physical manifestations and in relation to public discussions regarding the role and importance of island industries. Tourism in Jamaica provides one of the key earners of foreign exchange. At the same time, however, the island has been struggling with growing debt payments (Government Debt was 138.9 % of GDP in 2013 (Trading Economics 2014)), high levels of violent crime, and stereotypical media representations that oscillate between chaos and paradise (Thomas 2011; Mains 2008). In an effort to build on his widespread recognition, the Jamaica Tourist Board (JTB) has directly enlisted Usain Bolt to promote an idealized image of island life as part of a series of terrestrial, cable and internet based advertising campaigns. *The Times of India* (2012) commentary below, illustrates some of the expectations surrounding the outcomes of a Bolt endorsed tourism campaign.

> Jamaica expects the electrifying Olympic wins of the world's fastest sprinter Usain Bolt will enhance its image and help in wooing more foreign tourists to the picturesque Caribbean country.
>
> "The impact of Bolt's performance at the London Olympics will have an effect on the island for many years and will help us to attract more tourists," Jamaican Tourism Minister Wykeham McNeill said.
>
> McNeill said the interest in Jamaica, known for its serene butterscotch beaches with some of the best seafood eateries in the world, has reached its zenith since Bolt successfully defended in London his titles obtained in the 2008 Olympic Games in Beijing (*The Times of India*, 16/08/2012).

Building on an already established international profile and recognizable brand, Bolt's participation in tourism advertising was enthusiastically embraced by the JTB, culminating in a series of three advertisements on European outlets—*Pose*, *Speed* and *Stand Still*—released on 1 February, 2012 in the final run up to the games. Interestingly, these were not new commercials: each of the advertisements were already available in North America via television and online formats from 2010 (released as *Stop*, *Speed* and *Pose* as part of the "Once You Go, You Know" campaign (JTB 2010)), but the opportunity to re-release them in the run up to the London events apparently made strategic sense in terms of capitalizing on a popular promotion and making it available to a new regional market that would be tuning in to watch Bolt's races (to view the commercials see the links at Haagen 2010a, b, c).

The first of these promotions, *Pose*, travels around the island featuring a range of residents, including: musicians, a barber, a policewoman, fishermen, schoolchildren, and finally, Usain Bolt. All of the people are shown performing the "To di World" pose. The second advertisement, *Speed*, follows Usain Bolt as he runs around Jamaica (in 30 s!), through a range of dramatic landscapes, concluding with a brief pause to enjoy a burnished sunset. The final, and third video, *Stand Still* (also previously titled, *Stop*), features Bolt setting out running in pre-dawn light, his running "interrupted" by stopping at a range of striking locations throughout the course of the day—rivers, beaches, and waterfalls—then finishing with a commentary to camera: "On any other island, I'd probably be a marathon runner." Each of the advertisements features Bob Marley's *One Love* soundtrack, and the Jamaica Tourist Board tag line, "JAMAICA Once you go, you know."

The 2012 JTB advertising campaign provides an insightful example of the strategic ways in which celebrity and representations of place can be mobilized and disseminated in order to maximize a public profile and reach a wide international audience. In addition, although these images are largely targeted towards overseas Anglophonic audiences, they do provide an opportunity to more closely examine how celebrity, tourism and consumption are explicitly (and indirectly) intertwined. While Williams and Deslandes (2008) note that the island's infrastructure and proximity to the US, have given Jamaica a significant presence in terms of foreign direct investment in the tourism sector, the success and awareness of Jamaica as a "brand" (along with its associated landscapes and attractions) has also been seen as a critical factor. Jamaica is a relatively well "recognized" destination, regarded as punching above its weight in terms of international familiarity with the island for potential tourists who have never visited the Caribbean (even if that knowledge is fairly limited (e.g., via music (reggae), coastal landscapes (beaches and turquoise sea) and cuisine (tropical fruits, jerk chicken and rum)). Despite this recognition, there has been an increasing call for greater attention to be paid to illustrating a broader range of possible Jamaican experiences and destination attractions (particularly in relation to other Caribbean destinations) (Miller and Henthorne 2007). It has been argued that an awareness of potential tourist destinations is just one (if an important) component in the journey towards undertaking a physical trip to a new location, and building on familiarity with a unique "spin" is an additional crucial factor, which may lure a visitor to one island rather than another. As Qu et al. (2011, 465) state:

> Consumers are generally offered various destination choices that provide similar features such as quality accommodations, beautiful scenic view, and/or friendly people. Therefore, it is not enough for a destination to be included in the evoked set; instead the destination needs to be unique and differential to be selected as a final decision. From this perspective, the concept of destination branding is critical for a destination to be identified and differentiated from alternatives in the minds of the target market.

The use of Usain Bolt as a central figure in tourism advertisements produced on behalf of the JTB in recent years complements this notion of recognisability and familiarity, while adding the sought-after unique dimension to the destination story. Bolt becomes emblematic of Jamaica, the "frontrunner" in Caribbean tourism: he is internationally recognizable and associated specifically with the island, just as his frequently hailed precursor, Bob Marley, has also become part of that distinctive island narrative, reinforced through the advertisements' soundtrack, encouraging viewers to "feel alright." Indeed, music is central to the tone of the advertising campaign and once more connects to Bolt's publicly declared enjoyment and celebration of Jamaican musical artistes.

The over reliance on one or two specific celebrity athletes in order to represent more inclusive notions of nation and place is a problematic practice that has also been highlighted in relation to the 2000 Sydney Olympics (Gardiner 2003; Knight et al. 2007). During this 27th Olympiad held in Australia, Cathy Freeman, who won the gold medal for the 400 m, came under intense media scrutiny and veneration as being symbolic of both a "progressive" Australia that actively embraced Indigenous athletes (such as Freeman), and the potential for political protest supporting Indigenous calls for social justice (Elder et al. 2006). Just as Bolt has been promoted as part of the "Out of many one" Jamaican national narrative that alludes to racial harmony, Freeman has also been hailed as proof of a successful reconciliation between White Australians and largely marginalized Indigenous communities. As Elder et al. (2006, 182) point out, however, this re-imagined national story glosses over more contradictory tales of nationhood and mobility:

> In this context it is quite easy to see what was so important about Freeman's gold medal win: her sporting victory could be understood as a reflection of Australia as a progressive and liberal nation. However, what is less obvious is how, in the context of Australia hosting an international sporting event, it became possible for non-Indigenous people to figure themselves as pro-Indigenous or pro-reconciliation, and yet deploy discourses that further marginalized Indigenous peoples and their needs.

While Freeman and Bolt fit into desirable stories of success, diversity and belonging, the dominant narrators of these stories (athletics organisations, national governments, media corporations, tourist boards), systematically eclipse the stories of pervasive inequality, discrimination, resistance and social immobility on which they rely.

Returning to Jamaica, the recognisability of Usain Bolt and Jamaica as brands is also not without its limitations (which is significant given the JTB's active promotion of "Brand Jamaica"). Just as residents of other Caribbean islands critique depictions of Jamaica as representative of all islands in the region, an over emphasis on Usain Bolt may be seen as resulting in an under representation of other successful athletes. One particular example of this disparity can be seen in relation to Shelly-Anne Fraser-Pryce, a Jamaican athlete who also won the gold medal for the 100 m in both Beijing and the London Olympic Games (as well as additional medals in other events), but features in far less sports media coverage and high profile product endorsements. Although Fraser-Pryce does receive sponsorship from GraceKennedy, Digicel, Sagicor Jamaica and Nike (with a significant amount of that investment being put towards the *Pocket Rocket Foundation*, a non-profit organization she set up to support school-age athletes in Jamaica (Pocket Rocket Foundation 2013)), the scope of that sponsorship appears to be on a significantly smaller scale to that of Usain Bolt, and may also reflect wider gender inequalities in relation to income levels and media attention given to female sports stars relative to their male counterparts (McRobie 2012). Despite significant (if uneven) domestic support for female and male members of national sports teams, and the highly popular profile and visibility of Fraser-Pryce within Jamaican media, this has not been translated into a personalized global "brand" to the same extent.

Miller and Henthorne (2007) explain that although Jamaica is a country brand with a large public presence—i.e., a nation-state that has produced marketing icons that a wide audience easily associate with particular activities, landscapes, attitudes, a distinctive national "story"—more could be done by the national government to highlight its distinctive features and shift away from a more generalized (and somewhat vague) notion of the island's attractions (and one which could also incorporate greater gender diversity). This is a representational challenge that has been ongoing: for example, in relation to another recent online advertising campaign (in conjunction with *One love* Marley inspired advertisements) Miller and Henthorne (2007, 57) note:

> Jamaica is one of the most important tourism players in the region; a country whose customs, cuisine, and music are frequently imitated in the region; and a country often noted for the continuous innovation of its tourism product and the creativity of its marketing efforts. Perhaps Jamaica's slogan of "Explore Jamaica" is intended to impress the idea that there is much worth exploring while on vacation in Jamaica. If so, the message is subtle at best, and in contrast to alternative slogans for Jamaica that speak more.

While the desirability of focusing so much attention and financial investment into selling Jamaica to overseas visitors may be questionable from the point of view of protecting "hidden gems" for local residents and pandering to external markets, the problem of sweeping place stereotypes and ingrained repetitive landscape narratives is a relevant one for understanding the limitations of tourism-related media and mediated geographies (Mains 2014). The desire to create, represent and sell a unique tourist destination has dominated both media and material developments along Jamaica's coastal landscapes for many decades. Usain Bolt's insertion into this destination narrative may also be seen as an opportunity to shift away from simply depicting idyllic landscapes (e.g., through a bold personality and witty humour), but it remains to be seen whether this will translate into more inclusive and diverse Jamaican media geographies and/or a strengthening of Jamaica's tourist appeal.

In terms of the *Pose*, *Speed* and *Stand Still* advertisements, it is worth paying closer attention to their context and themes. Given that Usain Bolt is a central character, it is hardly surprising that mobility is a key theme. In *Speed* Bolt travels through Jamaica at an increasingly rapid pace—at "superhuman" speed—but even he must slow down to admire the striking sunset at the film's end. This idea of a transformative experience of time and place is also played on in the recent Puma campaign featuring Bolt and (the increasingly popular) Jamaican musicians, No-Maddz. Hailed as "The World's Fastest Band," No-Maddz appear in promotions as the characters Rocker, Flex and Groove, "from Jamaica, the world's fastest country" (Macleod 2011). The band is shown paralleling Bolt's commercial appearances: running with instruments in several adverts while promoting Puma's Faas running shoes (Bond 2011). Although at times juxtaposed with the slower beat of Marley's *One Love*, these images of tremendous speed also challenge the notion of Jamaica functioning at just one "laid back" tempo. *Stand Still*, however, reinforces the idea of a visit to Jamaica necessitating a "slow down" in order to "feel alright" and savour the view, an apparently enjoyable compulsion to which even superstar athletes must succumb.

Speed, and the pace of life depicted by tourist destinations such as Jamaica, represent contradictory media geographies and socio-spatial relations: images depicting the consumption of decelerated days in hidden spaces are seen as desirable commodities in marketing campaigns, while at the same time lively and exciting adventures in dramatic and/or highly visible beach front landscapes are also promoted. Such contradictory representations have been noted in other forms of tourism; Yiping (2003), for example, notes the conflicting sentiments between conservation and change in the context of heritage tourism in Singapore and Hong Kong, and points to the ongoing tensions, compromises and reinscriptions of destination spaces. A key problematic of these contradictions is the decision-making process around which representations, experiences and spaces are privileged and supported, and who gets to make that decision. For many Jamaicans this process has failed to include the broader population and has instead favoured a political and economic elite, often with questionable (or even dire) consequences (as can be seen in recent discussions about the Jamaican Government's purchase of an indebted tourist attraction, *The Outameni Experience* (Gordon 2014)).

Contradictions also permeate the process of representation and consumption in the context of tourism, enjoyment and labour. Through a Lacanian analysis, Kingsbury (2005, 125) explores the ways in which pleasure and pain are embedded in each other and the commodification and promotion of tourist- object-worker relations: "Tourism fantasies provide answers with a schema of desirable objects such as friendly Jamaicans, hotel managerial positions, exotic culinary fare, 'best employee' of the month awards, and white sandy beaches." When the fantastically "real" aspects of these emotive tourist experiences are ruptured by things like harassment from street vendors or muggings, Kingsbury points out a failure on the part of many observers, including the Jamaican Government, to engage with the intrinsically political and unequal nature of tourism and the production of destination-related interactions and desires. In a similar vein, when tourist promotions concentrate on idyllic beaches for tourist consumption—such as those shown in the *Stop* commercial—challenges, such as those via policy proposals from community organisations, street protests by craft vendors, critical documentary films and blogs, and investigative local news stories, provide important methods for chipping away at the cheerful sanitized depictions shown via the tourism industry—representations that are often actively hidden from overseas visitors. This unmapping and re-mapping of Jamaica's coastal landscapes highlights the numerous means by which local residents are excluded from "tourist" spaces (e.g., entry fees, fences, security guards, racial profiling), but are nonetheless an essential "absent" presence (see, for example, Virtue 2012). As well as affording spaces to pause, stretch and relax, Jamaican beaches, therefore, symbolize constraint, forced removal and frustration for a significant number of the island's population. In a landscape where the GNI per capita is $5,220 (World Bank 2013) and the average tourist income is several times that amount, for many island residents tourism advertisements—which are also accessible in Jamaica via widely available US-based television cable channels—depict a mythical landscape, mobility and tranquility that is "out of this—and their—world."

Usain Bolt also symbolizes some of these contradictions—his success is partly the result of talent, institutionalised athletics training through the Jamaican school system, parental support, international financial sponsorship and sustained hard work—he represents the pleasure of athletics success and the (usually hidden) pain of intensive training regimes. When he unsettles the dominant narratives of the relatively mute, disciplined, successful yet "unremarkable" athlete," e.g., by visiting nightclubs, publicly expressing an opinion or physically celebrating a race win, he faces the disciplinary gaze of popular media and (the inconsistent) chastisement of sports authorities (see, for example, Kessel 2009; Maguire 2008; McAllister 2008). Bolt maintains a tightly managed training schedule—which, despite being a media brand himself—excludes media interviews at critical points in the year (Paul 2012). At the same time as promoting the consumption of specific products (broadband, watches, holidays), Bolt himself is also a desirable object, subject and media presence. His charisma is viewed as endearing, but when viewed as being "off-message," is seen by some as becoming irresponsible (rather than refreshing). There is a tension between a desire for a sports celebrity with a new story and voice, and a range of narrowly defined narrative possibilities. Both Bolt and Jamaica, are represented as desirable sought after bodies, landscapes, and states of mind, yet also elusive in terms of speed, and in terms of "fitting" national and international narratives of success. These are brands that have become mapped onto each other, but at moments may be a slightly awkward fit. While Bolt's mobility enables him to "represent" Jamaica, in many ways he stands worlds apart from most Jamaicans who can rarely afford to stay in intensively marketed all-inclusive resorts or luxury villas found on the island. The Jamaican workers who feature in the *Pose* advertisement, briefly suggest the labour involved in the hidden tourism landscapes of salons, gardening maintenance, policing (and by association with Jamaica's current economic challenges: long shifts, below minimum wages, unaffordable housing, long commutes and erratic transportation). Despite the work places in *Pose* being central to the ongoing production of tourism landscapes and Jamaican brands, their time and spaces are often treated as less valuable than that of the tourist (and, therefore, more easily compromised) (UNEP 2014). Even in recent attempts at more socially aware "slow tourism"—to which these workers' labour is also integral—it is unclear to what extent the pace of work for waiters, villa room cleaners and taxi drivers (and the expectations to meet increasingly demanding service needs) is significantly reduced.

Although exquisite beach sunset views, tree framed country roads and imposing "great" houses illustrate the importance of place, the representation of time is key. Pace and speed of movement are played with in order to draw the viewer's attention towards a specific form of holiday experience, embodied by Usain Bolt and focused around a narrative of inevitable relaxation and rejuvenation (*aka* feeling alright). It is also movement and iconography of the male body that is central to this narrative (although the physical landscape is not an entirely passive backdrop, given its power to affect a slowing down of what appears to be unstoppable speed). The JTB advertisements also build on an ongoing fascination with Jamaican athletic success from overseas audiences, demonstrated through a range of documentary features, such as: *Why do Jamaicans Run so Fast*? (2009); *Usain Bolt*: *Running Man* (2012); and *Usain Bolt*: *The Fastest Man Alive* (2012).

Another, more recent example, that both reinforces and challenges the superhuman heroic figure of Usain Bolt, particularly the trademark solitary Lightning Bolt pose, is the 2014 Virgin Media advertisement, *Introducing the Bolt Family*, which was released via television, print press and online. In these advertisements (which were later banned in the UK due to unverified and overly vague claims about broadband speed (Reynolds 2014)), Bolt stars as several characters within the same family, including a grandfather online chatting with his (significantly younger) "girlfriends," an infant in a high chair and the mother who "awkwardly" states she can watch her favourite athlete—Mo Farah (this can be viewed at the link below attached to Weinstein 2014). The advertisement can be read as a humourous take on Bolt's ubiquitous media presence, and a nod towards the ways in which media convergence (such as the interconnection between television, print and online content, and the interweaving of sports, arts and musical celebrities), has facilitated this process. Although receiving a generally positive response, soon after the airing of the advertisements, several social media commentators expressed unease and more direct criticism of this perceived emasculation of Bolt's persona. Unhappy viewers couched the criticism within a context where Bolt was seen to be willing to sell his assured heterosexuality and (narrowly) defined masculine identity by being depicted as a woman wearing a dress in order to service the selling power of a multi-media corporation. This same corporation was being critiqued for failing to appropriately acknowledge the talents and successes of a global sports star, and Usain Bolt himself was attacked for having "sold his soul" in order to generate income: "the fastest man on the planet has since been blasted by those—most notably his fellow Jamaican countrymen—who claim the advert is offensive and that Bolt has 'sold out'" (Isokariari 2014). Although Usain Bolt's fame affords him a significant amount of social mobility, this was, and still is, publicly policed in relation to the means by which he performs gender (being viewed by some as overly feminized) and race (falling into the trap of black male celebrities mocking themselves to be seen as entertaining). Interestingly, the Jamaican newspaper, *The Gleaner* (2014) noted that to a certain degree Bolt's "entertaining" presence in the Virgin family advertisement overshadowed the product he was selling. These images could be hailed as falling into the limitations of Black identities being utilized as simply entertainment—what Kincaid calls, a "diversion"—to the extent that we lose the marketing motivation that has driven this representation and focus on the humorous element instead (Snell 1997, 2). Alternatively, this commercial highlights the seamless intertextual nature of media geographies; where both product, systems of representation, and products of personality/event promotion become inseparable.

Concerns about the ways in which Bolt's "cross-dressing" ideas were being enabled also suggests anxieties about the inexorability and interconnectedness of multi-modal media, the latter of which apparently facilitates the speed of change in attitudes towards gender, in tandem with the marketing of new media technologies (e.g., through multi-platform advertising campaigns utilized by companies like Virgin Media). This process of policing the performance of identity and humour, has not, however, gone unchallenged—*The Gleaner* (2014) sports commentary explains:

> But for some of us, Bolt in a dress is a bit too much so there has to be some sinister motive behind his decision to do so. What I have found is that there is a lot of insecurity among us as a people. The 'grandpa' Bolt is portrayed as a dirty old man with many girlfriends, yet we hear nothing about that. Instead we focus on the dress as if to suggest that Bolt in a dress defines who he is.

Such challenges, however, also point to the significant ways in which the intersection of local and national celebrities can coalesce and conflict with international representations and mediated geographies of those very same icons and the spaces with which they are associated (whether those are the track, the home, the beach, or a city intersection), and in the process the latter are also reproduced, recreated and transformed.

Go! Jamaica Haffi Run Wid It?

> Being in the spotlight, I too have had my fair share of run-ins with the media . . . This is one of the hazards of fame, but most of the time I'm fine with the attention. (Bolt 2010, 202)

Celebrities such as Usain Bolt, offer a window into, and a mirror reflecting, the ways in which we reify, reproduce and challenge notions of identity and place. Such figures become nodal points, and part of key places and events around which we celebrate national success stories, promote notions of unstoppable bodies, places to escape to, and create new global "living legends." These media geographies exist as part of street celebrations, internet advertising campaigns, and internationally broadcast sporting events. This chapter has been a warm-up run on the opening leg of a much longer journey towards critically engaging with the material, social and representational mediated geographies of fame and in particular the interconnected production of Usain Bolt and Jamaica as widely recognisable brands.

Usain Bolt is a hugely popular figure in Jamaica and beyond, and his incorporation into Jamaica's tourist campaigns, as well as other branding exercises, is strategic (and in many ways, appears to be inevitable). Representations of celebrities and the places with which they are associated, are not, however, straightforward and often become interwoven with tensions and conflicts around which success stories should be told and and/or challenged. Sports events and holidays are part of emotive events—celebrations, frustrations, unique adventures, time spent with families, a break from the routine of work in collective gatherings—all of which are reflected, reproduced, adapted and challenged via media geographies. Even someone as popular as Bolt—whose charitable organization, the *Usain Bolt Foundation*, has invested in numerous projects and training programmes for disadvantaged Jamaican youths, and material improvements for Jamaican schools (and has, to varying degrees, changed the material and social geographies of those specific locales)—finds his identity and relationship to Jamaica constantly policed and negotiated by a range of audiences. His highly visible brand, and his now significant economic wealth, has been accompanied with the opportunities and challenges of the public gaze(s). This monitoring has been noted above, for example, through the discussions of gender around the Virgin Media *Introducing the Family* advertisement (or in more recent media discussions around his alleged critique of the 2014 Glasgow Commonwealth Games (Williams 2014)). The Jamaican Government has also come under growing pressure to account for the ways in which Jamaica is depicted in overseas advertising, and through a range of media formats and forums has been urged to actively engage with a greater diversity of Jamaican experiences and spaces (e.g., Paul 2013; Figueroa and McCauley 2008). While on the one hand the process of branding can be seen as a declarative process, claiming identities and places and promoting them in particular ways, on the other it can be seen as an uncomfortable process that excludes identities, activities and regions who do not fit within narratives that are viewed as desirable. A recent call for research exploring the "Branding of Latin America," for example, notes that "Branding is also a painful act of marking, a declaration of possession and an enduring assignation of value"—it also often involves stereotyping of assumed place-based behaviours and national caricatures (Fehimovic 2014). While the corporate depictions of celebrities, tourism destinations and their associated brands, suggests a professionalization of managed landscapes, there is also an omnipresent "unofficial" subtext of violence via displacement (e.g., literally moving previous residents to allow for new tourist developments), silencing (e.g., ignoring or misrepresenting challenges to dominant nationalist narratives), and/or physical assault (e.g., through excessive training, police/public violence, work place fatalities, etc.). While this chapter has more closely explored a small number of specific examples associated with the more celebratory—and publicly visible—aspects of celebrity and place promotion in the context of Jamaica, these mediated geographies suggest the importance of further studies examining the interconnectedness of enjoyment, role models, mobility, disappointment, violence, recognition, dislocation and unequal power and economic relations associated with celebrity and tourism landscapes, and are important areas for future research. People, places and media move: a critical engagement with these mobilities is integral to a more nuanced and insightful understanding of the past, present and future landscapes of Jamaica.

References

Alderman, D. H. (2008). Martin Luther King, Jr. Streets in the south: A new landscape of memory. *Southern Cultures, 14*(3), 88–105.

Algeo, K. (2007). Teaching cultural geography with "Bend It like Beckham.". *Journal of Geography, 106*(3), 133–143.

Badenhousen, K. (2012). How Usain Bolt earns $20 million a year. *Forbes*, http://www.forbes.com/sites/kurtbadenhausen/2012/08/04/how-usain-bolt-earns-20-million-a-year/. Accessed 13 Nov 2014.

Benwell, M. C., Dodds, K., & Pinkerton, A. (2012). Celebrity geopolitics. *Political Geography, 31*, 405–407.

Bolt, U. (2010). *Usain Bolt 9:58: My story*. Kingston: Ian Randle.

Boyle, R. (2014). Television sport in the age of screens and content. *Television and New Media*. doi:10.1177/1527476414529167.

Brockington, D. (2014). *Celebrity advocacy and international development*. New York: Routledge.

Chacko, M. A. (2010). Bend it like Beckham: Dribbling the self through a cross-cultural space. *Multicultural Perspectives, 12*(2), 81–86.

Dittmer, J. (2012). *Captain America and the nationalist superhero: Metaphors, narratives, and geopolitics*. Philadelphia: Temple University Press.

Dyachkova, E. (2014). 'Champs' – the biggest deal in Jamaica. *International Association of Athletics Associations (IAAF)*, http://www.iaaf.org/news/feature/boys-girls-champs-jamaica-zharnel-hughes-issa. Accessed 13 Nov 2014.

Elder, C., Pratt, A., & Ellis, C. (2006). Running race: Reconciliation, nationalism and the Sydney 2000 Olympic games. *International Review for the Sociology of Sport, 4*(1/2), 181–200.

Fehimovic, D. (2014). *Branding Latin America: Call for papers*. http://www.academia.edu/8254545/Branding_Latin_America. Accessed 13 Nov 2014.

Gamson, J. (1994). *Claims to fame: Celebrity in contemporary america*. Berkeley: University of California Press.

Gardiner, G. (2003). Running for country: Australia print media representation of indigenous athletes in the 27th Olympiad. *Journal of Sport and Social Issues, 27*(3), 233–260.

Gordon, L. (2014 Nov 12). Investigate NHT purchase of Outameni. *The Gleaner*, http://jamaica-gleaner.com/gleaner/20141112/cleisure/cleisure2.html. Accessed 13 Nov 2014.

Hooks, B. (1992). *Black looks: Race and representation*. Boston: South End Press.

Hughes, G. (1998). Tourism and the semiological realization of space. In G. Ringer (Ed.), *Destinations: Cultural landscapes of tourism* (pp. 17–32). New York: Routledge.

Isokariari, M. (2014 Jan 7). Furore erupts over Usain Bolt dressing up as a woman for ad, *The Voice Online*, http://www.voice-online.co.uk/article/furore-erupts-over-usain-bolt-dressing-woman-ad. Accessed 13 Nov 2014.

Jamaica Gleaner (2008 Aug 17). Jamaica celebrates – it's dancing time! *Jamaica Gleaner*, http://jamaica-gleaner.com/gleaner/20080817/lead/lead4.html. Accessed 13 Nov 2014.

Jamaica Information Service (2008). *Big screen in St. William Grant Park for Olympics*. Jamaica Information Service, http://jis.gov.jm/big-screen-in-st-william-grant-park-for-olympics/. Accessed 13 Nov 2014.

JTB (2010). The Jamaica tourist board reveals new advertising campaign featuring "World's Fastest Man" Usain Bolt. *JTB News*, Jamaica: JTB.

Kessel, A. (2009 Apr 11). Usain Bolt still loves to party and still hates to train, *The Guardian*, http://www.theguardian.com/sport/2009/apr/11/usain-bolt-athletics-interview-records-100m. Accessed 13 Nov 2014.

Kingsbury, P. (2005). Jamaican tourism and the politics of enjoyment. *Geoforum, 36*, 113–132.

Knight, G., Neverson, N., MacNeill, M., & Donnelly, P. (2007). The weight of expectation: Cathy freeman, legacy, reconciliation and the Sydney Olympics – a Canadian perspective. *The International Journal of the History of Sport, 24*(10), 1243–1263.

Macleod, D. (2011). Puma Faas with Rocker, Flex & Groove. *Inspiration Room*, http://theinspirationroom.com/daily/2011/puma-faas-with-rocker-flex-groove/. Accessed 13 Nov 2014.

Maguire, S. (2008). Was Bolt celebration over the top? *Reuters*, http://blogs.reuters.com/world-wrap/2008/08/18/was-bolt-celebration-over-the-top/. Accessed 13 Nov 2014.

Mains, S. P. (2008). Representing cultural geographies of Jamaica: Rethinking tourism, place and identity. *Caribbean Geography, 15*(1), 1–13.

Mains, S. P. (2014). Fieldwork, heritage and engaging landscape texts. *Journal of Geography in Higher Education, 38*(4), 525–545.

McAllister, J. (2008). Coverage of Usain Bolt raises controversy in the community, *New York Daily News*, http://www.nydailynews.com/new-york/brooklyn/coverage-usain-bolt-raises-controversy-community-article-1.314101. Accessed 13 Nov 2014.

McCarthy, A. (2001). *Ambient television: Visual culture and public space*. Durham: Duke University Press.

McRobie, H. (2012). The gender equality Olympics: Medals and penalties so far. *50.50*, https://www.opendemocracy.net/5050/heather-mcrobie/gender-equality-olympics-medals-and-penalties-so-far. Accessed 13 Nov 2014.

Miller, M. M., & Henthorne, T. L. (2007). In search of competitive advantage in Caribbean tourism websites. *Journal of Travel and Tourism Marketing, 21*(2–3), 49–62.

Mugisa, K. (2012 Oct 10). Usain Bolt and the 'one, two, three' bring pride and joy to Jamaica. *The Guardian*, http://www.theguardian.com/sport/2012/aug/10/usain-bolt-jamaica-pride. Accessed 13 Nov 2014.

Paul, A. (2012). Profiling Usain Bolt... *Active voice*, http://anniepaul.net/2012/07/29/profiling-usain-bolt/. Accessed 13 Nov 2014.

Paul, A. (2013). Like diamonds and glass: The Barbados tourism authority vs the Jamaica tourist board. *Active Voice*, http://anniepaul.net/2013/03/18/like-diamonds-and-glass-the-barbados-tourism-authority-vs-the-jamaica-tourist-board/. Accessed 13 Nov 2014.

Papastergiadis, N., McQuire, S., Gu, X., Barikin, A., Gibson, R., Yue, A., Jung, S., Cmielewski, C., Yeong Roh, S., & Jones, M. (2013). *Theory Culture & Society, 30*(7/8), 325–341.

Pocket Rocket Foundation (2013). About the foundation. *Pocket Rocket Foundation*, http://www.pocketrocketfoundation.com/. Accessed 13 Nov 2014.

Qu, H., Hyunjung Kim, L., & Hyunjung Im, H. (2011). A model of destination branding: Integrating the concepts of the branding and destination image. *Tourism Management, 32*, 465–476.

Reynolds, J. (2014, September 17). Virgin media's Usain Bolt ad banned after BT and Sky complaints, *The Guardian*. http://www.theguardian.com/media/2014/sep/17/virgin-media-usain-bolt-ad-banned-bt-sky. Accessed 13 Nov 2014.

Serwer, J. (2012). LargeUp TV: Half way tree, Kingston reacts to Jamaica's 4×100-meter world record. *LargeUp.com*, http://www.largeup.com/2012/08/12/largeup-tv-half-way-tree-kingston-reacts-to-jamaicas-4x100-meter-world-record/. Accessed 13 Nov 2014.

Snell, M. B. (1997). Jamaica Kincaid hates happy endings. *Mother Jones*, (September/October) http://www.motherjones.com/politics/1997/09/jamaica-kincaid-hates-happy-endings?page=2. Accessed 13 Nov 2014.

Stanley Niaah, S. (2010). *Dancehall: From slave ship to ghetto*. Ottawa: University of Ottawa Press.

The Gleaner (2014, January 17). Why we can't see past Bolt's dress. *The Gleaner*, http://gleanerblogs.com/sports/?p=2304. Accessed 13 Nov 2014.

The World Bank. (2014). Mobile cellular subscriptions (per 100 people). *The World Bank*, http://data.worldbank.org/indicator/IT.CEL.SETS.P2/countries/1W-S3-JM?display=graph. Accessed 13 Nov 2014.

Thomas, D. A. (2011). *Exceptional violence: Embodied citizenship in transnational Jamaica*. Durham: Dike University Press.

Trading Economics (2014). Jamaica government debt to GDP. *Trading Economics*, http://www.tradingeconomics.com/jamaica/government-debt-to-gdp. Accessed 13 Nov 2014.

Turner, G. (2014). *Understanding celebrity* (2nd ed.). London: Sage.

UNEP. (2014). *Negative socio-cultural impacts from tourism*. United Nations Environment Programme (UNEP), http://www.unep.org/resourceefficiency/Business/SectoralActivities/Tourism/FactsandFiguresaboutTourism/ImpactsofTourism/Socio-CulturalImpacts/NegativeSocio-CulturalImpactsFromTourism/tabid/78781/Default.aspx. Accessed 13 Nov 2014.

Virtue, E. (2012 Jul 29). New beach policy seems lost at sea. *The Gleaner*, http://jamaica-gleaner.com/gleaner/20120729/lead/lead7.html. Accessed 14 Nov 2014.

Watts, J. (2014, July 7). Brazil 2014: World Cup where politics and social media invaded the pitch. *The Guardian*, http://www.theguardian.com/football/2014/jul/07/brazil-world-cup-politics-social-media-debate-fifa-football/print. Accessed 13 Nov 2014.

Williams, M. (2014). Stars defend Glasgow after Bolt embroiled in Games row. *The Herald*. http://www.heraldscotland.com/sport/commonwealth-games/cwg-news/stars-defend-glasgow-after-bolt-embroiled-in-games-row.24915517. Accessed 13 Nov 2014.

Williams, D. A., & Deslandes, D. (2008). Motivation for service sector foreign direct investments in emerging economies: Insights from the tourism industry in Jamaica. *The Round Table: The Commonwealth Journal of International Affairs, 97*(396), 419–437.

World Bank. (2013). *Data: Jamaica*. http://data.worldbank.org/country/jamaica. Accessed 14 Nov 2014.

Yiping, L. (2003). Heritage tourism: The contradictions between conservation and change. *Tourism and Hospitality Research, 4*(3), 247–261.

Filmography and Related Advertisement Links

Bond, F. (2011). *Puma Faas TV Spot #1: Blam! Blam!* US: Droga5. https://www.youtube.com/watch?v=Bp1KfEGUzZo. Accessed 13 Nov 2014.

Figueroa, E., & McCauley, D. (C-Dirs.) (2008). *Jamaica for sale*. Jamaica: Vagabond Media and Jamaica Environment Trust.

Galofré, M. (Dir.) (2009). *Why do Jamaicans run so fast?* Italy: Nice Time.

Haagen, J. (Dir.) (2010a). *Pose*. Jamaica: Jamaican tourist board. https://www.youtube.com/watch?v=aSBNrsYLLLw. Accessed 13 Nov 2014.

Haagen, J. (Dir.) (2010b). *Speed*. Jamaica: Jamaican tourist board. https://www.youtube.com/watch?v=1OJxRj9C3Gw. Accessed 13 Nov 2014.

Haagen, J. (Dir.) (2010c). *Stand still [stop]*. Jamaica: Jamaican Tourist Board. https://www.youtube.com/watch?v=u4GBEAzlifU. Accessed 13 Nov 2014.

Henzell, P. (Dir.) (1973). *The harder they come*. Jamaica: International Films Inc/Xenon Pictures.

Leiblang, G. (Dir.) (2012). *Usain Bolt: The fastest man alive*. UK: BBC Sport, http://www.bbc.co.uk/programmes/b0111xyh. Accessed 13 Nov 2014.

Martin, D. (Dir.) (2012). *Usain Bolt: Running man*. UK: BBC Sport, http://www.bbc.co.uk/sport/0/olympics/17352382. Accessed 13 Nov 2014.

Traktor.(Dir.) (2014). *Visa Europe: Bolt to Brazil*. UK: Saatchi & Saatchi, https://www.youtube.com/watch?v=5NF39wHAy9M. Accessed 13 Nov 2014.

Weinstein, B.(Dir.) (2014). *Virgin media: Introducing the Bolt family*. UK: BBH, http://www.campaignlive.co.uk/thework/1225874/. Accessed 13 Nov 2014.

Section VI

Transnational Media Production and Consumption

Development Communication, Popular Pleasure and Media Convergence

21

Julie Cupples

In 2007, the big question circulating in many villages, towns and cities across Latin America was whether one was with the *divinas* or *populares*. The question referred to a highly successful Argentinian telenovela,[1] *Patito Feo* (Ugly Duckling) about two groups of girls, one beautiful but mean (the *divinas*), the other "ugly" but good (the *populares*). It became a huge multimedia phenomenon, with many children and teenagers expressing their preference for the cruel *divinas* over the ethically minded *populares*. Experts of all kinds, including teachers, psychologists and feminist academics, expressed concern with its popularity and its perceived negative impacts. For these experts, the show gave out the wrong message, as it appeared to reinforce the idea that being successful meant behaving like the bullying *divinas* (see for example Bonavitta and de Garay Hernández 2011; Goy 2007).

Why such a show should resonate so strongly with young low-income girls across Latin America, who are forced to negotiate unequal relations of power in their everyday lives, requires careful analysis rather than a moral panic. Their interest in the show, which admittedly represents quite stereotypical versions of contemporary femininity and of the relationship between beauty, class and success as conventionally understood, should not be seen as evidence of their manipulation by a sexist and unequal society or as evidence that fans willingly participate in their

[1]The Latin American telenovela shares some similarity with the US or British soap opera but also has its own generic features. Many command large budgets, advertising rates and audiences. They get widely exported throughout Latin America and beyond. They are shown during the day and in prime time. Most run for 3–6 months and have "climactic, nation-paralyzing endings" (Soong 1999).

J. Cupples (✉)
Institute of Geography, School of Geosciences, University of Edinburgh, Drummond Street, Edinburgh EH8 9XP, UK
e-mail: julie.cupples@ed.ac.uk

S.P. Mains et al. (eds.), *Mediated Geographies and Geographies of Media*,
DOI 10.1007/978-94-017-9969-0_21

own ongoing subjectification. Indeed, much media dismissed as trashy or dangerous by elites can do quite important work in aiding subordinated populations to navigate the 'race', class and gender politics of their everyday lives, as a substantial body of scholarship in media and cultural studies has emphasized (see for example Ien Ang (1985) on *Dallas*, Lila Abu-Lughod (1995) on *Hilmiyya Nights*, Stephen Duncombe (2007) on Grand Theft Auto, Catherine Lumby (2003) on *Big Brother*, Kevin Glynn (2000) on tabloid television, and Yeddy Rivero (2003) on Colombian telenovela *Yo soy Betty la fea* (Ugly Betty)). Furthermore, media texts do not give out a single message, the hypodermic needle theory of media having long been discredited (see Gauntlett 1998), but tend to facilitate diverse readings and interpretations by active fans who bring different experiences, motivations and attitudes to bear on the text. Without doing empirical audience research or spending time engaging with social media, fan forums and bulletin boards, it is impossible for experts to know what it is viewers are doing with the media they consume and why they like it.

This media text circulated in the global South, where the question of development communication (often more recently referred to as information communication technologies for development or ICT4D) has been central to the development project since it began in the period after the Second World War. Since its emergence as a specific field of enquiry and development practice and policy, it has been subject to varying degrees of enthusiasm, suspicion and misunderstandings. In the 1950s and 1960s, when modernization theory was the dominant approach to development, communication was viewed by states and agencies as a means to change behaviour in line with modernizing and western agendas. Such an approach resulted in paternalistic forms of communication that told people how to behave and that were dismissive of traditional or indigenous knowledges. It was challenged in part by dependency theorists who began to question what they saw as the cultural imperialism in western media texts and their circulation in the global South (see McPhail 2009). Western media texts were viewed as instruments of domination that were capable of destroying local cultures. In an influential text, Ariel Dorfman and Armand Mattelart (1975) asserted for example that Disney movies were spreading an ideological message in the third world which posited US capitalist consumerism as both superior and inevitable. The cultural imperialism thesis has been significantly challenged for its failure to get to grips with the ways in which global media products are indigenized as they travel (Appadurai 1996; see also Tomlinson 1991). In more recent years, we have witnessed a clear shift from the diffusion model to an emphasis on interactive, multidirectional, participatory communication (Hemer and Tufte 2005a), but residues of the older models (both modernization and dependency varieties) continue to persist in different ways, in part because it is hard to dislodge the idea that media are powerful agents of social change, which they are but never in deterministic ways. While some development practitioners and scholars see media technologies as inherently or potentially beneficial, others believe that they should not be pursued until other development needs such as food, water, health care and education are met. This perspective is based on the idea that "the villager can't eat communication" (as cited in Hemer and Tufte 2005b): 11) and on a presumed separation between development goals. For Manuel Castells (2001: 269), the idea that Internet should be postponed until other development goals are secured, "reveals a profound misunderstanding of the current issues in development" as having access to the Internet can help people and communities to achieve education or clean water and participate in the development process more actively.

This chapter aims to connect the debates on development communication in the development studies literature with the debates on media convergence in the media and cultural studies literature. These different literatures have distinct emphases and bringing them into closer dialogue has the potential to produce new insights. The next section outlines these bodies of literature and then looks at recent scholarship that has called conventional understandings of the digital divide into question. I then discuss pleasure and entertainment and the extent to which these dynamics have been embraced in development communication practice. Finally, I discuss media convergence and show how this phenomenon should also be considered in a development context. This approach illuminates the need to place cultural politics at the centre of studies on the empowering potential of media and communication in and for development.

Media and Communication in/for Development

On one level, the value of media and communication in/for development is obvious. Although the presence of media technologies does not and cannot resolve poverty in a technologically deterministic way (see Warf 2001), it is not hard to observe that across Africa, Asia and Latin America, those parts of the world most identified with and transformed by the 20th project and concept of development, ordinary people are doing quite extraordinary things with radio, television, Internet and cell phones. Many low-income individuals, households and communities prioritize investment in media technologies for entertainment, information and in search of a route to improved quality of life, greater social mobility and enhanced political participation. Panos (2007) cites a World Bank survey of 40,000 poor people of their most pressing development needs, in which 'having a voice' was the second most common response after improved income and basic necessities.

Media technologies are therefore frequently sought and harnessed along with other material and discursive resources as a means to construct cultural citizenship. Of course, being able to communicate and express oneself safely and publicly is central to the development project. If your local water supply has dried up or become contaminated, if your land and food security are threatened by large agribusiness, or if you can observe elections being stolen, being able to communicate that state of affairs to a wider audience can be a life and death matter and fundamental to survival. Communication on one's own terms is also central to human dignity. The mainstream media are sometimes accused of framing marginalized groups in negative ways or silencing their perspectives in support of a discriminatory status quo. So if you belong to a social group or live in an area that is frequently stigmatized by elite groups in and through the media, having your own mediaspaces in which you can (try to) set the record straight is essential. Whether watching

CNN, Hollywood movies, soap operas and reality television is part of ICT4D is less obvious and less readily accepted by scholars in the field, although, as I argue, needs to be considered alongside and in interaction with alternative and bottom-up media forms.

Exactly how media and communication matter in development is not straightforward. In part, this is because the field is composed of quite disparate sets of literature that are not in adequate dialogue with one another. There are several bodies of literature that are all concerned with the question of media and communication for development to one degree or another, but they have quite different theoretical emphases and concerns and make different sets of assumptions. There is for example a substantial body of development-oriented literature on the digital divide, focused primarily on Internet connectivity. Although some recent geographical scholarship has produced much more nuanced and spatially sensitive accounts of the digital divide (see Warf 2001; Gilbert and Masucci 2005, 2011; Graham 2011), as Barney Warf (2001) writes, much of this scholarship, which generally calls for agencies to get the poor connected to the Internet, is frequently accompanied by a hyberbolic, teleological and neoliberalizing boosterism, along with identifiable residues of technological determinism. While the Internet is used for progressive and subversive ends all the time, as the Zapatista rebellion or the Iranian Twitter revolution more than reveals, much of the promotion of ICT4D is also insidiously aligned with corporate and neoliberal agendas. The creation of technologically literate compliant workers and willing consumers whose interests become aligned with those of for-profit corporations is seen as a key means to create and expand new markets. To some extent, the diffusion model has been dislodged by an equally problematic set of neoliberal governmentalities, in which the responsible development citizen uses the Internet to support the expansion of the market.

A second body of literature in this field is the scholarship focused on community, alternative or grassroots media or NGO-led edutainment. This literature is dominated largely by case studies of particular initiatives, showing how community media or NGO media productions are effective in producing new representations of key social issues and giving subordinated populations an effective political voice. While much of this work reveals that people need media that is not owned and controlled by private national or multinational corporations, some of it implicitly or explicitly suggests that corporate mainstream media is of no use or interest to those who have community media and that it cannot be put to alternative uses. It is a conclusion that is easily reached when the analytical focus privileges the conditions of production rather than the conditions of reception. While I share concerns about the contemporary global political economy of the media, in which a small number of very large corporations own the majority of newspapers, radio stations and television channels, it is important not to assume that economic dominance translates automatically into cultural dominance. In other words, in a valid celebration of community media, we must not explicitly or implicitly reproduce the line of thinking propagated by the Frankfurt School, which saw mass media as a mindless distraction that prevented people from engaging with or addressing political issues (see Hartley 2003), by viewing imported or corporate media as destructive to third world peoples and places or as an obstacle to development.

Finally, there is work in media and cultural studies, much of which does not identify with the question of development, but which is interested in popular pleasure and audience agency. As noted above, this body of scholarship is more likely to locate spaces of political possibility in commercial and mainstream media such as popular broadcast television than the literatures on the digital divide and on community media. John Fiske's (1987) work has emphasized how a media text can only become popular, and therefore profitable for media corporations, if it resonates with the lifeworlds of its viewers. For Fiske, active audiences do creative things with the media they have at their disposal. Media texts, even commercial ones, can be put to non-capitalist or anti-capitalist uses. It is therefore important to explore the cultural politics and the forms of civic engagement and sense-making which form around a media text rather than leaving the analysis with the (profit-seeking) intentions of the producer or television channel.

All of these bodies of work have their validity and there is much to be valued in their respective specializations and this chapter does not seek to deny the value of getting the poor connected to the Internet and also acknowledges that community and grassroots media forms can be powerful tools for political and social transformation and cultural revitalization. But a focus on the role played by media and communication for/in development needs to get to grips with the entangled, fluid and dynamic nature of the contemporary mediascape as it is navigated by the subjects of development in the third world and with the fact that media texts, consumers and producers and the discourses which they mobilize move across a range of media platforms. An incident filmed on a cell phone in a shopping mall might be a viral YouTube video a week later, television is watched on TV sets in domestic homes and on smartphones on the evening train commute, readers' comments on a news story in an online newspaper can take the story in a new direction, TV drama episodes are debated on social media, the blogosphere is a source of mainstream news, video mashups and memes are as much part of our everyday media environment as top rating soap operas or evening news programmes. These phenomena and many others constitute what is referred to as media convergence, defined by Klaus Bruhn Jensen (2010) "as a historically open-ended migration of communicative practices across diverse material technologies and social institutions".[2] Put simply, media convergence is about how different media interact with one another in ways that are mutually transformative and are leading to new ways of consuming and making media and therefore of making knowledge. Media convergence is associated with the creation of new communities and forms of belonging, along with new forms of fragmentation and exclusion. It is not just whole texts that are mobile, discourses and modes of sense-making which emerge from and around a media text also move across media platforms, facilitating

[2]Media convergence is a multifaceted and contested concept being taken up by a range of media studies scholars in a number of ways. To gain a sense of the different ways in which the question of media convergence is being approached, see Jenkins (2006), Jensen (2010), Meikle and Young (2011).

the multidirectional production and contestation of meaning. These dynamic media mobilities are therefore productive of new forms of discursive articulation and contestation. Their implications for how we imagine and remake the development project are paramount. In order to gain a deeper understanding of these media mobilities in development, development communication and ICT4D need to engage with the question of media convergence. To achieve this objective, it is useful to take both the geographic disruptions of the digital divide and the question of popular pleasure seriously.

Disrupting the Digital Divide

The question of the 'digital divide' has become one of the most salient development issues of the new millennium. As Mark Graham (2011: 212) writes: "hundreds of projects around the world are framed with the intention of solving or bridging a 'digital divide' and billions of dollars have been spent to achieve such goals", based on a belief that not having access to the Internet limits people's ability to participate in social and economic life, putting them at a distinct disadvantage. Broadly speaking, the digital divide is focused on differential access to information communication technologies, and there are many charts and maps which tell us the percentage of the population of a given nation that has access to the Internet (see for example Graham et al. 2012; internetworldstats.com). While having access to the Internet is undoubtedly better than not having access, just as literacy is better than illiteracy and clean water is better than dirty water, and the arrival of the Internet in a community can be as dramatic as the arrival of television was in domestic homes in the 1950s and 1960s (see Morley 2010), it is important to interrogate the idea of the digital divide as a simplistic binary between the "haves" and "have nots", as scholars such as Melissa Gilbert and Michelle Masucci (2005, 2011) have done. They question the assumption that 'the people stranded on the "have nots" side of the digital divide are undifferentiated, lacking in agency and employ similar frameworks for ICT use as the "haves" (Gilbert and Masucci 2011: 12). Graham (2011) disrupts the binary in another way by thinking about the Internet topologically rather than in terms of a divide in which one is simply in or out. He notes the multitude of temporal, infrastructural, technological, cultural, linguistic, political and economic barriers to digital interactivity, many of which persist after one is in cyberspace, having supposedly crossed an imagined physical divide. In many low-income urban neighbourhoods in Latin America, for example, Internet services are widely available, provided by NGOs, through commercial cybercafés or out of private homes, when the family computer is rented out to neighbours at an hourly rate. Many people lack the time, money or technological competence to access the Internet, but even those connected do not automatically find empowerment or a route out of poverty or structural disadvantage. Others could potentially find the time, money or skills required but the Internet is not perceived to be of interest, value or relevance in the context of their everyday lives. In part, this could be because other kinds of media seem more accessible, useful or entertaining to them.

Getting low-income neighbourhoods connected to the Internet can help people to achieve their development aspirations. But it is important to recognize the ongoing value of both radio and television both for intervening in and reimagining the development project. In this respect, I concur with David Morley's (2007, 2010) attempt to disrupt the emphasis on newness in much media studies. As Lisa Nakamura (2010: 28–9) writes, "often the most sophisticated and interesting forms of new media involve older instantiations of them ... precisely because they occupy lower bandwidth and are thus less bounded by particular infrastructural requirements". New forms of interconnectivity can also produce new forms of marginalization, especially if Internet is available but requires time, money and technological expertise or cybercafés or info-kiosks become male-dominated or youth-dominated spaces (Warf 2001). As Jensen (2010: 73) writes, while user-driven practices are often seen as more important than "old" style broadcasting, it can be "engaging and informative to go with the flow". Listening to local radio or watching popular television, what Jensen refers to as synchronous one-to-many communication, is much easier for most people than writing a blog or maintaining an active and influential Twitter account. Indeed, as Jesús Martín-Barbero (2004) observes with reference to Latin America, there is an important geography to television viewing that must be taken into account. Collective watching of a telenovela in one of the few homes in a neighbourhood to possess a television can be important in promoting or maintaining community cohesion and in providing people with a sense of belonging, factors which can be essential in tackling a range of development issues. Both "old" media and entertainment media must therefore be part of any consideration of media and communication for development.

Entertaining Development

Of course, one of the reasons that *Patito Feo* was so successful is that it was highly entertaining to children and teenagers in particular. A bit like *Glee* (although created and broadcast 2 years before *Glee*), it contains music, singing and dancing as the *divinas* and *populares* compete against one another in musical competitions. Díaz Somavilla et al. (2010) see its appeal in its combination of catchy music, simple choreography and the use of secrecy, intrigue and suspense, generic techniques common in melodramatic telenovelas. It is fair to say that many of those working in and with development communication or ICT4D would not see any development potential in *Patito Feo* and many would see it as detracting from real development needs. As Aswin Punathambekar (2010) has observed, the ICT4D community has failed to develop an analysis of popular pleasure in development communication and tends to assume that if something is pleasurable, it cannot be developmental. He notes how an ICT4D initiative to provide Internet access in info-kiosks in South India was removed when the project leaders learned that people were using the computers provided to watch Tamil or Hong Kong action films. Similarly, I have heard development practitioners in Nicaragua describe their frustration with shanty town dwellers who have prioritized the purchase of a television set while they and their children lack adequate shelter, nutrition or education.

The work of cultural theorists such as Néstor García Canclini and Jesús Martín-Barbero has provided us with an alternative way to understand the relationship between popular pleasure and development. For García Canclini (2001), media consumption cannot be divorced from the question of cultural citizenship. As noted above, activities such as television viewing provide a sense of belonging and provide resources that people can use in contesting the uneven and disempowering conditions of life in which they find themselves. Drawing on García Canclini, we can say that Hollywood movies, computer games, Facebook posts and telenovelas can be "good for thinking". For Martín-Barbero (2004), melodrama (central to telenovelas) is pivotal in the ways in which a collective cultural imagination is both constructed and maintained. He believes that media use and consumption cannot be divorced from the conditions of reception which means that popular cultural memories, indigenous ways of knowing and the practices of everyday life shape how telenovelas or other media texts are received. For George Lipsitz (1988: 100), subaltern and popular cultural traditions are not erased by commercial culture because commercial culture "expressly depends upon the residues of local popular narratives for its determinate forms and themes" and also generates "a new popular narrative response, one that draws upon both old and new forms of cultural creation". Watching *Patito Feo* or Hong Kong action films is a way in which people make sense of and interpret their conditions of life and produce new aspirational imaginaries. Yet as Punathambekar (2010) observes, most ICT4D initiatives "frame the poor as objects of the discourse of digital access, and they are rarely seen as the subject of digital imaginaries".

While I agree with this observation, which results in part from the continued influence of earlier theories of ideological domination as well as from limited scholarly interaction between development studies and cultural studies, the need for development communication to be entertaining in order to be effective has been accepted and embraced by some NGOs and development communicators. While retaining the idea that media can have positive developmental outcomes, they have realized that forms of communication that tell people what to do or TV programmes that are preachy, paternalistic and contrived are usually doomed to failure. For a development organization attempting to promote better reproductive health, it is crucial to recognize that a failure to insist on condom use during a sexual encounter is often not because of a lack of knowledge, education or information. The person might "know" that a condom will help to prevent pregnancy or HIV infection, but insisting on its use is often constrained by existing power relations, economic disadvantage, and cultural norms surrounding expressions of masculinity and femininity. Some organizations have therefore realized that producing entertaining television that is cognizant of the everyday cultural politics of the target population is much more effective at encouraging debate around and engagement with key development issues. In the so-called third world, there are many innovative examples of NGOs using media in an attempt to educate, inform and change behaviour. Nicaraguan feminist NGO, Puntos de Encuentro, has produced two very popular telenovelas,

Sexto Sentido and *Contracorriente*, set in Managua and which deal with questions of sex and sexuality, domestic violence and education as confronted by young low-income urban Nicaraguans (see Weinberg 2006; Howe 2008). In the post-apartheid era in South Africa, edutainment media has been central to the South African Broadcasting Corporation's broadcasting strategy. Hard-hitting and controversial TV dramas such as *Soul City* and *Yizo Yizo* or community reality makeover shows such as *Kwanda* have been both extremely popular and effective at encouraging debate around issues such as HIV infection, domestic violence, alcoholism and substance abuse and the crisis facing schools in the townships (see Kruger 1999; Tufte 2001; Barnett 2004). These media succeed in part because they do not aim to tell people what to do, but facilitate instead dialogic and interactive forms of engagement through local and familiar representations of place and are entertaining, in the way that commercially produced dramas and soap operas are.[3] They provide a space in which people can reflect on the everyday issues that both the characters and they are forced to negotiate and the media text makes it possible for people to discuss issues such as domestic violence or condom use in their neighbourhoods, schools and workplaces. In this way, the show might participate in the creation of strategies on behalf of viewers to manage these issues in their own lives in a more assertive or empowered way.

Media Convergence in/and Development

Let me use a fictional and composite character to demonstrate how media convergence is practised and experienced in the context of everyday life. Diego is a Central American man in his early twenties who has migrated from a rural area to the capital leaving family behind to farm the land. In the morning, he listens to farmers' union radio station to get information on the latest agricultural prices for his family's produce back home. He owns a cell phone and uses it regularly to text his sister who is working as a domestic servant in Costa Rica. He often gets hold of the daily newspaper in his workplace and if he has time for lunch, will read some of it. Most evenings he settles down with a group of people in the neighbourhood to watch the latest Brazilian or Venezuelan telenovela. If he has enough credit on his phone, he might vote in a reality television show. He accesses the Internet occasionally in a cybercafé but does so primarily to play games, although once he got caught up reading fan forums about a particular telenovela in which a character is diagnosed with HIV after intravenous drug use. He catches glimpses of CNN in a local store where he sometimes buys candy or cigarettes and he goes to a local bar every now and again to watch football or boxing. Most Saturday nights, he goes to a nightclub where the DJ plays mostly reggaeton, a Latin American blend of hip-hop, reggae and salsa. For a few months, he was involved in a community radio station as a volunteer

[3] Lesley Henderson (2007) shows how issues discussed in soap operas and other forms of TV fiction play an important role in shaping social attitudes and encouraging engagement.

and got to know a number of people in the neighbourhood who used it to make community announcements or express political grievances. Then one day, during a protest march against water privatization, he films police repression of protestors on his cell phone and uploads it to YouTube. The video is viewed by activists around the world who condemn the government for their brutality and international solidarity against water privatization and the criminalization of protestors grows and becomes politically effective.[4]

The relationship between development and Diego's media consumption and production is highly complex and it is not possible to definitively say which aspects are developmental. This difficulty results in part from the ways in which different media interact with one another and because any medium can be put to different uses. A cell phone used for maintaining social and familial networks can suddenly be used to disseminate and denounce an act of political repression. It is clear when we look at the contemporary media environment in a development context that the categories through which we talk about media for development (community media, mainstream media, the digital divide) are confounded and confused by the practices of everyday life. "Old" media and "new" media do not exist in parallel, but they remediate one another in multidirectional ways. Indeed, as Bolter and Grusin (1999: 55) argue, there is a need to disrupt the notion of a linear history in which new media such as Internet displace old media such as radio, as "older media can also remediate newer ones". The rapidity with which media content crosses media platforms and with which individuals move between bottom-up/community and top-down/corporate media blurs the boundaries between them and complicates how we understand the relationship between media, communication and development as well as the economic concentration of media ownership and the question of cultural imperialism. Development communication needs to take account of these dynamic interrelationships and their relational outcomes for improving standards of living or promoting self-esteem or empowerment.

In order to further probe the question of media convergence in/and development, it might be valuable to unravel analytically the tension between what Cowen and Shenton (1998) refer to as immanent and intentional development. Development and change come about through both intentional and unintentional ways. In this respect, both NGO-led entertainment and commercial soap opera can have potentially positive outcomes in terms of development. While the former clearly constitutes intentional development, the second comes about through a more diverse set of motivations, combining the immanent and the intentional. It is essential to recognize that producers, writers, editors, reporters and actors might also wish to change the

[4]This is a reference to a peaceful protest against water privatization in Suchitoto, El Salvador in 2007. It was brutally repressed by state forces and 14 protestors were charged with acts of terrorism. While such political violence is not uncommon in El Salvador, on this occasion the violence was filmed on cell phones and uploaded to You Tube. The images were spread rapidly around the world, and shared among activists and solidarity committees. In 2008, the charges against the protestors were dropped and there is no doubt that global media visibility played an important role in this outcome.

world, even though their products must return a profit to the companies for which they work. The immanent/intentional tension is also useful for thinking through what counts as development communication or ICT4D and how it does or does not differ from other forms of media consumption and entertainment. There might, therefore, be a need to include more immanent forms of development at work in media and communication. Of course, development communication includes communicating what we might define as development information to people who need it. Such information might include daily agricultural prices, warnings of an imminent hurricane or information of a local vaccination campaign. Increasingly, however, these services are much more user-driven than they were just a couple of decades ago. For example, risk communication after a major disaster is often dependent on volunteered geographic information or VGI provided by ordinary people which can be mapped on the Internet or communicated through radio, television, text messages and Twitter. These forms of risk or development communication can be crucial for survival and it is important for agencies to identify how best to reach people and engage with VGI effectively. But development communication also involves the mediation and potential rearticulation of development itself. If development is conceived in broad terms not just as a securing of basic needs but as a process of empowerment and construction of human dignity, then it is also about making sense of the conditions of life and imagining how things might be otherwise. Given that the development project has led to internalized feelings of inferiority among the targets or beneficiaries of development (Escobar 1995), doing better development or indeed making development obsolete[5] means that people need to feel positive about their lives, they need to cease feeling inferior to others and need to be able to envisage a more hopeful future for themselves and their communities, in which their ways of knowing are legitimated. Contemporary media are central to our imaginaries, as the media we produce and consume enable us to gain access to a range of different worlds, both real and fictional, both local and distant. Engagement with media might help people to connect the inadequate material conditions of life with imagined better futures, but in ways which make the latter possible and realizable. So any media which helps to imagine and aspire to improved well-being potentially has developmental dimensions. For Martín-Barbero (2004), everyday media consumption, including of telenovelas, is closely related to people's development aspirations and can even be conceptualized as a form of protest or assertion of human rights. While this has been the case since the advent of mass media, under conditions of media convergence the possibilities to imagine and aspire to a different future takes on new qualities.

In the convergent and increasingly user-driven media environment, the modes in which development can be appropriated proliferate, because people are able to engage with debates aired in media texts in a range of sites. Television and radio continue to be important not just as a site for critical engagement in themselves, but because they frequently provide a route into Internet connectivity and other media

[5]One of the key objectives of postdevelopment (see Escobar 1995).

forms. Consequently, television itself takes on new sets of meanings, as does radio. The vibrant Internet fan forums[6] which developed around Colombian telenovela *Yo soy Betty la fea* (Ugly Betty) enabled isolated Latin American women affected by racist and classed understandings of beauty and whiteness[7] to engage with the structures of domination that limit their mobility or prevent them from reaching their potential (see Rivero 2003). The Ugly Betty phenomenon was global. Not only was the Colombian telenovela widely exported and watched throughout Latin America and beyond, there were numerous international remakes of the show in Belgium, China, Croatia, Georgia, Germany, Greece, India, Russia, Turkey, the United States and many other countries. Betty's mediation across space and place and her presence in other media sites has enabled women who have experienced both racism and sexism in their everyday lives to speak back to power, and imagine collectively with others, using Betty as a vehicle, how things might be different for them. The US version dealt innovatively with the question of undocumented immigration by Latin Americans in the US and provided important resources for US-based Latinos with undocumented parents (Cupples 2013). *Patito Feo* also produced a substantial amount of Internet interactivity. A search on Google for "divinas y populares" produces more than five million hits, many of which are clips uploaded by fans to YouTube and which have attracted copious comments from users. While uploading, watching and commenting on these videos might appear to many to be a waste of time, and the discursive contestations around class, race and gender do not appear to be as significant as those which formed around Ugly Betty texts, the participatory online activity is developing key skills in users. Henry Jenkins' (2006) analysis of the 2004 presidential election shows how skills and forms of engagement learned in popular culture, what he calls "popular-culture-based strategies" (p. 209), could be applied to more overtly political contexts. In 2004, a blurring of fandom and political activism became evident, a phenomenon which was even stronger in the 2008 election when engagement with and use of social media platforms, especially by Obama supporters and campaigners, was central to the campaign and its outcomes. Through looking at and uploading images of cute cats online, a user might develop a set of transferable skills that might one day be used to achieve a different set of objectives. The potential might not be realized but it exists and is not to be dismissed as irrelevant to development.

All around the world, broadcast media (news and current affairs, election campaigns, entertainment television, radio programmes) are proving to be important routes into more user-driven practices, just as user-driven practices are changing the nature of broadcast media. To return to the question of the digital divide, it is clear that "old" media are providing routes to "new" media or Internet connectivity, without the former being replaced by the latter. Bluefields Stereo is

[6]It is important to recognize that this telenovela was being heavily discussed in face-to-face sites as well. Indeed, I heard many conversations about it during a 6 month stay in Nicaragua in 1999.

[7]The lighter-skinned you are in Latin America, the easier it is to get employment in jobs that require engagement with the public, such as retail assistants, bank tellers and receptionists.

a radio station operating out of Nicaragua's Caribbean Coast with the explicit aim of strengthening the region's political autonomy.[8] The region has a long-standing tradition of participatory radio (Glynn and Cupples 2011), and as such the channel is part of that well established media landscape, but more recently has been as active on Facebook as it is on the airwaves. The Facebook page was created in 2011 because at the time the station was unable to maintain its own website. Director, Wilder Wilson, told me that he felt a web presence was necessary to cover the information broadcast in the radio show *Autonomía en Marcha* (Autonomy in Progress), but to be able to add images and video, not possible in the radio format. It also allows people to engage with the content and add their own perspectives at a time that is convenient to them and they can do so whether or not they heard the radio broadcast, and whether or not they are in Bluefields. Since 2011, the newsfeed has broadened and includes not only local news stories, but announcement of key events, as well as images and photographs of Bluefields, some of which are historical. While many people in the region do not have access to the Internet, those that do are able to comment on all of these. In 2013, the station also began to broadcast live on Ustream (http://www.ustream.tv/channel/radio-bluefields-stereo) and many people comment on the show on the Ustream page too. At the time of writing, Bluefields Stereo began to upload videos to YouTube, promoting them through Facebook. Many people are using the Ustream and Facebook conversation threads simply to say hello to others, but this in itself is significant as it registers a presence on the web, a desire to be in a collective dialogue with others about the experience of autonomy and can in itself be a confidence- or skill-building exercise which might later lead to more politically significant interventions, as Jenkins' (2006) work has illustrated.

Citizens can engage with the question of autonomy and the development challenges facing Bluefields in a proliferating array of mediaspaces, and can often do so in a context which is entertaining. It is a way in which the Internet becomes relevant to citizens. And at each moment of engagement, of media consumption or production, there is the potential for a group or individual to catch a glimpse of political possibility or for a disempowering hegemonic discourse to be disrupted or rearticulated or for a counterhegemonic discourse to gain new traction. While mashup videos that go viral might make a visible political difference in a short space of time, it is also important to recognize the often slow, subtle and cumulative nature of discursive rearticulation, particularly when it pertains to deeply ingrained power relations. The dense interactions between broadcast and bottom-up digital media are not only making the distinction between them redundant, they potentially democratize the struggle for autonomy or for the making or dismantling of development.

---

[8] Nicaragua's Caribbean Coast is quite ethnically and linguistically distinct from the rest of Nicaragua as it is home to a number of indigenous peoples and ethnic communities who have often clashed with the Pacific region and the national political leadership in Managua. Political and military struggles in the region led to the passing of an autonomy law in 1987. Since then, local people have been attempting to strengthen the region's autonomy and make it more politically effective.

Conclusion

If we are interested in the potential of media and communication technologies for remaking development, it is important to start with the cultural politics of development, rather than with the media technology, which forces us to focus on the dynamics of convergence and how these are being reconfigured through everyday practices. People often use the media that they have available (which might be radio rather than Internet) and the media texts which already resonate with their everyday lives (which might be commercial soap opera rather than political blogs), but through that media use find routes into other media. Many move back and forth between "old" and "new" media, between community and corporate media, without distinguishing between these realms in the way that scholars or development practitioners might. Even for those who stick with commercial broadcast media, it will of course be with a reconfigured and convergent broadcast media, shaped by a range of transmedia activities. Where development communication starts and ends becomes impossible to discern. But in the spaces where the cultural politics of development are enacted and performed in collaboration with media texts and technologies, there is plenty for us to analyse.

Acknowledgement This research is supported by the Marsden Fund of the Royal Society of New Zealand, grant number MAU1108.

References

Abu-Lughod, L. (1995). The objects of soap opera: Egyptian television and the cultural politics of modernity. In D. Miller (Ed.), *Worlds apart: Modernity through the prism of the local* (pp. 190–210). London: Routledge.

Ang, I. (1985). *Watching Dallas*. London: Routledge.

Appadurai, A. (1996). *Modernity at large: Cultural dimensions of globalization*. Minneapolis: University of Minnesota Press.

Barnett, C. (2004). Yizo, Yizo: Citizenship, commodification and popular culture in South Africa. *Media, Culture and Society, 26*(2), 251–271.

Bolter, J. D., & Grusin, R. (1999). *Remediation: Understanding new media*. Cambridge: MIT Press.

Bonavitta, P., & de Garay Hernández, J. (2011). De estereotipos, violencia y sexismo: la construcción de las mujeres en los medios mexicanos y argentinos. *Anagramas, 9*(18), 15–30.

Castells, M. (2001). *The internet galaxy*. Oxford: Blackwell.

Cowen, M., & Shenton, R. (1998). *Doctrines of development*. London: Routledge.

Cupples, J. (2013). *Latin American development*. London: Routledge.

Díaz Somavilla, R., Canós Darós, L., & Magal Royo, T. (2010). El nuevo triunfo de la telenovela infantil-juvenil. El caso de *Patito Feo*. In M. F. Domènec (Ed.), *La calidad de los contenidos audiovisuales en la multidifusión digital* (pp. 125–138). Barcelona: Editorial UOC.

Dorfman, A., & Mattelart, A. (1975). *How to read Donald Duck: Imperialist ideology in the Disney comic*. New York: International General.

Duncombe, S. (2007). *Dream: Re-imagining progressive politics in an age of fantasy*. New York: New Press.

Escobar, A. (1995). *Encountering development: The making and unmaking of the third world*. Princeton: Princeton University Press.

Fiske, J. (1987). *Television culture*. London: Routledge.
García Canclini, N. (2001). *Consumers and citizens: Globalization and multicultural conflicts*. Minneapolis: University of Minnesota Press.
Gauntlett, D. (1998). Ten things wrong with the 'effects model'. In R. Dickinson, R. Harindranath, & O. Linné (Eds.), *Approaches to audiences: A reader* (pp. 120–130). London: Arnold.
Gilbert, M. R., & Masucci, M. (2005). Research directions for information and communication technology and society in geography. *Geoforum, 36*(2), 277–279.
Gilbert, M. R., & Masucci, M. (2011). *Information and communication technology geographies: Strategies for bridging the digital divide*. Vancouver: Praxis (e)Press.
Glynn, K. (2000). *Tabloid television: Trash taste, popular power and the transformation of American television*. Durham: Duke University Press.
Glynn, K., & Cupples, J. (2011). Indigenous mediaspace and the production of (trans)locality on Nicaragua's Mosquito Coast. *Television & New Media, 12*(2), 101–135.
Goy, M. (2007). La tira Patito Feo entró a las aulas. *El Litoral* 22 Nov http://www.ellitoral.com/index.php/id_um/26316/origen/ranking. Accessed 6 Mar 2013.
Graham, M. (2011). Time machines and virtual portals: The spatialities of the digital divide. *Progress in Development Studies, 11*(3), 211–227.
Graham, M., Hale, S., & Stephens, M. (2012). Featured graphic: Digital divide: The geography of internet access. *Environment and Planning A, 44*(5), 1009–10.
Hartley, J. (2003). *A short history of cultural studies*. London: Sage.
Hemer, O., & Thomas, T. (2005a). Introduction: The challenge of the glocal. In O. Hemer & T. Tufte (Eds.), *Media and glocal change: Rethinking communication for development* (pp. 13–22). Buenos Aires: CLACSO.
Hemer, O., & Thomas, T. (2005b). Foreword. In O. Hemer & T. Tufte (Eds.), *Media and glocal change: Rethinking communication for development* (pp. 11–12). Buenos Aires: CLACSO.
Henderson, L. (2007). *Social issues in television fiction*. Edinburgh: Edinburgh University Press.
Howe, C. (2008). Spectacles of sexuality: Televisionary activism in Nicaragua. *Cultural Anthropology, 23*(1), 48–84.
Jenkins, H. (2006). *Convergence culture: Where old and new media collide*. New York: New York University Press.
Jensen, K. B. (2010). *Media convergence: The three degrees of network, mass and interpersonal communication*. London: Routledge.
Kruger, L. (1999). Theater for development and TV nation: Notes on educational soap opera in South Africa. *Research in African Literatures, 30*(4), 105–126.
Lipsitz, G. (1988). Mardi Gras Indians: Carnival and counter-narrative in New Orleans. *Cultural Critique, 10*, 99–121.
Lumby, C. (2003). Real appeal: The ethics of reality TV. In C. Lumby & E. Probyn (Eds.), *Remote control: New media, new ethics* (pp. 11–24). Cambridge: Cambridge University Press.
Martin-Barbero, J. (2004). A nocturnal map to explore a new field. In A. del Sarto, A. Ríos, & A. Trigo (Eds.), *The Latin American cultural studies reader* (pp. 310–328). Durham: Duke University Press.
McPhail, T. L. (2009). Major theories following modernization. In T. L. McPhail (Ed.), *Development communication: Reframing the role of the media* (pp. 21–48). Malden: Wiley-Blackwell.
Meikle, G., & Young, S. (2011). *Media convergence: Networked digital media in everyday life*. Basingstoke: Palgrave Macmillan.
Morley, D. (2007). *Media, modernity and technology: The geography of the new*. London/New York: Routledge.
Morley, D. (2010). Domesticating dislocation in a world of "new" technology. In C. Berry, K. Soyoung, & L. Spigel (Eds.), *Electronic elsewheres: Media, technology and the experience of social space* (pp. 3–16). Minneapolis: University of Minnesota Press.
Nakamura, L. (2010). Avatars and the visual culture of reproduction on the web. In C. Berry, K. Soyoung, & L. Spigel (Eds.), *Electronic elsewheres: Media, technology and the experience of social space* (pp. 17–32). Minneapolis: University of Minnesota Press.

Panos. (2007). *The case for communication in sustainable development*. London: Panos. http://panos.org.uk/wp-content/files/2007/09/The-Case-for-Communication-in-Sustainable-Development.pdf (Accessed 12 Feb).

Punathambekar, A. (2010). What fandom could teach the IC4TD community. http://bollyspace.wordpress.com/. Accessed 5 Mar 2013.

Rivero, Y. M. (2003). The performance and reception of televisual "ugliness" in *Yo soy Betty la fea*. *Feminist Media Studies, 3*(1), 65–81.

Soong, R. (1999). Telenovelas in Latin America. *Zona Latina* 11 August. www.zonalatina.com/Zldata70.htm. Accessed 20 Mar 2013.

Tomlinson, J. (1991). *Cultural imperialism: A critical introduction*. London: Pinter.

Tufte, T. (2001). Entertainment-education and participation: Assessing the communication strategy of *Soul City*. *Journal of International Communication, 7*(2), 25–50.

Warf, B. (2001). Segueways into cyberspace: Multiple geographies of the digital divide. *Environment and Planning B: Planning and Design, 28*, 3–19.

Weinberg, C. (2006). This is not a love story: Using soap opera to fight HIV in Nicaragua. *Gender and Development, 14*(1), 37–46.

22 Indigenous Media: Linking the Local, Translocal, Global and Virtual

Pamela Wilson

Introduction

Indigenous cultures are, by their very nature, etched upon the land. Aboriginal cultural knowledge is mapped upon and situated in very specific places in the local geographic landscape, and cultural narratives are frequently ritually inscribed upon these places—even if those inscriptions are invisible to cultural outsiders. Indigenous cultures exist in all regions of the globe, in spite of the fact that, for various cultural political reasons, they may not all be labeled as such because of the particular histories of conquest and colonialism they have endured.

The tribal cultures of English-speaking settler nations, such as the United States, Canada, Australia and New Zealand, were among the first to be labeled as "indigenous." Those of other European colonies—in Central and South America, for example—soon followed. Slower to become recognized as part of the global community of indigenous peoples, but no less significant, have been those cultural groups of Africa, of Europe (the Sámi and the Basque, for example), of Asia (including the "Scheduled Tribes" or *Adivasi* of India, many of the *Minzu* minority groups of China, the aboriginal groups of Taiwan, Burma's hill tribes, the Ainu of Japan, the small-numbered peoples of Northern Russia, and more), of Indonesia and of Oceania.

Indigenous artists and activists have taken up the pen, the microphone, the camera and the computer to craft both nonfiction media pieces (to inform, arouse, and persuade a larger public through journalism, broadcasting, and documentary) and fictional narrative media such as literature, feature films, television series, and video games. And today, they are using electronic and digital structures of communication to serve their own needs for archiving and databasing as well as to

P. Wilson (✉)
Communication Program, Reinhardt University, Waleska, GA, USA
e-mail: PSW@reinhardt.edu

S.P. Mains et al. (eds.), *Mediated Geographies and Geographies of Media*,
DOI 10.1007/978-94-017-9969-0_22

distribute their cultural understandings and visions to audiences worldwide. In doing so, indigenous media producers help contribute to the creation of new geographies that translocally link the locally-based cultural knowledge, so inherently inseparable from the local physical landscape, with comparable bodies of indigenous knowledge deeply rooted in other landscapes around the globe.

The goal of this chapter is to elucidate the intersection between media studies, cultural anthropology, and cultural geography as a site in which the emerging hybrid discipline of indigenous media studies has taken hold and has emerged as a significant convergence point for understanding the role of electronic and digital technologies in creating and maintaining twenty-first-century indigenous geographies; mapping and remapping the cultural worlds, perspectives and causes of contemporary indigenous cultural groups onto the global stage; and providing a bridge between the local and the global. Indigenous media serve to link the grounded places of indigeneity not only to each other, translocally, but also to the more abstract (and often virtual) spaces of global culture.

One of the significant functions of indigenous media may be to create a greater understanding and exposure of those "layers of Indigenous history and understanding that are imbued in the land and the culture," as RDK Herman (2008) articulates, envisioning "the postcontact landscape as a thin veneer resting atop these other layers" (p. 86). Literary and visual media created by indigenous producers often serve as critically important means of sharing geographic and cultural knowledge—such as those iconic representations of place—with others beyond their immediate cultural group, as well as committing and archiving those representations in a new form for future generations. Indigenous media studies allow us access to the micro-processes of what Roland Robertson has famously coined 'glocalization'—in this case, the interpenetration of global media technologies with hyper-local needs, creatively adapted to work within and to sustain local cultures rather than to replace or homogenize them.

What are indigenous media, and why should they be of interest to geographers? Indigenous media may be defined as forms of media expression conceptualized, produced, and circulated by indigenous peoples around the globe as vehicles for communication: for purposes of cultural preservation, cultural and artistic expression, political self-determination, and cultural sovereignty. Indigenous media occupy a position on a spectrum with other types of minority-produced media that exist within and seek a distinct voice from larger political units, often sharing a kinship regarding many philosophical and political motivations with other minority media. What is distinctive about indigenous media, however, are those characteristics at the core of the indigenous or aboriginal experience. What defines a cultural group as indigenous? The current understanding in international law about indigeneity includes cultural groups that can claim to have occupied and used the resources of a specific territory prior in time to other known occupants, that self-identify as a distinct culture, that voluntarily perpetuate cultural distinctiveness, and that have experienced "subjugation, marginalization, dispossession, exclusion, or discrimination, whether or not these conditions persist" (Wilson and Stewart 2008, p. 14).

The discussion below will provide a summary of the rise of indigenous media studies as an academic field, an overview of the range of types and genres of non-print indigenous media, and an exploration of the potential of digital media to create virtual indigenous geographies for the future.

Rise of Indigenous Media Studies

The academic study of indigenous media has developed in parallel to the globalization process and, in particular, the international indigenous movement of the latter half of the twentieth century that was shepherded by agencies of the United Nations such as the UN Working Group on Indigenous Populations (1982–2006) and, since 2002, by the Permanent Forum on Indigenous Issues. The work of these international working groups led to the passing in September of 2007 of the Declaration of the Rights of Indigenous Peoples by the UN General Assembly.

Theoretical approaches to indigenous media as a distinct field of study developed first within anthropology during the 1980s, hand-in-hand with the radical shift in the subjectivity of social, cultural and linguistic anthropologists vis-à-vis the cultural groups that they studied; this epistemological paradigm change accompanied the introduction of a critical, anti-positivist, post-structuralist, post-colonialist approach to cultural study that tried to remove or minimize the power differential formerly embedded in the researcher relationship with those studied—as well as a growing reflexivity about social responsibility and a collaborative approach leading to advocacy and empowerment of the cultural actors and groups that had formerly been the passive objects of an often politically-detached ethnographic study. In terms of media production, this paradigm shift was marked by the transition from outsider-scholars creating ethnographic films to ethnographers acting as mentors who could enable and empower the "natives" to tell their own stories rather than needing a cultural interpreter.

In ethnographic film, the anthropological roots of "putting the cameras in the hands of the natives" dated back to Sol Worth's and John Adair's groundbreaking Navajo Project of the 1960s, which suggested that cultural differences might indeed inform new visual "languages" and aesthetic styles for filmmaking (Worth 1997). However, the mid-1980s saw the real beginnings of the indigenous media movement. In video and television production, American anthropologist Eric Michaels pioneered a new practical and theoretical approach in his collaborative work in the 1980s in Australia with Warlpiri Aboriginal broadcasting. Michaels (1984, 1986, 1994) was most concerned with Aboriginal use of television as a tool for maintaining culture, developing their own programming rather than "simply to observe the possible deterioration of a society whose choices had already been made for them" (Michaels 1984, p. 46). This was soon followed by the establishment of separate village-based video projects in Brazil by Vincent Carelli (1988) and Terence Turner (1990, 1992). In Aotearoa (New Zealand), Māori filmmaker Barry Barclay (1990, 2000) emerged as one of the most eloquent indigenous intellectuals and theorists of indigenous aesthetics and cinematic techniques.

The most unifying theoretical voice to establish indigenous media as a field has been that of American anthropologist Faye Ginsburg, whose essays (based upon her research in Australia), as well as her personal mentoring of graduate students and young scholars, have inspired new generations of emerging indigenous media scholars across disciplines. In her seminal work marking the official birth of indigenous media scholarship, Ginsburg (1991) articulated the "Faustian dilemma" of introducing technology to tribal peoples: Even as, on the one hand, the "invasion" of Western mass media technologies and media products has disrupted indigenous culture, language, social relationships, and relationships to traditional knowledge, on the other hand, the new media tools have proven to be valuable in constructing indigenous identities and challenging outside cultural domination.

Key recent theoretical contributions to the indigenous media studies field foreground the complex dynamics between local indigenous communities, their dominant nation-states, and global economic, cultural and political forces. Lorna Roth (2005) documents the intercultural negotiations to develop media policy and build infrastructures in Canada's indigenous North, resulting in the rise of discursive spaces and mediated public spheres both occupied and managed by Canada's First Peoples. Michael Robert Evans' (2008) ethnographic study of Igloolik, an Inuit community in Nunavut (Canada) that serves as a hotbed of indigenous media production, explores the work of the Isuma production group as a form of folk performance—as Inuit resistance to the political and economic domination of Canadian society, and as a production practice that honors the unique elements of Inuit tradition and storytelling, birthed in the arctic. Providing a rich ethnographic exploration of the role of media technologies in the remote Yolngu community of Gapuwiyak in Australia, Jennifer Deger (2006) challenges the generally unproblematized vision of media technologies as tools for indigenous resistance to imperialism, examining the ways that media technologies "mediate the spaces between cultures, between clans, between individuals, and between anthropologist and informant" (p. 32).

Bringing attention to indigenous struggles in Latin America, Mario Murillo (2010) provides insight into indigenous Colombians' alternative media movement for peace and social justice, focusing in particular on indigenous, community-based Radio Payu'mat, which operates surrounded by, and against, a mainstream hegemonic corporate media culture that excludes minority voices in a militarized environment of struggle for human and cultural rights. Freya Schiwy (2009) provides a comprehensive study of indigenous videomakers in the Andean region, including Bolivia, Colombia and Ecuador. Drawing on Latin American de-colonial literary theory, Schiwy argues that "the process [of indigenous media-making] is an example of how indigenous movements in the region are transforming dominant socioeconomic and political structures and largely hegemonic epistemologies" (p. 3). Jeff Himpele's study on Andean media (2007) incorporates indigenous-produced work into a circuit of national televisual media in Bolivia, focusing on the intersections of performed indigeneity with nation-building discourses in popular film and television.

This interdisciplinary field has also seen the rise of several indigenous media scholars who bring a personal cultural identity to bear upon their scholarship. Native American Studies scholar and documentary filmmaker Beverly Singer (2001) provides an account both personal and academic of the growth of indigenous filmmaking in Native America; this important work provided the seed for efforts to create a foundational body of critical literature on Native filmmaking as well as efforts to create infrastructures and networks that would nurture and support Native filmmakers. Candace Hopkins (2006), a Canadian Indigenous arts specialist and museum curator, provides penetrating insight into how new media are shaped by indigenous systems of circulating cultural knowledge through storytelling. With theoretical incisiveness, her work examines how indigenous knowledge systems shape the relational aspects of new media, focusing on the social and political role of the storyteller who adapts and personalizes cultural content into narrative as part of a cultural circulation of knowledge. Literary scholar Michelle Raheja (2007, 2011a, b) champions the concept of *visual sovereignty* from a Native American perspective: "the space between resistance and compliance wherein indigenous filmmakers and actors revisit, contribute to, borrow from, critique, and reconfigure ethnographic film conventions, at the same time operating within and stretching the boundaries created by these conventions" (2007, para. 4).

Characterizing Indigenous Media

From Representations 'of' to Self-Representations

The histories of indigenous media representation in the Anglophone settler nations share many commonalities, since such representations have been deeply rooted in the experience, ideology and vision of European colonialism. Until recently, most indigenous peoples of these nations—whether they are called Native Americans, American Indians, First Nations, Aboriginal Peoples, Aborigines, Indigenous, or by their specific tribal or cultural group names, such as the Maori—and their perspectives, were either effaced from the dominant literature and media, notable in their absence, or they were skewed, distorted and stereotyped into a limited range of archetypes that supported the vision of the dominant nation. Many of these tribes and communities have shared common historical traditions of removal from their native lands and resettlement onto reservations; residential boarding schools and forced acculturation; colonization of their knowledge and of artifacts by national institutions (such as universities and museums); stereotypical representations in mainstream film, television, and popular culture; and marginalization from the dominant societies and their social institutions, resulting in high levels of poverty and unemployment, poor education, and poor health care.

The 1960s and 1970s saw much indigenous rights activism in these settler nations, a transformation in agency growing out of larger civil rights movements. Self-representation movements became part and parcel of the struggle for self-determination, including land rights and title movements, protests and negotiations over the interpretation of treaties. Aboriginal rights activists in Australia in the 1960s–1970s, like the Red Power movement in the U.S., aspired to take control of their own cultural representations in the greater public sphere, especially through film and television as well as other forms of visual art, photography and literature (see Ginsburg and Myers 2006). Indigenous groups at a community level, as well, began to appropriate radio, and later television, broadcasting for their own uses. However, the introduction of broadcasting into remote communities generally required a technological and policy infrastructure as well as sources of funding, all to be negotiated with government bodies. The complexities of indigenous media require an examination of the policy and access issues as well how Aboriginal aesthetic and cultural sensibilities and belief systems have shaped distinctive forms and styles of media, both fiction and non-fiction, that serve indigenous cultural needs and will continue to do so in the future.

Ongoing activism, lobbying, and the establishment of distinctive relationships with nation-state policymakers and institutions in most of these nations have created infrastructures and funding for local and regional broadcasting, on the one hand, and the establishment of nationwide television in most of these settler nations, on the other—as well as national institutions supporting film development. With the exception of the United States, which has done very little, comparably, to support the development of Native American media production, the other three settler nations, as well as The Republic of China (Taiwan), have worked in negotiation with indigenous interest groups—sometimes for decades—to provide funding and infrastructures for indigenous media development. Canada, for example, has a strong tradition of state-supported filmmaking from the National Film Board of Canada (NFB) and was the first nation to develop a nationwide cable channel for indigenous programming (the Aboriginal Peoples Television Network, or APTN), launched in 1992, followed by Māori Television in New Zealand in 2004, Taiwan Indigenous Television in 2005 and National Indigenous Television in Australia in 2007. Having this assured outlet for broad distribution has sparked a great deal of Indigenous-produced program development in film and television around the globe, with exchange internationally.[1]

[1]The World Indigenous Television Broadcasters Network, WITBN [http://www.witbn.org/] is a global collaborative alliance of indigenous broadcasters established in 2008. Members include: TG4 (Ireland) [http://www.tg4.ie/]; Māori Television [http://www.Maori.television.com/default.aspx] and Te Reo [http://www.tereo.tv/] (both Aotearoa New Zealand); NITV (Australia) [http://www.nitv.org.au/]; TITV/PTS (Taiwan) [http://www.titv.org.tw/]; BBC Alba (Scotland) [http://www.bbc.co.uk/alba/]; NRK Sámi Radio and Television (Norway) [http://www.nrk.no/sapmi/]; S4C (Wales) [http://www.s4c.co.uk/hafan/c_index.shtml]; APTN (Canada) [http://www.aptn.ca/]; and 'Ōiwi TV (Hawai'i) [http://www.oiwi.tv/].

Indigenous Broadcasting

The rise of indigenous radio and television broadcasting has served many functions and enabled program producers to reach many audiences, both internal within the local indigenous community and external to local non-indigenous neighbors, regional and national audiences, translocal indigenous audiences, and global audiences of all types (often aided by digital streaming of program via the internet). Indigenous broadcasting and media production may be commercial or noncommercial, community-based or individual artistry. In nearly all cases, however, the growth of indigenous or aboriginal broadcasting has resulted in the creation of indigenous public spheres: not only spaces for discourse and dialogue that are controlled by native cultures (allowing for voices, perspectives and images rarely seen and heard via mainstream media), but also spaces that nurture counterhegemonic (and sometimes anti-hegemonic) social and political movements. Kevin Glenn and A. F. Tyson (2007) argue that contemporary mediascapes are sites where competing discourses and knowledge formations circulate and often clash on a global level; they see the presence of indigenous cultural voices in those mediascapes as significant in "inflect[ing] global cultural processes with localizing accents" (p. 206).

Radio has been the most pervasive media form to penetrate indigenous communities from the "outside world"; when adopted for indigenous use, radio technologies have provided local, even isolated, communities with the ability to communicate within their own communities as well as interculturally to a mixed regional audience. Radio is an important tool for sharing news, public affairs and dialogue on social/political issues—the serious business—as well as to convey cultural expression through traditional and popular music and other forms of aural entertainment.

Localized broadcasting for local indigenous audiences is widespread today, whether stations are state-sponsored or pirate radio; these range from projects such as the Inuit Broadcasting Corporation of Nunavut, Canada (Brisebois 1983) to the 'La Voz de la Montaña' radio emanating from Guerrero, Mexico (Rodríguez 2005), from the pioneering Warlpiri media experiment in Australia (Michaels 1994) to BilwiVision on Nicaragua's Mosquito Coast (Glynn and Cupples 2011). Community media in Africa, though rarely labeled as "indigenous media," occupy a similar status to indigenous community broadcasting projects in the Americas, Australia and New Zealand. Arising in the 1990s from an environment in which radio and television had been state-controlled prior to independence from colonialism, African community media have provided an avenue for civil society to develop an electronic voice beyond the control of government or commercial interests.

Community radio broadcasting has also been of central importance to the remote indigenous communities of Australia since the early 1970s, developing primarily through a system of regional media associations such as CAAMA (Central Australian Aboriginal Media Association), established in 1980. The launch of the Aussat communications satellite in 1985 enabled broadcasting of mainstream TV and radio into remote Australia; however, concern about the potential impact upon Aboriginal cultures led to the commissioning of reports by the federal government (see Michaels 1986), of the possible cultural impacts. This result in government formation of the Broadcasting for Remote Aboriginal Communities Scheme (BRACS) from 1987 to 1991, which encouraged creation of locally-produced content and reception of mainstream programming via satellite in 80 indigenous communities. Since 1996, the National Indigenous Radio Service has provided a 24-h feed of indigenous programming as well as an internet-based service across the country. Daniel Fisher's ethnographic work on Aboriginal radio (2005, 2009) examines the complex functions that radio serves in the social relationships of Aboriginal communities, diasporic communities, and as forms of cultural activism, while Michael Meadows (2009) reports upon and interprets the results of an extensive community media audience survey in Australia.

Although radio's reach has been broader and more pervasive in reaching indigenous communities globally, the development of indigenous-controlled television broadcasting has empowered aboriginal groups and established indigenous visual media producers as players in the larger mediascapes of nations within which they reside. Indigenous television stations and their programming may have local, regional, national, or—with the aid of the internet—global reach and impact. Even more than indigenous radio, while remaining deeply rooted in the local, aboriginal television has in many cases enabled translocal and pan-indigenous connections.

Television broadcasting also necessitates negotiations and involvement with the regulatory and licensing agencies and media policy-making bodies of the nation-state as well as funding from outside sources. Nowhere has this narrative been as well documented as in Australia and Canada. Michaels (1994) provides a distinctive insight into the pioneering establishment of broadcasting among the Warlpiri. Bell (2008) frames as a "battle" the efforts to obtain a remote television satellite license in order to establish the groundbreaking Imparja TV. Similarly, in Canada, the negotiations to establish local broadcasting corporations, as well as the nationwide APTN, are documented by Debbie Brisebois (1983), Lorna Roth (2005), and Ziggy Hafsteinsson and Marian Bredin (2010). From a very different angle, John Hartley (2004) provides a theoretical perspective on the way that indigenous television has reshaped the national mediascape—and in fact the sense of nation—in Australia, with implications for Canada, New Zealand, and other settler nations as well.

The media of the Sámi, an indigenous culture of far northern Europe, provide an excellent case study regarding the translocal and transnational nature of much indigenous media. Sari Pietikäinen (2008a, b) introduces issues of Sámi media and journalism, advocating for the importance of a public mediaspace for this indigenous group, creating an indigenous public sphere across several nation-states as well as throughout the Sámi diaspora. Since the Sámi homeland of Sapmi is transnational, Sámi media generally become fragmented by the imposition of state boundaries and national media institutions. Sámi Media/Sámi Radio in Finland is distinct from Sámi media in Norway—Norway's NRK Sápmi (TV Ođđasat) [http://www.nrk.no/sapmi/] is an entity within the Norwegian Broadcasting Corporation (NRK) that presents internet-based news as well as Sámi-language radio and television programming—and Sweden, where SR Sámi Radio [http://sverigesradio.se/sameradion/] is an entity within Sveriges Radio, and Sveriges Television has a Sámi component, SVT Sápmi. Beyond broadcasting, Sámi culture is thriving on digital media, social media, and through filmmaking, clearly providing a transnational, diasporic media space and public sphere for Sámi expression, politics and identity-building.

Indigenous Film and Video

Because Indigenous media circulate within and across local, national, regional and global spaces, through networks that are most often externally shaped by settler-state funding and distribution infrastructures, much of it tends to be produced within dominant genre categories.

Long the subject of documentaries by ethnographic filmmakers themselves, many native media producers use nonfiction to present their own cultural realities in their own voices, often arguing against social and political injustices (see Leuthold 2001). Documentary film and video often represents a "first-wave" of Indigenous media production, often created at the community level and frequently focusing on cultural heritage, but just as often associated with activism, galvanized by struggles for indigenous rights. In many cases, to provide funding, technology, and training, these productions are enabled through partnerships with state institutions; through university-based training programs; or with anthropologists, activists or NGOs that establish projects to encourage indigenous media production. Documentary videos are generally distributed either through local screenings, through regional festivals, or through broadcasting projects. Ginsburg (2005) distinguishes between (1) local or regional small-scale productions, generally focusing on local cultural and environmental issues, and (2) auteur-based feature filmmaking (introduced by settler colonialism and adapted to indigenous storytelling). Isuma TV [http://www.isuma.tv], a digital project developed by the Igloolik Isuma production company in the Canadian arctic, is the major online portal encouraging indigenous filmmakers to upload media productions for preservation, archiving and global sharing. Isuma.tv features documentary and nonfiction modes of storytelling in over 50 indigenous languages.

Accomplished indigenous storytelling practices and grassroots communication networks exist throughout Mexico and Central and South America. Indigenous video production in Latin America surfaced in the 1980s alongside participatory media projects inspired by Third Cinema, New Latin American Cinema, and experiences of grassroots, underground and radical community video. These paved the way for experimental or alternative video collectives in the 1990s (Salazar and Córdova 2008). The movement towards self-representation and for an "indigenous communication" (*comunicación indígena*) has grown alongside a history of resistance, the assertion of self-determination, and the defense of the rights of indigenous peoples, as summarized by Antoni Castells i Talens (2003), who theorizes how indigenous film may help consolidate historical indigenous resistance and support a move towards "cultural and linguistic normalization" in Latin America.

Exemplary community-based video projects in Latin America—such as Video nas Aldeias and the Kayapó Video Project, both in Brazil, and the Chiapas Media Project (Promedios) in Mexico—serve as one type of model for others worldwide. As anthropologists or activists brought technology into the villages in the mid-1980s (see Turner 1990), the indigenous villagers embraced video technology as a way to record and preserve their cultural practices (bypassing pen and paper), creating and circulating videos between and across communities. Villagers initially focused on documenting not only traditional knowledge and rituals but also records of intercultural political negotiations (Turner 1992). The non-literate Kayapó soon recognized the value of video as a tool to create social and political documents and to "fix" historical moments. They later found audiences in regional, national and even global publics. As Turner (1992) points out, these projects differ from government-sponsored indigenous broadcasting projects in other places since they are fairly low-tech (in the Amazon, video cameras and generator-powered editing decks) and are not accountable to outside governments.

Indigenous media-makers in Latin America have faced extremely limited funding for training, production and distribution, either relying on grants from advocacy organizations or state bodies, or opting for self-funding. Works in circulation are frequently developed in training workshops, revealing a democratizing drive in programming and exhibition as well as a commitment to support the indigenous training process.

The distribution and exhibition of indigenous media also involve much movement through geographical spaces to significant localized places. Indigenous media in their development have not followed conventional audiovisual production or distribution processes, nor do they generally circulate through conventional theatrical releases or in film and video circuits. While some works screen at major indigenous film and video festivals (Weatherford and Córdova 2010), most work is not in formal distribution. Rather, they inhabit a wide range of alternative venues, including film festivals, international meetings on indigenous rights, academic conferences, community television and local community screenings. Because they reflect various production methods, styles, and formats, their entrance into major broadcast spaces and, until recently, most mainstream film festivals, has been quite limited.

Indigenous cinematic fictional and narrative film has often developed around auteur filmmakers before coalescing (when and if it does) into "national cinema" movements representing indigenous nations. Ginsburg (2002) has emphasized that the recent international circulation of indigenous films on the global stage at festivals like Cannes and Sundance has often, ironically, made them representations of the cinema of their nation-states and has added cultural capital to those national cinemas as a whole, even though the initial impulse of most of the filmmakers is to create films that are counter-hegemonic, outside of the dominant film industries of their nations. Increasingly, First Nations features are included in the broader study of Canadian feature filmmaking. Recent critical approaches to the study of First Nations film and video have taken up the frameworks of decolonization (Knopf 2009), sovereignty (Taunton 2010), tribal epistemologies (Kelsey 2011), and transnationalism (Santo 2008). Others focus on the subversion of colonial records and frameworks (Hladki 2006) and the work of particular producers across broadcast and film media (White 2005).

Scholarly study of fictional and narrative indigenous films—primarily feature films, but also including shorts and docu-dramas—has expanded in step with global increase in production. Recent book-length studies indicate the new energy in the field and the particular dual emphasis upon national/continental cinemas and transnational/global comparative studies. For example, Singer (2001), Raheja (2011a), Knopf (2009), and Schiwy (2009) focus upon particular continents, geographic regions, or national borders as defining spaces for the production of particular kinds of films. Corinn Columpar (2010), Peter Limbrick (2010), and Houston Wood (2008), on the other hand, have produced more wide-ranging, expansive, trans-continental studies that take up theoretical models from post-colonial studies, settler studies, and other comparative frameworks to assess radically dispersed Indigenous cinema traditions.

Digital Natives: Creating Virtual Geographies for the Future

The convergence of ancient cultures and high-tech tools has created some very exciting possibilities for indigenous peoples worldwide. Katarina Soukoup (2006) retells[2] a lovely story about the need to translate the term (and concept) of the internet into the Inuit language, Inuktitut. The selected word was *ikiaqqivik*, or "traveling through layers": a concept describing an Inuit shaman's traveling through time and space to find answers or to locate living or deceased relatives. As Soukoup notes, this provides "an example of how Inuit are mapping traditional concepts, values, and metaphors to make sense of contemporary realities and technologies" (para. 1); she also suggests that, "like shamans in the digital age," indigenous digital media artists have adapted "cutting-edge technologies such as high-definition video and wireless broadband to 'travel through the layers' of time, geography, language, history, and culture... to present to the world a discourse from a distinctly Inuit point of view..." (para. 2).

Before indigenous communities and individuals are able to participate in digital culture-building, they must first have access to the technologies and the infrastructures. The majority of indigenous peoples, globally, reside in remote geographical areas that have generally been located beyond the scope of early technological development priorities by the dominant state and regional governing bodies. In addition to infrastructures, which are geographically and community-based, access also involves the need for adequate digital tools such as computer terminals, laptops or mobile devices, which require capital—a commodity that is often in short supply in many indigenous communities that may subsist on the lower end of the socioeconomic scale. New technologies also require digital skills training for digital literacy, which also requires funding. Gaining access to technologies

[2]Soukoup cites a 2005 article in *Nunatsiaq News* by Sara Minoque.

and infrastructure in remote communities may be the largest barrier to indigenous peoples seeking to become more digitally connected.

The use of digital media to maintain (and sometimes create) an indigenous public sphere at the translocal or regional level is an often-discussed theme regarding the indigenous adaptation of digital technology. Less discussed, though clearly as important, is the ability of digital media to bring indigenous groups into larger public spheres as equal discursive players at the translocal, regional, national, or even global levels. The development of a global pan-indigenous community has been enabled by digital media over the past few decades, as has the inclusion of formerly isolated indigenous voices in the democratic process and decision-making of nation-states.

In Nunavut, Canada, concerns and speculation about the integration of the Inuit into digital public sphere arose in the 1990s (see Savard 1998); 14 years later, the recently-launched Digital Indigenous Democracy project, *My Father's Land* [http://www.isuma.tv/did] (started in May 2012), provides the technology and high-speed infrastructure (community radio and television as well as the internet and social media) to enable Inuit citizens in remote communities to participate in a review of the human and environmental impact of a transnational iron mining project threatening their homeland. This enables Inuit citizens to use their traditional decision-making skills "to get needed information in language they understand, talk about their concerns publicly and reach collective decisions with the power of consensus" (DID Overview, para. 3).

Questions as to how to classify, manage, access, utilize, display and teach local and place-based indigenous cultural knowledge are deeply embedded in culturally-specific ontologies and in contested histories of colonial misappropriation and disrespect of cultural knowledges and cultural objects. As knowledge is power, so are paradigms for ways to appropriately deal with cultural knowledges that are imbricated in regimes of ideological power and histories of cultural politics. The more theorized scholars dealing with these issues (Michael Christie, Kimberly Christen, Ramesh Srinivasan and others) develop and build upon these not just in terms of technological adaptation but also regarding ideological issues emphasizing the need for local cultural control over how, to whom, and under what conditions cultural knowledge may be revealed or displayed. This awareness of the need for, and advocacy for, self-determination over representation by indigenous communities holds true and relevant for discussion of ways that digital information and communication technologies (ICTs) might serve local goals of storytelling (control over various types of narratives for various audiences), knowledge databasing and archiving, cultural heritage preservation and museum display, or education (both intra- and interculturally). Although these issues are not limited to the digital, they are essential central issues as more and more of these domains are becoming digitized. While questions of digitizing in any of these realms may on the surface appear to be primarily about technology, at the deepest levels they are about cultural ideologies and the increasing awareness of the need for local, community-generated, dialogic and culturally-specific solutions to how to structure any digital technology initiative.

Christie (2004) notes that digital architectures are not theory-neutral but rather arrive already encoded with specific politics of knowledge; Aboriginal databases don't just contain representations of Aboriginal knowledge but must be coded to reflect and reinforce Indigenous ontologies at every level. Elizabeth Povinelli (2011) theorizes the way configurations of power, structured into mainstream archiving systems, must be challenged through the construction of postcolonial archival systems. Christen (2005) reflects upon her work with digital collections in an Australian Aboriginal community, similarly suggesting that "local knowledge regimes redefine national and global debates concerning the preservation and production of indigenous traditional knowledge in the cultural commons. If the colonial idea of the archive was to collect and store the world's treasures for the betterment of mankind, this emerging Warumungu archive is part of an intimate set of kinship relations and a dynamic socioterritorial network that rubs up against national territorial boundaries and legal structures aimed at protecting indigenous culture" (p. 317). T.V. Padma (2005), in discussing indigenous knowledge archiving in South Asia, points out the need to develop laws to prevent misappropriation and commercialization of such knowledge by linking to the international patent classification system. International standards for registries of traditional knowledge were adopted by the World Intellectual Property Organization (WIPO) in 2003.

The international indigenous movement begun in the 1980s has brought together indigenous groups from around the world in pursuit of common goals of self-determination in governance, cultural intellectual property rights, and preservation of cultural lands and heritage. This has focused attention upon and has raised awareness of the deep hegemonic bias encoded into traditional mainstream museum representation and interpretation of indigenous cultural materials. Srinivasan et al. (2008a, b) note, "As institutions endowed with presenting the tangible and intangible cultural heritage of diverse populations, museums are confounded by the challenge of representation—in short, all of the many decisions that go into the selection of objects, their classifications, arrangements, accounts and the development of mechanisms by which the objects can circulate and interact with diverse stakeholder communities (such as indigenous groups, scholars, museum curators, students, etc.). These decisions generate a significant dilemma..." (para. 20). Together, this historical moment has thrown museums into crisis and given birth to a counter-hegemonic indigenous museum movement and "a growing dialogue between mainstream museums and indigenous people around museum representation" (Srinivasan et al. 2009, para. 5).

Digital heritage projects are the new trend as museum anthropologists seek partnerships with represented cultures in adopting protocols for cultural display. Notable museum and cultural heritage case studies include the Ara Irititja Project [http://www.irititja.com/] (see Hughes and Dallwitz 2007), The Mukurtu Project [http://www.mukurtu.org] (see Christen 2008), and the Creating Collaborative Catalogs Project [http://www.digital-diversity.org/?page_id=09] (see Srinivasan et al. 2008a, b).

Innovative digital artistic projects linking indigenous cultural space to cyberspace in dynamic and interactive ways have arisen in recent years. Aboriginal Territories in

Cyberspace (AbTeC) is working hard to carve out an autonomous Indigenous region of cyberspace in exciting new ways. Current projects include *TimeTraveller*™ [http://www.timetravellertm.com/] (a *machinima* production in *Second Life*), Skins (a video-game storytelling workshop), a video game called *Otsi*:! *Rise of the Kanien'kehá:ka Legends* [http://otsi.abtec.org/], and a project of performance art within *Second Life*. Active Pass to IR9 is a collaborative work of digital art by Kate Hennessey and Richard Wilson that maps indigenous cultural meaning onto geographical place through video. Vachiam Eecha is a collaborative web "Cuaderno" project by David Delgado Shorter and his Yoeme partners in Sonora, Mexico. These are just a few of many indigenous web projects available online today.

Closing

In conclusion, we have charted the rise of a distinct interdisciplinary field since the 1980s that engages the intersecting theoretical interests of scholars in cultural anthropology, media studies and cultural geography, among other fields, to focus upon the many ways that indigenous peoples are adopting digital and electronic communication technologies as tools for cultural expression, political assertion and cultural preservation. In doing so, these groups, both singly and collectively, are negotiating the forces of localism, nationalism and globalism in order to create novel types of indigenous public spheres and twenty first-century indigenous geographies. Indigenous media serve to map and re-map these dynamic indigenous cultural worlds, linking the grounded places of indigeneity—the homelands and sacred geographic spaces—not only to each other in a translocal remapping of the indigenous sphere but also to the more abstract (and often virtual) spaces of global geography, politics and culture. Indigenous digital architectures being constructed upon these geographies articulate distinctly indigenous knowledge systems: cultural ideologies, logics, sensibilities and aesthetics. Most significantly, the mastery and control over the technologies that enable such articulations also represents an empowerment expressed through self-representation that contributes ultimately to cultural and, in many cases, political sovereignty and, as Salazar (2009) points out, "self-determination in practice."

Acknowledgements I thank Oxford Press for permission to incorporate into this essay material from "Native America" and "Indigenous Media," articles I wrote and co-wrote for the Oxford Bibliographies Online project in Cinema Studies, edited by Krin Gabbard. I extend special thanks to Joanna Hearne and Amalia Córdova for their contributions to this essay.

References

Barclay, B. (1990). *Our own image*. Auckland: Longman Paul.

Barclay, B. (2000). *Mana Tuturu: Māori treasures and intellectual property rights*. Honolulu: University of Hawaii Press, 2006.

Bell, W. (2008). *A remote possibility: The battle for imparja television*. Alice Springs: IAD Press.

Brisebois, D. (1983). The Inuit Broadcasting Corporation. *Anthropologica (New Series), 25*(1), 107–115, Native North Americans and the Media: Studies in Minority Journalism.
Carelli, V. (1988). Video in the villages: Utilization of video tapes as an instrument of ethnic affirmation among Brazilian Indian groups. *Commission on Visual Anthropology Bulletin, 4*, 10–15.
Castells i Talens, A. (2003). Cine indígena y resistencia cultural. *Chasqui, 84*, 50–57, http://www.redalyc.org/pdf/160/16008407.pdf. Accessed 13 Sep 2013.
Christen, K. (2005). Gone digital: Aboriginal remix in the cultural commons. *International Journal of Cultural Property, 12*(3), 315–345.
Christen, K. (2008). Archival challenges and digital solutions in aboriginal Australia. *SAA Archaeological Record, 8*(2), 21–24, http://www.saa.org/Portals/0/SAA/Publications/thesaaarchrec/mar08.pdf. Accessed 13 Sep 2013.
Christie, M. (2004). Computer databases and aboriginal knowledge. *Learning Communities: International Journal of Learning in Social Contexts, 1*, 4–12.
Columpar, C. (2010). *Unsettling sights: The fourth world on film*. Carbondale: Southern Illinois University Press.
Deger, J. (2006). *Shimmering screens: Making media in an aboriginal community*. Minneapolis: University of Minnesota Press.
Digital Indigenous Democracy Overview. Isuma TV. (5 May 2012). http://www.isuma.tv/hi/en/DID/DIDoverview5May2012. Accessed 13 Sep 2013.
Digital Indigenous Democracy. Isuma TV. (4 May 2011). http://www.isuma.tv/hi/en/did/CMF. Accessed 13 Sep 2013.
Evans, M. R. (2008). *Isuma: Inuit video art*. Montreal/Kingston: McGill-Queen's University Press.
Fisher, D. T. (2005). "You Mob Listen": Cultural exchange and indigenous media in Northern Australia. Ph.D. dissertation, Department of Anthropology, New York University.
Fisher, D. T. (2009). Mediating kinship: Country, family, and radio in northern Australia. *Cultural Anthropology, 24*(2), 280–312.
Ginsburg, F. D. (1991). Indigenous media: Faustian contract or global village? *Cultural Anthropology, 6*(1), 92–112.
Ginsburg, F. D. (2002). Screen memories: Resignifying the traditional in indigenous media. In F. D. Ginsburg, L. Abu-Lughod, & B. Larkin (Eds.), *Media worlds: Anthropology on new terrain* (pp. 39–57). Berkeley: University of California Press.
Ginsburg, F. D. (2005). Blak screens and cultural citizenship. *Visual Anthropology Review, 21*(1/2), 80–97.
Ginsburg, F. D., & Myers, F. (2006). A history of aboriginal futures. *Critique of Anthropology, 26*(1), 27–45.
Glynn, K., & Cupples, J. (2011). Indigenous MediaSpace and the production of (Trans)locality on Nicaragua's mosquito coast. *Television and New Media, 12*(2), 101–135.
Glynn, K., & Tyson, A. F. (2007). Indigeneity, media and cultural globalization: The case of Mataku, or the Māori X-Files. *International Journal of Cultural Studies, 10*(2), 205–224.
Hafsteinsson, S., & Bredin, M. (Eds.). (2010). *Indigenous screen cultures in Canada*. Winnipeg: University of Manitoba Press.
Hartley, J. (2004). Television, nation and indigenous media. *Television and New Media, 5*(1), 7–25.
Herman, R. D. K. (2008). Reflections on the importance of indigenous geography. *American Indian Culture and Research Journal, 32*(3), 73–88, http://aisc.metapress.com/content/n301616057133485/. Accessed 13 Sept 2013.
Himpele, J. D. (2007). *Circuits of culture: Media, politics, and indigenous identity in the Andes*. Minneapolis: University of Minnesota Press.
Hladki, J. (2006). Decolonizing colonial violence: The subversive practices of aboriginal film and video. *Canadian Woman Studies, 15*, 83–87.
Hopkins, C. (2006). Making things our own: The indigenous aesthetic in digital storytelling. *Leonardo, 39*(4), 341–344.

Hughes, M., & Dallwitz, J. (2007). Ara Irititja: Towards culturally appropriate IT best practice in remote indigenous Australia. In L. E. Dyson, M. Hendriks, & S. Grant (Eds.), *Information technology and indigenous people* (pp. 146–158). Hershey: Information Science Publishing.

Kelsey, P. M. (2011). Condolence tropes and haudenosaunee visuality: It starts with a whisper and Mohawk girls. In D. K. Cummings (Ed.), *Visualities: Perspectives on contemporary American Indian film and art* (pp. 119–130). East Lansing: Michigan State University Press.

Knopf, K. (2009). *Decolonizing the lens of power. Indigenous films in north America*. Amsterdam: Editions Rodopi BV.

Leuthold, S. (2001). Rhetorical dimensions of native American documentary. *Wicazo Sa Review, 16*(2), 55–73.

Limbrick, P. (2010). *Making settler cinemas: Film and colonial encounters in the United States, Australia, and New Zealand*. New York: Palgrave Macmillan.

Meadows, M. (2009). Electronic dreaming tracks: Indigenous community broadcasting in Australia. *Development in Practice, 19*(4/5), 514–524.

Michaels, E. (1984). Ask a foolish question: On the methodologies of cross cultural media. *Australian Journal for Cultural Studies, 3*(2), 49–59.

Michaels, E. (1986). *The aboriginal invention of television in Central Australia, 1982–1986: Report of the fellowship to assess the impact of television in remote aboriginal communities*. Canberra: Australian Institute of Aboriginal Studies.

Michaels, E. (1994). *Bad aboriginal art: Tradition, media, and technological horizons*. Minneapolis: University of Minnesota Press.

Murillo, M. A. (2010). *Resistance on the air: Indigenous radio and the struggle for social justice in Colombia*. Boston: South End Press.

Padma, T.V. (10 Jan 2005). Digital library to protect indigenous knowledge. In *Science and Development Network*. http://www.scidev.net/en/news/digital-library-to-protect-indigenous-knowledge.html. Accessed 13 Sep 2013.

Pietikäinen, S. (2008a). Broadcasting indigenous voices: Sámi minority media production. *European Journal of Communication, 23*(2), 173–191.

Pietikäinen, S. (2008b). 'To Breathe Two Airs': Empowering indigenous Sámi media. In P. Wilson & M. Stewart (Eds.), *Global indigenous media: Cultures, poetics, and politics* (pp. 197–213). Durham: Duke University Press.

Povinelli, E. (2011). The woman on the other side of the wall: Archiving the otherwise in postcolonial digital archives. *d i f f e r e n c e s: A Journal of Feminist Cultural Studies, 22*(1), 146–171.

Raheja, M. H. (2007). Reading Nanook's smile: Visual sovereignty, indigenous revisions of ethnography, and Atanarjuat (The fast runner). *American Quarterly, 59*(4), 1159–1185.

Raheja, M. H. (2011a). *Reservation reelism: Redfacing, visual sovereignty, and representations of native Americans in film*. Lincoln: University of Nebraska Press.

Raheja, M. H. (2011b). Visual prophecies: Imprint and it starts with a whisper. In D. K. Cummings (Ed.), *Visualities: Perspectives on contemporary American Indian film and art* (pp. 3–40). East Lansing: Michigan State University Press.

Rodríguez Ramos, J. M. (2005). Indigenous radio stations in Mexico: A catalyst for social cohesion and cultural strength. *Radio Journal: International Studies in Broadcast and Audio Media, 3*(3), 155–169.

Roth, L. (2005). *Something new in the air: The story of first peoples television broadcasting in Canada*. Montreal: McGill-Queen's University Press.

Salazar, J. F. (2009). Self-determination in practice: The critical making of indigenous media. *Development in Practice, 19*(4/5), 513.

Salazar, J. F., & Córdova, A. (2008). Imperfect media and the poetics of indigenous video in Latin America. In P. Wilson & M. Stewart (Eds.), *Global indigenous media: Cultures, poetics, and politics* (pp. 39–57). Durham: Duke University Press.

Santo, A. (2008). Act locally, sell globally: Inuit media and the global cultural economy. *Continuum: Journal of Media and Cultural Studies, 22*(3), 327–340.

Savard, J.-F. (1998). A theoretical debate on the social and political implications of internet implementation for the Inuit of Nunavut. *Wicazo Sa Review, 13*(2), 83–97. Technology and Native American Culture.

Schiwy, F. (2009). *Indianizing film: Decolonization, the Andes, and the question of technology*. New Brunswick: Rutgers University Press.

Singer, B. R. (2001). *Wiping the war paint off the lens: Native American film and video*. Minneapolis: University of Minnesota Press.

Soukup, K. (2006). Report: Travelling through layers: Inuit artists appropriate new technologies. *Canadian Journal of Communication, 31/1*, 239–246, http://www.cjc-online.ca/index.php/journal/article/viewArticle/1769/1889. Accessed 13 Sept 2013.

Srinivasan, R., Becvar, K. M., Boast, R., & Enote, J. (2008a). Diverse knowledges and contact zones within the digital museum. *Science, Technology and Human Values, 35*(5), 735–768.

Srinivasan, R., Boast, R., Furner, J., & Becvar, K. M. (2008). Digital museums and diverse cultural knowledges: Moving past the traditional catalog. December 1, 2008. http://rameshsrinivasan.org/wordpress/wp-content/uploads/2010/04/1-SrinivasanetalTISBlobgects.pdf. Accessed 13 Sept 2013.

Srinivasan, R., Enote, J., Becvar, K. M., & Boast, R. (2009). Critical and reflective uses of new media technologies in tribal museums. *Museum Management and Curatorship, 24*(2), 161–181.

Taunton, C. (2010). Indigenous (Re)memory and resistance: Video works by Dana Claxton. *Post-Script, 29*(3), 44–57.

Turner, T. (1992). Defiant images: The Kayapó appropriation of video. *Anthropology Today, 8*(6), 5–16.

Turner, T. (Fall 1990). The Kayapó video project: A progress report. *Revue de la Commission d'Anthropologie Visuelle* (CVA). 7–10. http://www.yumpu.com/en/document/view/12313914/the-kayapo-video-project-a-progress-report. Accessed 13 Sep 2013.

Weatherford, E., & Córdova, A. (Spring 2010). Indigenous visual power: Film and media in Latin America. *American Indian*, 18–21.

White, J. (2005). Frozen but always in motion: Arctic film, video, and broadcast. *Velvet Light Trap: A Critical Journal of Film and Television, 55*, 52–64.

Wilson, P., & Stewart, M. (2008). Introduction: Indigeneity and indigenous media on the global stage. In P. Wilson & M. Stewart (Eds.), *Global indigenous media: Cultures, poetics and politics* (pp. 1–35). Durham: Duke University Press.

Wood, H. (2008). *Native features: Indigenous films from around the world*. New York: Continuum Press.

Worth, S. (1997). *Through Navajo eyes: An exploration in film communication and anthropology*. Albuquerque: University of New Mexico Press.

Salsation: Affective Encounters and the Right to the City

23

Tamara M. Johnson

The first thing I notice as I climb the old wooden stairs toward the faint sound of trumpets, congas, and clave growing steadily louder is the heat—it emanates from the warm glow of the restaurant's entrance and gently caresses my face, then envelops my body like a warm embrace. As I turn the corner and enter the restaurant, the vibrations of base and of rhythmic footsteps reverberate up from the wooden dance floor, entering my body through my heels, traveling up my legs to settle in my gut. I move through the gauntlet of warm smiles, damp hugs, superficial pecks on the cheek, and begin to navigate the precarious dance space of high heel shoes and flailing elbows in search of my first dance partner of the evening. I notice a woman strolling gracefully through the front doors wearing a spicy red dress that might be overdramatic on any other night but seems perfectly appropriate for this setting. A smartly-dressed guy in designer jeans and a black t-shirt, smooth and confident, enters the venue. His eyes find the girl with the magical shoes and he winks at her, approaches, and extends his hand. She gracefully accepts his request by taking his arm and leading him onto the dance floor. They sway together in a close embrace before he leads her to effortlessly execute triple spins, travelling turns, direction changes, and dramatic dips all with confidence, style, and a slight smile. I watch them briefly, engaged in a flawless performance that seems sensual and intimate, yet tinged with a hint of self-importance.

From a cozy chair in one corner of the restaurant, a casual observer can watch sweaty bodies twist and twirl to an Afro-Caribbean beat. At first glance, this observer might believe that the intensity of this zone of contact in one of Cape Town's hottest salsa clubs holds all of the promise of uninhibited social interaction and movement on the dance floor. In fact, many dancers describe the sense of

T.M. Johnson (✉)
Academic Planning and Analysis, Office of Academic Affairs, University of North Carolina at Charlotte, Charlotte, NC, USA
e-mail: tamaramargaret@gmail.com

S.P. Mains et al. (eds.), *Mediated Geographies and Geographies of Media*,
DOI 10.1007/978-94-017-9969-0_23

freedom that salsa offers. Others say it is the joyous and sensual nature of the dance and the self-expression possible that inspires them. Mostly, however, people claim to seek a community with their same passion for movement, expression, and feeling. Movements, communications, and interactions on the dance floor are driven by emotional responses to other participants and to the dance. Also, many participants in the Cape Town salsa scene are pleased with the multicultural or cosmopolitan nature of the scene and are attracted to the idea that salsa, for them, seems to transcend race. However, this sense of community is cultivated through passion and desire but is not unmarred by *schadenfreude*, jealousy, and the need for recognition. Investigating Cape Town's salsa scene reveals emotive processes of space-making in the city—creating networks, flows, and territories. These micro-political maneuverings and embodied negotiations are central to the struggle for and articulation of rights to the city. In this chapter, I argue that dance is a medium through which urban territories are negotiated. I discuss the dancing body as a lens for understanding the role of affective and emotional geographies in the active creation and contestation of spaces of exclusion and belonging.

Much like other forms of media (i.e. newspapers, advertising images, music, film), examining the dancing body can elucidate the complex intermingling of social and spatial interactions and the exchange of ideas, meaning, and emotion. According to geographer Paul Adams (2010), communications are "exchanges," or "arrangements that connect one to one, one to many, many to one, and many to many" (p. 39). Media, he argues, can "indicate any communication arrangement," whether in the form of material texts or expressed through social markers like language or code (p.39). I have argued previously that dance is a language of the body (Johnson 2011). The dancing body is often analyzed as a representation: the production of Irish heritage and identity (Leonard 2005), as a performance or an appropriation of black American culture (Johnson 2003; Laidlaw 2011) or as a way of memorializing events (Nelson 2008). The dancing body is also an affective entity; it is the vessel through which affections are transmitted and spaces are transformed through affective rendering. In this chapter, I examine the ways that dancing bodies and the spaces they create through affective rendering are defining territories, and providing avenues of urban participation for salsa dancers.

Dance holds social and political power in South Africa where crucial constructed categories of identity were visible on the body, and segregated landscapes were built and legislated according to State-ascribed racial and ethnic categories. In South African urban spaces, dance has a long legacy as a medium for defining geographies of exclusion. Apartheid-era restrictions prohibiting interracial dancing in public and private indicate that social dance practices have often been treated with suspicion, as sites of inappropriate social encounter (Johnson 2013). These restrictions highlight the significance that law- makers and gate-keepers of high society placed on social dancing as a dangerously intimate practice that could lead to devious social behavior and, most dangerously, to racial mixing. Social interaction, particularly sexual interaction, was a challenge to the delicate balance of minority rule premised on racial superiority (white supremacy) in order to establish and protect white political and economic power and social capital (Elder 1998, p. 156). One way to examine the intricacies of social contact and encounter in a post-apartheid city undertaking radical restructuring is to examine carefully the not-so-obvious spaces that bring together many different kinds of people into a relatively safe but unstable environment where rules evolve, roles change, and movement is not only expected but required. Social dancing can allow for the embodied expression of intense joy and communities form around a shared enthusiasm for the music and the desire to belong. However, social dance scenes are also landscapes through which struggles over control, legitimacy, and social and cultural capital are waged (see Johnson 2011). Studying dance and spaces of social dancing allows for an investigation of new avenues of participation, engagement, and movement, not only of and between bodies on the dance floor, but also within cities. In Cape Town, where difference and division were deliberately legislated and social (specifically sexual) interaction between racial groups explicitly forbidden, new social relationships are navigated, and urban spaces are negotiated and re-mapped through affective participation in dance scenes.

In this chapter, I use an ethnographic exploration of the salsa dancing scene in Cape Town, South Africa to explore affective movements as they shape interactions and stimulate the creation of urban spaces. Desires, hostilities, intentions are expressed through dance and communicated among dancers; affect is transmitted through the body, and translated to designate and define territory. I interrogate tension between dancers that are indicative of emotional struggles over territory and the power to create. Dance, as a medium, constitutes these social interactions in spaces. Urban spaces are written by everyday encounters with different ideas, backgrounds, values, and ways of living. Inherent in the processes of encounter are struggles over the rights to the city—the right to create spaces of belonging. I combine Henri Lefebvre's conceptualization of the 'right to the city' with literature on emotive and affective responses to urban space in order to engage with alternative understandings of urban participation.

Affective-Emotive Geographies, Dance, and the City

Geographers have recently emphasized the importance of affect and emotion when studying embodied negotiations of space. Nigel Thrift describes affect as "a form of thinking, often indirect and non-reflective true, but thinking all the same. And, similarly, all manner of the spaces which they generate must be thought of in the same way, as means of thinking and as thought in action. Affect is a different kind of intelligence about the world, but it is intelligence nonetheless" (2008, p. 187). Borrowing from Brian Massumi's work, Alan Latham and Derek McCormack describe affect as a "felt but impersonal, visceral but not neatly corporeal, force of intensive relationality" (2004, p. 706).

The body is the site of affective transmission, "emotional experience," and expression (Davidson and Milligan 2004, p. 523). Attempting to understand the transpersonal flow of affect or emotions as embodied practices has pushed research beyond the visual and textual into areas such as performance as non-representational geographies. Geographic studies dealing with affect, emotion and dance in particular, have emerged, in part, in response to Nigel Thrift's work on non-representational theory (Thrift 2008). The embodied and affective quality of dance contributes to what Thrift calls the "sixth kinesthetic proprioceptive sense," which is difficult to articulate (p. 139). In his explorations of vibrations and Jamaican dancehall scenes, Julian Henriques (2010) writes that movement and affect are embodied together, thus "to feel . . . is to feel moved" (p. 57). His article focuses on the ways in which affect is transmitted like vibrations between bodies on the dance floor. Dancers in Cape Town's salsa scene often described the "vibe" of a particular club. In these exchanges, vibe describes a feeling or energy transmitted between and among people and through space. The vibe is the energy generated by the sound and light attitudes as well as by the movement of dancing bodies. This description of vibe is reminiscent of the "sense of place" conceptualizations that emerged in the 1970s with the rise of humanist geography that put the human experience of place at the center of geographic inquiry. Influential human geographers examined the intimate, emotional relationships between place and the individual, including examinations of the pre-cognitive movements of people through place (Seamon 1979) the authenticity of place (Relph 1976), or the self as being-in-the-world (Tuan 1977). Drawing from this tradition, I describe vibe here as the dynamic, difficult to articulate, almost palpable affective rendering that attaches multiple sentiments to place and influences the dancing experience.

Affect and the Right to the City

Recently, prominent urban scholars have stressed the importance of examining affect to better understand the city (Thrift 2008; Morris 2004; Lantham and McCormack 2004). Edgar Pieterse recognizes the attention to affect as a promising epistemological contribution to African urban studies. He suggests the importance of examining affect/emotion and the related judgments, actions, responses, and intentions "because it is only through the redeployment of such registers that one can begin to fathom what is going on in the real city" (p. 14).

If affect is an important part of the ways in which cities in South Africa come to be understood, what exactly do empirical examinations of affect allow us to understand about lived urban experiences and the right to create urban space? Cities are sites where belonging is expressed, rights to space are negotiated, and identities are constructed and contested, all of which are everyday processes and practices (Secor 2004). In post-apartheid South Africa, many have embraced the language of rights in a continued struggle to create an equal and open society. However, in pondering the question of how to rebuild South African cities, prominent scholar Abdoumaliq Simone (2004) thinks that we have to go beyond viewing the right to the city as a right to basic services. For Simone, the right to the city indicates a right to pursue one's aspirations, a right which fundamentally relies upon the residents' abilities to create connections between the "infrastructures, spaces, populations, institutions, and economic activities of the city" (p. 323). In this interpretation, the right to the city cannot be legislated entirely by government and urban policy; urban residents must play an active role in the pursuit and realization of their aspirations.

Henri Lefebvre's idea of "right to the city" refers not only to the right to urban services but also the right to inhabit and transform urban space and thus to become a creator of the city. For Lefebvre, the city should be explored as a work of art constantly created and recreated in the everyday rituals and interactions of urban residents (Fenster 2005). The right to participation ensures that city dwellers become involved in decision-making processes that regulate, create, and maintain urban space (Lefebvre 1968 in Kofman and Lebas 1996, p. 174). According to Lefebrve, "the right to the city manifests itself as a superior form of rights: right to freedom, to individualization in socialization, to habitat and to inhabit" (1996, p. 173; quoted in Amin and Thrift 2002, p. 143). The right to the city "is not simply a right to consume, since human beings also have a need for creative activity, for the oeuvre . . ." (Parker 2004). According to Purcell (2002), producing space in the city requires planning, not only material spaces (roads, parks, buildings), but also constructing social relationships, attachments, sentiments, and an urban imaginary; it involves "producing and reproducing all aspects of urban life" (p.102). For Lefebvre, it is our legacy as human beings to create our own spaces; the freedom with which we are able to create this space is the crucial measurement of the quality of social life (Shields 1999). Likewise, Zayd Minty, Co-founder of Creative Cape Town, uses Lefebvre's right to the city to outline the role of culture and cultural expression in the cultivation of inclusivity in Cape Town. He argues that inhabitants often articulate an imagination of the city through the medium of performance and visual arts, forms of expression that can facilitate understanding and unity across socioeconomic and cultural boundaries (2008).

Methods

I conducted ethnographic research in Cape Town from August to January 2006, and from April 2009 to May 2011 that included 42 semi-structured, in-depth interviews with the city's dance club owners, dance instructors, dancers, musicians, and disk jockeys. I was also a semi-professional salsa dancer, performer, and salsa event organizer. I performed with most of the salsa dance companies in Cape Town as well as several of the local salsa musicians, taught workshops and private lessons, directed an all-female salsa dance company and have choreographed routines with prominent dancers in the scene. Through these collaborations, I experienced what salsa participants describe as "politics" among salsa stakeholders. During my fieldwork, I was accused of "messing" in other instructors' territory, stealing another dancer's choreography, and another dancer's dance partner. I made some missteps during the course of my research. Some I made knowing that by pleasing some parties, I would irritate others. It became almost impossible to become involved in the scene yet remain neutral in the world of salsa politics. This level of active participation allowed me not only to observe the emotional aspects of people and events but also to participate fully in them. People expressed feelings about actions or events to me in casual conversation on the dance floor that they would not repeat in a recorded interview. In addition to being expressed verbally, emotions such as frustration, excitement, and disappointment can be read through the body and are often subtly expressed through gestures, eye contact, posture, and in the lead and follow of social dancing.

Salsa and the Cape Town Context

In order to explore the context of social dancing in Cape Town, it is important to mention the legacy of apartheid in the city. Apartheid not only produced geographies of exclusion in the city, it cultivated a climate of distrust among urban residents (Nahnsen 2003). Apartheid was a carefully orchestrated system of racialized economic, social, and political domination in which a white minority controlled the mobility of non-white bodies as well as access to land, labor, and resources. Residential, institutional, and social spaces were legislatively and architecturally engineered to ensure racial separation (reflected in legislation such as the Group Areas Act of 1950 and the Separate Amenities Act of 1953), and one's existence within the Union of South Africa was determined by bodily appearance (reflected in the codification of race under the Population Registration Act of 1950). For scholar and philosopher, Achille Mbembe (2004), the end of apartheid raises the urgent question of how to inhabit the city because, for most South Africans, the city has been the dominant site of their exclusion from modernity. Exploring salsa dancing in Cape Town as an affective practice reveals new ways of understanding the urban habitation: interaction, participation, and creation.

Salsa's development and diffusion cannot be separated from a history of migration, first of Cuban and Puerto Rican music and musicians to New York City in the mid-twentieth century, entwining music and dance styles with African-American performance culture that was the product of a South-North migration of its own; and finally of the global circulation of the music that became known as salsa. As part of this global migration of music and dance, Cape Town's salsa scene developed relatively recently, beginning in 2001. During my preliminary research in the fall 2006, most of the salsa venues were located in the city center with occasional parties located in Camps Bay, Tygervalley, or Observatory. By 2011, many of the venues for salsa in the city center had closed, some had reopened, and new locations experimented with creating a salsa vibe. Although a few salsa instructors and venues have ventured out of the city center in order to host salsa events and classes, the city center is considered the home for salsa. A couple of the salsa instructors mention that salsa as an art form originated abroad, yet the scene in Cape Town was created primarily by South Africans. Salsa is therefore considered to "belong to everybody;" it is a "fresh platform" with a cosmopolitan vibe. Other instructors have mentioned salsa's global appeal, which connects dancers in Cape Town with people who practice salsa in other parts of the world, particularly when business or leisure travel brings salsa dancers to Cape Town. Dance can unite people despite language differences. As one dancer explains, economic sanctions and the cultural embargo implemented internationally to protest South Africa's apartheid policies left many South Africans of her generation (she is 55 years old) feeling isolated and cut off from the rest of the world. Participating in salsa has enabled her to feel part of an international phenomenon. For many salsa participants who have never left Cape Town, participating in a global art form is an engagement with a cosmopolitan sensibility that is not often accessible for many South Africans who have not yet had the opportunity to travel abroad.

Most salsa practitioners credit Buena Vista Social Café in Greenpoint with being the first venue to create and sustain a salsa vibe. As they climb the stairs, patrons were greeted with a Cuban flag, followed by a large photo of Fidel Castro, and a floor-to-ceiling mural depicting the cover of the Buena Vista Social Club documentary. The brick walls of the restaurant were peppered with framed and faded photographs of Havana street scenes. The atmosphere was one of revolutionary nostalgia. Mismatched furniture and old wooden floors created a cozy ambiance, yet a precarious dance floor. The lighting mimicked the soft warm glow of candlelight. A corner of the restaurant near the bar was cleared and dedicated to dancers after 9 pm; however after about 10 pm, the small space could no longer contain enthusiastic dancing bodies, and dancing couples began to spill out into the restaurant space, occupying the crevices in between tables. The best dancers muscled it out for dance space on the periphery, often inspired by the attention of spectators sitting at the tables or on the balcony, as well as for the room to attempt more intricate turn patterns on the dangerous dance floor. "Serious" salsa dancers refused to dance near the bar area in order to avoid careless cigarettes or the spilled beer that can ruin expensive suede-bottomed salsa shoes. To dance at Buena Vista, one was always aware of the eminent possibility of bodily harm caused by an ill-placed high heel or elbow. Waiters and waitresses carrying trays of nachos and drink specials were constantly dodging high-speed arms and hips in motion. But what drew both salsa dancers and non-dancers alike to Buena Vista week after week for 8 years was the energy of the place—this combination of body heat, rivers of sweat trickling shamelessly down the back, pervasive smell of cigar(ette) smoke and nacho cheese, infectious clave and conga beat, and non-stop movement that created a desirable "vibe." This affection was transmitted among Buena Vista patrons, fueling and being fueled by embodied responses to Latin beats. It was an affective circulation of exuberance, tension, sexual energy, anticipation, and reckless inhibition. It was this energy, this sexy vibe, some patrons say, that defines salsa dancing for them. For many salsa participants, the sensual social interactions that often occur on the dance floor would not have been possible under apartheid's legislated spatial and social segregation experienced in their lifetimes.

Emotional Politics and Territoriality in Cape Town's Salsa Scene

The salsa dancing scene is a window into the ways that people navigate, create, and territorialize spaces and networks, driven by feelings such as passion, desire, envy, and pride, as well as the need for recognition and belonging. Such collective dance practices necessitate the creation of venues to host dance classes and events. For some participants in Cape Town, the primary value of these spaces is economic: salsa venues provide opportunities for salsa instructors, event organizers, and venue owners to market themselves and earn money. For others, these venues provide opportunities for people to establish communities of like-minded individuals, to develop new dance skills or demonstrate mastery of dance steps. Salsa dancers discuss the sense of family they feel with other dedicated members of the salsa scene. For many, social dance venues offer escape from the stresses of everyday life. Therefore, creating venues to host social dance events is also about creating spaces of belonging in cities. Participation is not merely economic, but is also a social and micro-political activity. The ways that dancers negotiate, contest, collaborate in the creation of such spaces describes *participatory* actions.

In many casual conversations, members of the salsa community lament a sinister political element surrounding contestations over space and students on the part of salsa instructors that undermines a sense of community. On the surface, economic motivations drive territorial practices among salsa stakeholders in Cape Town. Dancers, DJs, and instructors often allude to market "saturation" and a "culture of scarcity" when describing many of the problems between members the salsa community: there are simply too many salsa instructors competing to profit from too few dancers willing to pay to attend salsa classes and parties. However, economic explanations do not fully explain these practices. Some instructors admit that collaboration among teachers is difficult yet economically necessary because all salsa instructors must encourage their students to attend salsa events, thereby simultaneously raising the profile and the profits of salsa in Cape Town.

As salsa promoters strive to recreate a Buena Vista vibe of their own in other venues, many of these endeavors failed: not enough members of the salsa crowd showed up and of those who attended the parties, not enough people spent money at the bar. Increasingly, there are few establishments interested in supporting salsa parties because venue managers do not feel as though they will make enough money from salsa. Many salsa dancers do not spend enough money on food and alcohol for the venue to recover the costs of hiring a salsa DJ and sound equipment. According to a salsa instructor operating in Cape Town's southern suburbs, venues do not support salsa parties because "salsa is not supporting the venue." As the venues willing to host salsa parties dwindle, scrambles for club venues for parties ensue among competing salsa instructors and schools, leading to territorial practices deriving from a "culture of scarcity." Salsa dancers claim the perception that there is not enough space or not enough students for each instructor to make a satisfactory profit results in highly competitive behavior. Dancers conjecture that, because the community of salsa dancers is relatively small numerically, instructors are pulling and tugging over the existing paying clients. Instructors are constantly bickering over party venues, class schedules, and students.

Despite the general perception among instructors of the economic benefits of collaboration, salsa events are approached with cold cooperation at best and deliberate sabotage at worst, resulting in poorly attended, lack-luster events. One dancer suggests that salsa schools do not collaborate to sponsor larger salsa events because of a suspicion that one school will make more money than the others. Therefore instructors withhold support or sabotage the events of other schools. Likewise, salsa studios territorialize salsa students, often treating them as objects without agency—as means to fulfill economic ends. Many salsa instructors accuse their colleagues of cajoling students away to join classes elsewhere. A language of possession and protection is so candid and pervasive among instructors that the notion of 'poaching' students, staff, and performers is explicitly written into and warned against in the Salsa South Africa code of conduct, a document that outlines appropriate behavior for members of the Salsa South Africa alliance. Cape Town salsa instructors formed the Salsa South Africa alliance to "bring scattered efforts together" to benefit all salsa schools. This alliance of existing salsa instructors, however, is also a product of territorial processes designed to exclude new initiatives and to protect existing instructors as the salsa elite. Salsa South Africa instructors insist that the organization of 'outsider' events threaten the economic viability of salsa schools, and competitive behavior is often justified as "just business" as schools protect their economic interests. The Salsa South Africa Alliance was intended to create spaces of inclusion and collaboration among salsa schools and participants; however, in its eventual corrupted form, members of Salsa South Africa created suspicion among instructors and zones of exclusion. This alliance was one element in a process of delineating and protecting territory. Even within the alliance, there is a general mistrust of intentions among instructors that makes collaboration disingenuous at best. One prominent instructor describes his distrust of the other instructors: "I'd like to say that salsa will come together . . . , but it's based around the teachers and every teacher has got . . . agendas, and agendas are never transparent as much as you might want them to be."

Instructors mention that the behavior behind accusations of stealing venues and poaching students can be excused as healthy competition. While economic arguments are often made to explain and justify territorial behavior, I argue that there is a deeper emotional motivation behind such maneuvering, particularly because instructors often act in ways that are counter to their economic interests to hinder the success of other schools. Instructors discuss the behavior of other salsa schools in terms of "betrayal" and "back-stabbing"—loaded words reserved for emotional sentiments. More than people's economic livelihoods are at stake—pride, prestige, status, respect, and recognition are entangled with salsa. What is at stake in these interactions is beyond economics; it is the power to create, influence and control space. The processes of territory-making in the salsa scene are not only about control of a social sphere and networks, it is also about the control of bodily movement, both on the dance floor and through the spaces and networks of the larger salsa scene. It is also about control of creativity—the right to make and remake space. Intense emotional involvement manifests itself in the politics of salsa dance scenes as instructors and dancers struggle for recognition and status. Therefore, emotional involvement influences behavior, shapes attachments, and inspires navigations and uses of urban space. Emotional involvement and bodily movement shape territory.

Shaping Space of Belonging Through Affective Rendering

As explained in the previous section, territorial maneuvers within the salsa scene challenge notions of community and shape spaces of exclusion and belonging. These practices of territorialization often generate frustration and general disgust among salsa participants, affecting the ways that people relate to a sense of community. The salsa community in Cape Town, as participants use the term, describes a core group of dancers and participants who are at most of the salsa events, take (or used to take) salsa courses, and communicate about salsa events and issues via blogs and Facebook. A sense of belonging, in contrast, describes an emotional connection, a feeling of comfort, home, or family within the salsa scene. For many salsa dancers, this sense of belonging is elusive because some dancers feel as though their contributions to the salsa community over time are not appreciated outside of an economic context. The economic policies, processes of venue selection, and the territorial maneuverings of instructors and event organizers also work to ensure that the salsa community is not as inclusive as it is narrated. Likewise, the seemingly petty conflicts and contestations over territory among the salsa elite disrupt the possibility of cosmopolitanism that the practice of salsa proposes. Often, hostility is made apparent in the salsa space, either communicated through the dancing body, or expressed in the noticeable absence of dance. This hostility creates a vibe that dancers describe as "cold," "competitive," or "uncomfortable."

In her essay on the science of territory, Andrea Mubi Brighenti defines territory as a practice rather than a physical space; it defines spaces through patterns of relationships of power among people (2010). Brighenti argues that it is the imagined nature of territory that enables distinction and recognition, and the act of boundary-drawing is implicated in the creation of ordered social relations. Importantly, territories are affective; they facilitate the spread of "moods, attitudes, desires, [and] beliefs" (p. 58). Respect plays an important role in territorial relationships because its expressive-affective nature organizes social structures internally while maintaining distance between those who are welcome inside territorial boundaries and outsiders. This desire for respect and recognition drives territorial practices and processes in Cape Town's salsa scene.

The establishment of networks and territory are all part of the transformation of mundane urban spaces through emotion into spaces of belonging and attachment. Cameron Duff (2010) explores the relationship between affect, belonging, and territory in youth social spaces in Vancouver. She explains the ways in which youth in Vancouver territorialize public spaces like parks, malls, beaches, and cafés in order to carve out spaces of belonging for themselves. Using Edward Casey's concept of "thick" and "thin" places, Duff seeks to explore the ways in which "thin" places are transformed through affective rendering into "thick" places imbued with memory, identity, and attachment. 'Thick' spaces are rich, meaningful, and provide for the individual a sense of belonging and an affective experience (Duff 2010). Duff suggests that "thin" places, in comparison, are unmemorable and indistinctive; they "offer nothing to hold the self in place" (p. 882). She explains how youth engage in "processes of navigation and memorialization" of places to construct a sense of belonging in the city (p. 888).

Borrowing from Duff, I argue that it is through affective rendering that Cape Town salsa dancers seek to create "thick" spaces of belonging on the dance floor. Years of affective labor—memories forged, conflicts mitigated, relationships formed and broken—have attached meaning and emotion to Buena Vista, transforming it into a "thick" space for many of Cape Town's salsa dancers. Importantly, the transformation of other salsa venues into "thick" spaces of meaning and belonging for dancers is disrupted by the constant opening and closing of salsa party venues. Salsa instructors and promoters, lured by the potential of recreating Buena Vista's energetic vibe and lucrative enterprise, attempt to launch regular salsa events at different bars and restaurants in Cape Town's city center. The success of these new venues is often thwarted by the territorial tactics of other salsa instructors often to the economic detriment of all salsa stakeholders. Therefore, many venues are not open long enough for dancers to develop an emotional attachment or to create spaces of belonging. In this case, an important relationship thus exists between emotional attachment to salsa spaces and economic commitment to salsa venues. Without economic commitment, instructors and club owners are unable to keep salsa venues open long enough for dancers to form emotional attachments to the new venues. Without emotional attachment to the venues, dancers are unwilling to support venues economically, wither through cover charges, dance lessons, or buying drinks at the bar. Furthermore, the culture of scarcity created when instructors feel economically pressured, fosters a negative vibe and fractures relationships, contributing to difficulty in forming emotional attachments to particular salsa spaces.

Of course, emotive and territorial maneuverings are not only about economics, but are often also about recognition, belonging, and the power to create and control space and interactions. Examining the role of dance and affect transmission allows us to see beyond the economic; if we examine the economic motivations behind conflicts among salsa stakeholders, we risk misunderstanding or underestimating the emotional and communicative power of dance to *move* people and, in doing so, reshape urban spaces. Urban spaces are created through feelings and actions driven by jealousy and *shadenfreude,* just as spaces of belonging are created through the transmission of energy, emotion and shared movement. It is in this way that dance inspires participation in the creation and contestation of urban space.

Conclusion: Toward an Understanding of Emotive Creation and the Rights to the City Through Dance

In his discussion of music and media geography, John Finn states that "music . . . comes into existence through movement" (2011, p. 1). The musicscape, he argues, "is the aural embodiment of myriad social and cultural forces, the sonic result of social rhythms resonating through space. The musicscape depends on movement, is defined by movement, come into being through movement" (p. 4). Dance is the body moving, feeling, and responding to rhythmic sounds, frequency, and energy. In this way, the spaces and relations of salsa dancing are mediated through intense emotions like joy, sensuality, jealousy, frustrations, and elation. In these scenes, music is more than a backdrop, it is the *raison d'être* for gathering. Therefore the mingling and movement of the body to the music is vital. It is the flow of energy and the transmission of affect among bodies that creates a particular vibe in a space. As with Buena Vista, this affective confluence of bodies in motion, energy, emotion, sound, and light creates vibrations, a vibe that attaches sentiment to place. It is the process of attaching sentiments to places that creates spaces of belonging in cities. Salsa dancing requires intimacy (because it involves close, often sensual contact), connection, and creativity. Facilitated by the music, creative, emotive energy is produced, exchanged, and channeled on the dance floor. Because the body facilitates the direct transmission of affect, energy, emotion, and desire through bodily contact, this connection can facilitate the creation of communities and spaces of belonging for people who share experiences. However, because of the intimate relationship between the body and the means of creative expression (dance), spaces and means of creative expression, when attacked or threatened, can also be fiercely defended as a means of self-preservation. Territorial maneuvers and the negotiation of tensions between salsa schools and dancers describe the processes of reclaiming, re-inventing, and re-creating urban space. Conflicts frequently arise, agreements are reached, alliances are formed and broken and new negotiations begin. These interactions are intensified by the embodied, intimate, and creative nature of social dancing, as well as though the affective rendering of spaces through music, movement, relationships.

As evident in the Cape Town salsa scene, encounters in these spaces might not always lead to resolutions, tolerance, mutual respect, or utopia, but these emotive encounters through dance represent a process in creating and imagining city spaces. In this way, the Cape Town salsa scene does not fail as a means to express allowing participants to express the right to the city. While the salsa scene may not live up to its cosmopolitan promise, it does provide a venue and opportunity for diverse people to come together through music and *create*. Individually and collectively, salsa dancers come together to create art, movement, relationships, networks, livelihoods, and spaces of belonging in their city. This speaks to the power of dancing bodies and dance scenes as communicative entities. Dance exemplifies the exchange that Paul Adams describes. Experiences in the establishment and decline of salsa venues are molded through emotional and economic encounters between dancers, instructors, DJs, and club owners. In this way, dance and dancing bodies communicate desire, displeasure, and territory. This process of contestation, collaboration, and negotiation, of space, bodies, and meaning describes acts of participation in the shaping of urban space; the rights to the city.

Dance allows us to examine urban participation beyond conventional notions of governance and civic responsibility such as voting activity, involvement in policy-making, and grassroots community organizing. These conceptualizations of the right to the city imply a *direct* and *intentional* engagement with the state or structures of governance. Examinations of participation that only investigate these activities do not explain the important affective and emotional maneuverings and negotiations that are part and parcel of urban participation. I expand the definition of participation to include the everyday negotiations and contestations inherent in social dance practices and integral to creation of social space, thereby exploring the notion of participation as shaping spaces for interaction, creativity, and expression in the city. According to David Harvey (2008), "the freedom to make and remake our cities and ourselves is … one of the most precious yet most neglected of our human rights" (p. 23). Harvey argues Lefebvre's concept of the right to the city is far more than the individual liberty to access urban resources: it is a right to change ourselves by changing the city (2008). Therefore, participation in the conflicts and negotiations that I highlight are struggles over the right and the power to create not just spaces of economic empowerment, or even of belonging, but also to explore and ultimately shape the *self*.

References

Adams, P. (2010). A taxonomy for communication geography. *Progress in Human Geography, 35*(1), 37–57.

Amin, A., & Thrift, N. (2002). *Cities: Reimagining the urban*. Cambridge: Polity Press.

Anderson, B. (2006). Becoming and being hopeful: Towards a theory of affect. *Environment and Planning D: Society and Space, 24*, 733–752.

Brighenti, A. M. (2010). On territoriology: Towards a general science of territory. *Theory Culture and Society, 27*, 52.

Davidson, J., & Milligan, C. (2004). Embodying emotion, sensing space: Introducing emotional geographies. *Social and Cultural Geography, 5*(4), 523–532.

Duff, C. (2010). On the role of affect and practice in the production of place. *Environment and Planning D: Society and Space, 28*, 881–895.

Elder, G. (1998). South African body politics: Space race, and heterosexuality. In H. Nast & S. Pile (Eds.), *Places through the body* (pp. 114–122). New York: Routledge.

Fenster, T. (2005). The right to the gendered city: Different formations of belonging in everyday life. *Journal of Gender Studies, 14*(3), 217–231.

Finn, J. (2011). Introduction: On music and movement. *Aether: The Journal of Media Geography, 7*(20), 1–11.

Harvey, D. (2008). The right to the city. *New Left Review, 53*, 23–40.

Henriques, J. (2010). The vibrations of affect and their propagation on a night out on Kingston's dancehall scene. *Body and Society, 16*(1), 57–89.

Johnson, E. P. (2003). *Appropriating blackness: Performance and the politics of authenticity*. Durham: Duke University Press.

Johnson, T. M. (2011). Salsa politics: Desirability, marginality, and mobility in North Carolina's salsa nightclubs. *Aether: The Journal of Media Geography, 7*(20), 97–118.

Johnson, T. M. (2013). Some dance to remember: The emotional politics of marginality, reinvention, embodied memory, and all that (Cape) Jazz. In A. Pine & O. Kuhlke (Eds.), *Geographies of dance*. Lanham: Lexington Books.

Laidlaw, A. (2011). *Blackness in the absence of blackness: White appropriations of rap music and hip-hop culture in newcastle upon Tyne, explaining a cultural shift*. Ph.D. dissertation Loughborough University.

Lantham, A., & McCormack, D. (2004). Moving cities: Rethinking the materialities of urban geographies. *Progress in Human Geography, 28*(6), 701–724.

Lefebvre, Henri. (1995 [1968]). The right to the city. In E. Kofman and E. Lebas (Eds and Trans.) *Writing on cities*. Oxford: Blackwell, 63–181 (originally published as *Le Droit a la Ville*. Paris: Anthropos).

Leonard, M. (2005). Performing identities: Music and dance in the Irish communities of Coventry and Liverpool. *Social and Cultural Geography, 6*(4), 515–529.

Massumi, B. (2002). *Parables for the virtual: Movement, affect, sensation*. Durham: Duke University Press.
Mbembe, A. (2004). Aesthetics of superfluity. *Public Culture, 16*(3), 373–406.
Minty, Z. (2008). *Culture and the right to the city: Diversity in the cultural ecology of Cape Town*. Cape Town: Isandla Institute.
Morris, B. (2004). What we talk about when we talk about 'walking in the city'. *Cultural Studies, 18*(5), 675–697.
Nahnsen, A. (2003). Discourses and procedures of desire and fear in the re-making of Cape Town's central city: The need for a spatial politics of reconciliation. In C. Haferburg & J. O. βenbrügge (Eds.), *Ambiguous restructurings of past-Apartheid Cape Town: The spatial form of socio-political change*. London: Transaction Publishers.
Nash, C. (2000). Performativity in practice: Some recent work in cultural geography. *Progress in Human Geography, 24*(4), 653–664.
Nelson, C. T. (2008). *Dancing with the dead: Memory, performance, and everyday life in postwar Okinawa*. Durham: Duke University Press.
Parker, S. (2004). *Urban theory and the urban experience: Encountering the city*. New York: Routledge.
Pieterse, E. (2009). *African cities: Grasping the unknowable. Inaugural lecture*. Cape Town: The University of Cape Town.
Purcell, M. (2002). Excavating Lefebvre: The right to the city and its urban politics of the inhabitant. *GeoJournal, 58*, 99–108.
Relph, E. (1976). *Place and placelessness*. London: Pion.
Revill, G. (2004). Cultural geographies in practice: Performing French folk music: Dance, authenticity and nonrepresentational theory. *Cultural Geographies, 11*, 199–209.
Robinson, J. (1997). The geopolitics of South African cities: States, citizens, territories. *Political Geography, 16*(5), 365–386.
Saldanha, A. (2005). Trance and visibility at dawn: Racial dynamics in Goa's rave scene. *Social and Cultural Geography, 6*(5), 707–721.
Seamon, D. (1979). *A geography of the life-world: Movement, rest and encounter*. London: Croom Helm.
Secor, A. (2004). There is an Istanbul that belongs to me: Citizenship, space, and identity in the city. *Annals of the Association of American Geographers, 94*(2), 352–368.
Shields, R. (1999). *Lefebvre, love & struggle: Spatial dialectics*. London: Routledge.
Simone, A. (2004). The right to the city. *Interventions, 7*(3), 321–325.
Thrift, N. (2008). *Non-representational theory: Space, politics, affect*. New York: Routledge.
Tuan, Y. F. (1977). *Space and place*. Minneapolis: University of Minnesota Press.

Section VII

Media Practices and Pedagogies

Mediated Geographies Across Arizona: Learning Literacy Skills Through Filmmaking

24

Chris Lukinbeal, John C. Finn, John Paul Jones III, Christina Beal Kennedy, and Keith Woodward

This chapter explores the project *Mediated Geographies: Critical Pedagogy and Geographic Education,* which combined the geography departments from the three Universities in Arizona (Northern Arizona State University, Arizona State University, and the University of Arizona). In the pedagogical project we encouraged students to both critically evaluate the vast amount of visual information in their daily lives and become literate in technologies related to digital media. Our four goals were as follows: (1) create a series of integrated geography courses across three universities; (2) have students work in groups on semester-long projects to produce digital video documentaries or multi-media photo essays; (3) use learner-centered education principles combined with critical pedagogy to enhance

C. Lukinbeal (✉)
School of Geography and Development, The University of Arizona, Harvill Building 437B, 85721 Tucson, AZ, USA
e-mail: chris.lukinbeal@arizona.edu

J.C. Finn
Department of Sociology, Social Work, and Anthropology, Christopher Newport University, Newport News, VA, USA
e-mail: john.finn@cnu.edu

J.P. Jones III
School of Geography and Development/College of Social and Behavioral Sciences, The University of Arizona, Tucson, AZ, USA
e-mail: jpjones@email.arizona.edu

C.B. Kennedy
Department of Geography, Planning and Recreation, Northern Arizona University, Flagstaff, AZ, USA
e-mail: tinakennedyster@gmail.com

K. Woodward
Geography Department, University of Wisconsin–Madison, Madison, WI, USA
e-mail: kwoodward@wisc.edu

S.P. Mains et al. (eds.), *Mediated Geographies and Geographies of Media*,
DOI 10.1007/978-94-017-9969-0_24

geographic media literacy within the courses; and (4) have students communicate what they learned to fellow students in a formal, conference setting. Elsewhere, we examine how critical pedagogy and learner-centered education strategies were used to engage students in these projects and how students communicated what they learned in a conference setting (Lukinbeal et al. 2007). In this paper we first review the student documentaries that were created and offer web links to these productions, before turning to our assessment of students' geographic media literacy skills. We conclude by noting the problems, difficulties, and successes of our project and by making suggestions on how to better assess and implement geographic media literacy skills in pedagogic practice.

Lukinbeal and Craine (2009, 176) argue that while "geographic literacy is widely written about and discussed in geography … media literacy is not." In an era of increased globalization, media saturation and mediated technologies, along with their continued rapid rate of changes, have become the new norm. Lukinbeal (2014, 41) defines geographic media literacy as "the ability to locate, evaluate, effectively use and produce geographic information." It incorporates four basic literacies: visual literacy, information technology literacy, information literacy, and media literacy.

Visual literacy was a term created by John Debes to reference "a group of vision-competencies" that "enable a visually literate person to discriminate and interpret the visible actions, objects, symbols, natural or [sic] man-made, that he encounters in his environment" (Fransecky and Debes 1972, 7). According to the National Research Council, information technology literacy references three skill sets: (1) up-to-date skills, concentrating on the application of contemporary knowledge; (2) foundational concepts, covering the principles of technology; and, (3) intellectual capabilities, that allow one to apply skills and concepts to solve problems (National Research Council 1999, 1).

Information technology literacy in geography is often positioned only within the domain of geospatial technologies. However, as Lukinbeal (2014) argues, we need to not only position GIS as media (Sui and Goodchild 2001) but we also "need to expand geographic information technology skills across the broad array of media and communication technologies that deliver geographic information" (Lukinbeal 2014, 43). The Association of College and Research Libraries and Resources produced information literacy standards that were endorsed by the American Association for Higher Education (October 1999) and the Council of Independent Colleges (February 2004). These standards "recognize when information is needed, having the ability to locate, evaluate, and effectively use information, an understanding of the economic and social issues of information use, and knowing how to access and use information ethically and legally" (Lukinbeal 2014, 43). According to Kimsey and Cameron (2005, 17), information literacy is important to geography because students "must learn to find, evaluate, and manipulate information in almost any context." Further, majors, graduates and alumni need good research skills and an understanding of the need for lifelong learning (Kimsey and Cameron 2005). Buckingham (2003, 36) defines media literacy as "the knowledge, skills and competencies required using and interpreting media." According to Hobbs (1998) there is a consensus amongst Western educators that media literacy is constituted by five key elements: "(1) mediated information is always constructed; (2) mediated information is the product of the social and historical milieu from which it was created; (3) mediated information affects how people understand their lived world; (4) meaning is an interactive process between the reader, text, and culture; and, (5) media has a unique language and system of communication" (Lukinbeal 2014, 44).

In this chapter we first provide an overview and background about the *Mediated Geographies* project, then review the student documentaries that were produced. All of the documentaries are available online for viewing and represent the major outcome of this project: that a learned-centered education approach facilitates deeper learning through praxis. Following this we discuss the methodology deployed to assess learner outcomes including pre- and post-course essays and film clip assessments where students watched three different short film clips about human-environment interactions. Finally, we turn to a discussion about the assessment process and conclude with some thoughts about geographic media literacy and the *Mediated Geographies across Arizona* project.

Project Overview

In 1999, in an effort to renovate pedagogic practices and thereby change social conditions, the Arizona Board of Regents Regents (ABOR) adopted a work plan focused on learner-centered education (LCE) principles that place students at the center of the learning process, with instructors as facilitators. According to ABOR, LCE is expected to provide a more efficient and flexible college education that promotes deeper, more lasting, and more transferable learning. We created three courses that followed the principles of LCE to serve as vehicles for this project. Each course focused on a different grade level and theme: a sophomore-level course in Cultural Geography; a junior-level course in U.S. Geography; and a senior-level special topics course entitled *Cinematic Geographies*. Prior to teaching the courses, assessment matrices and LCE materials for each course were produced. While the same assessment matrices were used in each course, professors dealt with individual thematic content and geographic media literacy materials differently. In all three courses group projects required students to engage in either creating digital video documentaries or multi-media photo essays in PowerPoint that included voice-over narration and music.

Finally, as part of this project, documentary filmmaker Ari Palos was hired as a consultant. He assisted students at one of the universities on a regular basis; students at the other university that created video documentaries had more limited access. All students were brought together at the end of the semester for a conference held during the final weekend of the semester. The conference provided a professional venue where students from the three universities presented and discussed their projects. It was also a forum for peer evaluation of the student projects, and for evaluation of the tri-university project as a whole.

Student Documentaries

In this section we highlight seven different student productions. Although more were created, the documentaries presented here are the ones that do not violate copyright issues. The documentaries are central to this paper and show how geographic media literacy pushes the traditional means of publishing research.

The documentary *Mediated Geographies*, by Ann Fletchall and Kristy Smith, provides background information related to the inception and goals of the project *Mediated Geographies: Critical Pedagogy and Geographic Education.* The documentary includes interviews with the project's Principal Investigators Chris Lukinbeal, Tina Kennedy, and John Paul Jones III. It then follows the journey of two groups of students in Chris Lukinbeal's Cinematic Geographies class at Arizona State University from their unique classroom learning environment to the creation and impending completion of two documentaries on a local community in transition. Due to copyright restrictions the second half of the documentary does not include David Bowie's song *Under Pressure*, which provided the basis for the montage sequence. Copyright issues around the use of music became a central means through which to address information literacy skills. The documentary can be viewed at http://projects.sbs.arizona.edu/projects/clukinbe/Mediated_Geographies.mp4 (Fig. 24.1).

The documentary *Civic Engagement and Revitalization in Garfield*, by Katie deVriese, James Wagner, Chris Lukinbeal, Marilyn Dantico, John Finn, Natalie Lopez, Sarah Bongiovanni, and Stuart Bricker (with music by Peter deVriese), explores civic engagement, neighborhood revitalization and community building in the Garfield neighborhood of Phoenix, AZ. One of the oldest inner-ring neighborhoods in Phoenix, according to the 2010 Census, Garfield's population is 86 % Hispanic. It experienced white middle-class flight to the suburbs in the 1950s and increased physical deterioration and rising crime in the 1970–1980s (Price et al. 2011; Lopez and Lukinbeal 2010). Funded by the Arizona Humanities Council,

Fig. 24.1 Mediated geographies across Arizona student documentary

Fig. 24.2 Student documentaries about the Garfield neighboorhood in Phoenix Arizona

the Department of Political Science, and the School of Geographical Sciences at Arizona State University, the documentary uses interviews with community leaders, public officials, clergy, professors, and everyday citizens to explore the interactions that residents have with the institutions designed to serve them as they seek a common goal. The documentary incorporates and builds upon the student documentary project: *Garfield: The Rise, Fall and Revitalization* which was a part of the *Mediated Geographies: Critical Pedagogy and Geographic Education.* The documentary can be viewed at http://projects.sbs.arizona.edu/projects/clukinbe/Garfield_Revitalization.mp4 (Fig. 24.2).

The documentary *For Life* is by Sam Herr, Chelsea Kappeler, and Stephanie Lipple, who state, "we have put our blood, sweat and tears into the creation of this documentary." The documentary is the story of Shanti Sellz and Daniel Strauss, two *No More Deaths* volunteers who were arrested in 2005 for providing humanitarian aid in the Sonoran Desert. The documentary follows the lives of Shanti and a small contingent of other humanitarian aid workers and sympathizers over the last couple of months leading up to their trial. The lawsuit would later be thrown out in 2006. This documentary can be viewed at http://projects.sbs.arizona.edu/projects/clukinbe/For_Life.mp4 (Fig. 24.3).

The documentary *Porque es Barato/Because it's Cheaper*, by Claire Kleese, Chris Bentley, John Holden, and Nicole Disante, examines the market of tourist goods in Nogales, Mexico and Mexicans crossing the border to shop in Nogales, Arizona at big stores like WalMart. This documentary focuses on the exchange of goods between the two countries sharing Nogales. Both groups state that 'it is cheaper' on the other side. This documentary can be viewed at http://projects.sbs.arizona.edu/projects/clukinbe/Because_its_Cheaper.mp4 (Fig. 24.4).

The documentary *Crossing the Line*, by Kirby Brady, Nell McCallum, Mary Perry, and Nick Sexton, examines the relationship between the US/Mexico border and the people of Douglas, Arizona. Douglas seems to be an example of common, small-town America, with a main street and many parks within the town's

Fig. 24.3 Student documentary, *For Life*

Fig. 24.4 Student documentary, *Because Its Cheaper*

boundaries. However, being located directly on the border with Mexico has many social and economic benefits and consequences. Having a wall dividing the town puts strain on the people of both sides. Conflict has developed over what is best for both the town of Douglas and the United States as a country. With excessive growth on the Mexican side, the town of Douglas has changed in a way that does not necessarily benefit the people of Douglas or the people of Mexico. Through interviews with locals, humanitarian groups, and the border patrol, the film gives the viewer an understanding of ways the town and the border coexist and interpenetrate one another, for better and worse. This documentary can be viewed at http://projects.sbs.arizona.edu/projects/clukinbe/Crossing_the_Line.mp4 (Fig. 24.5).

The documentary *Changes…Route 66,* by Katie Smith, Shuko Ogi, Carson Cherland, and Mark Poland, takes us on a tour of Arizona's Route 66 near Flagstaff, Arizona. Route 66 was developed for one reason and one reason only: transportation from one U.S. coast to the other. Construction resulted not only in the highway, but also in towns and businesses. Towns developed due to the mere fact that they were

Fig. 24.5 Student documentary, *Crossing The Line*

Fig. 24.6 Student documentary, *Route 66*

located along Route 66. Over time, Route 66 was not large enough to accommodate the increased traffic from the country's growth. Some portions of a new highway were recreated on top of the original route; in others, completely new routes were created. Some sections of the new highway were only 100 ft away from Route 66's original path. Landscapes changed along with the road. New developments sprouted, while old businesses died. This documentary can be viewed at http://projects.sbs.arizona.edu/projects/clukinbe/Route_66.mp4 (Fig. 24.6).

The documentary *Mexican Pharmaceuticals at the Border… OR, Where the Pills At?*, by Tad Barker, Michael Wyman, and Ellis Harper, is about US/Mexico border pharmacies and their clientele. The film shows people that frequent these pharmacies, what they purchase, why they purchase medications there, their opinions on prescription drugs from both the US and from Mexico, and the legal circumstances in which they buy prescription drugs in Mexico. Interviewees included a senior citizen tour group on a day trip to Algodones, Baja California (which is about 15 min away from Yuma), a local expert on border health, Howard Eng, a University of Arizona student who was caught crossing the Nogales port of entry with controlled drugs without a prescription, as well as a Mexican pharmacy owner. Alongside the many Arizona college students who use controlled prescription drugs as a form of recreation, there are the thousands of senior citizens who come to Arizona each year with the similar intent of buying drugs in Mexico. The migration

Fig. 24.7 Student documentary, *Mexican Pharmaceuticals at the Border OR, Where the Pills At?*

of these people to and from the border for this commerce is an important aspect of border culture. This documentary can be viewed at http://projects.sbs.arizona.edu/projects/clukinbe/Mexican_Pharmaceuticals.mp4 (Fig. 24.7).

Having students work on semester long video projects put geographic media literacy into praxis and allows for deeper learning through doing. As Buckingham (2003) notes, media literacy is not just about the interpretation of texts, but also embraces the field of praxis through the use of media. As technologies become more easy to use and more pervasive, this style of teaching will spread. Both Garrett (2011) and Jacobs (2013) have noted the rise of videographic geographies and/or filmic geographies. This new style of applied geographic media literacy positions education in a learner centered environment. According to Jacobs (2013, 715), "When we actively engage with the production of a moving audio-image, place—as a historical, social, economic, gendered, ethnic and political space—is put firmly on screen, and other senses are able to enter the ontology of film and digital video." Both Jacobs (2013) and Garrett (2011) argue for geographers to begin producing videos and documentaries as a means to educate and as a new outlet for research. Making videos aligns well with engaging in public discourse and informal science education outside of the classroom.

Methodology: Assessment Matrix

In this project we used a mixed methodological approach with several assessment tools in order to measure progress in students' geographic media literacy resulting from the courses. First, at the beginning of the semester, we asked students to write a short autobiographical essay providing information on why they were taking the class, what they hoped to gain from it, their educational and family background, places and environments in which they had lived, and the role of media in their daily life. These autobiographical essays also asked for basic demographic data such as age, gender, number of years in school, college major, and whether or not they had taken a geography or critical thinking class before. At the end of the course we again asked students to write a self-reflective statement based on a number of specific questions such as, "What was the one most useful or meaningful thing about visual media you learned in this course?" and, "What would you like to learn further about this subject/discipline?"

Our second assessment tool looked at changes in geographic media literacy skills as it related to levels of cognition (discussed below). We showed short clips from three documentary films at the beginning of the semester to establish a baseline and again at the end of the semester to detect changes. Each clip was related in some way to human-environment relationships, and each had a different narrative style. The first clip focused on the construction of the Hoover Dam to control flooding on the Colorado River. The narrator, a white male with a tweed suit, glasses, and a British accent, tells a triumphant story of American ingenuity and power in the control of nature. The message is unabashedly one of modernity and progress. The second clip focused on the relocation of cane toads from Hawaii to Australia to help mitigate an agricultural pest. This clip subverts classic documentary narrative conventions by destabilizing authority figures, making the cane toad a central character in the *mise en scène*, and using music in a humorous fashion. The third clip, which focused on tourism, was a low budget film that used dramatic music, "home movie" style shots, aesthetically pleasing imagery of landscapes, and an omnipresent narrator to promote the Oregon Coast as a tourist destination.

Fifteen questions related to the film clips were used to evaluate the geographic media literacy skills of the 40 total students in the 3 courses (a total of 600 pre- and 600 post-course entries to evaluate). The clips were shown on the first and last day of class and were not discussed or used in classroom exercises in order to avoid tainting assessment results. Each question was ranked by Bloom's (1956) taxonomy of (1) knowledge, (2) comprehension, (3) application, (4) analysis, (5) synthesis, and (6) evaluation. The first three levels relate to lower-order cognition: "knowledge" refers to the ability to recall or describe information; "comprehension" relates to meaning, understanding, and interpretation of instructions; and "application" is where one seeks to use a concept in a new manner. Levels 4–6 relate to higher-order cognition: "analysis" tests one's ability to separate ideas into component parts to reveal an organizational structure; "synthesis" is where one can assemble the component parts into a whole, thus producing new meaning; and "evaluation" allows one to make value judgments about information. In our assessment two questions were classed as knowledge based questions (80 total responses) and two were categorized as comprehension based questions (N = 80). As there were no application based questions, lower order cognitive responses totaled 160. Three questions were categorized as analysis (N = 120), while only two questions were categorized as synthesis (N = 80). The bulk of the questions (6) were evaluative (N = 240). In all, 26.7 % of the questions related to the lower order cognitive domain while 73.3 % related to the higher order cognitive domain (N = 440), reflecting our interest in assessing higher order cognitive domain skills.

Film clips were assessed using two matrices: first, a five-point Likert scale, and second, a categorical classification of responses. We assigned a Likert score of one for the most incomplete, least detailed, and least thoughtful answers, while a score of five was given for the most complete, detailed, and thoughtful responses. By comparing changes in individual scores from pre-course to post-course responses we were able to assess geographic media literacy improvement. There was an 85.8 % response rate to all pre- or post-questions (515 out of 600). Lower order cognitive domain questions had a response rate of 93.1 % (149 out of 160); higher order cognitive domain questions had a response rate of 83.2 % (366 out of 440).

Because we were concerned that the subjective nature of converting qualitative responses to a Likert scale might privilege positive outcomes over negative ones, we conducted a secondary analysis that compared pre- to post-course responses and assigned each response pair to one of the following categories: blank, same, different (but neither better nor worse), worse, or better. All 600 pre- and post-course pairs were used for analysis (N = 600). Anything coded as same, worse, different, or better required the presence of both a pre- and post- response. The blank category was used where there was no response to either pre- or post- questions.

Finally, content analysis of the students' written assessments allowed us to look for trends in responses, as well as to provide confirmation of the quantitative analysis. Qualitative results from the students' pre- and post-course essays, along with quantitative results from the assessment matrices, are presented below.

Pre- and Post-course Essays

Responses collected in the pre-course autobiographies proved useful for analysis in three ways: they gave important insights into the types of students in the courses, provided information on issues of interest to students, and detailed the role of visual media in students' lives. The three courses consisted of freshmen (1), sophomores (6), juniors (11), seniors (16) and graduate students (6). Their ages ranged from 19 to 39, with an average of 23. The gender of students was more or less evenly split, with 47.5 % female (19) and 52.5 % male (21). Non-geography majors outnumbered geography majors nearly two to one.

Despite the differences in course level and content many students expressed common interests in international travel and related this interest to their previous personal experiences. They showed a curiosity about cross-cultural issues, social justice, immigration, and the US/Mexico border, and many displayed a desire to learn and apply digital media skills. As one student explained, "I am eager to dive into the photo essay because it not only allows me creative control over my work, but photojournalism and photography in general are areas of considerable interest to me" (student 27)[1].

[1]To maintain confidentiality, all students that participated in this project were assigned a number. Responses cited in this paper reference this coding procedure.

The pre-course autobiographies also highlighted significant ways that media influences and impacts students' daily lives. One sentiment common to all the students was the media's omnipresence, with television in particular being seen as having had the greatest effect on them. A small number of students also expressed a more negative outlook on media and its role in society, citing its manipulative and/or addictive qualities.

Post-course reflective essays were administered on the last day of class. The first question asked was: "What was the one most useful or meaningful thing about visual media you learned in this course?" Most responses focused on the need to critically examine messages. According to one student,

> The most meaningful thing I learned this semester about visual media is that one should not sit back and absent-mindedly take it all in. Nearly every message sent through visual media has intent or a bias behind it and one can be negatively affected by that message if he or she is not careful. While I do not think it is necessary to write a complete, detailed analysis about the thousands of messages we receive every day, I do think it is important to take a second to step back and think about where the message is coming from and what the creator might be trying to say. In essence, I have learned that it is important to take a more active role in receiving visual media rather than passively let yourself be bombarded with sounds and images all day (student 38).

This statement highlights many key aspects of geographic media literacy including examining intent and bias, being an active rather than passive receiver of information, examining the source of information, and being conscious of the effect visual media has on an individual. By learning how media assembles audiovisual material to tell a story, sell a product, or transmit knowledge, students were better able to differentiate between the affective nature of media and the quality of information received from various sources.

Another critical aspect of geographic media literacy that students acknowledged was how the source of information affected the content. This interrelates two issues in media literacy: first, that sources of information are culturally and historically situated and impose particular perspectives on viewers, and second, that receivers will differentially interpret the information based on their own subjectivity and characteristics (Zonn 1990; Zube and Kennedy 1990).

While in the pre-course essays only one student mentioned identifying manipulation in mediated messages, this number grew substantially by the post-course essays, where more than 50 % of the students said that by engaging in visual media production they learned firsthand how form and context influence information. Many students further remarked on the importance of being active, critical consumers in the wake of proliferating forms of media. We are not suggesting a 'manipulation of the masses' Frankfurt School critique of media literacy here, rather, we would argue geographic media literacy is nuanced, open, and an ongoing process of meaning deliberation between producers, products and consumers.

The final question of the post-course reflective essays asked students what they would like to learn further about this subject/discipline. While at the beginning of the course many students questioned the integration of producing videos, media literacy exercises, and geography, by the end of the class students seemed to grasp the interrelationships.

As a whole, the pre- and post-course essays reveal significant improvement in students' ability to critically evaluate media. Specifically, students became more conscious about intent, bias, and the construction of information content, and how presentation techniques have an effect upon information. Furthermore, many students gained a better appreciation for how a different perspective, based on social characteristics of individuals, affects information presentation and reception. Key visual literacy skills highlighted in student post-reflective essays included being critical about the source of information and being judicious when it comes to selecting information sources.

Film Clip Assessment

The pre- and post-course film clip assessments provided valuable insights into students' digital media literacy before and after the courses. In this section we outline the most important findings based on qualitative content analysis of pre- and post-course film clip assessments before turning to the two methods we used to quantify this shift.

Based on content analysis of the pre- and post-course film clip assessment, students began to link empirical examples to deeper theoretical issues. By the end of the course many students had developed a more critical eye and were able to move from descriptive accounts to more critical commentaries on human-environment relations, thus demonstrating an improvement in their geographic media literacy by inadvertently changing lower-level comprehension questions into upper-level evaluative questions. By the end of the course students were also able to use technical terminology related to digital photography and filmmaking, showing that they were better able to analyze the packaging of messages, narrative conventions, camera angles and shots, color and lighting, and style.

We were further able to distill important trends by quantifying pre-and post-course film clip assessment responses in two different ways. First, after ranking all pre- and post-course responses on a 5-point Likert scale, we found that pre-course scores tended to be higher for questions in lower domain cognitive skills (average score 3.7) and lower for higher domain cognitive skills (average score 2.89). While slight improvements occurred in both lower and higher domain cognitive skills, gains were better on the higher domain cognitive skills questions. We presume that this result was partially due to pre-course scores being high on lower domain cognitive skills questions. Without having taken the course, read any of the material, or completed the digital media project, students had an easier time answering questions focused on knowledge and comprehension rather than those focused on analysis, synthesis and evaluation.

In order to answer higher order questions students must understand and interpret meaning, separate material into component parts for more profound understanding, distinguish between fact and inference, and make judgments about the value of ideas or materials (Bloom 1956). At the beginning of the semester most students had not yet engaged in this type of evaluation regarding media and thus wrestled with questions such as, "Do you think the documentary will have a happy ending?" or, "What contributes to the narrator's authority in this clip?" By emphasizing geographic media literacy skills in geography courses, higher domain cognitive skills regarding media can be improved; however, improvement is slow and difficult to quantitatively measure.

Discussion

The three courses involved in this study varied in class level, size, and approach to geography (regional, cultural, thematic), as well as in the nature of the visual media project (documentary film, multi-media photo essay) and the pedagogic approaches and styles of the instructors. Still, all three were integrated through their focus on learner-centered education and geographic media literacy, the same assessment material, interactions between faculty and teaching assistants, and a student conference.

The course that required students to produce multi-media photo essays in PowerPoint showed the greatest improvement. Pre-course assessment may have been lower in this course because only 4 of 18 were geography majors. Further, with a less technologically intensive final project, there was more time for evaluative geographic media literacy exercises. The potential drawbacks of highly intensive technology projects were mentioned in self-reflection essays. Some students even suggested that the format of the course be changed to either focus on evaluative media literacy exercises or praxis: "If I were to do the entire class over again, I would recommend not trying to combine content and production into one course. We needed substantially more time to edit our projects" (student 9). However, we found that the best method for improving geographic media literacy skills was applying learner-centered education through project-based exercises and discussion.

Students' personal attitudes and feelings also affected their responses to questions; indeed, only a few students could separate their personal views from critical evaluations. For instance, for the *America by Design* film, we asked, "Who do you imagine the filmmakers had in mind as their intended audience?" Whereas one student (student 2) saw the film's audience consisting of "environmentalists," another saw it as constituted by "Republicans" (student 28). For the same film clip we asked if the documentary would have a happy ending. In their pre-course response one student (student 7) said, "No because it ends describing a misunderstanding between humans and nature." That student's post-course response remained the same, but had greater detail: "It doesn't seem to have a happy ending. The film started with the notion that Americans don't take care of the land they have been 'given,' but rather exploit every bit of it for welfare and prosperity." Another pre-course response states, "Happy endings are relative. *In my opinion*, no this will not have a happy ending because the narrator seems to be arguing that 'conquering' or exploiting resources for the benefit of mankind is a positive thing, and *I do not agree*" (emphasis added, student 8). This student's post-course response remains the same: "Happy endings are subjective. *In my opinion*, no—because it seems to promote the Hoover Dam as necessary" (emphasis added).

Many student responses show how media provokes an affective response that forestalls a critical evaluation of information. In short, rather than challenging a person's opinion, values, attitudes, or feelings, media's affect may reinforce them (Shaw and Warf 2009). The act of reinforcing personal opinion maintains the consensus group that individuals perceive themselves to belong to, while simultaneously curtailing their ability to critically judge media information from positionalities beyond such standpoints. However, these processes are always in a state of flux and becoming, a never ending shift of meaning and identity creation and interpretation.

While the quantitative assessment matrices showed mixed results on students' ability to improve higher order cognitive skills, the qualitative results appeared to suggest that these skills did improve. This leads us to believe that constructing quantitative measures that assess geographic media literacy skills can be problematic. In our efforts to focus on questions that address human-environment relations and higher order cognitive skills we ended up with no taxonomic questions in the application area. Further, as was shown, a Likert-based analysis worked well to report improved skills (however small), but the categorical analysis seemed to negate part of these findings. The manner of implementing a quantitative analysis had its problems also, as a number of factors could have influenced the results. In contrast, the qualitative assessment appeared to allow students to be more open with regards to their answers and to express their opinions, values, beliefs, and knowledge by going beyond what was asked for in a question to expound upon ideas that were of concern to their personal lives.

Conclusion

As Westernized nations move further into a postindustrial world where culture is increasingly commodified and packaged into mediated information, students need to be able to critically evaluate and differentiate not only between data, information, and knowledge, but also between evidence, meaning, stereotypes, rhetoric, and ideology. Some key geographic media literacy issues exposed through this project include the following: omnipresence of media, the medium's effect on content, intent and bias, being active consumers, differential reception, and the importance of hands-on practical work in enhancing visual media literacy.

If representations such as those portrayed in documentary films are already subjective (Natter and Jones 1993), and if the reception of such information is also deeply subjective, then why is our ability to critique media so important? The answer lies in the need to improve our ability to assess the quality, source, and content of visual media, as well as the need to understand that the author(s), the medium, and the positionality (including both biography and wider socio-cultural context) of the viewer are always in a co-constituting process of meaning and identity negotiation.

Developing skills to critically evaluate visual media should include exercises that analyze specific texts and generate technical understanding of the techniques of narration within a visual text. Textual analysis exercises can involve everything from content analysis and technical-stylistic assessment to decoding or deconstructing the meanings embedded in images. It can also include situated analysis of the socio-spatial, cultural, and historical contexts of representations and their receptions. Critical thinking is not just for evaluative purposes; it can also probe the affective nature of media. Becoming conscious of how media affects our emotions, behaviors, and values is just as important as learning skills with which we can critically analyze media.

References

Bloom, B. (1956). *Taxonomy of educational objectives, handbook I: The cognitive domain*. New York: David McKay Company.

Buckingham, D. (2003). *Media education: Literacy and contemporary culture*. Cambridge: UKL Polity/Blackwell.

Fransecky, R., & Debes, J. (1972). *Visual literacy: A way to learn, a way to teach*. Washington, D.C.: AECT Publications.

Garrett, B. (2011). Videographic geographies: Using digital video for geographic research. *Progress in Human Geography, 35*(4), 521–541.

Hobbs, R. (1998). The seven great debates in the media literacy movement. *Journal of Communication, 48*(1), 16–32.

Jacobs, J. (2013). Listening with your eyes: Towards a filmic geography. *Geography Compass, 7*(10), 714–728.

Kimsey, M., & Cameron, S. (2005). Teaching and assessing information literacy in a geography program. *Journal of Geography, 104*(1), 17–23.

Lopez, N., & Lukinbeal, C. (2010). Comparing police and residents' perception of crime in a Phoenix neighborhood using mental maps and qualitative GIS. *Yearbook of the Association of Pacific Coast Geographers, 72*, 33–55.

Lukinbeal, C. (2014). Geographic media literacy. *Journal of Geography, 113*(2), 41–46.

Lukinbeal, C., & Craine, J. (2009). Geographic media literacy: An introduction. *GeoJournal, 74*, 175–182.

Lukinbeal, C., Kennedy, C. B., Jones, J. P., III, Finn, J., Woodward, K. A., Nelson, D., et al. (2007). Mediated geographies: Critical pedagogy and geographic education. *Yearbook of the Association of Pacific Coast Geographers, 69*, 31–44.

National Research Council. (1999). *Being fluent in information technologies*. Washington, D.C.: National Academies Press.

Natter, W., & Jones, J. (1993). Pets or meat: Class, ideology, and epace in *Roger and Me*. *Antipode, 25*(2), 140–158.

Price, P., Lukinbeal, C., Gioioso, R., Arreola, D. D., Fernández, D., Ready, T., et al. (2011). Placing Latino civic engagement. *Urban Geography, 32*(2), 179–207.

Shaw, I., & Warf, B. (2009). Worlds of affect: Virtual geographies of video games. *Environment and Planning A, 41*(6), 1332–1343.

Sui, D., & Goodchild, M. (2001). GIS as media? *International Journal of Geographical Information Science, 15*(5), 387–390.

Zonn, L. (1990). Tusayan, the traveler, and the IMAX theater: An introduction to place images in media. In L. Zonn (Ed.), *Place images in media: Portrayal, experience, and meaning* (pp. 1–5). Savage: Rowan and Littlefield.

Zube, E., & Kennedy, C. B. (1990). Changing images of the Arizona environment. In L. Zonn (Ed.), *Place images in media: Portrayal, experience, and meaning* (pp. 183–203). Savage: Rowan and Littlefield.

Film References

America by Design: The Shape of the Land. (1987). Director of Photography, Tim Housel. Guggenheim Productions. 3121 South Street Northwest, Washington D.C. 2007. http://www.gpifilms.com/.

Cane Toad: An Unnatural History. (1988). Directed by Mark Lewis. A Film Australia Production. http://www.screenaustralia.gov.au/.

The Oregon Coast: An Adventure Awaits. (2003). EMA Video Productions Inc. 1706 NW Glisan Street, Suite 7, Portland, Oregon 97209. http://www.emavideo.com/.

Indigenous Media and Postcolonial Pedagogy

25

Laurel C. Smith

Introduction

Indigeneity is an inherently political concept that is commonly culturally configured and connected to demands for social and environmental justice (Cameron et al. 2009, 2014). For many people who identify themselves, their families, communities, and ways of being as Indigenous, Indigeneity merits capitalization. The capital "I" signifies recognition of continuities such as heritage, territorial dispossession, and environmental practice, as well as more emergent global movements and international laws recognizing rights and responsibilities. But the conceptual category of I/indigeneity arose out of colonialist practices of representation and repression based on racialized social hierarchies. So formulations of indigeneity (without the capital I) also operate: historically, contemporaneously, and often—but not always—beyond Indigenous aegis.

Images and ideas about Indigenous men, women and children abound in EuroAmerican humanities and arts. In the Americas, indigeneity figures prominently in national narratives of identity, usually as ghostly symbols, sometimes idyllic and other times horrific and abject. When synchronized with scientific explanations, such symbolism has helped fix formerly more fluid community relations into the categories operationalized by the colonial and then state institutions to structure territorial dispossession and the 'development' of a 'disappearing people.' Geographical imaginations of indigeneity also profitably brand commodities such as artwork, travel and entertainment experience, real estate, franchises, and tobacco. In response to these caricatures and economic opportunities, diverse Peoples have appropriated and reworked state and scholarly categories to identify communities, political struggles, and cultural heritage.

L.C. Smith (✉)
Department of Geography and Environmental Sustainability, University of Oklahoma, Norman, OK, USA

S.P. Mains et al. (eds.), *Mediated Geographies and Geographies of Media*,
DOI 10.1007/978-94-017-9969-0_25

Most recently, historically marginalized social groups have rallied relatively new multimedia and broadband technologies to articulate alternative (to massive, commercial, hegemonic, 'whitestream') visualizations of the(ir) world (cf. Downing 2011; Rodriguez et al. 2010). This chapter suggests materials and strategies for teaching about the aims, content and production of Indigenous videos. How and by whom are video technologies accessed and used to mediate Indigenous geographies? How can Indigenous videos help decolonize geographic knowledge production? How about geographic curricula? With the concept of visual sovereignty this chapter emphasizes lessons that center the ways Indigenous media makers re/claim territory, place and agency in settler colonial spaces. Students are invited to explore how Indigenous videos help disrupt the colonialist hierarchies of knowledge production that have historically shaped both scholarly and popular knowledge about Indigenous geographies.

Decolonizing Indigenous Geographies

In pursuit of scholarly and commercial enterprise, explorers relied on the knowledge and experience of the diverse Peoples they encountered. But these white authors wrote their sources out of the texts and established their geographic authority (Barnett 1998). This colonialist tradition lingers. It conflates Indigenous peoples with notions of nature, which makes it hard to see them as agents of authoritative knowledge production (Willems-Braun 1997). Ample research demonstrates how legal and other institutional assumptions about expertise have hindered and continue to complicate Indigenous actors' participation in technology-mediated geographic knowledge production in relation to land tenure, resource management, and government policy (e.g., Bravo and Triscott 2011; Sletto 2009; Wainwright and Bryan 2009; Hale 2006; Palmer and Rundstrom 2012).

Although partial progress has been made in terms of including women in the field, there is no avoiding the fact that geography remains a white, English-speaking enterprise.[1] The President of the Association of American Geographers, Eric Sheppard, recently observed, "the proportion of African-American, Hispanic and American Indian students remains distressingly small." He went on to note that, "[p]erversely, we are more successful recruiting African and Hispanic scholars from outside the U.S. than from within." To change this situation, Sheppard suggests we diversify the "elite academic spaces" where geographers work. He notes that a key step toward this socio-cultural transformation is valuing "under-resourced communities, organizations, and activists" as research partners instead of study subjects (Sheppard 2012). Making space for broader participation in the production

[1]For overviews of geography's Whiteness, see Berg (2012), Delaney (2002), Pulido (2002), and Tímar (2004), who examines the dominance of the English language.

of knowledge has the potential to make geographic inquiry and pedagogy more relevant to the communities involved.[2]

Journals such as *Geografiska Annaler, Geographical Research, American Indian Culture and Research Journal, Journal of Cultural Geography, cultural geographies*, and *Cartographica* recently published special issues focused on Indigenous geographies and featuring Indigenous authors (see Shaw et al. 2006; Johnson et al. 2007; Berry 2008; Sletto 2009; Louis et al. 2012; Larsen and Johnson 2012; Cameron et al. 2014). Indigenous scholarship richly informs recent reports on Indigenous geographies in *Progress in Human Geography* (Coombes et al. 2011a, 2012), as well as chapters focused on Indigeneity in disciplinary handbooks, companions, and guides (e.g., Coombes et al. 2011b; Cameron et al. 2009; Howitt and Suchet-Pearson 2003). A key characteristic of this academic work is the call to decolonize the production of Indigenous geographies.

Coombes et al. (2011b) argue that decolonizing geography requires a relational approach to knowledge production. Anti-essentialist and reflexive, a relational approach draws attention to the co-production of Indigenous geographies. This analytical angle draws on post-structural, postcolonial, and feminist traditions to resist the temptation to "treat Indigeneity as if it is a fixed and absolute construction." It also "demands openness to shifts in the positionality of researchers." A relational approach to geographic knowledge asks how the field of inquiry "contributes to, and is transformed by, engagement with Indigenous geographies" (Coombes et al. 2011b, 485). This question prompts reflection on selfhood and institutional relations. It asks that geographers "learn to recognize our own stories and the ways in which they sustain power relations," and to figure out how to "give room for speaking back by those whose stories have been silenced" (Coombes et al. 2011b, 486).

Responding to the call to decolonize geography with a relational approach, this chapter proposes a postcolonial pedagogy that emphasizes how and why Indigenous videos "give room for speaking back." First I more fully flesh-out a postcolonial approach to geographic pedagogy. Next I introduce the concept of visual sovereignty and suggest how to teach about it with resources that examine the influence of Hollywood, documentary and experimental cinema on the production of Indigenous films made in Canada. Afterward I recommend ways to introduce a different, but connected, form and process of Indigenous media—*video indígena*, which arose in southern Mexico at the end of the twentieth century. My recommendations dwell on the roles played by academic advocates and state agencies in the making, moving, and viewing of Indigenous video. Both the Canadian and the Mexican parts of this postcolonial pedagogy mobilize videos and related reading assignments to illustrate and foster discussion about the geographies of Indigenous media. The key lesson

[2]Motta (2012) discusses the need for transgressing traditional subjectivities with critical pedagogies. See also the special issue of *Canadian Geographer* that is devoted to community-based participatory research involving Indigenous peoples in Canadian geography and introduced by Castleden et al. (2012).

is that Indigenous videos help decolonize geographical imaginations by unsettling neo/colonialist hierarchies of geographic authority and authorship. They disrupt geopolitics of representation that position Indigenous peoples, places, and practices solely as subjects of study and state control.

Proposing a Postcolonial Purpose

Ample debate swirls around definitions of postcolonial and post-colonial. Rather than review these arguments, I describe how the concepts are operationalized here. A temporal notion of post-colonial that demarcates a time before or after European colonization is not very useful to me because my research has focused on the Americas, where direct colonial rule gave way centuries ago to still-present settler society rule and state-led internal colonialism. Instead I use postcolonial theory as a tool to critically intervene in the production of authoritative knowledge. Postcolonial scholars read with and against the grain to detect the agency of colonized peoples who are sometimes called subalterns. Their analyses of colonial and state archives confirm the oppression of subaltern populations while also noting the patchy nature of power. Neo/colonialist forces never were and are not monolithic or unchallenged; they confront(ed) friction all over the world (Tsing 2005).

In solidarity with anti-colonialist struggles, many postcolonial theorists strive to post, or leave behind, colonialist practices of representation that position subalterns as eternally passive and needy. But they also recognize the lingering value of colonialist categories of analysis and administration. These categories help identify, discuss, and—whenever possible—*redress* the cultural violence, inequitable political economies and devastated ecologies that have been initiated, legitimated, and sustained by neo/colonialist discursive practices. A postcolonial focus draws attention to intersections, appropriations, and interdependencies. This makes postcolonial theory especially useful for deconstructing how hegemonic imaginations of nations, states, and other communities—including the discipline of geography—selectively incorporate *and* exclude Indigenous peoples (and other marginalized populations, e.g., women).

Informed by these deconstructions, postcolonial geographers work toward reconfiguring the field of geography in terms of demographics, epistemologies, and ontologies. Similar to recent work concerned with Indigenous geographies described earlier, postcolonial geography aims "to decolonize the production of geographical knowledge both in and beyond the academy" (Blunt and McEwan 2002, 1). One way to work toward this goal is a postcolonial pedagogy. A postcolonial pedagogy historicizes today's educational institutions and dominant knowledge production practices; it "gives the present a past" (Kanu 2006, 8–16). Postcolonial *visual* pedagogy invites students to examine visibility as well as knowledge and power (Jones 2011). It aims to foster a critical media literacy that looks for and listens to voices speaking from marginalized standpoints "in order to gain multiple perspectives on issues and phenomena that appear as common sense" (Kellner and Share 2005, 371). Because they amplify historically marginalized voices, Indigenous media are ideal resources for working toward this goal of decolonizing curricula and the production of authoritative geographic knowledge more generally.

I use Indigenous videos in the classroom to challenge the ways academic discourse obscures ontological pluralism (Howitt and Suchet-Pearson 2003). Indigenous videos provide a glimpse of different ontologies—i.e., culturally- and place-specific ways of conceptualizing existence. Learning to recognize and respect ontological difference encourages students "to account for the ways colonial power has shaped their approaches to knowledge production while inscribing the process of self-production" (Kincheloe 2006, 184). These are not easy lessons. Prompting students to reflect on their self-production is an emotional as well as intellectual intervention. It tasks individuals with situating themselves and their education within uneven power dynamics, ambiguous identities, and shifting relations. Postcolonial pedagogy troubles categories of difference that we tend to take for granted (Brydon 2004). Such "pedagogy of discomfort" may be disorienting, but it is also invaluable. It connects privilege to disadvantage and suggests students have an obligation to listen to and learn from the testimonies of those who have been historically and violently marginalized (Boler 1999).

In the following section, I propose a 16-week semester-long course designed to operationalize a postcolonial visual pedagogy centered on Indigenous media geographies. I have not yet had the opportunity to teach this precise course. But more than 15 years of using Indigenous videos to teach geography informs its structure and aims (see Smith 2002, 2006). The course objectives are shaped by the fact that I have mostly taught medium-sized geography classes at big state universities in the U.S.A. where the student body is overwhelmingly white. I strive to get my students to notice who is and who is not involved in the production of authoritative geographic knowledge, and to ask why uneven participation exists. Asking such "why questions" helps white students think about how repositioning their analytical perspective offers insight into how structural violence and cultural continuities not only shape Indigenous geographies, but also geographic (and other forms of) education, and indeed their own lives (Sleeter 1995).

Teaching with and About Indigenous Videos

In what follows, I describe a 16-week semester-long course designed to operationalize a postcolonial visual pedagogy centered on Indigenous media geographies. The class features two parts, both of which include an array of video screenings and reading assignments (see Table 25.1). The first part focuses on Canada and the second on Mexico. Although each part could stand on its own, incorporating elements of both will allow for a regional comparison of Indigenous media geographies. It will also highlight the transborder entanglements that connect the media makers and productions from both regions. Instructors can adjust the amount of material assigned according to educational level (e.g. upper- and lower-level undergraduate or graduate).

Table 25.1 Resources for teaching with and about Indigenous video

Canada	
Title	Type
Rich Hall's Inventing the Indian (BBC Four 2012)	Documentary program
Reel Injun: On the Trail of the Hollywood Indian (Diamond 2009)	Documentary film
Reservation Reelism: Redfacing, Visual Sovereignty, and Representations of Native Americans in Film (Raheja 2010)	Book written by Native media scholar
Atanarjuat The Fast Runner (Igoolik Productions 2002)	Indigenous film available online
Isuma TV	Website showcasing Indigenous media, especially Inuit work
Sovereign Screens: Aboriginal Media on the Canadian West Coast (Dowell 2013)	Book written by anthropologist
Nikamowin (Burton 2007)	Indigenous video available online
Mexico	
Title	Type
Indigenous Media in Mexico: Culture, Community, and the State (Wortham 2013)	Book written by anthropologist
"Mobilizing Indigenous Video: The Mexican Case" (Smith 2006)	Article written by geographer
"Gaining Ground: Indigenous video in Bolivia, Mexico and Beyond" (see Himpele 2004a)	Visual anthropology collection of essays published in *American Anthropologist*
Ojo de Agua Comunicación	Website of an Indigenous media organization
"The Search for Well-Being: Placing Development with Indigenous Identity" (Smith 2003)	Article written by geographer
"Locating post-colonial technoscience: through the lens of Indigenous video" (Smith 2010)	Article written by geographer
"Visualizing Indigenous women in Oaxaca: Mexico at the end of the twentieth century" (Smith 2012b)	Article written by geographer
Mujeres del mismo valor/Women of Equal Worth (Monteforte 2000)	Indigenous video available online
"Decolonizing hybridity: Indigenous video, knowledge, and diffraction" (Smith 2012a)	Article written by geographer
Dulce convivencia/Sweet Gathering (Gómez 2005)	Indigenous video available online
"Interview with Guillermo Monteforte" (Cousineau 2012)	Interview with member of Ojo de Agua Comunicacíon available online

Resources featured in the postcolonial pedagogy for teaching with and about Indigenous videos that is discussed here. They are listed in the order they appear in this chapter

Confronting the 'Reel Injun' with 'Reservation Reelism'

One way to begin a classroom conversation about the importance of Indigenous videos is to review the ways in which Indigenous peoples have been portrayed in Hollywood cinema. Numerous books detail the stereotypes reified in film, but films addressing these issues are especially effective for conveying such critical lessons. Students might first watch the imperfect, but still useful, BBC Four "All American" program titled *Chris Hall's Inventing the Indian*. Broadcast in 2012, this hour and half long documentary puts the irascible American comedian Rich Hall in dialogue with Dallas Goldtooth, a Mdewakanton Dakota and Diné filmmaker who is a member of the comedy troupe the 1491s.[3] Together Hall and Goldtooth deconstruct literary and cinematic representations of American Indians, ridicule racist tropes, highlight how Indigenous actors have subverted them, and (among other things) underscore Native diversity in the face of absurdly essentialist assumptions about the Peoples now called Indians.

The BBC's *Inventing the Indian* serves as a gateway vehicle that sparks useful discussion about the geopolitics of representation with university students who may have never considered such things before. In my experience, many students find Hall's deadpan, rambling monologues accessible and amusing. They also enjoy how Goldtooth's superb sense of timing and calm, but sharp, humor often serve as a fine foil to Hall's fuming tirades about injustice. Laughing helps students learn to recognize the violence of mobilizing images of Others as symbols. Yet not all students will appreciate all of the humor; sometimes it approaches "bro humor," which is quick to demean and dismiss. For instance, at one point Hall rants about the pop star Miley Cyrus's "dream catcher" tattoo and declares a desire to punch her. Collectively "unpacking" this and other scenes from this program can inform life-changing lessons that sensitize all sorts of students to the trouble with popular representations of Indigenous peoples (and women).

I recommend following *Inventing the Indian* with Cree filmmaker Neil Diamond's 2009 film *Reel Injun: On the Trail of the Hollywood Indian*. Designed as a road movie, this documentary follows Diamond as he drives a "rez car" from his community Waskaganish, Quebec on the James Bay coast in northern Canada across the U.S. to Hollywood. He searches for and unsettles the mythical images of Native peoples that he and other Indigenous people encounter regularly. Humor also plays a key role in this film. Diamond and other Native filmmakers locate, laugh at and lament the lens through which they and other Indigenous peoples are so commonly seen. The 'Reel Injun' is the image of an Indian who lives in a tepee and races across the prairie on his war pony that came to dominate cinema and globally shape geographical imaginations about the Indigenous peoples of the Americas. Film clips, travel footage and interviews with Tribal Elders and activists,

[3]To learn more about Rich Hall see http://www.offthekerb.co.uk/rich-hall/ and http://en.wikipedia.org/wiki/Rich_Hall. You can find Dallas Goldtooth on myspace, Twitter, and Facebook and you can enjoy the brilliant work of the 1491s here: https://www.youtube.com/user/the1491s.

filmmakers, actors, and scholars reveal the origins of this cinematic illusion and illustrate some of its impacts. Indigenous filmmakers note how they once pined to be "The Cowboy" because he always wins. Meanwhile non-Indigenous peoples enjoyed countless films featuring Indians, often played "The Nobel Injun" in films, and started longing to be "The Groovy Injun" in the late 1960s. As an antidote to the Reel Injun, the final portion of *Reel Injun* celebrates how a "Renaissance" of Native cinema speaks back to and/or ignores Hollywood typecasting with more human and geographically specific representations of Indigenous families, communities, and nations. The concluding scene of *Reel Injun* features Ojibway film critic Jesse Wente on the verge of tears as he describes the innovations and importance of the critically acclaimed 2001 Inuit feature film *Atanarjuat: The Fast Runner*.

Viewing and discussing *Inventing the Indian* and *Reel Injun* prepares students to engage with Michelle Raheja's book *Reservation Reelism: Redfacing, Visual Sovereignty, and Representations of Native Americans in Film*. Raheja's book does many things. It examines Native Americans' historical involvement in and influence on the earliest days of Hollywood cinema. It emphasizes the invisibility of Native American women in cinema historiography. And then it highlights how contemporary Native American performers make marginalized individuals (and by extension, communities) more visible by creatively retelling their stories. Raheja underscores how these efforts necessarily confront "a representational field that has already been defined by the dominant culture and often does not reflect the lived experiences of people who have lived most of their lives in Native American communities" (Raheja 2010, 143). And she theorizes contemporary Indigenous films as a response to this necessity by drawing on Edward Soja's operationalization of Henri Lefevre's "trialectics of space" to disrupt the binary between the real and the reel. With Soja's overlapping trio of social spaces (perceived, conceived, and lived) she describes how film serves as "the virtual reservation" that mediates Indigeneity. Recognizing the reservation as a place that can be seen as both homeland and an embodiment of genocidal intentions, Raheja describes the virtual reservation "as a recombinant, fluid re-reading of space that exists within and in between geographical territory; the past, present, and future; the Internet; film; everyday lived experience; the possible, and other such sites of Indigenous production and practice" (Raheja 2010, 150). Through the lens of the virtual reservation, Indigenous films look like a third space that rejuvenates cultural autonomy and reworks geographical imaginations.

Raheja argues Indigenous films create fertile ground for representing and reproducing sovereignty. The concept of sovereignty is paradoxical. For many Indigenous peoples it describes how and why place-centered communities live in relation to other communities, both human and non-human. At the same time it also references "European notions of nation-to-nation political sovereignty" (Raheja 2010, 198). Raheja makes good use of this geopolitical hybridity to analyze Indigenous films in terms of "visual sovereignty." Indigenous filmmakers select, critique and refashion cinematic conventions, blending them with other ways of visualizing the past, present and future with which they are familiar, such as prophecy. They "suggest ways of decolonizing knowledge and methods of interpretation through putting seemingly discordant discourses in conversation with each other" (189).

To illustrate how such a dialogic process refreshes and fortifies Indigenous traditions, Raheja—like Diamond's documentary *Reel Injun*—examines *Atanarjuat: The Fast Runner*. She highlights how "the *Atanarjuat* filmmakers strategically adjust and reframe the registers on which Inuit epistemes are considered with the twin, but not necessarily conflicting, aims of operating in the service of their home communities and forcing viewers to reconsider mass-mediated images of the Arctic" (Raheja 2010, 193). Her reading of *Atanarjuat* demonstrates how the virtual reservation of Indigenous film provides a powerful, multi-faceted medium for re/claiming Indigenous geographies. In addition to reading Raheja's analysis, students could view *Atanarjuat* online via Isuma TV, the website of Igloolik Isuma Productions, the organization based in Igloolik, Nunavut that made the film. They might also explore this Internet site through which they can view thousands more Indigenous videos. Of particular note is the work of Arnait Video Productions, an Inuit women's media organization also based in Igloolik.[4]

Kristin Dowell's book *Sovereign Screens: Aboriginal Media on the Canadian West Coast* complements Raheja's book and its analysis of *Atanarjuat*. Dowell examines how Aboriginal media makers use video to represent themselves and their communities in the ways they wish their historical and contemporary triumphs, struggles, and everyday lives to be represented. They create "image nations" and "decolonize the screen" in the pursuit of a deeply politicized cultural sovereignty. *Sovereign Screens* differs from Raheja's book because it is based on extensive ethnographic fieldwork. Dowell's research centered on a media organization in Vancouver staffed by Aboriginal activist-artists and non-Aboriginal advocates—including Dowell, who work together to produce and screen Aboriginal films.

Dowell participated in and studied the "the off-screen, behind the scenes, social practices of Aboriginal media production" (Dowell 2013, xii). Her book-length study elucidates how the social geographies generated through Aboriginal media making galvanize and fortify a diverse urban community. It demonstrates how media organizations shift over time as leaders and mentors come and go. *Sovereign Screens* introduces students to the key roles that women and their families play. It details how state funding can both enable and hobble the production and exhibition of Indigenous media art. And like *Reservation Reelism*, it features thoughtful analyses of several Indigenous videos. Ideally, students should watch one or more of the videos Dowell discusses. Doing so offers them an opportunity to critically consider the film's aesthetics, the filmmakers' techniques, and Dowell's analyses. For example, *Nikamowin* is an award-winning experimental film available for viewing online.[5] This 11-min video showcases the ways Aboriginal media makers

[4]See http://www.isuma.tv/. On this website you can view and/or download the film *Atanarjuat: The Fast Runner* at http://www.isuma.tv/isuma-productions/atanarjuat-the-fast-runner. You can also access the website of Arnait Video Productions: http://www.isuma.tv/arnaitvideo. Additionally or alternatively students can read Michael Evan's 2010 book *The Fast Runner: Filming the Legend of Atanarjuat*.

[5]See http://www.beatnation.org/kevin-lee-burton.html#null.

utilize innovative audio and editing strategies to artistically (re)create cultural traditions such as language, song and storytelling. Dowell argues that Burton "created a Cree mediascape" and that social actors like him are "negotiating the interface between digital technologies and indigenous knowledge through the media production process" (Dowell 2013, 171).

Situating Indigenous Video in Mexico

The second component of this Indigenous media course focuses on the geographies of Indigenous video in Mexico. It begins with the book *Indigenous Media in Mexico: Culture, Community, and the State* by Erica Cusi Wortham. Like *Sovereign Screens*, this book is based on the author's extensive fieldwork, institutional advocacy, and organization of festivals. Wortham examined the social practices of Indigenous media by supporting, spending time and talking with the media makers who are "making culture visible" in the face of "hegemony on steroids" (Wortham 2013, 3 and 5). Her book details the invention of *video indígena*, a documentary style of video making that was disseminated through a series of video production workshops orchestrated by mostly non-Indigenous reformers working in the *Instituto Nacional Indígena* (INI-National Indigenous Institute), a federal agency targeting Mexico's Indigenous population. Although this mode of media making emerged out of a state institution, video indígena has provided a platform for the documentation and revalorization of cultural traditions historically denigrated by state policy and public education. Wortham's book shows students how much of the cultural work initiated and informed by aims of video indígena fortifies connections between cultural production, political action and social organization, connections formerly and currently severed by neo/colonialist practices of representation.

Wortham explores the organizational relationships and cultural programs that preceded, informed, and emerged out of this state-supported technology transfer program. She describes how many community-based media collectives sought, and seek, to work beyond state sponsorship and sometimes struggle(d) to explain their efforts to their neighbors, relatives, and community authorities. She also explains how a small independent media organization, Ojo de Agua Comunicación, arose out of the ashes of a national Indigenous Video Center that had been established in Oaxaca at the climax of state investment in Indigenous media production.[6] Refusing to tell a simple tale of triumph, Wortham also probes the ways Indigenous videos made in Oaxaca risk replicating state discourses of a depoliticized, male-dominated and folkloric cultural patrimony. To make her case, Wortham compares video indígena to more revolutionary but still related (Smith 2006) videos the Chiapas Media Project produced with Zapatista communities. This scholarly strategy effectively illustrates a spectrum of Indigenous media operating in southern Mexico at the dawn of the twenty-first century.

[6] See also http://www.ojodeaguacomunicacion.org/.

Recent research on Indigenous videos made in Oaxaca, Mexico also spotlights academics who have worked to reform the educational and research institutions that employ them, as well as the scholarly venues where they publish, in order to enable and foster greater Indigenous participation in the visualization of authoritative knowledge about Indigenous geographies. This angle of analysis helps draw students' attention to the contributions of Indigenous media makers to scholarly knowledge production. To do this, instructors might assign a special collection of short articles titled "Gaining Ground: Indigenous video in Bolivia, Mexico and Beyond" that was published in a 2004 issue of *American Anthropologist*. It features an introduction by the journal's Visual Anthropology editor, Jeff Himpele (2004a), an interview Himpele did with Bolivian media makers (Himpele 2004b), an article about a particular Indigenous media organization in Oaxaca by Wortham (2004), and an interview with Juan José García, a Zapotec media maker from Oaxaca (Brígido-Corachán 2004). García was interviewed in April 2003, when he was president of Ojo de Agua and attending the Taos Talking Picture Festival, where Ojo de Agua received a lifetime achievement award. The interview allows students to hear what García has to say about the transition from the state-sponsored Indigenous Video Center to the independent organization Ojo de Agua. Emphasizing the interview's publication in the flagship journal of the American Anthropological Association provide a starting point for a conversation about how Indigenous videos can and should contribute to scholarly discussions about Indigenous geographies, and the ways that scholars can also contribute to the making and moving of Indigenous videos.

Such a classroom conversation about the decolonization of geographic knowledge production could be fortified by assigning an article (Smith 2003) that considers the contributions of Stefano Varese, a Peruvian anthropologist who worked in Mexican cultural institutions before relocating to U.S.A., where he is now the Director of the Indigenous Research Center of the Americas at the University of California-Davis. In the early 1980s, Varese spearheaded state supported popular culture programming that hired representatives of community-centered collectives to undertake initiatives such as theater and traditional knowledge about medicinal plants. Positioned as organic intellectuals, these individuals and the Indigenous collectives associated with them accessed media technologies, as well as some support for using them. Varese and several other academics that initiated institutional reform also used this programming to share and rethink critical theories of ethnicity and development. These institutional, intellectual and technological exchanges are evident in videos made by particular community-based Indigenous organizations, especially when the videos are theorized as the result of postcolonial technoscientific practices (see also discussion of Gustavo Esteva in Smith 2010).

The coproduction of Indigenous videos is also evident in a project made possible by Josefina Aranda, an anthropologist who helped establish a statewide coffee cooperative organized along the lines of community governance traditions that distinguish the Indigenous regions of Oaxaca where coffee is grown. Aranda worked with Ojo de Agua to make a video called *Mujeres del mismo valor* (*Women of*

Equal Worth), which is available online for viewing.[7] Smith (2012b) argues this 27-min video can be viewed not only as a multi-purpose postcolonial archive that illuminates Indigenous women's new forms of participation in development and community governance, but also as an illustration of Aranda's research on the gendered inequalities characterizing coffee and other forms of agricultural production. Introducing students to this kind of analytical perspective encourages them to think about how and why Indigenous videos intervene in neo/colonialist and androcentric geographical imaginations of Indigeneity. It can also prompt discussions about the relevance of geographic research undertaken with emancipatory aims.

This emphasis on particular bodies involved in the coproduction of Indigenous media is also central to an examination of, and personal entanglements with, another Indigenous video made in Oaxaca with the assistance of Ojo de Agua: *Dulce convivencia* (*Sweet Gathering*) by Filoteo Gómez Martínez. Smith (2012a) provides a tripartite analysis of this video's production and meaning that neatly intersects with the conclusion of Wortham's book, where she discusses the same video (Wortham 2013, 210–218). Both examinations of *Dulce convivencia* emphasize the ways Indigenous media makers and differently located audiences understand Indigenous media in sometimes drastically different ways. In addition to reading and comparing both analyses, students can view *Dulce convivencia* by accessing it online on Isuma TV.[8] These three resources—Smith (2012a), *Dulce convivencia*, and Isuma TV—link Indigenous media made in Oaxaca with Indigenous media made in and/or circulated online from a server in Igloolik (see Table 25.1). Instructors might ask students to think (or write) about these connections in terms of a relational approach to Indigenous geographies.

To expand on this lesson about tranborder geographies of Indigenous media, I suggest wrapping up this second part of a class on Indigenous video by assigning a 20-min interview with Guillermo Monteforte, an Italian-Canadian documentary maker who was one of the architects of the state technology transfer program and founding director of the related Indigenous Video Center, as well as a key figure in the establishment and maintenance of Ojo de Agua. Like *Atanarjuat* and *Dulce convivencia*, this interview is available on an Isuma TV. It is found on a channel titled "Making Connections," which was created by Marie-Hélen Cousineau, a non-Indigenous media maker who has worked with the all-women Arnait Video collective in Igloolik since the early 1990s. Cousineau interviewed Monteforte during a 12-day tour in 2009 that brought the four members of the original Arnait video crew (including Cousineau) and others to Oaxaca.[9] Her conversation with

[7]The video *Mujeres del mismo valor/Women of Equal Worth* is available for viewing on channel Ojo de Agua Comunicación's Vimeo site, specifically the channel features the group's work in English. See http://vimeo.com/ojodeaguacomunicacion.

[8]To view *Dulce convivencia/Sweet Gathering* online see http://www.isuma.tv/mixe/sweet-gathering.

[9]The interview with Guillermo Monteforte is found here: https://www.isuma.tv/en/making-connections.

Monteforte not only provides insight into the formation and continuing goals of Ojo de Agua, but it also discusses the possible impacts of the cultural exchanges characterizing Arnait's visit. This interview encapsulates some of the far-flung relationships that constitute Indigenous media geographies. It also reveals the invaluable cultural work of non-Indigenous advocates who dedicate their lives to creating conditions for exchange and access that will enable Indigenous peoples to mediate Indigenous geographies.

Conclusion

Wortham closes her book by asking readers to consider how "the study of indigenous media constitutes a collaborative space in which researchers work alongside other participants" (2013, 221; see also Ginsburg 1997). While this collaborative space fails to rectify asymmetrical geopolitics, it does allow participants and a growing range of audiences to reimagine Indigenous peoples' relationships with states, and—I would add—scholars. Wortham argues that, "the struggle for self-determination is being waged on the fields of media production and consumption. Indigenous media production and consumption act as catalysts, engendering discussions that lead to empowering awareness." While this "project of decolonizing our imagination is far from accomplished … the accomplishments of making culture visible brings us further along" (Wortham 2013, 221). Indigenous videos are powerful pedagogical tools that provide mobile grounds for establishing visual sovereignty. They educate viewers about Indigenous geographies, as experienced by Indigenous peoples. Indigenous videos also catalyze valuable conversations about exclusion and inclusion.

To fully understand the power of Indigenous video, students and other scholars must learn to situate them in relation to the commercial, cultural, political and institutional settings from which they arise and through which they operate. So my two-part dream class begins with lessons about the hegemonic celluloid imaginations of Indigenous peoples. Then it introduces theoretical tools, ethnographic research, and particular videos for studying the production and circulation of Indigenous media in Canada and Mexico. The overarching objective of the two-part pedagogical plan outlined above is to reflect on how the conversations catalyzed by Indigenous video can help change the disciplinary practices of geography. Indigenous videos should be mobilized to teach about particular Indigenous peoples, places, and practices. Students can be encouraged to witness these videos in terms of visual sovereignty, which suggests how the videos do not just embody, but also interpret and analyze, Indigenous geographies. Ascribing this sort of recognition and respect for new forms of mediating authorship reconfigures geographic authority. It includes formerly excised voices, illuminates alternative perspectives, and might even facilitate the recruitment and retention of students from historically marginalized communities. Indigenous videos decolonize geography by prompting geographers to ask who participates (or not) in their own geographic visualizations. They also provide a template for visualizing pluricultural conversations in which no one cultural perspective dominates the conversation.

References

Barnett, C. (1998). Impure and worldly geography: The Africanist discourse of the royal geographical society, 1831–73. *Transactions of the Institute of British Geographers, 23*, 239–251.

Berg, L. (2012). Geographies of identity I: Geography – (neo)liberalism – White supremacy. *Progress in Human Geography, 36*(4), 508–517.

Berry, K. A. (2008). Introduction: Mainstreaming indigenous geography. *American Indian Culture and Research Journal, 32*(3), 1–3.

Blunt, A., & McEwan, C. (Eds.). (2002). *Postcolonial geographies*. New York/London: Continuum.

Boler, M. (1999). *Feeling power: Emotions and education*. New York: Routledge.

Bravo, M. T., & Triscott, N. (2011). *Arctic geopolitics and autonomy* (Arctic Perspective Cahier No. 2). Ostfildern: Hatje Cantz.

Brígido-Corachán, A. (2004). An interview with Juan Jose García President of Ojo de Agua Comunicación. *American Anthropologist, 106*(2), 368–373.

Brydon, D. (2004). Cross-talk, postcolonial pedagogy, and transnational literacy. *Situation Analysis, 4*, 70–87.

Cameron, E., de Leeuw, S., & Greenwood, M. (2009). Indigeneity. In N. Thrift & R. Kitchin (Eds.), *International encyclopedia of human geography* (pp. 107–123). London: Elsevier.

Cameron, E., de Leeuw, S., & Desbiens, C. (2014). Indigeneity and ontology. *Cultural Geographies, 21*(1), 19–26.

Carter, S., & McCormack, D. P. (2006). Film, geopolitics and the affective logics of intervention. *Political Geography, 25*, 228–245.

Castleden, H., Mulrennan, M., & Godlewska, A. M. C. (2012). Community-based participatory research involving indigenous peoples in Canadian geography: Progress? An editorial introduction. *The Canadian Geographer, 56*(2), 155–159.

Coombes, B., Gombay, N., Johnson, J. T., & Shaw, W. S. (2011a). The challenges of and from Indigenous geographies. In V. J. Del Casino Jr., M. E. Thomas, P. Cloke, & R. Panelli (Eds.), *A companion to social geography* (pp. 472–489). London: Blackwell.

Coombes, B., Johnson, J. T., & Howitt, R. (2011b). Indigenous geographies I: Mere resource conflicts? The complexities in indigenous land and environmental claims. *Progress in Human Geography, 36*(6), 810–821.

Coombes, B., Johnson, J. T., & Howitt, R. (2012). Indigenous geographies II: The aspirational spaces in postcolonial politics – Reconciliation, belonging and social provision. *Progress in Human Geography, 37*(5), 691–700.

Delaney, D. (2002). The space that race makes. *The Professional Geographer, 54*(1), 6–14.

Dowell, K. (2013). *Sovereign screens*. Lincoln/ondon: University of Nebraska Press.

Downing, J. (Ed.). (2011). *Encyclopedia of social movement media*. Los Angeles/London/New Delhi/Singapore: Sage Publications.

Evans, M. R. (2010). *The fast runner: Filming the legend of Atanarjuat*. Lincoln/London: University of Nebraska Press.

Ginsburg, F. (1997). 'From little things, big things grow': Indigenous media and cultural activism. In R. G. Fox & O. Starn (Eds.), *Between resistance and revolution: Cultural politics and social protest* (pp. 118–144). New Brunswick: Rutgers University Press.

Hale, C. R. (2006). Activist research v. cultural critique: Indigenous land rights and the contradictions of politically engaged anthropology. *Cultural Anthropology, 21*(1), 96–120.

Halkin, A. (2008). Outside the indigenous lens: Zapatistas and autonomous videomaking. In P. Wilson & M. Stewart (Eds.), *Global indigenous media: Cultures, practices and politics* (pp. 160–180). Durham: Duke University Press.

Himpele, J. (2004a). Introduction. *American Anthropologist, 106*(2), 353–354.

Himpele, J. (2004b). Packaging Indigenous media: An interview with Invan Sanjinés and Jesús Tapia. *American Anthropologist, 106*(2), 354–363.

Howitt, R., & Suchet-Pearson, S. (2003). Ontological pluralism in contested cultural landscapes. In K. Anderson et al. (Eds.), *Handbook of cultural geography* (pp. 557–569). London: Sage Publications.

Johnson, J. T., et al. (2007). Creating anti-colonial geographies: Embracing Indigenous peoples' knowledges and rights. *Geographical Research, 45*(2), 117–120.

Jones, R. B. (2011). *Postcolonial representations of women: Critical issues for education*. New York: Springer.

Kanu, Y. (2006). Introduction. In Y. Kanu (Ed.), *Curriculum as cultural practice: Postcolonial imaginations* (pp. 3–29). Toronto: University of Toronto Press.

Kellner, D., & Share, J. (2005). Toward critical media literacy: Core concepts, debates, organizations, and policy. *Discourse: Studies in the Cultural Politics of Education, 26*(3), 369–386.

Kincheloe, J. L. (2006). Critical ontology and Indigenous ways of being: Forging a postcolonial curriculum. In Y. Kanue (Ed.), *Curriculum as cultural practice: Postcolonial imaginations* (pp. 181–202). Toronto: University of Toronto Press.

Larsen, S. C., & Johnson, J. T. (2012a). In between worlds: Place, experience, and research in indigenous geography. *Journal of Cultural Geography, 29*(1), 1–13.

Larsen, S. C., & Johnson, J. T. (2012b). Towards an "open" sense of place: Geography and the question of being. *Annals of the Association of American Geographers, 102*(3), 632–646.

Louis, R. P., et al. (2012). Introduction: Indigenous cartographies and counter-mapping. *Cartographica, 47*(2), 77–79.

Motta, S. C. (2012). Teaching global and social justice as transgressive spaces of possibility. *Antipode, 45*(1), 80–100.

Palmer, M., & Rundstrom, R. (2012). GIS, internal colonialism, and the U.S. Bureau of Indian Affairs. *Annals of Association of American Geographers, 103*(5), 1142–1159.

Pulido, L. (2002). Reflections on a white discipline. *The Professional Geographer, 54*(1), 42–49.

Raheja, M. H. (2010). *Reservation reelism: Redfacing, visual sovereignty, and representations of native Americans in film*. Lincoln/London: University of Nebraska Press.

Rodriguez, C., Kidd, D., & Stein, L. (Eds.). (2010). *Making our media. Vol. 1: Creating new communication spaces*. Cresskill: Hampton Press.

Schiwy, F. (n.d.). *Who's laughing now? Indigenous media and the politics of humor*. Unpublished manuscript.

Shaw, W., et al. (2006). Encountering indigeneity: Re-imagining and decolonizing geography. *Geografiska Annaler B, 88*(3), 267–276.

Sheppard, E. (2012). Beyond the U.S. academy. *AAG Newsletter, 47*(8), 3.

Singer, B. R. (2001). *Wiping the war paint off the lens: Native American film and video*. Minneapolis/London: University of Minnesota Press.

Sleeter, C. E. (1995). Reflections on my use of multicultural and critical pedagogy when students are white. In C. E. Sleeter & P. L. McLaren (Eds.), *Multicultural education, critical pedagogy, and the politics of difference* (pp. 399–413). Albany: State University of New York Press.

Sletto, B. (2009). "Indigenous people don't have boundaries": Reborderings, fire management, and productions of authenticities in indigenous landscapes. *Cultural Geographies, 16*, 253–277.

Smith, L. C. (2002). The "cultural turn" in the classroom: Two examples of pedagogy and the politics of representation. *The Journal of Geography, 101*, 240–249.

Smith, L. C. (2003). The search for well-being: Placing development with indigenous identity. In T. Cresswell & G. Verstraete (Eds.), *Mobilizing place, placing mobility: On the politics of representation in a globalized world* (Intersecting: Place, sex, and race series, pp. 87–108). Amsterdam: Rodopi Press.

Smith, L. C. (2006). Mobilizing indigenous video: The Mexican case. *The Journal of Latin American Geography, 5*(1), 113–128.

Smith, L. C. (2010). Locating post-colonial technoscience: Through the lens of indigenous video. *History and Technology, 16*(3), 249–277.

Smith, L. C. (2012a). Decolonizing hybridity: Indigenous video, knowledge, and diffraction. *Cultural Geographies, 19*(3), 329–348.

Smith, L. C. (2012b). Visualizing Indigenous women in Oaxaca: Mexico at the end of the twentieth century. *Historical Geography, 40*, 61–83.

Tímar, J. (2004). More than 'Anglo-American', it is 'Western': Hegemony in geography from a Hungarian perspective. *Geoforum, 35*, 533–538.

Tsing, A. L. (2005). *Friction: An ethnography of global connection*. Princeton/Oxford: Princeton University Press.

Wainwright, J., & Bryan, J. (2009). Cartography, territory, property: Postcolonial reflections on indigenous countermapping in Nicaragua and Belize. *Cultural Geographies, 16*, 153–178.

Willems-Braun, B. (1997). Buried epistemologies: The politics of nature in (post)colonial British Columbia. *Annals of the Association of American Geographers, 87*(1), 3–31.

Wortham, E. C. (2004). Between the state and indigenous autonomy: Unpacking video indígena in Mexico. *American Anthropologist, 106*(2), 363–368.

Wortham, E. C. (2013). *Indigenous media in Mexico: Culture, community and the state*. Durham: Duke University Press.

Owning and Sharing Experiences of Adventure: Tourism, Video and Editing Practices

26

Eric Laurier

Introduction: The Car Wreck

In Harvey Sacks's (1992) lecture 'Storyteller as "witness;" Entitlement to experience' he investigates the different ways in which knowledge and experience circulate in everyday conversation. Sacks reminds his students that we can pass a fact from one person to another (e.g. water boils at 100 °C) and they can then pass that item of knowledge on to another member of society and they can then pass that on to another and so on, relatively easily.[1] Sacks then turns to experiences which seem to have a much more limited circulatory logic in conversation between members of society. Experiences quickly run out of persons who can tell them and persons who will listen to them. Sacks draws upon the example of conversations where one person – 'Ethel[2]' – has witnessed the wreck from a recent car crash. Having had this experience, Ethel can tell her friend 'Betsy' about seeing the wreck because it is recognisably a tell-able event. It is also Ethel's experience, she possesses it; it happened to her and she felt certain things in response to it. Having been told of what Ethel witnessed, Betsy might just about be able to persuade her husband to hear about Ethel's experience that day but Betsy's husband would hardly be able to tell his colleagues at work about how his wife's colleague had felt on witnessing a car wreck. In tracing out the short distance Ethel's experience can travel before it runs out of tellers and audiences, Sacks identifies that there are constraints

[1] Elsewhere in his lectures Sacks examines jokes which are equally to pass around and as he reveals also distribute and rely upon knowledge.

[2] I will call the persons here 'Ethel' and 'Betsy' so we can more easily keep track of them.

E. Laurier (✉)
School of GeoSciences, University of Edinburgh, Edinburgh, Scotland, UK
e-mail: eric.laurier@ed.ac.uk

S.P. Mains et al. (eds.), *Mediated Geographies and Geographies of Media*,
DOI 10.1007/978-94-017-9969-0_26

on how a second party can feel in response to the first party's experiences and there 'are even sharper limits on the good feeling that they can give to a third' (1992: 244).

The manner in which an ordinary member of society's daily personal experiences become increasingly un-interesting and un-tellable beyond their immediate acquaintances is a feature of conversation we all become familiar with through sharing our daily experiences and the experiences of others. For Sacks, what this raises is that the communicative geography of experience is tied to each teller's *entitlement* to relate a story of such an experience and their entitlement to have certain emotions in relation to it. What we begin to realise from his work is that stories that turn on the experience of events do not circulate in quite the same way that other stories do. When Betsy receives the story from Ethel she does not thereby acquire Ethel's experiences as hers; for Betsy they are *indirect* experiences. Nor is Betsy entitled to feel the same way in relation to the original event because she did not come upon the car crash herself. Betsy ought not to faint or burst into tears on hearing about the crumpled car and police cordon that Ethel passed by on the motorway that morning.

It might then seem that experience can only have a very limited dispersal through those with direct access to the events and through being told in conversations that form part of close relationships to one another, as family or friends or colleagues. However, in his lecture, Sacks goes on to pick out cases where an indirectly experienced (or reported) event does allow for a widespread entitlement to an affectual response. Using the then current examples of the Vietnam War and the assassination of JFK, Sacks shows that these were events where unacquainted members of US society were entitled to talk about what they felt and, for some, to become motivated to further action be it political protests or mass mourning. In leaping to the national scale and to media events Sack's ideas connect to both media studies and cultural geography. The former has been concerned with the production and reception of events through the idea of experience (Scannell 1996, 2004). Cultural geography has had a more recent shift in its focus toward the circulation of emotion (Davidson et al. 2007) and affect (Thrift 2008). Each form of mediated experience moves through the world through different sets of rights and distributional properties than knowledge (Jay 2005). My interest in this chapter is in taking up Sacks's writings on the circulation of experience through stories told in conversation, to examine how experience circulates through home movies made by adventurers. Adventurers that have much in common with tourists while also being distinct from them. To arrive at the adventurer and their video practices will require traversing, firstly, earlier studies of the circulation of experience in broadcast media and, secondly, the relationship between tourism and adventure.

Mediated Geographies of Experience

Social scientists trying to record and understand the experiential have given photographic cameras to groups in order to document their experiences of, for instance, being refugees in social housing, seniors visiting the countryside, children at school or residents in a particular neighbourhood of the city. Video cameras, as they have become cheaper and easier to use, have been handed out following similar principles of auto-documentation of, for instance, children in the car (Noy 2012). Video, then, has become a further method for bringing social scientists closer to the experience of 'being someone' or 'doing something' (Garrett 2011). This has come at a time when there has been an explosion of new communicative practices via video-sharing sites, Youtube being the most well known (Burgess & Green 2009). To understand how experiences are shared I am suggesting we can then turn toward the production and consumption of amateur videos on Youtube and in turn this requires us to touch upon the production cultures of the home movie (Thornton Caldwell 2008).

In media studies, Scannell (1996) returned to Sacks's material on ordinary members of society witnessing a car crash, to point out that 'Ethel' checks on the newsworthiness of her experience by, quite literally, turning to consider whether the local radio 'news' had reported on it. In her selection of the local news channel, she demonstrates her assessment of the event she experienced as not being of national news significance. The news, then, becomes an objective measure to which witnesses to an event can compare their subjective experience of it. Picking up Sacks's concern with the shared national experience of the death of JFK, Myers (2000) and Scannell (2004) examine how modern media events such as the death of Princess Diana and 9/11 have changed, firstly, the circulation of eyewitness testimony and, secondly, the emotional responses of members of the public. In terms of the latter, Myers underlines that entitlement to feeling upset about the death of a public figure, known only indirectly, turns upon a number of justifications: how that person featured in their lives, a sense of similarity of role (e.g. as members of a family losing one of its members) and as being spokespersons for the feelings of all other ordinary people (in relation to the loss of this particular public figure). Where Scannell and Myers explored the broadcast media and members of the public what I will turn toward in this chapter is the distinct geographies of mediation and circulation of experience of home movies and adventure tourists.

Were Sacks researching the experience of encountering a car wreck now, it is not impossible that Ethel would have pulled her smartphone out of her handbag to record what she saw. Nor is it all that unlikely that Betsy would have then learnt about Ethel's experience by seeing the videoclip through Facebook. Nor that the videoclip might (or might not) lead eventually to a conversation about the car wreck on the phone after all (Miller 2011). The circulation of experience is changing as social media alter both the geography and temporality of how we keep up with the events in our friends' lives. In others words, where and when we expect our friends and their (and our) events to show up is different from previous generations. Simultaneously, the media that we expect them to share with us have multiplied. This is not to say that the social actions that we are trying to accomplish with them are unfamiliar. Friends and families are still gossiping, forgiving, updating, flirting, telling stories, remembering and, here, sharing experiences (Miller 2011).

My concern in this chapter is not to explore the direct parallel case of bystander videos of dramatic events and how they are shared. My interest is in a form of home movie that has a longer history: the tourist's holiday movie (Nicholson 2004). Its extended lineage usefully downplays what might otherwise seem, from the account above, like an epochal transformation in the sharing of experiences. Rather than examine all forms of tourism and their video documentation, in this short chapter, it is the adventure-holiday and its video documentation. The adventure-tourist, as we will see later, is not a conventional tourist, even if they are not quite an adventurer either. As a genre, the adventure-tourist videos of snowboarding, mountain-biking and so on, are common on Youtube and, moreover have, alongside videos of children and parents at play, become part of the advertising campaigns of home movie editing apps. In part, the popularity of adventure tourist videos resides in their spectacle but it has also emerged in the desire to share extraordinary experiences.

Tourism, Experience and Media

Tourism has always been bound up with the experience of other places and in his classic book on tourism, Urry identified the centrality of the gaze of the tourist (Urry 1990). The geography of the mass tourist industry has grown up around supporting, servicing and presenting itself for that gaze. As Urry (1990) documented, the tourist gaze has varied by period and continues to vary across different societies. The travels of tourists parallel the pilgrimages to holy sanctuaries in search of religious experiences (MacCannell 2002), thereby allowing religious experience to circulate. The tourist industries exploit this logic of experience by reminding us that the only way to have and possess the experience of other places is to travel to them and spend time there. Understanding the pursuit of those experiences has classically been understood in terms of a search for authenticity in the face of commodification (Wang 1999) rather than how a more varied and less pilgrimmatic set of experiences are distributed.

Unlike Ethel who, as an ordinary person with a concern with ordinariness (Sacks 1984), happened upon a grim car crash, adventure tourists are a combination of pilgrims and Simmel's adventurer. The latter goes in search of experiences that have 'something alien, untouchable, out of the ordinary' (Simmel 2006). There is a desire to participate in and produce the extraordinary and to escape both the job of 'doing being ordinary' and the way of reporting on experiences that accompanies it (Sacks 1984). While the adventurer is a distinct figure from the pilgrim, the adventure-tourist finds themselves the target of businesses trying to commodify their adventures in sacred or spectacular sites and charging for the opportunity to have extraordinary experiences in these places. The locations for adventure-tourists are often beyond the beaten tracks of mass tourism (Cloke and Perkins 2002; Kane 2012).

Not only are places sought out that provide suitable spectacular locations for adventure, sites are developed and adapted for adventurers. Bungee jumping, for example, has had platforms purpose-built in dramatic locations. Bungee jumping also captures the centrality of embodied experience to adventures. As Cloke and Perkins put it: 'from fear to adrenaline-filled exhilaration – from 'AARH' to

'YEEHAA' – is the essence of commodified adventure' (2002: 538). Bungee jumping serves the commodification critique well since it does indeed appear to be a diminished adventure and one that is, in turn, a target of critique by the adventure tourists that we spoke to during the larger research project that this chapter arises out of[3] (capturing that spirit see also (Heywood 1994)). However bungee jumping also helps us appreciate the distance the adventure tourist lies from the tourist who is primarily looking at the place they are visiting (Urry and Larsen 2011). Compared to the sedate tourist photographing the Eiffel Tower there are sharply distinct sensoriums in adventure tourism, for example in pursuits of scuba diving (Merchant 2012) or rock climbing (Lewis 2000).

Moving on to the role of media in adventure-tourism, the sites travelled to, are places that have been anticipated through pre-visiting them. Both mass tourism and adventure tourism are mediated through magazines, TV holiday programs, brochures and, more recently, through online social media such as Youtube, Flickr, Yelp, Tripadvisor etc. Correspondingly, and in a reflexive relationship with the media that precede them, tourists' experience of places are dominated by visual practices (Crang 1997). The role of photography in shaping the experience of place at the time but also in sharing experiences later has been written about extensively (Crang 1997, Urry 1990). The relationship between cinema, television and the representation and consumption of place has become a substantial domain of empirical work and theorising by human geographers and others (Clarke 1997; Lukinbeal 2004). In a recent collection on the relationship between the idea of the hotel and its uses in cinema, tourism, of course, figures centrally (Clarke 2009). Finally, from studies of social media, tourist's creative consumption of multiple traditional media before and during their visits has been inquired into (e.g. on Rosslyn Chapel, Månsson (2011). Central to my examination here, amateur Youtube videos made by adventure-tourists are drawn upon by other adventure-tourists to guide their adventures.

While the production of holiday home-movies has been examined predominantly through historical studies of their representations of place (Nicholson 2004) and their role in the preservation of alternative archives of past events (Ishizuka and Zimmermann 2008), the spread of video cameras through mobile phones and other devices, with the parallel rise of Youtube and other video sharing systems, have lead to a rising interest in contemporary video cultures (Buckingham and Willett 2010). A variety of alternative and extreme sports (e.g. snowboarding, mountain biking), alongside urban subcultures (e.g. street-dancing, parcour, urban exploration) have, from their outset, had video as part of their equipment (Fogarty 2010; Garrett 2011). The history and practices of most of these new activities have grown up around their recording, viewing and, indeed, the spread of their very techniques via video (Booth 1996; Woermann 2012). The video produced by these groups are usually heavily

[3] 'Assembling the line: amateur and professional work, skill and practice in digital video editing practices'. A 3 year ethnographic study of video editing practices funded by the ESRC RES 062-23-0564

edited with jump cuts, slow motion, accelerated motion, titling, cuts to musical rhythms and so on. The adventure tourist's video practice sits somewhere between the conventional tourist's steady gaze upon touristic places with a video camera and the jump-cut videos of extreme and alternative sports. Each practice, be it of the mass tourist at the Eiffel Tower, the trial biker traversing several impossible walls (Spinney 2010) or, as we will in the next section, the adventure tourist climbing cliffs in the Mediterranean, has a concern with capturing their experiences on video for sharing with others.

Making Home Movies of Adventure

The home movie emerges from a longer history than the digital video cultures and social media described above. Home movies have been used by friends and family to share their experiences in ways that resemble Sacks's description of storytelling in conversation. The maker of the home movie was present at the events, be it a wedding or a holiday, and they subsequently share their experience with others in the form of a video. They witnessed those original events but, more than witnessing them, they also documented them with their video camera. When compared to the speed of story-assembly in conversation, what is distinct to home movie making is the time, effort and resources required to construct the medium that will tell the story (Moran 2002; Zimmermann 1995). While many of the studies of the home movie have been interested in what it can tell us about home and family (Chalfen 1987; Moran 2002) what I am treating it as here, is as a site of production. In doing so

I want to shift into considering media production in an amateur mode and this toward a wider array of media production practices:

> there is an increasingly wide range of amateur practices that go well beyond the focus on domestic life. Indeed, the home mode may itself be evolving, as both technology and the forms of family life have changed. These different practices have their own rules and traditions, and their own modes of social organisation, and they cannot be simply collapsed together (Buckingham 2010: 46)

Home-mode movie makers are producing the extreme sports videos from above but also video blogs (Harley and Fitzpatrick 2009; Laurier 2014), 'how to' videos for Youtube, amateur natural history video, dance videos, monitoring of police practices (Jones and Raymond 2012) and many other new video practices (see also Broth et al. 2014). There is, in other words, an ever-increasing proliferation of forms emerging from the home-mode, many of which are borrowed and adapted from broadcast genres (and then borrowed back again). The growth of the amateur mode of production is moreover a dispersal of video as a medium akin to the spread of the writing technologies of pens, paper and the mail service. Amateurs put video to work for different purposes than the professional broadcast media as we will see in more detail below.

An important part of the amateur mode is the relationship between the form of the video and its maker's experience, which more firmly reconnects my discussion of amateur video with Sacks's original consideration of the telling of experiences in conversation. What I will now do is shift briefly from this literature-based account to the practices of one of the home-movie editors from our project. In the project I spent a number of evenings with each home-movie maker in an 'edit-along' (borrowing from the now common 'go-along' method of Kusenbach 2003). This involved joining editors, usually during an evening when they would have been editing anyway and then sitting alongside the home-movie makers while they edited together their videos. Like the 'go-along' it is then a way of finding out about editing practices that sits somewhere between hanging-out and interviewing.

Editor of His Experiences

In the evenings, after a weekend away climbing, kayaking or mountain-biking, James assembled the footage recorded by himself and his friends into videos that are in an experiential mode. They were almost always set to music and usually shared afterward among his climbing, cycling and sailing companions. A handful of these videos are posted on Youtube, which is where I first saw them. The movie-making that I edited-along was of a recently completed sailing holiday with five friends. Over Easter they had hired a yacht, navigated along the south coast of France and dropped anchor in a scattering of isolated bays in order to climb limestone cliffs.

With a hard-drive based camcorder, the editing of James's videos actually began during the events themselves. Because he could easily select which shots to delete and which to keep, James disposed of clips on camera, something that had been too laborious and time-consuming with the tape-based cameras he had owned in the past. Editing in situ, as he told me, helped him stay close to what the experience of the event was like at the time. As we talked beside the library of clips for his project, many showing images of blue sea, James explained his technique for editing as soon as possible after the event:

> *When you've just been somewhere and you're surrounded by the mountains, or, the diver's just come out of the water, you have this sort of vivid recollection of the moment which really captured that last ten minutes of your life. And so, you go to the camera looking for that moment. Whereas, back here in a cold city, or tired at the end of a day, or whatever, if I look at a diver in turquoise waters, it could be the most inanimate moment of that minute and forty seconds, but here it's like 'Oh great! Get that in there'.* James

In trying to capture the experience of being there, he is looking for what clips to keep, then and there. By editing at that moment and in that place he can compare the record of the event with the event just after he experienced it. The closeness in time between the event and the editing of the video produces an experiential connection by their very proximity. The preservation of the experience of the event is accomplished by editing it at that moment, rather than editing it as part of a later recollection from a distance. While leaving the camera running should be closer to what the experience was like, the amateur editor works in the aftermath of the history of the unedited amateur holiday movie, which was more often a test of friendship through suffering than a gripping 2–3 min video of climbing or cycling.

Although it seems that editing is after the event, the larger project of the adventure-holiday within which those experiences are found, remains ongoing. The editing is being carried out before he goes home, before Andrew is back in his everyday routines and places. Echoing Simmel (1911), when the adventure is over it becomes disconnected from the existence that Andrew had within it, taking on a dreamlike quality. Like a dream that we wake from, it fades on our return from it and, more troublingly, the adventure is left outside of our usual experience. What it was to be in the adventure becomes increasingly distant to us and removed from the mundane existence we return to afterwards. For professional media production, live-editing is the solution to how we can be part of an event and, even though we are seemingly distant spectators of TV coverage (Auslander 2012; Scannell 2009) liveness is destroyed once the event is edited into highlights. What Andrew does while in the midst of his adventure is an amateur variation of live-editing practices, a rolling edit that will help him recapture his own experience of his adventure.

There is also a significant departure here from the relationship between the professional documentary-maker and their object: it is the person editing the footage that also experienced the events that are recorded, they are not the editor of the experiences of others (Scannell 1996). Consequently securing the video as the experience that belongs to him requires James to maintain proximity and directness between his experience of the event and his editing of the event. By contrast, the renowned feature film editor Walter Murch tries to keep himself at a distance from the shooting of the films that he edits (Koppelman 2005). For Murch, what was appreciated in situ as a great performance by one or more of the cast on set, distracts the professional editor from establishing the best shot in the editing suite. Murch thus identifies the opposite problem, that the editor would find themselves searching for the footage of that remembered brilliant performance and setting aside the superior takes as they are viewed in the editing suite.

Up until this point it appears as if the home-movie maker, until they share their edited video, is editing their experiences not only by themselves but, for themself. And yet that is not the case, their video mediated experiences are not in an individualized register, the experiences are collective in three senses:

(a) For each climb, they undertake it, to adapt a phrase from Goffman, as a climbing-together (see this adapted to cycling in McIlvenny 2015);
(b) They edit drawing upon the collective knowledge of climbers of the looks of climbing
(c) The videos are built to be shared in the future with others

Concentrating on the third elements of its collective nature, even though the video is not being produced for broadcast, James's editing is directed toward sharing it with others and it is akin to a public remembering (Middleton and Brown 2005) of each climb he and his friends have undertaken. At a simple level his editing attends to the inclusion of shots of each person from the adventure holiday but as James noted, he tries to capture what happens during climbs as their collective experiences. In his editing's orientation to his friends, the work of assembling and sharing his videos also bears similarities with the family photograph album (Rose 2010). While there is memory work going on during the editing, as there is when the family photographs are edited, there is a future orientation in this work toward future occasions when these will be collective personal, familial and friendly histories climbing.

> *I'm not making money out of this, I'm not entertaining people on this, it is, it is because an experience for me is partly in the moment. But it's probably about just as much in the moment as it is in the anticipation and in the years and years afterwards that, you know, you're telling family and you're reminiscing and you're getting together with friends.* James

Because these are James's experiences he will be entitled to be 'telling family' in this future and, again, there is a marked similarity with parents assembling family photo albums for later tellings of the histories and geographies of the family. Or, as in his alternative anticipation of the future, he will be together with his friends again and the videos will be resources for reminiscing together. In this future orientation we can discern something of the distinctiveness of the amateur where the production orients toward these future viewings. These are viewings enmeshed in the relationships of family and friends rather than broadcasts, publics and viewers.

By contrast to the long project of learning to climb and climbing as an ongoing endeavor, bungee jumping is emblematic of an extreme experience easily acquired. Any person can turn up on the day, a trained staff is there to sort out the details for them in advance and the tourist's only requirement is to have the courage to jump. Moreover the bungee jumping company usually have already organised photographing or filming the jump for the adventure tourists involved. It may be that more serious climbers, surfers and the like, do go bungee jumping for fun or become involved in planning and engineering it. Equally bungee jumping can be a first step toward a life of adventure and so one would not want to dismiss it entirely. Nevertheless, the tourists that bungee-jump remain only weakly entitled to begin to describe what exceptional demands twanging themselves off a bridge put upon them, what unimaginable sufferings or joys they endured.

Because they are climbers, James and his fellow climbers gain the rights to pursue what shots best depict the climbs that they do. They do this both through 'subject-side' entitlements and 'object-side' entitlements to describe and assess the experience and the video of the experience of each climb (Edwards 2005; Stokoe and Edwards 2007). It is not only thus that their experiences have to be located in being experiences of the ease or difficulty of a climb (the object-side of the experience) it is that such an assessment is based in their expertise in climbing (the subject side of the experience). When both shooting with the camera and editing the footage these entitlements then generate criteria for assessing the video:

> *Now climbers always say that when you look back at climbing footage and climbing photography it doesn't look as hard and so: the skill therefore is to capture the difficulty of it. The gymnastic and the, the, the vertigo of it.* (James, while editing the series of clips shown in the panels)

As Scannell (2001) notes, we accept that certain kinds of members of society are the ones who can assess their experience of how difficult or easy any project, such as a climb, actually was. Indeed it is only those categories of persons that *can* see the difficulty of the climb on the video since they see the climb in terms of doing it themselves. On Youtube, there is a wider community of climbers beyond those who were on the same climbing holiday that are drawn to and can appreciate James's videos. James's desire though remains to show 'the gymnastic' and 'the vertigo of it'. Consequently he selects, firstly, clips of climbers contemplating the climb above and, then, of muscle-straining shifts upwards that document the remarkable agility and strength required to scale limestone cliffs of this formation. Secondly, he selects and intercuts clips of the downward view of the climber (see the second and fifth panels of the sequence of panels) that document the feeling of vertigo that the climber might suffer on looking down. By cutting in this way his expectation is that a non-climber will then also be drawn into the experience of climbing through the video, even if their rights to claim it as their experience remain limited. It is, however, one of the qualities of cinema that it can allows us to escape spectating and help us enter into the embodied experiences of others (Sobchack 2004). For James, if his viewers are able to enter into the climber's gymnastic moves or feel vertigo on viewing then he has passed along something of the experience of climbing in these mountains.

Concluding Remarks

In this chapter I have begun to outline the entitlement to experience that the home movie maker orients toward during the editing process. Film production, be it professional or amateur, is a practice where the story-yet-to-come is assembled in an ongoing editing of camera shots (rather than in forms of talk). Departing from Sacks's description of ordinary members missing or finding what anyone could see in the extraordinary (see also Antaki 2004) in order to continue 'doing being ordinary', I have begun to describe the logics of circulation of extraordinary experiences by extraordinary members of society. They are, in short, doing being extraordinary where telling of the extraordinary is part of that same accomplishment and cannot be reduced to the ordinary. Exploring extraordinary life required distinguishing the adventurer from the tourist. The ongoing differentiation between adventure and tourism, between the extraordinary and the ordinary, inhabits the video-making practices of the adventure-tourist. Should the adventurer's hard won experience in dangerous and difficult climbing conditions become a quite ordinary holiday video then it would begin to undermine the very form of life that produced it. Adventure is a form of life that should, according to Simmel (1913), tear ordinary life completely out of itself. In common with other social figures from Simmel, such as the soldier, one of their problems is then how to return to ordinary life having been outside of it.

The circulation of extraordinary experiences is through the changing and volatile networks of social media and mediated social relationships. The adventurer's videos of their experiences are sometimes only shared amongst their fellow adventurers, sometimes shared amongst wider friends and family, sometimes amongst other enthusiasts, sometimes they are picked up by the national or international media. Quite how these experiences are received by their varied audiences (to use a more media studies term) has been beyond the scope of this brief chapter. However for the makers themselves, part of the circulation of experience is toward a projected future occasions within a life course ahead. Home videos of extraordinary experiences are to be sent onward to a later occasion of re-viewing by their maker and their future friends and family, there to be re-assessed as a shared, mediated and familial history of experience.

Acknowledgements The hospitable and thoughtful James. Barry Brown and Ignaz Strebel. ESRC for funding the research that this chapter draws upon – RES-062-23-0564. The review team for the edited collection.

References

Antaki, C. (2004). Reading minds or dealing with interactional implications? *Theory & Psychology, 14*(5), 667–683.

Auslander, P. (2012). *Liveness*. London: Routledge.

Booth, D. (1996). Surfing films and videos: Adolescent fun, alternative lifestyle, adventure industry. *Journal of Sport History, 23*, 313–327.

Broth, M., Laurier, E., & Mondada, L. (2014). Introducing video at work. In M. Broth, E. Laurier, & L. Mondada (Eds.), *Studies of video practices: Video at work* (pp. 1–131). New York: Routledge.

Buckingham, D. (2010). A commonplace art? Understanding amateur media production. In D. Buckingham & R. Willett (Eds.), *Video cultures media technology and everyday creativity* (pp. 23–50). Basingstoke: Palgrave Macmillan.

Buckingham, D., & Willett, R. (2010). *Video cultures media technology and everyday creativity*. Basingstoke: Palgrave Macmillan.

Burgess, J., & Green, J. (2009). *YouTube*. Cambridge: Polity.

Chalfen, R. (1987). *Snapshot versions of life*. London: Popular Press.

Clarke, D. B. (1997). *The cinematic city*. London: Psychology Press.

Clarke, D. B. (2009). *Moving pictures/stopping places: Hotels and motels on film*. Plymouth: Lexington Books.

Cloke, P., & Perkins, H. C. (2002). Commodification and adventure in New Zealand tourism. *Current Issues in Tourism, 5*(6), 521–549.

Crang, M. (1997). Picturing practices: Research through the tourist gaze. *Progress in Human Geography, 21*(3), 359–373.

Davidson, J., Smith, M., & Bondi, L. (Eds.). (2007). *Emotional geographies*. Aldershot: Ashgate.

Edwards, D. (2005). Moaning, whinging and laughing: The subjective side of complaints. *Discourse Studies, 7*(1), 5–29.

Fogarty, M. (2010). Learning hip hop dance: Old music, new music and how music migrates. In: *Crossing conceptual boundaries II* (pp. 1–20). London: University of East London.

Garrett, B. L. (2011). Videographic geographies: Using digital video for geographic research. *Progress in Human Geography, 35*(4), 521–541.

Harley, D., & Fitzpatrick, G. (2009). YouTube and intergenerational communication: The case of geriatric 1927. *Universal Access in the Information Society, 8*, 5–20.

Heywood, I. (1994). Urgent dreams: Climbing, rationalization and ambivalence. *Leisure Studies, 13*(3), 179–194.

Ishizuka, K. L., & Zimmermann, P. R. (2008). *Mining the home movie*. London: University of California Press.

Jay, M. (2005). *Songs of experience*. London: University of California Press.

Jones, N., & Raymond, G. (2012). "The Camera Rolls": Using third-party video in field research. *The Annals of the American Academy of Political and Social Science, 642*(1), 109–123.

Kane, M. J. (2012). Professional adventure tourists: Producing and selling stories of "authentic" identity. *Tourist Studies, 12*(3), 268–286.

Koppelman, C. (2005). *Behind the seen. How Walter Murch edited cold mountain using Apple's final cut pro and what this means for cinema*. Berkeley: New Riders.

Kusenbach, M. (2003). Street phenomenology. *Ethnography, 4*(3), 455–485.

Laurier, E. (2014). Dissolving the dog: The home made video. *Cultural Geographies, 21*(4): 1–12.

Lewis, N. (2000). The climbing body, nature and the experience of modernity. *Body and Society, 6*(3–4), 58–80.

Lukinbeal, C. (2004). The map that precedes the territory. An introduction to essays in cinematic geography. *GeoJournal, 59*(4), 247–251.

MacCannell, D. (2002). *Empty meeting grounds*. London: Routledge.

Månsson, M. (2011). Mediatized tourism. *Annals of Tourism Research, 38*(4), 1634–1652.

McIlvenny, P. (2015). The joy of biking together: Sharing everyday experiences of Vélomobility. *Mobilities, 10*(1), 55–82.

Merchant, S. (2012). Negotiating underwater space: The sensorium, the body and the practice of scuba-diving. *Tourist Studies, 11*(3), 215–234.

Middleton, D. J., & Brown, S. D. (2005). *The social psychology of experience*. London: Sage.

Miller, D. (2011). *Tales from Facebook*. Oxford: Polity Press.

Moran, J. (2002). *There's no place like home video*. London: University of Minnesota Press.

Myers, G. (2000). Entitlement and sincerity in broadcast interviews about Princess Diana. *Media Culture and Society, 22*(2), 167–185.

Nicholson, H. N. (2004). At home and abroad with cine enthusiasts: Regional amateur filmmaking and visualizing the Mediterranean, ca. 1928–1962. *GeoJournal, 59*(4), 323–333.

Noy, C. (2012). Inhabiting the family-car: Children-passengers and parents-drivers on the school run. *Semiotica, 191*(1/4), 309–333.

Rose, G. (2010). *Doing family photography*. Farnham: Ashgate.

Sacks, H. (1984). On doing "being ordinary". In J. M. Atkinson & J. Heritage (Eds.), *Structures of social action* (pp. 413–431). Cambridge: Cambridge University Press.

Sacks, H. (1992). *Lectures on conversation* (Vol. 2). Oxford: Blackwell.

Scannell, P. (1996). Authenticity. In P. Scannell (Ed.), *Radio, television & modern life* (pp. 93–106). Oxford: Blackwell.

Scannell, P. (2001). Authenticity and experience. *Discourse Studies, 3*(4), 405–411.

Scannell, P. (2004). What reality has misfortune? *Media Culture and Society, 26*(4), 573–584.

Scannell, P. (2009). Review essay: The liveness of broadcast talk. *Journal of Communication, 59*(4), E1–E6.

Simmel, G. (2006). The adventurer 1911. In J. Cosgrave (Ed.), *The sociology of risk and gambling reader* (pp. 215–224). London: Routledge.

Sobchack, V. C. (2004). *Carnal thoughts*. London: University of California Press.

Spinney, J. (2010). Performing resistance? Re-reading practices of urban cycling on London's South Bank. *Environment and Planning A, 42*(12), 2914–2937. doi:10.1068/a43149.

Stokoe, E. H., & Edwards, D. (2007). 'Black this, black that': Racial insults and reported speech in neighbour complaints and police interrogations. *Discourse and Society, 18*(3), 337–372.

Thornton Caldwell, J. (2008). *Production culture: Industrial reflexivity and critical practice in film and television*. London: Duke University Press.

Thrift, N. J. (2008). *Non-representational theory: Space | politics | affect* (pp. 1–336). London: Routledge.

Urry, J. (1990). *The tourist gaze: Leisure and travel in contemporary societies*. London: Sage.

Urry, J., & Larsen, J. (2011). *The tourist gaze 3.0*. London: Sage.

Wang, N. (1999). Rethinking authenticity in tourism experience. *Annals of Tourism Research, 26*(2), 349–370.

Woermann, N. (2012). On the slope is on the screen: Prosumption, social media practices, and scopic systems in the freeskiing subculture. *American Behavioral Scientist, 56*(4), 618–640.

Zimmermann, P. R. (1995). *Reel families*. London: Indiana University Press.

About the Authors

Paul C. Adams is Associate Professor of Geography at the University of Texas at Austin. He received his PhD from the University of Wisconsin-Madison in 1993. His research addresses place images in the media, the historical geography of communication technologies, virtuality, geopolitical discourses, and the ways in which communication technologies are integrated into places. He has published articles in the Annals of the Association of American Geographers, Progress in Human Geography, Political Geography, and Communication Theory, among other journals. His books include: The Ashgate Research Companion to Media Geography (co-edited with Jim Craine and Jason Dittmer, Ashgate 2014), Geographies of Media and Communication (Wiley-Blackwell 2009), Atlantic Reverberations (Ashgate 2007), The Boundless Self (Syracuse University Press 2005), and Textures of Place (co-edited with with Steven Hoelscher and Karen E. Till, University of Minnesota Press 2001). He is the founder of the Communication Geography Specialty Group of the Association of American Geographers.

Stuart C. Aitken, PhD is Professor of Geography and June Burnett Chair at SDSU. He directs the Center for Interdisciplinary Studies of Young People and Space (ISYS). Stuart's research interests include critical social theory, qualitative methods, children, families and communities. His recent books include The Ethnopoetics of Space and Transformation (Ashgate 2014), The Fight to Stay Put (Verlag 2013), Young People. Border Spaces and Revolutionary Imaginations (Routledge 2011), Qualitative Geographies (Sage 2010) and The Awkward Spaces of Fathering (Ashgate 2009). Stuart has published over 200 papers in academic journals and edited book collections.

Andrew Boulton, PhD is a Technology Consultant and front-end developer with Humana, Inc., a Fortune 500 well-being and health insurance company based in Louisville, Kentucky. He is interested primarily in the art and science of user experience/user interface design. As such, his doctoral research examined the intersections between landscape, locative media and neogeographies in the context of an "ordinary" residential neighborhood. Andrew obtained a BSc in Geography from the University of Bristol, before completing his Master's and Doctoral studies

S.P. Mains et al. (eds.), *Mediated Geographies and Geographies of Media*,
DOI 10.1007/978-94-017-9969-0

at the University of Kentucky. His published works are focused primarily on the imbrications of "virtual" with "material" cultural landscapes, and the critical potential of reformulating cultural landscape studies to consider these intrinsic relationalities.

Brett Christophers is associate professor at the Institute for Urban and Housing Research at Uppsala University in Sweden. A geographer, he has researched and published widely on the political and cultural economies of capitalism in historical and contemporary perspectives. The media and cultural industries are one of his main areas of interest, and are the subject of his 2009 monograph Envisioning Media Power: On Capital and Geographies of Television (Lexington Books). Other areas of topical focus include money, finance and banking; housing and housing policy; urban political economy; markets and pricing; accounting, modelling and other calculative practices; and competition and intellectual property law.

James W. Craine is an associate professor of Geography at California State University, Northridge. He specializes in the geography of media and also works applying geovisualization theory to digital and analog cartography. His most recent books include The Ashgate Research Companion to Media Geography (co-edited with Paul Adams and Jason Dittmer) and The Fight To Stay Put: Social Lessons Through Media Imaginings of Urban Transformation and Change (co-edited with Giorgio Hadi Curti and Stuart Aitken). He is co-editor of Aether: The Journal of Media Geography (http://www.aetherjournal.org) and the current editor of the Yearbook of the Association of Pacific Coast Geographers.

Julie Cupples is Reader in Human Geography and Co-Director of the Global Development Academy at the University of Edinburgh. She joined the University of Edinburgh in 2013 after more than a decade working at the University of Canterbury in New Zealand. Her research interests span cultural geography, development studies and media and cultural studies. Since the early 1990s, she has been working in Central America, primarily Nicaragua, and has published widely on questions of development/postdevelopment; revolution and conflict; neoliberal economic policies; gender; disasters and environmental risk; electoral geographies and municipal governance. She is currently engaged in a collaborative research project exploring the geographic dimensions of media convergence, funded by the Marsden Fund of the Royal Society of New Zealand. This project is exploring the democratizing and decolonizing dimensions of the new media environment, with a focus on mainstream entertainment media as well as community and indigenous media operations. She is the author of Latin American Development (Routledge, 2013) and the co-author of Media/Communications/Geographies: Interdisciplinary Perspectives (forthcoming with Routledge).

Michael Curtin is the Duncan and Suzanne Mellichamp Professor of Global Studies in the Department of Film and Media Studies at the University of California, Santa Barbara. He is also co-founder and Director of the Media Industries

Project at the Carsey-Wolf Center. Curtin's research and teaching focus on media globalization, cultural geography, industry studies, and creative labor. His books include Distribution Revolution: Conversations about the Digital Future of Film and Television (2014), Reorienting Global Communication: Indian and Chinese Media Beyond Borders (2010), The American Television Industry (2009), and Playing to the World's Biggest Audience: The Globalization of Chinese Film and TV (2007). Curtin is currently at work on Media Capital: The Cultural Geography of Globalization and Precarious Creativity: Global Media, Local Labor. He is co-editor of the Chinese Journal of Communication and the International Screen Industries book series of the British Film Institute.

John C. Finn, PhD, is an Assistant Professor of Geography in the Department of Sociology, Social Work, and Anthropology at Christopher Newport University. He obtained his PhD in Geography from Arizona State University in 2011. His research is situated at the nexus of race, place, and struggles over identity, especially in Latin America. Specifically, his research is focused in three areas: (1) music, race, and nationalism in Brazil; (2) food and food culture in Revolutionary Cuba; and (3) the production of racial and racializing landscapes. He has conducted extensive fieldwork in Brazil and Cuba.

Ann Fletchall, PhD, is a Visiting Assistant Professor at Mississippi University for Women. She obtained her BA from Gustavus Adolphus College and received her PhD in geography from Arizona State University in 2009. Her primary research interest lies in the geographies of television, especially the role that television plays in the creation and experience of place. She also has research interests in tourist landscapes, film tourism, beer tourism, and the impact of these on place identity.

Mike Gasher is director of the Concordia Centre for Broadcasting and Journalism Studies as well as a professor and MA program director in the Department of Journalism at Concordia University in Montreal. A former newspaper reporter and editor, his principal research area is news geography, which posits journalism as a form of cartography. His interest in questions of space and place permeate his studies of media practices and forms. He is currently writing a book on the geography of news and edited a special volume of the online journal Aether pertaining to news geography. He is co-author of the textbook Mass Communication in Canada, now in its eighth edition, and he has also published in the areas of international news flows, journalism education, media representation, health reporting, cultural policy, and questions of space and place in Canadian cinema.

Kevin Glynn teaches media studies at Massey University in Wellington, Aotearoa/New Zealand. He received his Ph.D. from the Media and Cultural Studies program in the Department of Communication Arts at the University of Wisconsin–Madison. His research is situated at the intersection of media studies and cultural studies, informed by post-Marxism and poststructuralism, and centrally concerned with politics of/and contemporary media cultures. He is author of Tabloid Culture: Trash Taste, Popular Power, and the Transformation of American Television (Duke University Press). His work has also appeared in many leading international journals such as Television & New Media; Cultural Studies; Camera Obscura: Feminism, Culture and Media Studies; International Journal of Cultural Studies; Communication Studies; Comparative American Studies: An International Journal; Annals of the Association of American Geographers; Gender, Place and Culture: A Journal of Feminist Geography; Antipode: A Radical Journal of Geography; and others. His recent publications have explored issues such as Indigenous peoples' media and globalization, digital media and convergence culture, popular and political cultures of the Americas, popular geopolitics, decolonial struggle in the new media environment, and intersections between media & cultural studies and geography. He is a Primary Investigator on a collaborative research project entitled "Geographies of Media Convergence: Spaces of Democracy, Connectivity and the Reconfiguration of Cultural Citizenship," which is funded by the Marsden Fund of the Royal Society of New Zealand.

Mark Graham is an Associate Professor and Senior Research Fellow at the Oxford Internet Institute, a Research Fellow at Green Templeton College, and an Associate in the University of Oxford's School of Geography and the Environment. He has published articles in major geography, communications, and urban studies journals, and his work has been covered by the Economist, the BBC, the Washington Post, CNN, the Guardian, and many other newspapers and magazines. He is an editorial board member of Information, Communication, and Society, Geo: Geography, and the Environment, and Big Data and Society. He is also a member of DFID's Digital Advisory Panel and the ESRC's Peer Review College. In 2014, he was awarded a European Research Council Starting Grant to lead a team to study 'knowledge economies' in Sub-Saharan Africa over 5 years. This will entail looking at the geographies of information production, low-end (virtual labour and microwork) knowledge work, and high-end (innovation hubs and bespoke information services) knowledge work in fifteen African cities.

Leon Gurevitch is the Deputy Head of School, Royal Society Research Scholar and Senior Lecturer at Victoria University of Wellington's School of Design. He is also Associate Editor of Animation: In Interdisciplinary Journal and has published his work in The Journal of Television and New Media, Convergence, Animation, Senses of Cinema and more. His current research on Weta Digital is a major 3-year project to study digital image industry work cultures. Leon currently lectures graduate and postgraduate courses in photographics, visual culture and software studies.

Ken Hillis is Professor Media and Technology Studies in the Department of Communication Studies at the University of North Carolina at Chapel Hill. He is author of Digital Sensations: Space, Identity, and Embodiment in Virtual Reality (1999); and Online a Lot of the Time: Ritual, Fetish, Sign (2009). He is also co-author of Google and The Culture of Search (2012) and co-editor of Everyday eBay:

Culture, Collecting, and Desire (2006); and Networked Affect 2015. His research focuses on the relationships among networked cultures, affective identity formations and the technological assemblages upon which they rely.

Tamara M. Johnson received her doctorate in Geography from The University of North Carolina at Chapel Hill in 2011. Her selection of Cape Town as a research site for a project on dance and urban change was driven by her broader research interests in land and dispossession, urban contexts marked by rapid and significant change transition, and the ways in which continued conflicts over changes in the material and symbolic infrastructure of the city play out in urban social spaces. Tamara has been dancing salsa socially and semi-professionally for over a decade, and has performed salsa, samba, mambo, tango, and Afro-Cuban rumba in the United States, Canada, and South Africa. Currently, she is Research Associate for Academic Planning and Analysis in the Office of Academic Affairs at The University of North Carolina at Charlotte, where her work focuses on student success initiatives and high impact practices with a special emphasis on best practices for global community engagement and international undergraduate research experiences for students. She is also an adjunct faculty member in the Department of Global, International and Area Studies at UNC Charlotte.

John Paul Jones III is Professor in the School of Geography and Development and Dean of the College of Social and Behavioral Sciences at the University of Arizona in Tucson. He previously worked at the University of Kentucky where he co-founded and co-directed the institution's Committee on Social Theory. His interests span most areas of geographic theory and methodology. In addition to work on the variability of parameters in spatial regression models, his interventions in critical geography include cultural analyses and deconstructions of spatial representations; poststructuralist theories linking identity and space; explorations of the effects of spatial epistemologies on geographic thinking; and assessments of the concepts of scale and globalization in socio-spatial explanation. With colleagues he has just completed a project on art-science collaborations jointly funded by the U.S. National Science Foundation and the U.K. Arts and Humanities Research Council. He serves as the editor for the Wiley-Blackwell series, Critical Introductions to Geography, and as the co-editor of the Sage journal, Dialogues in Human Geography.

Christina Beal Kennedy is Professor Emeritus from the Department of Geography, Planning and Recreation at Northern Arizona University in the United States. A geographer, she has researched and published on our relationship to landscapes and place as well as on resource management issues. It was her focus on landscape that brought her to study film. Her writings on film include a 2010 chapter "Science Fiction/Fantasy Films, Fairytales and Control: Landscape Stereotypes on a Wilderness to Ultra-Urban Continuum", coauthored by her daughters Mèlisa and Tiànna, in *Cinema and Landscape*, Graeme Harper and Jonathan Rayner eds. (Bristol, UK) and in 2008, "Living with Film: An Autobiographical Approach" chapter in *The Geography of Cinema – A Cinematic World*, eds. Christopher Lukinbeal and Stefan Zimmerman.

Sara Beth Keough, Ph.D., is an Associate Professor of Geography at Saginaw Valley State University in Michigan, USA. She earned her doctorate degree in Geography from the University of Tennessee, an M.S. in Geography from Virginia Tech, and a B.S. in History and a B.A. in Spanish from Jacksonville University in Florida. She serves as the Editor of the scholarly journal, *Material Culture.* Her research examines the connections between media, migration, and cultural identity in Canada, specifically in Newfoundland and northern Alberta. Her doctoral dissertation on music, radio broadcasting, and cultural identity in Newfoundland won the 2007 Outstanding Dissertation Award from the Association for Canadian Studies in the United States. Sara Beth also studies culture and social relations surrounding water and water venders in West Africa, particularly in the country of Niger. Her research, which often employs qualitative methods such as ethnography, has been published in scholarly journals including the *Journal of Cultural Geography*, *The American Review of Canadian Studies*, and *FOCUS on Geography,* as well as the Oxford University Press bibliography series on Geography where she wrote the entry on the cultural landscape. She was the 2013 recipient of the American Geographical Society's McColl Fellowship, and she sits on the review board for the *American Review of Canadian Studies*.

Eric Laurier is Reader in Geography and Interaction at the School of GeoSciences, University of Edinburgh. He was the principal investigator on 'Assembling the line: amateur & professional work, skills and practice in digital video editing', funded by the ESRC, working together with Barry Brown & Ignaz Strebel. The chapter in this book emerges out of that project. Currently he is researching the uses of smartphones in producing social relationships and haggling with machines.

Chris Lukinbeal holds a B.S., M.A., and Ph.D. in Geography. He is an Associate Professor and Associate Director in the School of Geography and Development at The University of Arizona and the Director of Geographic Information Systems Technology programs and of the Geospatial Innovation Science and Technology center. He has published over 50 academic papers and book chapters related to topics ranging from representation, media, GIS, remote sensing, cartography, urban geography, mental maps and GIS, landscape studies, and education.

Susan P. Mains is a Lecturer in Human Geography at the University of Dundee. Her work explores transnational identities and media representations of mobility, borders, and security in the context of Caribbean migration, creative industries in Jamaica and Scotland, and heritage tourism. She has published in a range of international journals including *Social and Cultural Geography*, *GeoJournal*, *Journal of Geography in Higher Education*, *The Singapore Journal of Tropical Geography*, *Small Axe: A Caribbean Journal of Criticism*, and *Caribbean Geography*. Her previous work has been funded through the University of the West Indies, the

Association of American Geographers and the American Geographical Society. Currently she is the coordinator for a Carnegie Trust funded project, "Ties to the Tay," which utilises participatory filmmaking as part of an interdisciplinary project between a group of academics at the University of Dundee, the Nethergate Writers, artists, and community organisations (further information can be found here: https://tiestothetay.wordpress.com/). She is also a partner in a Royal Society of Edinburgh funded collaborative project exploring representations of place and landscape connections between the Scottish Highlands and the Caribbean.

Jonathan Mendel is Lecturer in Human Geography at the University of Dundee. His research focuses on networks, conflict, security and policy and has been published in the Journal of Cultural Economy and Science as Culture. His recent work has looked at science blogging, human trafficking and online networks, and urban violence.

Lisa Parks, Ph.D. is Professor and former Department Chair of Film and Media Studies at UC Santa Barbara, where she is currently the Director of the Center for Information Technology and Society. Parks has conducted research on uses of media and information technologies in different national contexts, most recently in Zambia. Her work is highly interdisciplinary and engages with fields such as geography, international relations, communication, and art. Parks is the author of Cultures in Orbit: Satellites and the Televisual (Duke UP, 2005), Vertical Mediation and the War on Terror (forthcoming), and Mixed Signals: Media Infrastructures and Cultural Geographies (in progress). She is co-editor of: Signal Traffic: Critical Studies of Media Infrastructures (U of Illinois, 2015), Down to Earth: Satellite Technologies, Industries and Cultures (Rutgers UP, 2012), and Planet TV: A Global Television Reader (NYU, 2003). Parks has held visiting appointments at the Institute for Advanced Study (Wissenschaftskolleg) in Berlin, McGill University, University of Southern California, and the Annenberg School of Communication at the University of Pennsylvania. She has delivered invited lectures in more than 25 countries and is currently a PI on two major research grants, one from the National Science Foundation and another from the US State Department.

Michael Petit is Director of Media Studies and the Joint Program in New Media Studies at the University of Toronto Scarborough. He is author of Peacekeepers at War (Faber and Faber 1986); co-editor of Everyday eBay: Culture, Collecting and Desire (Routledge 2006), co-author of Google and the Culture of Search (Routledge 2013), and co-editor of Networked Affect (MIT Press 2015). His research focuses on technological affect and the intersection of digital media and pedagogy.

Hauke Riesch is a lecturer in Sociology at the Department of Social Sciences, Media & Communications at Brunel University, London. He has a PhD in Science and Technology Studies and has specialised on the public understanding of science, environmental policy and risk. He is currently interested in citizen science, science blogs and representations of science in popular culture, especially in comedy.

Laura Sharp is a Ph.D. student and Research Assistant in the School of Geography and Development at The University of Arizona. Her research looks at the application of information communication technologies to on-location filming practices in the United States. Laura is managing editor of *Aether: The Journal of Media Geography* and (co)editor of the 2014 issue of *you are here: the journal of creative geography*.

Laurel C. Smith is an associate professor in the Department of Geography and Environmental Sustainability at the University of Oklahoma, where she early earned a MA in history of history of science before obtaining her PhD in geography from the University of Kentucky. She studies the organizational geographies that enable Indigenous actors to use video technologies for participating in the production of authoritative knowledge about socio-ecological worlds they inhabit. Her research has focused on the production and circulation of Indigenous videos made in Oaxaca, Mexico and the ways this kind of digital storytelling can reconfigure the cultural geographies of technoscience. It has been published in journals such as Historical Geography, History and Technology, and cultural geographies. She is currently Principal Investigator on a project that brought together a team of Indigenous and non-Indigenous researchers and media artists. This project resulted in the production of Listening for the Rain, http://vimeo.com/91082165. This video aims to start a pluricultural conversation in which some Indigenous people who live in the central United States of America discuss their observations and understandings of, as well as responses to, climate change and variability. Her hope is that this kind of digital storytelling will help decolonize institutions of higher learning by fostering community engagement.

Lynn Spigel is a Professor at Northwestern University in Chicago. She is author of Make Room for TV: Television and the Family Ideal in Postwar America University of Chicago Press, 1992); Welcome to the Dreamhouse: Popular Media and Postwar Suburbs (Duke University Press, 2001); TV By Design: Modern Art and the Rise of Network Television (University of Chicago Press, 2009), and TV Snapshots: An Archive of Everyday Life (forthcoming, Duke University Press).

Monica Stephens, PhD, is an Assistant Professor of GIScience in the Department of Geography at the University at Buffalo (SUNY Buffalo) in New York. She obtained her doctoral degree from the University of Arizona in 2012, worked as a visiting scholar at the University of Kentucky and an Assistant Professor at Humboldt State University in California. Her research integrates methodologies in Geographic Information Science (GIS) with Social Network Analysis (SNA) and Big Data. She harnesses and critiques these methodologies with data from social media and user-generated content to trace inequalities across gender, race and economic status.

Daniel Sui is currently a Professor of Geography and Distinguished Professor of Social & Behavioral Sciences at the Ohio State University. He also serves as Chair of Geography (since July 2011) and Director of the Geographical Analysis Core, Institute of Population Research (IPR) (since September 2012). He served as

Director of the Center for Urban & Regional Analysis (CURA) (2009–2012). Sui holds adjunct professorship at the John Glenn School of Public Affairs, Knowlton School of Architecture (City & Regional Planning Program), and College of Public Health at OSU. Prior assuming his current position at OSU in July 2009, Daniel Sui was a professor of geography (1993–2009) and holder of the Reta A. Haynes endowed chair (2001–2009) at Texas A&M University. He holds a B.S. (1986) and M.S. (1989) from Peking University and Ph.D. from University of Georgia (1993). His research current interests include GIScience theory, open/alternative GIS, spatial-temporal synergetics, volunteered geographic information, and crowdsourcing geographic knowledge production. Sui was a 2009 Guggenheim Fellow, 2006 winner of the Michael Breheny Prize for best paper in environment and planning, and 2014 recipient of the distinguished scholar award from the Association of American Geographers. He also served on the U.S. National Mapping Science Committee for two terms (2007–2013) and currently serves as editor-in-chief for GeoJournal.

Matthew W. Wilson is an associate professor in the Department of Geography at the University of Kentucky and a visiting scholar at the Center for Geographic Analysis at Harvard University. He co-founded and co-directs the New Mappings Collaboratory which studies and facilitates new engagements with geographic representation. His research in critical GIS draws upon STS and urban political geography to understand the development and proliferation of location-based technologies, with particular attention to the consumer electronic sector. He has previously taught at the Harvard Graduate School of Design, and his current research project focuses on the founding of the Laboratory for Computer Graphics at Harvard in 1965, a catalyzing moment in the advent of the digital map. His work has been published in leading journals and collections including, Society & Space, Landscape & Urban Planning, Qualitative GIS (Sage), Geoforum, The Professional Geographer, the Annals of the Association of American Geographers, Cartographica, Social & Cultural Geography, Gender, Place & Culture, and Environment & Planning A.

Pamela Wilson is on the Communication faculty at Reinhardt University in Georgia, USA. She is the co-editor of Global Indigenous Media (2008, Duke UP) and numerous articles and chapters on the cultural politics of media at the intersection of racial, ethnic, gender or regional subcultures and identities and the mainstream.

Keith Woodward is an Assistant Professor of Social Theory in the Geography Department at the University of Wisconsin–Madison. His work focuses upon affective dimensions of social struggle and social change, particularly in the context of organizing and direct action. He is also the co-author, with Sallie A. Marston and John Paul Jones III, of several articles on Site Ontology.

Bo Zhao is currently pursuing a Ph.D. in geography at The Ohio State University. He holds a B.S. (2006) and M.S. (2008) from Nanjing University, China. He has worked as a GIS programmer and spatial analyst in Center for Urban and Regional Analysis at The Ohio State University (since 2012), and Center for Geographic Analysis at Harvard University (2009–2010). During the gap period (2011–2012), he was temporarily away from the academia and served as a CTO of a major GIS company (http://www.trgis.com) located in China. He has led a team to design GIS software such as cluster-based web mapping server as well as spatial database for mobile devices, and also undertaken GIS related projects in multiple areas, such as urban planning, municipal administration, tourism, and archaeological excavations. His current research interests lie in the intersection of social media, VGI, GIScience and society. He is a major member of a NSF-funded project (A GIScience Approach for Assessing the Quality, Potential Applications, and Impact of Volunteered Geographic Information, Award Number: 1048100): he has designed a cluster based web crawler for harvesting Sina Weibo and Twitter feeds, and written several programs to process, manipulate and visualize (geo) big data like the social media feeds.

Leo Zonn, PhD (1975), is Professor of Geography in the Department of Geography and the Environment at the University of Texas at Austin. He has a been a member of the Geography faculty at Arizona State University (1975–1986), East Carolina University (1986–1997), the University of North Carolina at Chapel Hill (1997–2004) and the University of Texas at Austin (2004–present). He has served 19 years as a department chair—11 years at ECU, 5 at UNC, and 3 at UT—along with a variety of other administrative duties. Professor Zonn is interested in issues of geographic representation that can be associated with a variety of media, including landscapes, literature, and travel writing, but his special interest is in cinema. This curiosity has usually been within some of the more classic frames of textual analysis, although more recently he has become especially interested in cinematic exhibition. As such he is concerned with the complex network that frames the integration of technology, production, audience, cinematic text, and the site of exposition into a place-based filmic experience. This means that the drive-in, home screening room, traveling film theater, portable DVD player, and the many standard forms of the walk-in movie theater, as examples, provide geographic experiences worthy of consideration. His overall research agenda is not informed by any one conceptual structure, but instead draws from a rich and varied set of mostly social-theoretical views, while even humanist and experiential place-based influences can be found blended into the mix.

Matthew Zook is a Professor of Geography at the University of Kentucky and study how code, space and place interact as people increasingly use of mobile, digital technologies to navigate through their everyday, lived geographies. In the recent past he has served as a Fulbright Scholar in Estonia, been a Visiting Fellow at the Oxford Internet Institute and helped found the research blog/website FloatingSheep.org

Index

A
AAG. *See* Association of American Geographers (AAG)
ABOR. *See* Arizona Board of Regents (ABOR)
Aboriginal Territories in Cyberspace (AbTeC), 379–380
Abuse, online, 246–248
Admiral Broadway Revue (TV program), 44
Advertising, 74–75
  news, 134
  radio, 274, 276
  tourism, 337–338, 340, 343
Aerial images, 162
Aerial reconnaissance, 160
Aesthetic geosophy, 9
Affect
  and cities, 388–389
  definition, 387
  and film viewing, 29
  mediated space, 86–88
  and space, 394–396
  virtual, 85
Afghanistan, US wars in, 159, 161, 163, 166–168
  reconstruction, monitoring, 170–171
Afghanistan Oil and Natural Gas Viewer, 170
Afrikaans Idol (TV program), 121
Alamo Drafthouse
  Celebrity Guest, 151
  construction, 144–146
  Food and Film, 149
  Heckle-Vision, 149
  Master Pancake Theater, 150
  Open Screen Night, 150–151
  Quote-Along, 148
  Ritz, 146, 149–150
  Rolling Roadshow, 151
  Sing-Along, 147
  Videoke, 150
Al Baghdadiya, 169
Al Iraqiya, 169
All in the Family (TV program), 114
Al Sharqiya, 169
America by Design (film), 413
American Idol (TV program), 124
America's Got Talent (TV program), 116
Anderson, Benedict, 128–129
Anti-feminism, and television, 50
Apartheid, 386, 390, 391
Apollo 17, 177
Apparatus theory, 28
The Apprentice (TV program), 116
Arab Idol (TV program), 121
Ara Irititja Project, 379
ArcGIS, 169, 192
Arizona Board of Regents (ABOR), 403
Armchair travel, 48–50
Association of American Geographers (AAG), Environmental Perception and Behavioural Geography (EPBG) specialty group, 10–11
Atanarjuat: The Fast Runner (film), 425
ATR. *See* Author-text-reader (ATR) model
Attention economy, 74–75
Audience
  Alamo Drafthouse, 146–147, 153–154
  broadcasting, 373
  expansion, in cultural industries, 68, 71
  as market, films, 27
  news, 129, 134, 135
  as object of spectatorship, 46
  participation in stage/films, 141
  power, 74
  radio, 275
  TV program, 122–123

S.P. Mains et al. (eds.), *Mediated Geographies and Geographies of Media*,
DOI 10.1007/978-94-017-9969-0

Augmented realities, 224
critical geographies of, 237–238
crowd-sourced, 223–224
Osama bin Laden's Hideout Compound in Abbottabad, 231–237
power in (*see* Power, in augmented reality)
social media and, 224
socio-spatially nuanced concept of, 224–225
Auteurship, 22
Author-centered approach, 21–24, 30
Author-text-reader (ATR) model, 21–22, 31
Autonomía en Marcha radio show, 363

B
Badscience blogging community, 252–253
uncivil comments on sites, 253–254
Badscience blog network, 243, 246, 249
Baywatch (TV program), 116
BBC. *See* British Broadcasting Corporation (BBC)
Beckham, David, 331
Bedroom, television in, 42
Below-the-line interactions, 246
Below the line labor, 23
Below-the-line spaces, 243, 252, 257
Bend it like Beckham (film), 331
Bewitched (TV program), 117
BFI. *See* British Film Institute (BFI)
Big Brother (reality show), 56, 114, 116
Big data, 203, 212, 214
Billy's Perv Website, 266
Bing, 214
Bing Maps, 193
Biondi, Dick, 275
Black Mirror (TV series), 56
Blade Runner (film), 84, 86
Blogging community. *See also* Science blogging
development, 245
interaction, 243–244
UK-based, 249
The Blue Marble (photograph), 177–179
Bolt, Usain, 329–345
Border Patrol policing, 7
Borders, 6, 7
actual *vs.* cinematic, 25
science blogging, 250
Boundary marking, and television, 42
BRACS. *See* Broadcasting for Remote Aboriginal Communities Scheme (BRACS)
Brand, Stuart, 177
Breaking Bad (TV program), 120
Brighenti, Andrea Mubi, 394
British Broadcasting Corporation (BBC), 78
Inventing the Indian, 423
British Columbia (Canada) cinema, 23
British Film Institute (BFI), 7–8
Broadcasting, 47, 77–78, 99
aesthetics of liveness, 52
indigenous, 373–375
radio, 274–277
satellite radio, 279, 280, 282–283
use of geoweb, 193
user-driven practices, 362
Broadcasting for Remote Aboriginal Communities Scheme (BRACS), 374
"Broadcast Yourself!", 269–271
Broadway Music channel, 282
Buena Vista Social Café, 391
Buena Vista Social Club, 391
Buena Vista vibe, 392, 395
Bush, George W., 184

C
CAAMA. *See* Central Australian Aboriginal Media Association (CAAMA)
California Proposition 187, 7
Campbell, Bill, 283–284
Canadian Idol (TV program), 113
Cape Town, salsa dance, 388
emotional politics and territoriality, 391–393
Capitalism
accumulation, 98
and advertising, 75
copyright capitalism, 73
and cultural production, 69
political economy, 66, 75–77
Caribbeanness, 8
Cartography, 89–90, 180, 216–217
and journalism, 127–132
and power, 226, 230
Casey, Edward, 394
Celebrity
geographies, 330
mediascapes, 331
and popular culture, 331
tourism, 334, 338, 344, 345
Central Australian Aboriginal Media Association (CAAMA), 373
Centralization
of code power, 230
in cultural industries, 70–71

Chiapas Media Project, 376, 426
Chicago Daily News (newspaper), 129
Children, and bedroom television, 42
China Central Television, 105
Chinese cinema
 contested capitals and identities, 102–106
 historical geography of, 23
 history, 96
 independent films, 105
 media capital, 97–100
  characteristics of, 100–102
 television medium, 105
 transfer of Hong Kong, 103
Chris Hall's Inventing the Indian program, 423
Cinema. *See also* Film geography
 borders, 25
 in China (*see* Chinese cinema)
 and emotion/affect, 87
 landscapes, 10, 11
 sites of exhibition, 143, 152–153
 venues, 27–28, 151, 152
Circulation network, newspapers, 136
Civic Engagement and Revitalization in Garfield, 404
Classical political economy, 66–67
Clique percolation, 321
Clustering
 of creative labor, 98–99
 racial clustering in social media, 215
 spatial clustering, 67
Code power, 229–231, 236–237
Code space, 224
Coffeecam, 263
Cold Mountain (film), 23–24
Collaborative Catalogs Project, 379
Collective viewing of television, 45
Colonialism
 archives, 379
 de-colonial literary theory, 370
 and indigenous media, 371
 postcolonial pedagogy, 419–420
Commercialization of journalism, 135–136
Commodification
 cultural, 70–71, 76
 of radio, 276, 283
 and tourism, 436–437
Communication
 geography, Nobel Prize ceremony, 313–326
 indigenous communication, 375
 infrastructure sites, geospatial images of, 166–167
 interpersonal communication, 52, 54
 map-based, 216–217
 political economy of, 66, 76
 power, 228–229, 231, 234–236
 role in development, 353–356
 role in space-and place-making, 195–196
Communities of dissensus, 303
Community-based video projects, 376
Community building, and media, 13
Community media, 354, 373, 374
Community of journalism, 127–128
Community radio broadcasting, 373
Competition-based TV formats, 119
Consciousness, during film viewing, 29
Consolidation, in cultural industries, 70–71
Consumer demand, in cultural production, 75
Copyright capitalism, 73
Copyrights, 72–73
 territorialization of, 73
Countervisualization, 303
Creative autonomy, 69
Creative capacity, and media capitals, 101
Creative labor, 69, 98
 and media capitals, 101
Creative migration, trajectories of, 98–99
Criado-Perez, Caroline, 247
Crisis of representation school, 83
Critical dystopia, 56
Critical geography, 212
 of augmented realities, 237–238
Critical theories, 83–84, 427
Crossing the Line (documentary), 405
Crowd-sourced augmented realities, 223–224
Cultural commodities
 distinctiveness of, 66
 distribution of, 66, 71
 integration into circulation, 75
 as public good, 70
 realization of values, 75
 reproduction of, 68
 satellite radio as, 278, 284
Cultural discount, 112
Cultural economy, 23–24, 30, 71, 72, 74, 99
Cultural geography, 144, 331, 368, 434
 new cultural geography, 10, 11
Cultural heritage, television for expressing, 41
Cultural homogenization, of TV formats, 118–121
Cultural imperialism, 68
 thesis, 352
Cultural industries, 66
 attention economy, 74–75
 audience expansion in, 68
 consolidation and centralization, 70–71
 creative labor, 69
 globalization in, 68–69
 monopoly rents, 69–70

Cultural industries (*cont.*)
  rationalization, 71
  role of state, 77–78
  scarcity
    copyrighting, 72–73
    territorializing, 73–74
Cultural narratives, 367
Cultural production
  and capitalism, 69
  in cultural industries, 66
  geographies of, 67–68
  insertion of labor, 75
Cultural studies, 6, 12, 14
  and film geography, 23–24, 30
Cultural texts, 23–24
Cultures of media convergence, 300–301
Cyberspace, 85, 88, 203, 224, 248, 297

D
DAB. *See* Digital audio broadcasting (DAB)
Dallas (TV program), 122
Dancing body, 386, 396
Deal or No Deal (TV program), 116
De-colonial literary theory, 370
Decolonization, of indigenous geographies, 418–420
Demand management, and advertising, 74–75
Democratic totalitarianism, 294, 301–304
Deterritorialization, 249–250
  of radio spaces, 281
Development communication, 353–356
Dialectics, approach to real/real binary, 22, 26, 30
Diamond, Neil, 423
Diaspora, and television, 50
DigiPlace, 230, 256
Digital annotations, and interactions, 223
Digital artistic projects, 379–380
Digital audio broadcasting (DAB), 277
Digital divide, 203–204, 354
  disruption, 356–357
Digital Earth, 180
Digital Globe, 160, 161, 166, 168, 169, 173, 179–184
  future eventualities, 187–188
Digital heritage projects, 379
Digital Indigenous Democracy project, 378
Digital media, 378
  effects, 85–86
Digital natives, 377–380
Digital screens, 55
Digitization of news, 128n, 135
Disc jockies (DJs), 275, 282
Disney movies, 352
Distributed power, 227–228, 231, 233–234
Divinas, 351
DJs. *See* Disc jockies (DJs)
Documentary film/video, 375
Domestic space, and television, 39–43
Dowell, Kristin, 425
The Dr. Oz Show (TV program), 118
Dulce convivencia, 428
Duplicate code, 224–226
Dynamic media mobilities, 356
Dystopias, media, 55–56

E
Economic geography, 22, 30
  of Hollywood film industry, 23
Economics of film production, 22–23
Ecosystem of earth, 178, 179
Emotions
  and dance, 386
  and film-viewing, 29, 87
  and place, 136, 195
  in salsa dance, 393
  and space, 387, 395
Endemol, 116, 117
Entertainment
  and development, 357–359
  film as, 25, 30
  radio as, 275
Environmental representation, Google Earth, 185–187
Epistemology, real/real binary at, 25
ESRI
  ArcGIS.com, 192
  MapStudio, 193
Essentialism, 27
Ethnographic film, 369
Ethnographic research
  in Cape Town, 389
  of television, 41, 42
Ethnographies of film consumption, 27
European school *vs.* American school, of media geography, 10
Everybody Loves Raymond (TV program), 117, 118, 121
Executives, television, 119
Exhibitions, film, 27–28, 31, 143, 152–153
Exporting Raymond (film), 121
Extreme Makeover: Home Edition (TV program), 116, 120

F

Facebook, 193, 212, 214, 215, 262, 267, 284–286
Fahey, Morgan, 13
Family intimacy, and television, 39–43
"Faustian dilemma", 370
FCC. *See* Federal Communication Commission (FCC)
Fear Factor (TV program), 116
Federal Communication Commission (FCC), 275, 276
 licensing agreements, 282
Federal Radio Commission, 275
Female viewer identity, 29
Feminism, and television, 50n
Fictional television programming, 112
Film clip assessment, 412–413
Film festivals, 27, 376
Film geography
 application of transactionalism in, 11
 author-centered approach, 22–24
 reader-centered approach, 27–29
 text-centered approach, 24–27
Film-going, 143–144, 150
Film production, economics of, 22–23, 30
Film theory, 21
First Run Cinema Eatery (FRCE), 142
Flexible accumulation, 7, 84
Flexible specialization, in Hollywood film industry, 23
Football World Cup, 334
Footprints, 323–325
For Life, 405, 406
Format radio, 275
Formats, television, 111, 113–116
 active audiences and place, 122–123
 agreement, 113
 cost effectiveness of, 115
 homogenization/localization, 118–121
 international distributors, 115–118
Fraser-Pryce, Shelly-Anne, 339
FRCE. *See* First Run Cinema Eatery (FRCE)
Freeman, Cathy, 339
Freeways, 38, 52
FremantleMedia, 116, 117
Frozen River (film), 24–25
Fugard, Athol, *Sizwe Bansi is Dead*, 6

G

Gaia theory, 178
Game shows, 112, 114, 116, 117, 119–120
García Canclini, Néstor, 358
Gay/queer men, 261–267
Gender, 13
 and online abuse, 246–248
 and television, 40, 42, 50, 54
Geo-engineering, 185, 187
GeoEye (satellite operator), 160, 161, 168
GeoEye-1 satellite, 163
The Geographical Magazine, 82, 83
Geographical political economy, 65
Geographic information systems (GIS), 161, 191–192, 197, 202–203
 maps as media, 215–217
 mediated relationships, 212–215
 paradigm shift, 193
 as science, 211–212
 as social media, 192–193, 211–212
 and social network analysis, 215
 as tool, 211–212
Geographic media literacy, 11–12, 402, 411
GEOINT, 162, 167–168, 172
Geolocation systems, satellite-based, 161
Geopolitics, popular, 14, 25
Geosophy, 9
Geospatial imagery, 159, 161
 microphysics of, 162–167
Geospatial industry, 160
Geospatial intelligence, 162
Geospatial web, 192
Geoweb, 192, 193, 214
 Adam's tetradic framework, 194–196
 convergence of traditional mass media with, 194
 media in places, 201–202
 media in spaces, 202–204
 places in media, 197–199
 spaces in media, 199–200
Gift economies, 73
Ginsburg, Faye, 370
Girls Around Me (social networking app), 202
GIS. *See* Geographic information systems (GIS)
GIScience, 212, 214
Glasgow University Media Unit (GUMU), 5–6
The Gleaner (newspaper), 343, 344
Global connectivity, 48
Globalization, 22, 23, 248
 media, 68–69
 and news geographies, 134–135
 and television, 38
Global media landscape, satellite radio in, 280
Global village, 76, 139, 178
Globe. *See* Digital globe
Goldacre, Ben, 246, 250–252
Goldacre's blog, 246

Google, 160, 168, 214
Google Earth, 160, 161, 167–168, 170, 173, 178–179, 181, 193
  "child of ten" rule, 183
  Crisis in Darfur layer, 185
  environmental representation, 185–187
  history, 182–184
  Iraq Media Mapping layer, 169, 170
  Outreach program, 186
  prediction of future terrorist attacks, 172
Google Latitude, 214
Google Map Maker, 232
  bin Laden compound, 233–237
Google Maps, 193, 229, 231
  and territorial disputes, 197, 199
Google's goggles, 261, 262, 267
Google's PageRank, 231
Gore, Al, 183–184
GPS, 161
  cameras, 170
  tracking, 194, 225
Group Areas Act of 1950, 390
Guido, Michael, 277
GUMU. *See* Glasgow Unisity Media Unit (GUMU)
Gun permit holders, online mapping of, 201–202

H

Hacking, television, 47
Half Way Tree screenings, 333, 335, 336
Haptics, approach to real/real binary, 22, 26–27, 29, 30
Hard-drive based camcorder, 439
The Harder They Come (film), 336
Hermeneutics, of cultural texts, 23–24
Heteronormativity, 51, 264–268
Himpele, Jeff, 370
Hollywood East, 96, 104
Hollywood film industry
  economic geography of, 23
  film rating system, 104
  globalization, 68–69
  runaway productions of, 23, 24
  viewing subjects, 28–29
Home movies, 4–5
  of adventure, 438–439
Home theaters, 43–47
Homogenization, of TV formats, 118–121
The Honeymooners (TV program), 120
Hong Kong cinema, 23, 102–104
  and PRC government, 104
Housing
  and home theatres, 46–47
  inequality, television as compensation for, 44
Hoyle, Fred, 177
Human agency, 52, 88
Human geography, 4, 12, 14, 15, 331
Humanist geography, 9–10, 388
Hunters International (TV program), 49
Hyper-commercialization of journalism, 135–136
Hyperproduction, in Hong Kong cinema, 103

I

Ibsen, Henrik, 316
ICT4D community, 357
ICTs. *See* Information and communication technologies (ICTs)
I Dream of Jeanie (TV program), 117
IED. *See* Improvised explosive device (IED) attacks
Igloolik (Inuit community), 370
Igloolik Isuma Productions, 375, 425
Ikonos satellite, 163
IMAGEINT, 162
Imagination, 9
  of indigeneity, 417, 423, 428, 429
  popular, 44
Imagined communities, 127–128
  of satellite radio, 285–286
Immigration policies in media, 7
Improvised explosive device (IED) attacks, 171
Indigeneity, 368, 417, 423, 428, 429
Indigenization, of TV formats, 119–120, 122
Indigenous artists/activists, 367
Indigenous broadcasting, 373–375
Indigenous communication, 375
Indigenous communities, 377
Indigenous-controlled television broadcasting, 374
Indigenous culture, 367
Indigenous film/video, 375–377
Indigenous geographies, 368, 425, 427, 429
  decolonizing, 418–420
Indigenous media
  academic study of, 369
  complexities of, 372
  distribution and exhibition of, 376
  functions of, 368
  interdisciplinary field, 371

movement, 369
producers, 368
from representations 'of' to self-representations, 371–372
theoretical approaches to, 369
theoretical contributions to, 370
theoretical voice to establish, 370
Indigenous Media in Mexico: Culture, Community, and the State (Wortham), 426
Indigenous rights activism, 371
Indigenous videos, 375–377, 418
in Mexico, 426–429
teaching with/about, 421–422
Industrialisation, and digital globe, 181
Information and communication technologies (ICTs), 192, 378
Information space, 196
Information technology (IT)
infrastructure, 203–204
literacy
in geography, 402
skill set, 402
Infrared geospatial images, 163
Institutional structure, of media capitals, 100–101
Instituto Nacional Indígena, 426
Insurgent movements, monitoring, 171
Intellectual property, 168, 171, 185
rights, 72, 73, 379
Intercultural negotiations, and media policy, 370
Interdisciplinary research, 7
television in, 39
International distributors, of television formats, 115–118
International Society for Digital Earth, 184
Internet, 106, 203
-based file-sharing, 72
convergence of television with, 42–43
cross-border distribution of cultural products, 74
and development communication, 354, 356–357, 363
digital divide, 203–204, 354, 356–357
science blogging (*see* Science blogging)
Interpersonal communication, through television, 52, 54
Iraq, US wars in, 159, 161, 163, 166–168
Iraq Media Mapping layer, 169, 170
Isuma TV, 375, 425
IT. *See* Information technology (IT)

J
Jamaican dancehall scenes, 388
Jamaican tourism, 337–344
Jamaica Tourist Board (JTB), 337–339
Jardetzky, Paul, 263
Jasbits.com, 268
JenniCam, 264
Jeopardy (TV program), 114
Journalism. *See* News geographies
JTB. *See* Jamaica Tourist Board (JTB)

K
Kayapó Video Project, 376
Keith, Michael C., 286
Keyhole. *See* Google Earth
KH-11 satellite, 163
KH-12 satellite, 163
The King of Queens (TV program), 117
Kolstad, Arnulf, 318–319
KSAN radio station, 276
KSFR radio station, 276

L
Labor
creative, 69, 98, 101
cultural, modes of insertion, 75
television industry, 23
Landscapes, 10, 85
cinematic, 10, 11
of Hollywood film industry, 23
vernacular, 10
and visual imagery, 10
Language
and communication power, 228
and film analysis, 24
role in news stories, 132
Lar Dolce Lar (TV program), 120
Large screens, television, 55–57
The Last of the Mohicans (film), 23–24
LCE. *See* Learner-centered education (LCE)
League, Kerry, 145–146
League, Tim, 145–146
Learner-centered education (LCE), 403
Lefebvre, Henri, 389
Lewis, Helen, 247
Liberal Conspiracy blog, 252
Likert-based analysis, 414
LinkedIn, 193
Liveness, television, 51–52, 54
LMAs. *See* Local marketing agreements (LMAs)
Localization, of TV formats, 118–121

Localization economies, 67
Local marketing agreements (LMAs), 276
Location-based social media, 193–194
Location-based social networking, 194
Location-based spatial media, 214
Logic of accumulation, 97–98
London Olympic Games, 339
Loose Change (film), 305–307

M
Malls, 52
Managerial revolution, 98
ManyEyes, 199
Map communication model (MCM), 216–217
Map Maker, 232, 234
Mapping parties, 192–193
Maps. *See also* Cartography
  making, and news reporting, 130–133
  as media, 215–217
  news, 127–128, 137
Market(s)
  cultural products, 72, 73
  failure, 78
  and news geographies, 134
Marley, Bob, 338
Married With Children (TV program), 121
Marshall, Alfred, 67
Martín-Barbero, Jesús, 358
Marx, Karl, 66, 67, 97
Marxism, 10, 30, 74, 84
Material affects, 86–88
Material boundaries, 250
Material space, 52, 57, 224, 267, 268, 331
MCM. *See* Map communication model (MCM)
Meanings
  creation, 21, 22, 25, 28, 30, 83, 85
  of language, 132
Media archives, 7–8
Media capital, 97–100
  characteristics of, 100–102
  contours of socio-cultural variation, 99
  logic of accumulation, 97–98
  trajectories of creative migration, 98–99
Media convergence, 14, 355
  cultures of, 300–301
  in/and development, 359–363
Media economics, 65
Media environment, geography of new, 308–309
Media events, 293–294
  and image insurgency, 301–304
  modernity and visual, 294–297
Media literacy, 402, 408
Media Matters, 302
Media-rich bedroom, 42
Mediascapes, 169, 196, 274, 331, 373, 374
Mediated Geographies: Critical Pedagogy and Geographic Education project
  assessment matrix, 408–410
  discussion, 413–414
  film clip assessment, 412–413
  overview, 403
  pre-and post-course essays, 410–412
  student documentaries, 404–408
Mediated relationships, 212–215
Mediated spaces, 4, 265, 266, 331
  as affective geovisualization, 87–88
  affectiveness of, 86–88
Media theory, 85
Mediating communities, 134–137
Mental maps, 88, 132
Mercy Corps, 170
Metastasis (TV program), 120
Metropolis (TV program), 55
Metz, Christian, 21
Mexican Pharmaceuticals at the Border : : : OR, Where the Pills At? (documentary), 407, 408
Michaels, Eric, 369
Microphysics of geospatial imagery, 162–167
Middle kingdom, in China, 95
Migration, and television, 43–44
Military-industrial-media-entertainment network, 161
Military metaphors, for television, 45–46
Mind/body dualism, and film viewing, 29
Minorities, television as boon for, 44
Minty, Zayd, 389
Mircophysics of power, 164
Mirror stage, 28
Mobile phones, 53–54
Mobile privatization, 47–51, 53
Mobile technologies, personal TV, 42–43
Mobile television, 54–55
Modern family symbol, television as, 40–41
Modernity, 87
  audiovisual media as endgame of, 49
  and visual, 294–297
Modernization theory, 352
Modern Times (1936), 55
MoMo (social networking app), 202
Monopoly rents, 69–70
"Morning Edition" radio news program, 273–274
Movie theaters, 141–144. *See also* Alamo Drafthouse
  behavior, 152

Mujeres del mismo valor, 427
Mukurtu Project, 379
Multi-media photo essays, 413
Multiscreen environments, 55
Multispectral sensing, 163
Murch, Walter, 440
Music-by-decade channels, 277
Mutual learning, in media capitals, 98, 101
My Father's Land project, 378

N

The Nanny (TV program), 117, 121
Narration, film, 25, 26, 31
Narrative conventions, 84
National Counterterrorism Center, Worldwide Incidents Tracking System, 171
National Film Board of Canada (NFB), 372
National Geographic Magazine, 82, 83
National Geospatial-Intelligence Agency (NGA), United States, 160, 161, 167
National Imagery and Mapping Agency (NIMA), United States, 160
National Indigenous Radio Service, 374
Nationalism, in China, 96
National Public Radio (NPR) station, 273
National Reconnaissance Office (NRO), United States, 159–161, 167
Natural resources, scouting, 169–170
Navajo Project (1960), 369
Neo/colonialist forces, 420
Neogeography, 192, 214
Neoliberalism, 185–186
New cultural geography, 10, 11
News geographies, 194
  and geoweb, 193
  importance of, 138–139
  journalism as practice of cartography, 127–130
  maps, 137
  mediating community, 134–137
  news reporting and map-making, comparison, 130–133
Newspapers
  circulation network of, 136
  framing of subject matter in, 131
  role in creating imagined communities, 128–129
NFB. *See* National Film Board of Canada (NFB)
NGA. *See* National Geospatial-Intelligence Agency (NGA), United States
Nightmares, television, 55–56
Nikamowin (film), 425
NIMA. *See* National Imagery and Mapping Agency (NIMA), United States
9/11 attacks, 298, 299, 302
9/11 truth movement, 294, 304–308
1984 (novel), 55
Nobel Prize ceremony, communication geography, 313–326
Non-human phenomena, 164–165
Non-material studies, 10
Non-places, 38, 57
Nonrivalry, in cultural industries, 72
Non-spaces, 52, 53
Non-textual author-centered approach, 22
Norwegian Broadcasting Corporation (NRK), 322, 374
Nostalgia, 254, 255
Novels, role in creating imagined communities, 128
NPR. *See*National Public Radio (NPR) station
NRK. *See* Norwegian Broadcasting Corporation (NRK)
NRO. *See*National Reconnaissance Office (NRO)

O

Objectivity, 83, 132
Objects, geospatial imagery, 164–165
The OC (TV program), 122–123
Ocularcentrism, 295
The Office (TV program), 120–121
Olympic Games, Beijing, 329–332
Online abuse, gender and, 246–248
Online social networking, and social relations, 212
On-location filming, 11
Ontology
  of GIS, 210–212
  object-oriented, 165, 166
  real/real binary at, 25
  tele-travel, 49
  television, 56
OpenStreetMap (OSM), 192–193
Optics, and film viewing, 29
Ordinary landscape, 10
Osama bin Laden's Hideout Compound in Abbottabad, augmented realities of, 231–233
  code power, 236–237
  communication power, 234–236
  distributed power, 233–234
  timeless power, 237
OSM. *See* OpenStreetMap (OSM)
Overhead images, 162
Overview effect, 178

P
Palos, Ari, 403
Pandora, music genome project, 281
Panhumanity, 178
Patito Feo (film), 357, 358, 362
Penny, Laurie, 246
Persona 4 (video gema), 56
Personal perception, 9
Personal TV, 42–43
Person-environment transactions, 11
Phillips, Mark, 299–300
PHOTOINT, 162
Pieterse, Edgar, 388
Place(s), 13. *See also* Space(s)
  in media, 195, 197–199
  media in, 196, 201–202
  news geography, 127–128
  non-places, 38
  science blogging, 250–252
  and spatial media, 214
  televisual, 122–123
  trolling, 256
Pleasantville (1999), 56
Pocket Rocket Foundation, 339
Political autonomy, and media capitals, 101–102
Political economy, 22, 30, 65
  of communication, 66, 76
  of public-service broadcasting, 77–78
  of time and space, 76–77
Political geographers, 331
Political orientations, 83–84
Politics, and media, 5–6
Politics of verticality, 160
Poltergeist (1982), 56
Pop Idol (reality show), 114
Populares, 351
Popular geopolitics, 14
  and cultural studies, 25
Population Registration Act of 1950, 390
Porque es Barato/Because it's Cheaper (documentary), 405, 406
Portable TV, 53
Pose, Speed and *Stand Still* (advertisement), 332, 337, 340
Postcolonial geography, 420–421
Postcolonial theory, 420
Postcolonial visual pedagogy, 420, 421
Postmodern geography, 37
Postmodernism, 82, 86, 89
Postmodern space, 85
Postmodern sublime, 297
Postmodern theory, 84
Post-structuralist theories, and textual metaphor, 24
Power
  audience, 74
  in augmented reality, 226–227
    code power, 229–231, 236–237
    communication power, 228–229, 231, 234–236
    distributed power, 227–228, 231, 233–234
    timeless power, 230–231, 237
  -bearing dimensions of visuality, 301
  of images, 297–300
  mircophysics of, 164
  in radio broadcasting, 274–277
Predictive analytics, 171
Privacy
  and smart homes, 47
  and television, 45–46
Private bubbles, 53
Private rooms, television in, 42
Private space, and satellite television, 45, 50
Privatization
  Google Earth, 185–186
  media, 169
  mobile, 47–51, 53
Privatized mobility, 53
Project Glass, 209, 216
Proximity, news value of, 136–137
PSB. *See* Public-service broadcasting (PSB)
Psychoanalysis, of film viewing, 28–29, 31
Public community, definition, 129
Public culture
  and mobile phones, 53
  and television, 45
Public events, broadcasting, 52
Public good, cultural commodity as, 70
Public-service broadcasting (PSB), 78–79
Public spaces
  and satellite television, 50
  television in, 55, 56

Q
Queer space, 267
  and television, 51
QuickBird satellite, 163
QuickCam, 263

R
Racist geographies, and television, 43–44
Racist systems of governance, 6
Radical geography, 211

Radio broadcasting, historical evolution of power in, 274–277
Radio Corporation of America (RCA), 37
Radio spaces, deterritorialization of, 281
Raheja, Michelle, 424
Rapkin, Mickey, 271
RCA. *See* Radio Corporation of America (RCA)
Reader-centered approach, 21, 27–29, 31
  viewing subject, 28–29
Reality
  augmented. *See* Augmented realities
  film as, 26
  of geospatial imagery objects, 165
  social construction of, 132
Reality TV, and formatting, 114–115
Real/reel binary, 22, 24–25, 30
  approaches to, 26–27
Reconnaissance, aerial/satellite, 159–160
Red Power movement, 372
Reel Injun: On the Trail of the Hollywood Indian (film), 423–426
Re-imagineering, 6
Relativism, 27, 31
Remote sensing, 159, 160, 162–163, 296
Representation, definition, 132
Reservation Reelism: Redfacing, Visual Sovereignty, and Representations of Native Americans in Film (Raheja), 423–426
Resource speculation, 169–170
Rewards, in TV formats, 119
Rising Star (TV program), 124
Rocky Horror Picture Show (film), 147
Roger and Me (film), 26
Roosevelt, Franklin D., 275
Route 66, 406, 407
Routinized dispersion, 42
Royal Geographical Society (with IBG) guide, science blogging, 244
Royalties, for creative labor, 69
RTL (TV network), 123
Runaway productions, of Hollywood film industry, 23, 24

S

Sacks, Harvey, 433–435
Salsa dance, Cape Town, 390–397
  salsa community, 394
  Salsa South Africa Alliance, 393
Sámi media, 374
Satellite radio, 274
  broadcasting, 282–283
  as commodity, 283–285
  as landscape, 279–280
  new imagined communities of, 285–286
  placing, 278
  SiriusXM, 274, 277–279, 282–286
Satellites
  images, 162
    processing, 163–164
  reconnaissance, 159–160
Satellite television, 50
Scale, 12
Scale-free networks, 212
Scarcity of cultural commodities, 70, 71
  copyrighting, 72–73
  territorializing, 73–74
Scene footprints, in satellite images, 168
Science blogging, 244–245
  badscience blogging community, 252–253
  badscience blog network, 243, 246, 249
  borders, 250
  geographies of, 248–249
  moron baiting, 253
  place, 250–253
  space, 250, 252–253
  territory, 249–250
  uncivil comments, 253–254
Science fictions, 86
SDI. *See* Spatial data infrastructure (SDI)
Search engine, 226, 256
Separate Amenities Act of 1953, 390
Shame, television as site of, 41
Shange, Ntozake, 6
Sheppard, Eric, 418
Shutter control, of satellite imagery, 168
Simmel, Echoing, 440
Simulacra, approach to real/real binary, 22, 26, 30–31
Sina WeiBo, 199–200
SiriusXM satellite radio, 274, 277–279, 282–286
Sitcoms, formatting of, 120–121
Site-seeing *vs.* sight-seeing, 26
Situatedness-in-displacement, 267–269
  "Broadcast Yourself!", 269–271
  postscript, 271
Situatedness of film reception, 27
Skepticism, 249, 294, 304, 307
Small films, 8
Smart homes, 47, 49
Smart television, 43
SNA. *See* Social network analysis (SNA)
Snyder, Donald, 266
Social check-in sites, 194
Social class status, of television, 41, 42

Social congregation, use of mobile phones in, 54
Social construction of reality, 132
Social dancing, 387
  in Cape Town, 390
Social etiquette, and television, 42
Social inequality, and television, 44
Socialism, and television, 41
Social marginalization, 264
Social media
  and augmented realities, 224
  Facebook, 193, 212, 214, 215, 224, 262, 267, 284–286
  Foursquare, 224
  GIS as, 192–193
  impact on everyday life, 213
  location-based, 193
  and maps, 216
  pervasiveness of, 225–226
  Twitter, 193, 214, 224, 254–255, 284
Social network analysis (SNA), 212–215, 262
Social networking, 262
  location-based, 194
The Social Network (film), 261
Social representation, film as, 26
Social review sites, 194
Social scheduling/events sites, 194
Social spaces
  and dance, 390, 394, 397
  and news, 134
  of spatial data, 226
  and television, 45
Socio-cultural variation, contours of, 99
Sony Pictures Television (SPT), 116–118
South African Idol (TV program), 119
Sovereign Screens: Aboriginal Media on the Canadian West Coast (Dowell), 425
Space(s). *See also* Place(s)
  blogging community, 250
  in classical physics, 87
  gay/queer, 267
  hyperreality of, 85
  in media, 196, 199–200
  media in, 196, 202–204, 274
  news geography, 130
  political economy of, 76–77
  and television, 39–43
Spatial clustering, in cultural industries, 67
Spatial data infrastructure (SDI), 203
Spatial de-realization, television, 52–53
Spatial homogenization, 38
Spatialities, 89–90, 274
Spatialization, 76–77
Spatial media, 192
  impact on everyday life, 214
  and maps, 216, 217
Spatial science, 211
Spatial stories, 84
Spectator amusements, impact of television on, 44
Spectatorship, 27, 31, 143
  and public sphere transformation, 151–152
Spectator theory, 28
Spotify.com, 281
SPT. *See* Sony Pictures Television (SPT)
Stafford-Fraser, Quentin, 263
State, role in cultural industries, 77–78
Stereotypes, 5, 423
Structuration, 76
Student documentaries, 404–408
Sub-contracting, in Hollywood film industry, 23
Subject matter, of news stories, 131, 132
Subject/object dualism, in film viewing, 29
Suburban towns, and television, 43–47
Summer Olympic Games, Beijing, 329–332
Super-formats, 114, 115
Super 8mm cine films, 4
Supermodernity, 38
Surgical strikes, 161
Surplus matter of geospatial imagery, 166
Surveillance medium, television as, 45–47
Survivor (TV program), 49, 114
  Japanese version, 119
Sydney Olympics 2000, 339
Symbolism
  content of media products, 99
  of television, 40–42

T

Technogenesis, 85
Telecommunications, 47, 48
Telecommunications Act of 1996, 280
Telefetish, 265, 266, 269–271
Telematic culture, 85, 89
Telenovelas, 13, 69, 112–113, 120, 351, 352, 357–359, 362
Telepresence, 51, 54, 265, 267, 269
Television (TV)
  broadcasting, 374
  in China, 105, 106
  domestic space and family intimacy, 39–43
  home theaters and suburban towns, 43–47
  as interdisciplinary object, 39
  liveness, telepresence, and spatial de-realization, 51–57
  mobile, 53

mobile privatization and virtual travel, 47–51
portable, 53
Radio Corporation of America (RCA), 37
spectator as couch potato, 50
stasis of, 49, 50
studies, 57–58
Televisual places, 122–123
Terrestrial radio, 277
Terrestrial signals, of radio, 274
Territory
blogging community, 249–250
disputes, and online mapping, 197, 199
Text
-centered approach, 21, 24–27, 30–31
approaches to real/reel binary, 26–27
and context, 24
film as, 143
messaging, 13
practices of journalism, 131, 133
Textual analysis, 21, 143
Textual metaphor
in film geography, 21
and language, 24
relationship of real/real binary to, 25
That Girl (Sitcom), 51
Theatre studies, 6
Thrift, Nigel, 387, 388
Till Death Us Do Part (TV program), 114
Time
in classical physics, 87
political economy of, 76–77
Timeless power, 230–231, 237
The Times of India (newspaper), 337
Time-space compression, 37, 77, 84
TimeTravellerT, 380
Totalitarianism, democratic, 294, 301–304
Tourism, 8, 436–438
Jamaican, 337–344
Transactionalism, 10–11
application in film geography, 11
Trend surfaces, 323–325
Tribal cultures, 367
Trolling, 251–252
comments, 253–254
methodologies, 255–256
places, 256
on Twitter, 254–255
The Truman Show (film), 56
Truth, as moment of falsehood, 82–83
Turner, Victor, 268
TV. *See* Television (TV)
TV Dinner Date (TV program), 44
Twin Towers, attacks, 299
Twitter, 193, 214, 284
trolling on, 254–255

U
Ubik (novel), 56
Ugly Betty phenomenon, 13
UK national territory, 249–250
Ukorrigerte Meninger, 319
United States Agency for International Development (USAID), 170
United States television
abroad, 112–113
formats (*see* Formats, television)
reasons for dominance, 112
ubiquity of, 111–113
Urban agglomerations, in cultural industries, 67
Urbanization economies, 67
Urban spaces, 387, 395
Urban theory, 84
Usain Bolt Foundation, 344
US Geological Survey (USGS), 169

V
Vachiam Eecha, 380
Values, news, 136–137
Venues, cinematic, 27–28, 151, 152
Verdens Gang newspaper, 314, 320–322
Vernacular landscape, 10
Vertical mediation
Afghanistan and Iraq, 167–168
geospatial imagery, 162–167
mediascaping, 169
monitoring reconstruction, 170–171
predictive analytics, 171
resource speculation, 169–170
VGI. *See* Volunteered geographic information (VGI)
VG-peaceprize-index, 321, 322
Video indígena, 419, 426
Viewing subjects (film theory), 28–29
Virgin Media (advertisement), 343, 345
Virtual boundaries, 250
Virtual camera, 186, 187
Virtual community, of online-mapping sites, 192
Virtual digital environment, 88–89
Virtual geographies, 85–86, 377–380
Virtual places, 197, 251
Virtual space, 88, 224, 245, 249, 250, 256, 264, 265
Virtual travel, 47–51

Virtual worlds, 54
Visual credibility, 297
Visual imagery, and landscapes, 10
Visuality, power-bearing dimensions of, 301
Visual literacy, 402
Visual media, 293, 368
  producers, 374
  theory, 85
Visual sovereignty, 371, 424
The Voice (TV program), 124
Volunteered geographic information (VGI), 192, 203, 214–215
The Voronins (TV program), 121
Vowell, Sarah, 278
Voyeurism, 26, 29, 31
  of television, 46

W
Warlpiri Aboriginal broadcasting, 369
Washington Redskins game, 285
Webcams-R-Us, 263–267
Weiderman, Jason, 268, 269
West African Idol (TV program), 119
What's My Line (TV program), 114
Wheel of Fortune (TV program), 112, 114, 119
When Old Technologies Were New (Marvin), 263
White flight, 44
Whole earth photograph, 177–178
  as cultural artefact, 182, 186–187
Who's the Boss (TV program), 117
Who Wants to Be a Millionaire? (TV program), 114, 119–121
Wife Swap (TV program), 120
Williams, Jay, Jr., 286
Williams, Sean Patrick, 264–265, 269–270
Wilson, Wilder, 363
WIPO. *See* World Intellectual Property Organization (WIPO)
WITBN. *See* World Indigenous Television Broadcasters Network (WITBN)
Women
  and home theaters, 45
  household labor, and television, 40
World Indigenous Television Broadcasters Network (WITBN), 372n
World Intellectual Property Organization (WIPO), 379
Worldmaking, journalism as, 129, 130
World Trade Center (WTC) attacks, 299, 300, 303, 305–306
WorldView-1 satellite, 163
WorldView-2 satellite, 163
Wortham, Erica Cusi, 426
Writer's Guild Strike, 114
WTC. *See*World Trade Center (WTC) attacks

X
The X Factor (TV program), 116
Xiaobo, Liu (Nobel Prize ceremony, 2010), 313–320

Y
Yahoo, 214
Yo Soy Betty, la Fea (TV program), 120
Your Face Sounds Familiar (TV program), 117–118

Z
Zuckerberg, Mark, 261, 262

The manufacturer's authorised representative in the EU is Springer Nature Customer Service Centre GmbH, Europaplatz 3, 69115 Heidelberg, Germany. If you have any concerns regarding our products, please contact ProductSafety@springernature.com

Printed and bound by CPI Group (UK) Ltd, Croydon, CR0 4YY
08/01/2026
02031440-0011